Communications and Control Engineering

Communications and Control Engineering is a high-level academic monograph series publishing research in control and systems theory, control engineering and communications. It has worldwide distribution to engineers, researchers, educators (several of the titles in this series find use as advanced textbooks although that is not their primary purpose), and libraries.

The series reflects the major technological and mathematical advances that have a great impact in the fields of communication and control. The range of areas to which control and systems theory is applied is broadening rapidly with particular growth being noticeable in the fields of finance and biologically-inspired control. Books in this series generally pull together many related research threads in more mature areas of the subject than the highly-specialised volumes of *Lecture Notes in Control and Information Sciences*. This series's mathematical and control-theoretic emphasis is complemented by *Advances in Industrial Control* which provides a much more applied, engineering-oriented outlook.

Publishing Ethics: Researchers should conduct their research from research proposal to publication in line with best practices and codes of conduct of relevant professional bodies and/or national and international regulatory bodies. For more details on individual ethics matters please see:
https://www.springer.com/gp/authors-editors/journal-author/journal-author-helpdesk/publishing-ethics/14214

More information about this series at http://www.springer.com/series/61

W. Murray Wonham · Kai Cai

Supervisory Control of Discrete-Event Systems

Springer

W. Murray Wonham
Systems Control Group, Department of Electrical and Computer Engineering
University of Toronto
Toronto, ON, Canada

Kai Cai
Department of Electrical and Information Engineering
Osaka City University
Osaka, Japan

ISSN 0178-5354 ISSN 2197-7119 (electronic)
Communications and Control Engineering
ISBN 978-3-030-08458-5 ISBN 978-3-319-77452-7 (eBook)
https://doi.org/10.1007/978-3-319-77452-7

Mathematics Subject Classification (2010): 93C65

This Springer imprint is published by the registered company Springer Nature Switzerland AG
The registered company address is: Gewerbestrasse 11, 6330 Cham, Switzerland

Preface

This monograph is based on the first author's lectures at the University of Toronto during the sessions 1987–88 through 2017–18, as well as at Washington University (St. Louis) in May 1988, the Indian Institute of Technology (Kanpur) in February 1989, Bilkent University (Ankara) in May 1989, the Universidade Federal de Santa Catarina (Florianopolis) in February 1993, the Centro de Investigación y de Estudios Avanzados (Guadalajara, Mexico) in February 1997, the University of Stuttgart in May 1998, the Banaras Hindu University in February 2000, December 2001, and February 2006, Zhejiang University in May–June 2006–2011, the Sapienza Università di Roma in October 2011, and Xidian University (Xi'an) in May 2013 and 2015. The material on control theory originated with the U. of T. doctoral theses of Peter Ramadge (1983), Feng Lin (1987), and subsequently many others, together with joint publications with the first author.

The material has also been used by the second author for his lectures at Osaka City University during 2015–2017.

The software package *TCT* (for untimed DES) has been developed most recently by Renyuan Zhang, while the package *TTCT* for timed DES was developed by Christian Meder and Ali Saadatpoor.

Toronto, Canada
Osaka, Japan
January 2018

W. Murray Wonham
Kai Cai

Contents

1 Algebraic Preliminaries 1
1.1 Posets 1
1.2 Lattices 4
1.3 Equivalence Relations 7
1.4 Equivalence Kernel and Canonical Factorization 12
1.5 Application: Internal Model Principle 23
1.6 Notes 42
References 43

2 Linguistic Preliminaries 45
2.1 Languages 45
2.2 Nerode Equivalence and Right Congruence 46
2.3 Canonical Recognizers 48
2.4 Automata 57
2.5 Generators 60
2.6 Regular Expressions 66
2.7 Causal Output Mapping and Hierarchical Aggregation 69
2.8 Chains of Regular Languages 78
2.9 Notes 84
References 84

3 Supervision of Discrete-Event Systems: Basics 85
3.1 Introduction 85
3.2 Representation of Controlled Discrete-Event Systems 86
3.3 Synchronous Product, Shuffle, and Meet 89
3.4 Controllability and Supervision 100
3.5 Supremal Controllable Sublanguages and Optimal Supervision 104
3.6 Implementation of Supervisory Controls by Automata 109
3.7 Design of Supervisors Using *TCT* 114

3.8 Forced Events ... 125
3.9 Supervisory Control for Reconfiguration ... 130
3.10 Mutual Exclusion ... 136
3.11 Supervisory Control by Relabeling ... 138
3.12 Supervisor Reduction ... 141
3.13 Notes ... 142
References ... 144

4 Decentralized and Distributed Supervision of Discrete-Event Systems ... 147
4.1 Introduction ... 147
4.2 Conjunction of Supervisors ... 148
4.3 Naive Decentralized Supervision: Deadly Embrace ... 150
4.4 Decentralized Supervision: Small Factory ... 153
4.5 Decentralized Supervision: Big Factory ... 154
4.6 Decentralized Supervision: Transfer Line ... 157
4.7 Decentralized Supervision: AGVs in a Manufacturing Workcell ... 163
4.8 Decentralized Supervision by Natural Projection ... 184
4.9 Reasoning About Nonblocking ... 190
4.10 Synchronization and Event Hiding ... 196
4.11 Distributed Supervision by Supervisor Localization ... 198
4.12 Notes ... 201
References ... 202

5 Hierarchical Supervision of Discrete-Event Systems ... 205
5.1 Hierarchical Control Structure ... 205
5.2 Two-Level Controlled Discrete-Event System ... 207
5.3 High-Level Control Structure ... 210
5.4 Command and Control ... 215
5.5 Hierarchical Consistency ... 220
5.6 Hierarchical Supervision of Transfer Line ... 224
5.7 Hierarchical Supervision with Nonblocking ... 228
5.8 Notes ... 239
References ... 254

6 Supervisory Control with Partial Observations ... 257
6.1 Natural Projections and Normal Languages ... 257
6.2 Observable and Relatively Observable Languages ... 271
6.3 Feasible Supervisory Control ... 280
6.4 Infimal Closed Observable Sublanguages ... 295
6.5 Supervisory Control and Normality ... 300
6.6 Control of a Guideway ... 310
6.7 Nondeterminism, Quasi-congruences, and the Observer Property ... 315

6.8 Efficient Coordination in Decentralized Control 331
6.9 Notes 337
References 337

7 State-Based Control of Discrete-Event Systems 339
7.1 Introduction 339
7.2 Predicates and State Subsets 339
7.3 Predicate Transformers 341
7.4 State Feedback and Controllability 343
7.5 Balanced State Feedback Controls and Modularity 347
7.6 Dynamic State Feedback Control 349
7.7 Notes 352
References 358

8 Supervision of Vector Discrete-Event Systems 361
8.1 Introduction 361
8.2 Vector Discrete-Event Systems 362
8.3 VDES Modeling 365
8.4 Linear Predicates 367
8.5 State Feedback and Controllability of VDES 368
8.6 Reachability and Loop-Freeness 371
8.7 Loop-Freeness and Optimal Control 375
8.8 Example: FACT#5 377
8.9 Memory and Dynamic State Feedback Control for VDES 381
8.10 Modular Dynamic State Feedback Control for VDES 382
8.11 Example: FACT#2 383
8.12 Modeling and Control of a Production Network 386
8.13 Representation of Optimal Control by a Control VDES 392
8.14 Notes 401
References 409

9 Supervisory Control of Timed Discrete-Event Systems 411
9.1 Introduction 411
9.2 Timed Discrete-Event Systems 412
9.3 Example 1 416
9.4 Example 2 418
9.5 Time Bounds as Specifications 419
9.6 Composition of TDES 420
9.7 Example 3 421
9.8 Controllability of TDES 421
9.9 Supremal Controllable Sublanguages and Optimal
Supervision 428
9.10 Example 4: Endangered Pedestrian 430
9.11 Example 5: Timed Manufacturing Cell 433

9.12 Modular Supervision of Generalized TDES 439
9.13 Notes 443
References 443

Appendix A: Supervisory Control of a Mine Pump 445

Bibliography 461

Index 481

Introduction

The control of discrete-event systems (DES) is a research area of current vitality, stimulated by the hope of discovering general principles common to a wide range of application domains. Among the latter are manufacturing systems, traffic systems, database management systems, communication protocols, and logistic (service) systems. The contributing specialities are notably control, computer and communication science and engineering, together with industrial engineering and operations research.

With this variety of supporting disciplines, it is no surprise that the DES research area embraces a corresponding variety of problem types and modeling approaches. It is fair to say that no single control-theoretic paradigm is dominant, nor is it necessarily desirable that matters should be otherwise.

From a formal viewpoint, a DES can be thought of as a dynamical system, namely an entity equipped with a state space and a state transition structure. In particular, a DES is discrete in time and (usually) in state space; it is asynchronous or event-driven: that is, driven by events other than, or in addition to, the tick of a clock; and it may be nondeterministic: that is, capable of transitional 'choices' by internal chance or other mechanisms not necessarily modeled by the system analyst.

The present monograph is devoted to a simple, abstract model of controlled DES that has proved to be tractable, appealing to control specialists, and expressive of a range of control-theoretic ideas. It was introduced in the control literature by P. J. Ramadge and the first author in 1982; in this monograph it will be referred to as *Supervisory Control Theory* (SCT). SCT supports the formulation of various control problems of standard types, like the synthesis of controlled dynamic invariants by state feedback, and the resolution of such problems in terms of naturally definable control-theoretic concepts and properties, like reachability, controllability, and observability. SCT is automaton-based, or dually language-based, depending on whether one prefers an internal structural or external behavioral description at the start. Briefly, a DES is modeled as the generator of a formal language, the control feature being that certain events (transitions) can be disabled by an external controller. The idea is to construct this controller so that the events it currently disables depend in a suitable way on the past behavior of the generating DES. In this way the DES can be made to behave optimally with respect to a variety of criteria, where 'optimal' means 'in minimally restrictive fashion'. Among the criteria are 'safety' specifications like the avoidance of prohibited regions of state space, or the enforcement of service priorities; and 'liveness' specifications, at least in the weak sense that distinguished target states always remain reachable.

As a practical matter, in the present state of SCT software technology, DES and their specifications and controllers must be representable by finite transition structures (FTS), although there is no intrinsic restriction to FTS in the theory itself. In the FTS setup the computations for many of the control solutions entail only polynomial effort in the model's state size. However, complex controlled DES are directly modeled as product structures of simpler components; so each time a new component is adjoined (with state space size N, say) the state size of the product FTS is multiplied by N; and thus the size of the model increases exponentially with the number of components. The situation is actually worse in the case of control with partial observations: in most versions of this problem, the computational effort is exponential (rather than polynomial) in the model size itself, for instance owing to the necessity of converting from a nondeterministic FTS to its deterministic counterpart in terms of 'uncertainty subsets'.

While exponential complexity is not inevitably disastrous (after all, salesmen continue to travel) it is surely a strong incentive to refine the approach. For this, two well-known and universal systemic strategies can be invoked, each of them already familiar in control theory. The first is to create suitable architecture: that is, to exploit horizontal and vertical modularity, or in this context decentralized and hierarchical control. The second is to exploit internal regularity of algebraic or arithmetic structure if it happens to be present.

Thus a specialization of the SCT base model to 'vector' DES (VDES) allows the exploitation of vector-additive arithmetic state structure: for instance, when dealing with sets of similar objects of which the number in a given state may be incremented or decremented by various events (machines in a factory workcell, entering a busy state or a breakdown state). For modeling and analysis in this domain Petri nets have been widely utilized, especially by computer and communications specialists, but there seems little need to adopt the arcane terminology of nets to treat what after all are just standard control problems in a specialized state framework. Of course, insights from the extensive literature on Petri nets can and should be exploited whenever advantageous.

Taking a different approach, one may seek to generalize the SCT model in directions of greater realism and modeling flexibility. For instance, a generalization to 'timed transition models' incorporates real-time features along with modeling enhancements like program variables, their transformations, and transition guards. Another, deep and rather technical, generalization in the spirit of temporal logic (due to both Peter Ramadge and John Thistle) brings in languages over sets of infinite strings and addresses issues of 'eventuality', or liveness in the long run.

While the present text does not cover all the topics just listed, it provides an introduction to, and preparation for research in, control of DES in the style described. Two software packages are available as project tools: *TCT*, for untimed DES, and *TTCT*, for both timed and untimed systems. These are linked to the Web site

http://www.control.utoronto.ca/DES

It should be stressed that these tools are used only as auxiliary teaching aids; the theoretical content of the text is independent of them. They provide basic procedures for the systematic computation of examples, but could be replaced by any other tools of comparable functionality that the reader might prefer.

Numbering Convention: In each section, theorems, propositions, examples, exercises, etc., are numbered sequentially: e.g. Theorem 1, Proposition 2, Example 3, Exercise 4. When referring to a theorem (proposition, etc.) in a different section, chapter#.section#.theorem# is cited; for instance, Theorem 3.5.8 refers to Theorem 8 in Section 5 of Chapter 3.

Introductory Exercise: Elevator Modeling and Simulation

Part 1: Write a simple simulation program for a single-cage elevator in a six-story building. The elevator should respond to both external and internal calls, in a 'reasonable' way which matches your experience. No special background knowledge is assumed about DES or elevators.

Simulation runs are generated by allowing calls to occur randomly. The elevator state can be taken to be the floor at which the cage is currently located, together with the current pattern of unserviced calls and such other information that you deem relevant. Cage transitions between adjacent floors can be assumed instantaneous.

The display may be quite simple, say in the form of a table, as shown. Here the cage is at Floor 3, and there are unserviced calls to Floors 1 and 5 that have originated inside the cage, along with external up-calls at Floors 0, 1, 4 and down-calls at Floors 2, 4.

Floor	Cage location	Incalls	$\text{Excalls}_{\text{up}}$	$\text{Excalls}_{\text{dn}}$
5		x		
4			x	x
3	x			
2				x
1		x	x	
0			x	

In the presence of unserviced calls the cage location should change by one level at each stage, following which new x's in the CALLS columns may appear and old ones disappear.

Include a brief description of your approach and your control logic.

Part 2: Develop an automaton model of the system in Part 1, including a complete specification of the state set and transition structure. For instance, the state set could take the form

$$Q = F \times H \times (U0 \times \cdots \times U4) \times (D1 \times \cdots \times D5) \times (I0 \times \cdots \times I5)$$

where $F = \{0, 1, \ldots, 5\}$ is the floor set, $H = \{up, rest, down\}$ is the cage heading set, and Ui, Di, Ii represent two-state switches ('buttons') with state sets $\{set, reset\}$ for external up-calls, external down-calls, and inside-cage-calls. Thus $ui = set \in Ui$ indicates the presence of an up-call at floor i; ui will be switched back to $reset$ when the call is serviced. The state size is $|Q| = 6 \times 3 \times 2^5 \times 2^5 \times 2^6 = 1179648$. Write $f \in F$ for the current floor, $h \in H$ for the current heading, u, d, i for the button vectors (thus $u = (u0, \ldots, u4) \in U0 \times \cdots \times U4$, and $calls = (u, d, i)$. Then

$$h_{\text{next}} = \delta_H((calls)_{\text{next}}, f, h)$$

$$f_{\text{next}} = \delta_F(h_{\text{next}}, f)$$

for suitable functions δ_H, δ_F. Define $calls_{\text{next}}$ as a suitable (in part, random) function of the current values $(calls, f, h)$, so the computation sequence is

$$(calls, f, h) \mapsto calls_{\text{next}}$$

$$(calls_{\text{next}}, f, h) \mapsto h_{\text{next}}$$

$$(h_{\text{next}}, f) \mapsto f_{\text{next}}$$

Part 3: Check systematically whether or not your simulation code from Part 1 is an implementation of your automaton model in Part 2. If the automaton model were developed first, would it be helpful in writing the code?

Part 4: Discuss possible performance specifications for your elevator (e.g. 'Every call is eventually serviced'). Sketch a proof that your automaton model satisfies them.

TCT: General Information

TCT is a program for the synthesis of supervisory controls for discrete-event systems. Generators and recognizers are represented as standard DES in the form of a 5-tuple

[Size, Init, Mark, Voc, Tran]

Size is the number of states (the standard state set is {0, ..., Size−1}), Init is the initial state (always taken to be 0), Mark lists the marker states, Voc the vocal states, and Tran the transitions. A vocal state is a pair [*I*, *V*] representing positive integer output *V* at state *I*. A transition is a triple [*I*, *E*, *J*] representing a transition from the exit (source) state *I* to the entrance (target) state *J* and having event label *E*. *E* is an odd or even nonnegative integer, depending on whether the corresponding event is controllable or uncontrollable.

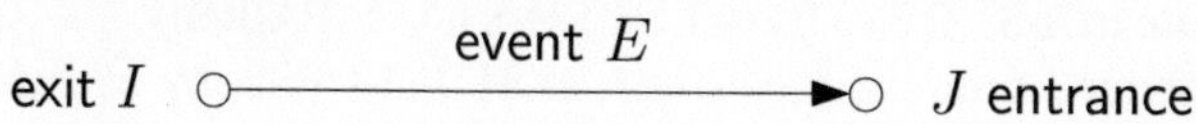

All DES transition structures must be deterministic: distinct transitions from the same exit state must carry distinct labels.

Synthesis Procedures

DES = **create**(DES)

is a new discrete-event system (DES). Option 1 allows fast user input via a sequence of prompts, resulting in direct creation of a .DES file. Option 2 allows the user to create a text (.ADS) file with any ASCII text editor; this file can be converted to a .DES file using the *TCT* procedure **FD**.

DES2 = **selfloop**(DES1,[SELF-LOOPED EVENTS])
is DES1 augmented by a selfloop (q, E, q) for each listed event E, at each state q where a transition (q, E, q') is not already defined. The event E can be chosen freely and need not belong to the alphabet of DES1.

DES2 = **trim**(DES1)
is the trim (reachable and coreachable) substructure of DES1.

DES = **sync**(DES1,DES2,…,DESk)
is the (reachable) synchronous product of DES1,DES2,…,DESk. DES need not be coreachable. Not for use with vocalized DES.

DES = **meet**(DES1,DES2,…,DESk)
is the meet (reachable cartesian product) of DES1,DES2,…,DESk. DES need not be coreachable. Not for use with vocalized DES.

DES3 = **supcon**(DES1,DES2)
is a trim generator for the supremal controllable sublanguage of the marked legal language generated by DES2 with respect to the plant DES1. DES3 provides a proper supervisor for DES1. Not for use with vocalized DES.

DES2 = **force**(DES1,[FORCIBLE EVENTS],[PREEMPTIBLE EVENTS], TIMEOUT EVENT)
is DES1 modified by the insertion of a new timeout event whose disablement forces an event in the forcible list to preempt every event in the preemptible list. Not for use with vocalized DES.

DES3 = **mutex**(DES1,DES2,[EXCLUDED-STATE-PAIRS])
is formed from the product of DES1 and DES2, by excluding state pairs listed as [[I1,J1],[I2,J2],…], plus all state pairs from which they are reachable along an uncontrollable path; and then taking the reachable substructure of the result. DES3 is reachable and controllable, but need not be coreachable. For the corresponding control data, compute DES = **sync** (DES1,DES2), then DAT = **condat**(DES,DES3). If DES3 is trim, it provides a proper supervisor for the mutual exclusion problem; if not, a solution is SUP = **supcon**(DES,DES3). Not for use with vocalized DES.

DAT2 = **condat**(DES1,DES2)
returns control data DAT2 for the supervisor DES2 of the controlled system DES1. If DES2 represents a controllable language (with respect to DES1), as when DES2 has been previously computed with **supcon**, then DAT2 will tabulate the events that are to be disabled at each state of DES2. In general **condat** can be used to test whether the language represented by DES2 is controllable with respect to DES1: just check that the disabled events tabled in DAT2 are themselves controllable (have

odd-numbered labels). To show DAT call **SA**. Not for use with vocalized DES.

DES3 = **supreduce**(DES1,DES2,DAT2)
is a reduced supervisor for plant DES1 which is control-equivalent to DES2, where DES2 and control data DAT2 were previously computed using **supcon** and **condat**. Also returned is an estimated lower bound slb for the state size of a strictly state-minimal reduced supervisor. DES3 is strictly minimal if its reported state size happens to equal the slb.

{LOC1,LOC2,...,LOCm} = **localize**(PLANT,{PLANT1,...,PLANTm},SUPER)
is the set of localizations of SUPER to the *m* independent components PLANT1,...,PLANTm of PLANT. Independence means that the alphabets of PLANT1,...,PLANTm must be pairwise disjoint. Optionally, correctness of localization is verified and reported as ControlEqu(...) in MAKEIT.TXT. **localize** is mainly for use when SUPER is a decentralized supervisor with authority over PLANT1,...,PLANTm, and PLANT is their synchronous product. Not for use with vocalized DES.

DES2 = **minstate**(DES1)
is a minimal-state DES that generates the same closed and marked languages as DES1, and the same string mapping induced by vocalization (if any). DES2 is reachable, but not coreachable unless DES1 is coreachable.

DES2 = **complement**(DES1, [AUXILIARY-EVENTS])
is a generator of the marked language complementary to the marked language of DES1, with respect to the extended alphabet comprising the event labels of DES1 plus those in the auxiliary-event list. The closed behavior of DES2 is all strings over the extended alphabet. The string mapping induced by vocalization (if any) is unchanged.

DES2 = **project**(DES1, [NULL/IMAGE EVENTS])
is a generator of the projected closed and marked languages of DES1, under the natural projection specified by the listed NULL or IMAGE events. In decentralized control, DES2 could be an observer's local model of DES1. Not for use with vocalized DES.

DES2 = **relabel**(DES1, [OLD-NEW EVENT LABEL PAIRS])
is a generator for the relabeled closed and marked behaviors of DES1, where the relabeling of languages maps each listed DES1 event label (alphabet element) into a specified label of DES2; unlisted DES1 labels are unchanged, while some DES2 labels may coincide with unlisted DES1 labels. Not for use with vocalized DES.

DES2 = **vocalize**(DES1, [STATE-OUTPUT PAIRS])

has the same closed and marked behaviors as DES1, but with user-selected state output at the entrance state corresponding to each selected (exit state, event input) pair.

DES2 = **outconsis**(DES1)

has the same closed and marked behaviors as DES1, but is output-consistent in the sense that nonzero state outputs are unambiguously controllable or uncontrollable. A vocal state with output V in the range 10,…,99 may be split into siblings with outputs, respectively, V1 or V0 in the range 100,…,991.

DES2 = **hiconsis**(DES1)

has the same closed and marked behaviors as DES1, but is hierarchically consistent in the sense that high-level controllable events may be disabled without side effects. This may require additional vocalization together with changes in the control status of existing state outputs. **hiconsis** incorporates and extends **outconsis**.

DES2 = **higen**(DES1)

is defined over the state-output alphabet of (vocalized) DES1 and represents the closed and marked state-output (or 'high-level') behaviors of DES1. For instance, starting with a 'low-level' vocalized model GLO, the sequence

OCGLO = **outconsis**(GLO)
HCGLO = **hiconsis**(OCGLO)
HCGHI = **higen**(HCGLO)

returns a DES pair (HCGLO, HCGHI) that is hierarchically consistent: controllable languages in HCGHI can be synthesized, via the state-output map, as controllable languages in HCGLO.

DES3 = **supnorm**(DES1,DES2,[NULL/IMAGE EVENTS])

is a trim DES which represents the supremal sublanguage of the legal language represented by DES2, that is normal with respect to the marked behavior of the plant generator DES1 and the natural projection specified by the NULL/IMAGE event list. Not for use with vocalized DES.

DES3 = **supscop**(DES1,DES2,[NULL/IMAGE EVENTS])

is a trim DES which represents the supremal normal solution to the Supervisory Control and Observation Problem (SCOP), corresponding to the plant DES1, legal language DES2, and specified natural projection. In this solution, only observable controllable events may be disabled. Not for use with vocalized DES.

DES3 = **suprobs**(DES1,DES2,[NULL/IMAGE EVENTS])

is a trim DES which represents the supremal relatively observable sublanguage of the language represented by DES2, with respect to the plant DES1, and specified natural projection. Not for use with vocalized DES.

DES3 = **supconrobs**(DES1,DES2,[NULL/IMAGE EVENTS])

is a trim DES which represents the supremal controllable and relatively observable sublanguage of the legal language represented by DES2, with respect to the plant DES1 and specified natural projection. In this solution, any controllable event, observable or not, is subject to disablement. Not for use with vocalized DES.

DES2 = **sup(s)qc**(DES1,[NULL/IMAGE EVENTS])

is a possibly nondeterministic DES with 'silent' transitions (labeled 'e') which represents DES1 reduced by canonical (i.e. supremal) (strong) quasi-congruence with respect to the specified natural projection. The user may select whether or not to print the corresponding state partition in MAKEIT.TXT; the printout omits singleton cells. Not for use with vocalized DES.

DES2 = **allevents**(DES1)

is a marked one-state DES self-looped with all the events of DES1. DES2 can be used as specification for the supervisory control of DES1 with respect to the sole criterion of nonblocking.

(DES3,DES4) = **natobs**(DES1,DES2)

returns a natural projection DES3 of DES1, where the projection is a natural observer, for the marked behavior of DES1, with (**allevents**) image representation DES4. The event list of DES4 is an economical extension of the Seed event list supplied by the user in the form of an **allevents** representation DES2. For **natobs** to be valid DES1 should be trim.

DES2 = **uncertmod**(DES1,[NULL/IMAGE EVENTS])

returns the uncertainty model of DES1 obtained by projection modulo uncertainty sets, corresponding to the specified natural projection on strings.

true/false = **nonconflict**(DES1,DES2)

tests whether DES1, DES2 are nonconflicting, namely whether all reachable states of the product DES are coreachable. Not for use with vocalized DES.

true/false = **isomorph**(DES1,DES2)

tests whether DES1 and DES2 are identical up to renumbering of states (but with initial state held fixed at 0); if so, their state correspondence can be displayed.

true/false = **(s)observ**(DES1,DES2,[NULL/IMAGE EVENTS])
tests whether the language represented by DES2 is (strongly) (DES1,P)-observable, for the specified natural projection P. If observability fails, a diagnostic is provided with state(s) and event(s) where failure occurs. Not for use with vocalized DES.

Utilities

DES2 = **BFS**(DES1)
is DES1 with its state set recoded by breadth-first search from state 0. BFS recoding can facilitate drawing a transition diagram, or in compensating for previous recodings introduced by **outconsis** or **hiconsis**.

DES2 = **edit**(DES1)
is obtained from DES1 by user-selected modifications.

ScreenDisplay = **show**(DES)
SE displays an existing DES, **SA** a DAT (**condat**) table, **SX** a TXT (text) file. Tables can be browsed with page keys. The file MAKEIT.TXT keeps a record of user files as they are generated.

TextFile = **FE**(DES)/**FA**(DAT)
is an ASCII text file PDS/PDT for printing or offline conversion to another format.

DESFile = **FD**(ADS)
converts an ADS file or group of files, formed using **create** (option 2), to the corresponding DES file(s).

UserDirectory = **UD**
is a listing of the current user subdirectory.

GIF = **CE**(DES)
converts a DES file to a GIF file for graphical display.

LabeledTransitionGraph = **DE**(GIF)
displays a GIF file obtained from **CE**, as a labeled transition graph; marker states are denoted by a double circle.

Chapter 1
Algebraic Preliminaries

This chapter provides a brief introduction to partially ordered sets (posets), lattices, the lattice of equivalence relations, and derived concepts such as the factorization of a function through its equivalence kernel. These ideas capture, in a primitive but useful way, the general notion of 'information' in a feedback control loop together with that of 'dynamic observer' by which such information is extracted. An application is sketched to the Internal Model Principle of regulation theory, in a relatively unstructured setting of plain sets and functions.

1.1 Posets

Partially ordered sets or 'posets' play a basic role in system theory; for instance, such structures are needed to support certain concepts of optimization. Posets lead to lattices, of which the lattice of subsets (of a fixed set) and the lattice of equivalence relations (on a fixed set) will be of key importance in the theory of formal languages.

Let X be a set. A *binary relation* on X is a subset of $X \times X$, the cartesian product of X with itself. Let $\leq$ be a binary relation on X. We use infix notation (as usual) and, for $x, y \in X$, write $x \leq y$ to mean that the ordered pair $(x, y) \in X \times X$ belongs to the relation $\leq$. The relation $\leq$ is a *partial order (p.o.)* on X if it is

reflexive: $(\forall x \in X)x \leq x$
transitive: $(\forall x, y, z \in X)x \leq y \;\&\; y \leq z \Rightarrow x \leq z$
antisymmetric: $(\forall x, y \in X)x \leq y \;\&\; y \leq x \Rightarrow x = y$

Elements $x, y \in X$ are *comparable* if either $x \leq y$ or $y \leq x$. A p.o. is a *total ordering* if every two elements of X are comparable. In a p.o. in general, it need not be the case

W. M. Wonham and K. Cai, *Supervisory Control of Discrete-Event Systems*, Communications and Control Engineering,
https://doi.org/10.1007/978-3-319-77452-7_1

that two arbitrary elements $x, y \in X$ are comparable; if x, y are not comparable, we may write $x <> y$.

If $\leq$ is a p.o. on X, the pair $(X, \leq)$ is a *poset* (or *partially ordered set*). If $\leq$ is understood, one speaks of 'the poset X'.

Example 1:
(i) Let $X = \mathbb{R}$ (the real numbers), or $X = \mathbb{N} := \{0, 1, 2, \ldots\}$ (the natural numbers), or $X = \mathbb{Z} := \{\ldots, -1, 0, +1, \ldots\}$ (the integers), with $\leq$ the usual ordering.
(ii) Let $X = \mathbb{N}^+ := \{1, 2, 3, \ldots\}$ and define $x \leq y$ iff[1] $x|y$ (x divides y), namely $(\exists k \in \mathbb{N}^+) y = kx$.
(iii) Let $X = \mathbb{Z} \times \mathbb{Z}$. Let $x = (x_1, x_2)$ and $y = (y_1, y_2)$ belong to X. Define $x \leq y$ iff $x_1 \leq y_1$ and $x_2 \leq y_2$. Thus (7,-2) $\leq$ (9,-1), but (7,2) $<>$ ($-$10,3).
(iv) Let A be a set and let $X = Pwr(A)$ be the set of all subsets of A (the *power set* of A). Thus $x, y, \ldots \in X$ are subsets of A. Define $x \leq y$ iff $x \subseteq y$.
(v) With n fixed, let $X = S^{n \times n}$ be the set of $n \times n$ symmetric matrices with real elements. For $P, Q \in X$ define $P \leq Q$ iff the matrix $Q - P$ is positive semidefinite.
(vi) Let X, Y be posets. Define a relation $\leq$ on $X \times Y$ by the recipe:

$$(x_1, y_1) \leq (x_2, y_2) \text{ iff } x_1 \leq x_2 \text{ in } X \text{ and } y_1 \leq y_2 \text{ in } Y$$

Exercise 2: Verify that the definition in Example 1(v) really does turn X into a poset. By considering P and Q as quadratic forms, interpret the relation $P \leq Q$ in geometric terms. What is the picture if $P <> Q$?

Exercise 3: In Example 1(vi) check that $(X \times Y, \leq)$ is actually a poset. It is the *product poset* of X and Y.

From now on we assume that X is a poset. Let $x, y \in X$. An element $a \in X$ is a *lower bound* for x and y if $a \leq x$ and $a \leq y$. An element $l \in X$ is a *meet* (or *greatest lower bound*) for x and y iff

$$\begin{aligned} l \leq x \ \& \ l \leq y \quad & \text{[i.e. } l \text{ is a lower bound for } x \text{ and } y] \\ \& \ (\forall a \in X)\ a \leq x \ \& \ a \leq y \Rightarrow a \leq l & \\ & \text{[i.e. } l \text{ beats every other lower bound for } x \text{ and } y] \end{aligned}$$

Exercise 4: Check that if l, l' are both meets for x and y then $l = l'$: a meet, if it exists, is unique. If it exists, the meet of x, y is denoted by $x \wedge y$.

Dually, an element $b \in X$ is an *upper bound* for x, y iff $x \leq b$ and $y \leq b$. An element $u \in X$ is a *join* (or *least upper bound*) of x and y iff

$$x \leq u \quad \& \quad y \leq u \quad \& \quad (\forall b \in X) x \leq b \ \& \ y \leq b \Rightarrow u \leq b$$

If the join of x and y exists it is unique, and is written $x \vee y$.

[1] 'iff' stands for 'if and only if'.

Example 5:
(i) Let $X = Pwr(A)$ and $x, y \in X$. Then $x \wedge y = x \cap y$ (set intersection) and $x \vee y = x \cup y$ (set union). Thus the meet and join always exist.
(ii) Let $X = \mathbb{Z} \times \mathbb{Z}$, and let $x = (x_1, x_2)$, $y = (y_1, y_2) \in X$. Then

$$x \wedge y = (\min\{x_1, y_1\}, \min\{x_2, y_2\}),$$
$$x \vee y = (\max\{x_1, y_1\}, \max\{x_2, y_2\}).$$

Again the meet and join always exist.
(iii) Let $X = S^{n \times n}$. In general $P \vee Q$ and $P \wedge Q$ do not exist.

Exercise 6: Explain the situation in Example 5(iii) with a 2×2 counterexample, and draw the picture.

Hint: Consider $P = 0$, $Q = \begin{bmatrix} 0 & 1 \\ 1 & 0 \end{bmatrix}$.

Exercise 7: Investigate the existence of meet and join for the poset $(\mathbb{N}^+, \cdot|\cdot)$ defined in Example 1(ii).

The following extensions of our notation are often useful. Write $x \geq y$ for $y \leq x$; $x < y$ for $x \leq y$ & $x \neq y$; $x > y$ for $x \geq y$ & $x \neq y$. Notice that, in general, the negation of $x \leq y$ is (either $x > y$ or $x <> y$). Also let $\perp$ stand for *bottom element*

Fig. 1.1 Hasse diagram: $X = Pwr(A)$

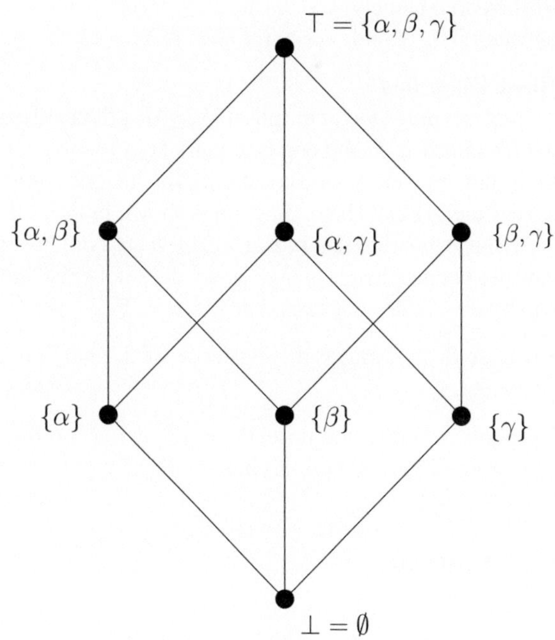

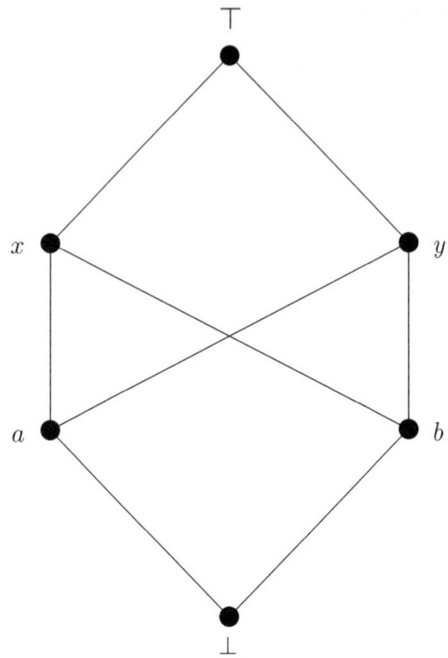

Fig. 1.2 Hasse diagram: $X = \{\top, x, y, a, b, \bot\}$

(if it exists): namely $\bot \in X$ and $\bot \leq x$ for all $x \in X$. Similarly let $\top$ stand for *top element*: $\top \in X$ and $\top \geq x$ for all $x \in X$.

Hasse Diagrams

A *Hasse diagram* for the poset $(X, \leq)$ is a directed graph with nodes corresponding to elements $x \in X$ and edges to pairs (x, y) with $x < y$. Edges are drawn as rising lines and are usually displayed only for 'neighboring' x, y. For $A = \{\alpha, \beta, \gamma\}$ and $X = Pwr(A)$ the Hasse diagram is shown in Fig. 1.1.

In the Hasse diagram of Fig. 1.2, a, b are both lower bounds for x, y, but x and y have no greatest lower bound: $x \wedge y$ does not exist. However, $a \wedge b$ exists and is $\bot$. Dually $a \vee b$ does not exist, but $x \vee y = \top$.

Exercise 8: Investigate the existence of $\bot$ and $\top$ in $(\mathbb{N}, \leq)$, $(\mathbb{N}^+, \cdot|\cdot)$, $(\mathbb{Z}, \leq)$ and $(Pwr(A), \subseteq)$.

Exercise 9: Define the poset $X = (\mathbb{N}, \cdot|\cdot)$ according to $x \leq y$ iff $x|y$, i.e. $(\exists k \in \mathbb{N}) y = kx$. Thus $x|0$ $(x \in \mathbb{N})$ but *not* $0|x$ if $x \neq 0$. Show that $\bot = 1$ and $\top = 0$.

1.2 Lattices

A *lattice* is a poset L in which the meet and join of any two elements always exist; in other words the binary operations $\vee$ and $\wedge$ define functions

$$\wedge : L \times L \to L, \quad \vee : L \times L \to L$$

It is easy to see that, if $x, y, z \in L$ and if $\star$ denotes either $\wedge$ or $\vee$ consistently throughout, then

$$\begin{array}{ll} x \star x = x & (\star \text{ is } \textit{idempotent}) \\ x \star y = y \star x & (\star \text{ is } \textit{commutative}) \\ (x \star y) \star z = x \star (y \star z) & (\star \text{ is } \textit{associative}) \end{array}$$

So for any k one can write $x_1 \star x_2 \star \cdots \star x_k$, say, without ambiguity, namely the meet and join are defined for arbitrary nonempty finite subsets of elements of L.

In addition one has the easily verified relationships

$$\begin{array}{ll} x \wedge (x \vee y) = x \vee (x \wedge y) = x & (\text{traditionally called 'absorption'}) \\ x \leq y \text{ iff } x \wedge y = x \text{ iff } x \vee y = y & (\text{'consistency'}) \end{array}$$

Exercise 1: Verify the above relationships.

Exercise 2: In any lattice

$$\begin{array}{ll} y \leq z \Rightarrow (x \wedge y \leq x \wedge z) \,\&\, (x \vee y \leq x \vee z) & (\wedge \text{ and } \vee \text{ are } \textit{monotone}) \\ x \wedge (y \vee z) \geq (x \wedge y) \vee (x \wedge z) & \\ x \vee (y \wedge z) \leq (x \vee y) \wedge (x \vee z) & (\textit{distributive} \text{ inequalities}) \\ x \leq z \Rightarrow x \vee (y \wedge z) \leq (x \vee y) \wedge z & (\textit{modular} \text{ inequality}) \end{array}$$

Exercise 3: Investigate the lattices $X = Pwr(A)$ and $X = \mathbb{Z} \times \mathbb{Z}$ to see whether the distributive inequalities, or the right side of the modular inequality, can be strengthened to equality.

If in a given lattice, the distributive inequalities are actually always equalities, the lattice is *distributive*; if the right side of the modular inequality is actually always equality, the lattice is *modular*. Clearly every distributive lattice is modular.

Let $(L, \wedge, \vee)$, or simply L, be a lattice and let S be a nonempty, and possibly infinite, subset of L. To generalize the notion of meet (greatest lower bound) of the elements of S, define $l := \inf(S)$ to mean that l is an element of L with the properties:

$$(\forall y \in S) l \leq y \quad \& \quad (\forall z \in L)((\forall y \in S) z \leq y) \Rightarrow z \leq l$$

Notice that it is not required that l belong to S. Similarly, the notion of join is generalized by defining an element $u = \sup(S)$ in dual fashion. It should be clear that $\inf(S)$, $\sup(S)$ are always unique. If S is finite, then $\inf(S)$ and $\sup(S)$ reduce to the meet and join as defined above for a finite number of elements of L, and hence always exist because L is a lattice; but if S is an infinite subset, it need not be true that $\inf(S)$ or $\sup(S)$ exist. The lattice L is *complete* if, for any nonempty subset S of L, both $\inf(S)$ and $\sup(S)$ exist (as elements of L). Thus one easily verifies that $L = (Pwr(A), \cap, \cup)$ is complete, but that $L = (\mathbb{Z} \times \mathbb{Z}, \wedge, \vee)$ is not complete.

Exercise 4: Let V be a finite-dimensional linear vector space and let $X = \mathcal{S}(V)$ be the set of linear subspaces of V. For subspaces x and y of V, define $x \leq y$ iff $x \subseteq y$ (subspace inclusion). Verify that $(X, \leq)$ is a complete lattice, where $\wedge$ is subspace intersection and $\vee$ is subspace addition (i.e. vector addition extended to subspaces). Show that X is modular but not distributive.

Exercise 5: $L = (\mathbb{Q}[0, 1], \inf, \sup)$, the rational numbers in $[0, 1]$ with the usual real analysis definitions of inf and sup, is not complete; while $L = (\mathbb{R}[0, 1], \inf, \sup)$, the real numbers in $[0, 1]$, is complete.

Exercise 6: If L and M are lattices, show that the product poset $L \times M$ is a lattice as well. It is the *product lattice* of L and M. Show that $L \times M$ is complete iff L, M are both complete.

Whether or not L is complete, if $\sup(L)$ (or $\top$) happens to exist then the empty subset $S = \varnothing \subseteq L$ can be brought within the scope of our definition of $\inf(S)$ by the convention

$$\inf(\varnothing) = \sup(L)$$

Similarly, if $\inf(L)$ (or $\bot$) exists then one may define

$$\sup(\varnothing) = \inf(L)$$

These odd-looking conventions are, in fact, forced by 'empty set logic', as can easily be checked.

Exercise 7: Adjoin to the (incomplete) lattice $(\mathbb{Z}, \leq)$ two new elements $-\infty, +\infty$ to form $\bar{\mathbb{Z}} := \mathbb{Z} \cup \{-\infty, +\infty\}$. Extend $\leq$ to $\bar{\mathbb{Z}}$ according to

$$\begin{aligned} x < +\infty \quad &\text{if} \quad x \in \mathbb{Z} \cup \{-\infty\} \\ -\infty < x \quad &\text{if} \quad x \in \mathbb{Z} \cup \{+\infty\} \end{aligned}$$

Show that $(\bar{\mathbb{Z}}, \leq)$ is complete and identify $\bot, \top$.

Exercise 8: Show that, if $\inf(S)$ exists (in L) for every subset $S \subseteq L$, then L is complete. **Hint:** Let $S^+ := \{x \in L | (\forall y \in S) x \geq y\}$. Show that $\sup(S) = \inf(S^+)$.

Let $L = (X, \leq)$ be a lattice and $Y \subseteq X$. We say that $M := (Y, \leq)$ is a *sublattice* of L if Y is closed under the meet and join operations of L.

Exercise 9: Referring to Exercise 5, show that $\mathbb{Q}[0, 1]$ is an incomplete sublattice of the complete lattice $\mathbb{R}[0, 1]$.

Exercise 10: For $L = (X, \leq)$ a lattice and $Y \subseteq X$ an arbitrary subset, show that there is a (unique) smallest subset $Z \subseteq X$ such that $Y \subseteq Z$ and $M = (Z, \leq)$ is a sublattice of L. M is the *sublattice of L generated by Y*. **Hint:** First show that the intersection of an arbitrary nonempty collection of sublattices of L is a sublattice.

1.3 Equivalence Relations

Let X be a nonempty set, and $E \subseteq X \times X$ a binary relation on X. E is an *equivalence relation* if

$$
\begin{array}{ll}
(\forall x \in X) x E x & (E \text{ is } \textit{reflexive}) \\
(\forall x, x' \in X) x E x' \Rightarrow x' E x & (E \text{ is } \textit{symmetric}) \\
(\forall x, x', x'' \in X) x E x' \,\&\, x' E x'' \Rightarrow x E x'' & (E \text{ is } \textit{transitive})
\end{array}
$$

Instead of $x E x'$ we shall often write $x \equiv x' (\text{mod } E)$.

For $x \in X$ let $[x]$ denote the subset of elements x' that are equivalent to x:

$$[x] := \{x' \in X | x' E x\} \subseteq X$$

The subset $[x]$ is the *equivalence class* (or *coset*) of x with respect to the equivalence relation E. By reflexivity $x \in [x]$, i.e. every coset is nonempty. The following proposition states that any two cosets either coincide or are disjoint.

Proposition 1

$$(\forall x, y \in X) \quad \text{either} \quad [x] = [y] \quad \text{or} \quad [x] \cap [y] = \varnothing$$

Proof

Let $x, y \in X$. Assuming $[x] \cap [y] \neq \varnothing$, let $u \in [x] \cap [y]$, so that $u E x$ and $u E y$. We claim that $[x] \subseteq [y]$. If $x' \in [x]$ then $x' E x$. Since $x E u$ (by symmetry) we have $x' E u$ (by transitivity) and so (again by transitivity) $x' E y$, namely $x' \in [y]$, proving the claim. Similarly $[y] \subseteq [x]$, and therefore $[x] = [y]$. □

We write $|E|$ for the cardinality (number of cosets) of E.

Let $\mathcal{P}$ be a family of subsets of X indexed by α in some index set A:

$$\mathcal{P} = \{C_\alpha | \alpha \in A\}, \quad C_\alpha \subseteq X$$

The family $\mathcal{P}$ is a *partition* of X if

$$
\begin{array}{ll}
(\forall \alpha \in A) C_\alpha \neq \varnothing & (\text{each subset } C_\alpha \text{ is nonempty}) \\
X = \bigcup \{C_\alpha | \alpha \in A\} & (\text{the subsets } C_\alpha \text{ cover } X) \\
(\forall \alpha, \beta \in A) \alpha \neq \beta \Rightarrow C_\alpha \cap C_\beta = \varnothing & (\text{subsets with distinct indices are} \\
& \text{pairwise disjoint})
\end{array}
$$

The second of these conditions could be written $(\forall x \in X)(\exists \alpha \in A) x \in C_\alpha$. The subsets C_α are the *cells* of $\mathcal{P}$.

Thus for the equivalence relation E we have proved that the collection of distinct cosets $[x]$ (each of them indexed, say, by some representative member x) is a partition of X. Conversely for a partition $\mathcal{P}$ of X, as above, we may define an equivalence relation E on X by the recipe

$$xEy \quad \text{iff} \quad (\exists \alpha \in A) x \in C_\alpha \;\&\; y \in C_\alpha$$

namely x and y are equivalent iff they belong to the same cell of $\mathcal{P}$. It is easily checked that E as just defined really is an equivalence relation on X. With this correspondence in mind we shall often speak of equivalence relations and partitions interchangeably, and use the term 'cell' in preference to 'coset' or 'equivalence class'.

Let $\mathcal{E}(X)$, or simply $\mathcal{E}$, be the class of all equivalence relations on (or partitions of) X. We shall assign a p.o. to $\mathcal{E}$ in such a way that $\mathcal{E}$ becomes a complete lattice, as follows:

$$(\forall E_1, E_2 \in \mathcal{E}) E_1 \le E_2 \quad \text{iff} \quad (\forall x, y \in X) x E_1 y \Rightarrow x E_2 y$$

In other words $E_1 \le E_2$ iff, whenever $x \equiv y (\text{mod } E_1)$ then $x \equiv y (\text{mod } E_2)$; that is, every cell of E_1 is a subset of some (and therefore exactly one) cell of E_2. If $E_1 \le E_2$ one may say that E_1 *refines* E_2, or E_1 is *finer* than E_2, or E_2 is *coarser* than E_1. If $E_1 \le E_2$, clearly $|E_1| \ge |E_2|$.

Exercise 2: Verify that $\le$ really does define a p.o. on $\mathcal{E}$; that is, $\le$ is reflexive, transitive and antisymmetric.

Proposition 3

In the poset $(\mathcal{E}, \le)$ the meet $E_1 \wedge E_2$ of elements E_1 and E_2 always exists, and is given by

$$\begin{aligned}(\forall x, x' \in X) x &\equiv x' (\text{mod } E_1 \wedge E_2)\\ &\text{iff} \quad x \equiv x' (\text{mod } E_1) \;\&\; x \equiv x' (\text{mod } E_2)\end{aligned}$$

Proof (Outline)

Write $E := E_1 \wedge E_2$ as just defined. Then E really is an equivalence relation on X, that is $E \in \mathcal{E}$. Next, $E \le E_1$ and $E \le E_2$. Finally if $F \in \mathcal{E}$ and $F \le E_1$, $F \le E_2$ then $F \le E$. □

Exercise 4: Supply all the details in the above proof.

The meet may be described by saying that $E_1 \wedge E_2$ is the coarsest partition that is finer than both E_1 and E_2. In general $|E_1 \wedge E_2| \le |E_1| \cdot |E_2|$. The situation is sketched in Fig. 1.3.

The definition of join is more complicated to state than the definition of meet.

Proposition 5

In the poset $(\mathcal{E}, \le)$ the join $E_1 \vee E_2$ of elements E_1, E_2 always exists and is given by

$$\begin{aligned}&(\forall x, x' \in X) x \equiv x' (\text{mod } E_1 \vee E_2)\\ &\quad \text{iff } (\exists \text{ integer } k \ge 1)(\exists x_0, x_1, \ldots, x_k \in X) x_0 = x \;\&\; x_k = x'\\ &\qquad \& \; (\forall i) 1 \le i \le k \Rightarrow \big(x_i \equiv x_{i-1} (\text{mod } E_1) \text{ or } x_i \equiv x_{i-1} (\text{mod } E_2)\big)\end{aligned}$$

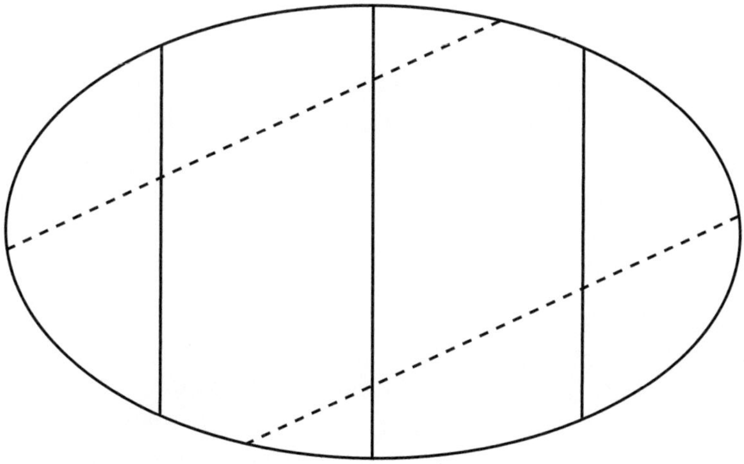

Fig. 1.3 Meet of two partitions
———————— boundaries of E-cells
- - - - - - - - - - - - - - boundaries of F-cells
$(E \wedge F)$-cells are formed by intersecting E-cells and F-cells

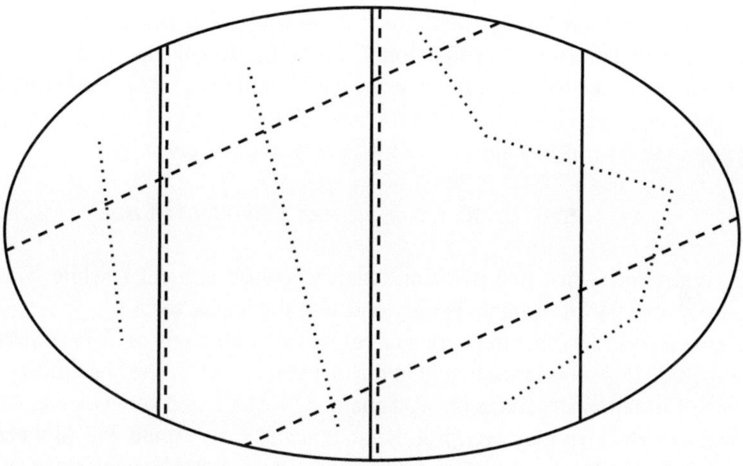

Fig. 1.4 Join of two partitions
———————— boundaries of E-cells
- - - - - - - - - - - - - - boundaries of F-cells
$(E \vee F)$-cells are formed by zigzag linkage of E-cells and F-cells

Exercise 6: Prove Proposition 5.

The definition of join amounts to saying that x and x' can be chained together by a sequence of auxiliary elements $x_1, \ldots, x_{k-1}$, where each link in the chain represents either equivalence (mod E_1) or equivalence (mod E_2). In case $k = 1$, either $x \equiv x'(\text{mod } E_1)$ or $x \equiv x'(\text{mod } E_2)$. The join may be described by saying that $E_1 \vee E_2$

is the finest partition that is coarser than both E_1 and E_2. The situation is sketched in Fig. 1.4.

We have now established that $(\mathcal{E}, \leq)$ is a lattice, the *lattice of equivalence relations* (or *partitions*) on X. Finally we show that the lattice $\mathcal{E}$ is complete.

Proposition 7

Let $\mathcal{F} \subseteq \mathcal{E}$ be a nonempty collection of equivalence relations on X. Then $\inf(\mathcal{F})$ exists; in fact

$$(\forall x, x' \in X) x (\inf(\mathcal{F})) x' \text{ iff } (\forall F \in \mathcal{F}) x F x'$$

Also $\sup(\mathcal{F})$ exists; in fact

$$\begin{aligned}
&(\forall x, x' \in X) x (\sup(\mathcal{F})) x' \\
&\quad \text{iff } (\exists \text{ integer } k \geq 1)(\exists F_1, \ldots, F_k \in \mathcal{F})(\exists x_0, \ldots, x_k \in X) \\
&\qquad x_0 = x \ \&\ x_k = x' \ \&\ (\forall i) 1 \leq i \leq k \Rightarrow x_i \equiv x_{i-1} (\text{mod } F_i)
\end{aligned}$$

Proof (Outline)

As defined above, $\inf(\mathcal{F})$ and $\sup(\mathcal{F})$ are indeed equivalence relations on X, and have the properties required by the definitions of inf and sup given in Sect. 1.2. □

Exercise 8: Supply the details.

The definition of $\inf(\mathcal{F})$ says that x and x' are equivalent with respect to each of the equivalence relations in the collection $\mathcal{F}$. The definition of $\sup(\mathcal{F})$ says that x and x' can be chained together by a finite sequence of elements, where each link of the chain represents equivalence (mod F) for some $F \in \mathcal{F}$.

In particular $\bot = \inf(\mathcal{E})$ and $\top = \sup(\mathcal{E})$ exist and are given by

$$x \equiv x' (\text{mod } \bot) \text{ iff } x = x', \quad x \equiv x' (\text{mod } \top) \text{ iff } \mathit{true}$$

Thus $\bot$ is the finest possible partition (each singleton is a cell), while $\top$ is the coarsest possible partition (there is only one cell: the whole set X).

We have now shown that the lattice of equivalence relations on X is complete.

To conclude this section we note that the elements of $\mathcal{E}(X)$ may be crudely interpreted as information structures on X. Thus if $E \in \mathcal{E}(X)$, and some element $y \in X$ is 'known exactly', we may interpret the statement '$x \equiv y (\text{mod } E)$' to mean that '$x$ is known to within the cell that contains y'. [As a metaphor, consider an ideal voltmeter that is perfectly accurate, but only reads out to a precision of 0.01 volt. If a voltage may be any real number and if, say, a reading of 1.23 volts means that the measured voltage v satisfies $1.225 \leq v < 1.235$, then the meter determines a partition of the real line into cells (intervals) of length 0.01. What is 'known' about any measured voltage is just that it lies within the cell corresponding to the reading.] On this basis a given partition represents more information than a coarser one. The element $\bot \in \mathcal{E}$ stands for 'perfect information' (or 'zero ignorance'), while $\top$ stands for 'zero information' ('complete ignorance'). If $E, F \in \mathcal{E}(X)$ then $E \wedge F$ represents 'the information contributed by E and F when present together, or cooperating'. [Let X be the state space of an electrical network, and let E and F represent an ideal ammeter and an ideal voltmeter.] Dually $E \vee F$ might be interpreted as 'the information that E and F produce in common'. [With X as before, suppose the state $x \in X$ of the network can be perturbed, either by a shove or a kick. Assume that shoves and kicks are known to define partitions S and K belonging to $\mathcal{E}(X)$: a shove can only perturb x to a state x' in the same cell of S, while a kick can only perturb x to a state x' in the same cell of K. If initially x is measured with perfect accuracy, and the network is subsequently perturbed by some arbitrary sequence of shoves and kicks, then the best available information about the final state y is that $y \equiv x (\text{mod } S \vee K)$.]

Exercise 9: In a certain Nation, regarded as a network of villages, it is always possible to make a two-way trip from any village to itself or to one or more other villages, by at least one of the modes: canoe, footpath, or elephant. Show that, to a Traveler restricted to these modes, the Nation is partitioned into Territories that are mutually inaccessible, but within each of which every village can be reached from any other.

Exercise 10: For $X = \{1, 2, 3\}$ present $\mathcal{E}(X)$ as a list of partitions of X and draw the Hasse diagram. Repeat for $X = \{1, 2, 3, 4\}$.

Exercise 11: Let $E, F, G \in \mathcal{E}(X)$. Investigate the validity of proposed distributive identities

$$(?) \qquad E \wedge (F \vee G) = (E \wedge F) \vee (E \wedge G)$$

and

$$(?) \qquad E \vee (F \wedge G) = (E \vee F) \wedge (E \vee G)$$

If either one is valid (i.e. holds for arbitrary E, F, G), prove it; otherwise provide a counterexample. **Hint:** Examine the first Hasse diagram of Exercise 10.

Exercise 12: Investigate whether $\mathcal{E}(X)$ is modular. That is, either prove that $\mathcal{E}(X)$ is modular, or show by a counterexample that it is not. **Hint:** Examine the second Hasse diagram of Exercise 10.

Exercise 13: Given two elements $E, F \in \mathcal{E}(X)$, say that elements $x, x' \in X$ are *indistinguishable*, and write xIx', if either $x \equiv x' (\text{mod } E)$ or $x \equiv x' (\text{mod } F)$, or possibly both. It can be checked that the binary relation I is reflexive and symmetric but not necessarily transitive. Such a relation is a *tolerance relation*. Provide examples of tolerance relations that are not transitive. If R is any binary relation on X (i.e. $R \subseteq X \times X$), the *transitive closure* of R is the smallest (in the sense of subset inclusion in $X \times X$) transitive binary relation that contains R. Show that the transitive closure of I is an element of $\mathcal{E}(X)$ and compute this element in terms of E and F. **Hint:** First show that the transitive closure of R is given by

$$R^* := \{(x, x') | (\exists k)(\exists x_0, \ldots, x_k) x = x_0, \ x' = x_k, \ (\forall i = 1, \ldots, k) x_{i-1} R x_i\}$$

Exercise 14: For $|X| = n$ let $p_n := |\mathcal{E}(X)|$ $(n = 1, 2, \ldots)$ and let $p_0 := 1$. Show that

$$p_{n+1} = \sum_{k=0}^{n} \binom{n}{k} p_k$$

Deduce that (i) $p_{2n} \geq n!2^n$, and (ii) $p_n \leq n!$. Write a program to compute p_n ($1 \leq n \leq 10$), and calculate p_n approximately for $n = 20, 30, \ldots$ (e.g. $p_{50} \simeq 1.86 \times 10^{47}$).

1.4 Equivalence Kernel and Canonical Factorization

Let X and Y be sets. In this section we shall write π, ρ etc. for elements of $\mathcal{E}(X)$. It will be convenient to think of the elements of $\mathcal{E}(X)$ as partitions of X. Let $\pi \in \mathcal{E}(X)$ and let $\bar{X}$ denote the set of (distinct) cells of π. Alternatively $\bar{X}$ can be taken to be a set of labels or indices for these cells. Often $\bar{X}$ is written X/π, read 'X mod π'. Denote by P_π the surjective function mapping any x to its cell (or cell label):

$$P_\pi : X \to X/\pi : x \mapsto [x]$$

where $[x]$ is the cell of x (mod π). We call P_π the *canonical projection* associated with π.

Let $f : X \to Y$ be a function with domain X and codomain Y. With f we often associate the induced function $f_* : Pwr(X) \to Pwr(Y)$ taking subsets of X into their f-images in Y:

$$f_*(A) := \{f(x) | x \in A\}, \qquad A \in Pwr(X) \text{ (i.e. } A \subseteq X)$$

Thus $f_*(A) \subseteq Y$. In particular $f_*(X)$ is the *image of f*, denoted im f. Usually we do not distinguish notationally between f and f_*, simply writing $f(A)$ for $f_*(A)$.

With $f : X \to Y$ we also associate the*inverse image function* $f^{-1} : Pwr(Y) \to Pwr(X)$ according to

$$f^{-1}(B) := \{x \in X | f(x) \in B\}, \qquad B \in Pwr(Y) \text{ (i.e. } B \subseteq Y)$$

Thus $f^{-1}(B) \subseteq X$. Notice that f^{-1} is always well-defined, even if f does not have an 'inverse' in the ordinary sense. If f happens to be bijective, then the ordinary inverse of f (strictly, its induced map on subsets of Y) coincides with the inverse image function f^{-1}.

Next, with $f : X \to Y$ associate an equivalence relation in $\mathcal{E}(X)$ called the *equivalence kernel* of f and denoted by ker f, as follows:

$$(\forall x, x' \in X) x \equiv x' (\text{mod ker } f) \text{ iff } f(x) = f(x')$$

For instance, ker $P_\pi = \pi$. The cells of ker f are just the subsets of X on which f assumes its distinct values in Y and are sometimes called the *fibers* of f. [For illustration consider the function $f : \mathbb{S}^2 \to \mathbb{R}$ that maps points on the earth's surface into their elevation above sea level in m. Then $f^{-1}(\{100\})$ is the cell of ker f consisting of those points whose elevation is 100 m.] The partition corresponding to ker f consists of the subfamily of nonempty subsets of X formed from the family of subsets

$$\{f^{-1}(\{y\})|y \in Y\}$$

Intuitively, f 'throws away more or less information according to whether its kernel is coarser or finer'. [Consider the ideal voltmeter as a function $f : \mathbb{R} \to \mathbb{R}$. Compare the kernels of two voltmeters, respectively reading out to the nearest 0.01 v. and 0.001 v.]

Exercise 1: Let $f : X \to Y$. Show that

(i) $(\forall B \subseteq Y) f(f^{-1}(B)) \subseteq B$;
equality holds for B iff $B \subseteq \text{im } f$; and equality holds for all $B \subseteq Y$ iff im $f = Y$, i.e. f is surjective.

(ii) $(\forall A \subseteq X) f^{-1}(f(A)) \supseteq A$;
equality holds for A iff ker $f \leq \{A, X - A\}$; and equality holds for all A iff ker $f = \perp$, i.e. f is injective.

(iii) Let $\{A_\alpha\}$, $\{B_\beta\}$ be arbitrary families of subsets of X, Y respectively. Then

$$f(\cup_\alpha A_\alpha) = \cup_\alpha f(A_\alpha), \quad f(\cap_\alpha A_\alpha) \subseteq \cap_\alpha f(A_\alpha)$$

$$f^{-1}(\cup_\beta B_\beta) = \cup_\beta f^{-1}(B_\beta), \quad f^{-1}(\cap_\beta B_\beta) = \cap_\beta f^{-1}(B_\beta)$$

Illustrate strict inclusion in the second of these distribution relations.

(iv) For all $A \subseteq X$, $B \subseteq Y$, $f(A \cap f^{-1}(B)) = f(A) \cap B$.

Let $f : X \to Y$, and let $P_f : X \to X/\ker f$ be the canonical projection. Then there is a unique function $g : X/\ker f \to Y$ such that $f = g \circ P_f$. ($\circ$ denotes composition of functions). Indeed if $z \in X/\ker f$ and $z = P_f(x)$ for some $x \in X$, define $g(z) := f(x)$. This definition is unambiguous because if, also, $z = P_f(x')$ then $P_f(x') = P_f(x)$ and therefore $f(x') = f(x)$ since $\ker P_f = \ker f$. Uniqueness of g follows by the fact that P_f is surjective. In this way we obtain the *canonical factorization* of f through its equivalence kernel. The situation is displayed in the commutative diagram below.

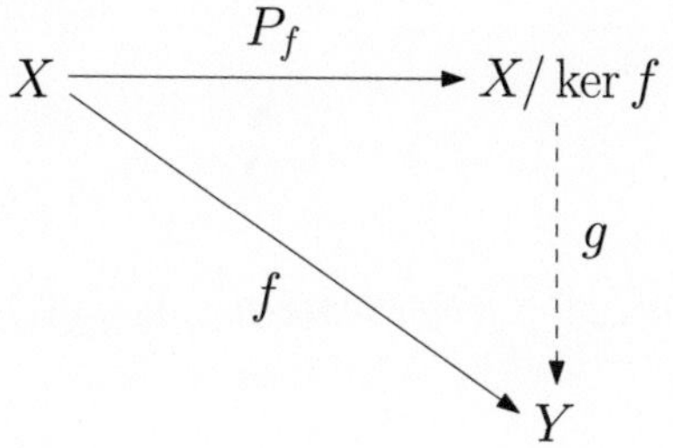

In a canonical factorization $f = g \circ P_f$ the left factor g is always injective. For suppose $z, z' \in X/\ker f$ and $g(z) = g(z')$. If $z = P_f(x)$ and $z' = P_f(x')$ then

$$f(x) = g \circ P_f(x) = g(z) = g(z') = g \circ P_f(x') = f(x')$$

namely $x \equiv x' (\text{mod } \ker f)$, so $x \equiv x' (\text{mod } \ker P_f)$, i.e.

$$z = P_f(x) = P_f(x') = z'$$

as claimed.

If $\pi \in \mathcal{E}(X)$ we may write $P_\pi : X \to X/\pi$ for the canonical projection; thus $\ker P_\pi = \pi$.

The following propositions offer variations on the foregoing theme; their proof will be left as an exercise.

Proposition 2

Suppose $f : X \to Y$ and let $\rho \in \mathcal{E}(X)$ with $\ker f \geq \rho$. There exists a unique map $g : X/\rho \to Y$ such that $f = g \circ P_\rho$.

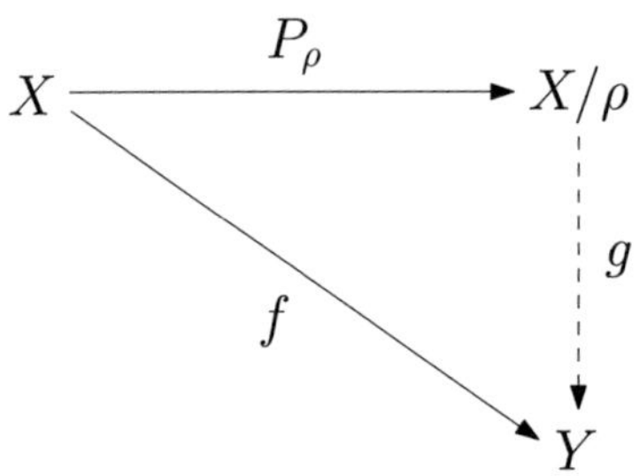

$\ker f \geq \rho$

Proposition 3

Suppose $f : X \to Y$ and $g : X \to Z$ and let $\ker f \geq \ker g$. Then there exists a map $h : Z \to Y$ such that $f = h \circ g$. Furthermore h is uniquely defined on the image $g(X)$ of X in Z; that is, the restriction $h|g(X)$ is unique.

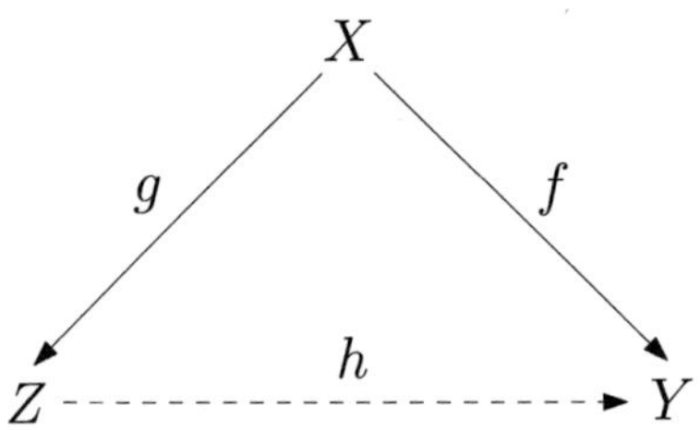

$\ker f \geq \ker g$

In this situation f is said to *factor through* g. Intuitively, 'g preserves enough information to calculate f (via h)'.

Proposition 4

If $\pi, \rho \in \mathcal{E}(X)$ and $\pi \leq \rho$, there is a unique function $f : X/\pi \to X/\rho$ such that $P_\rho = f \circ P_\pi$.

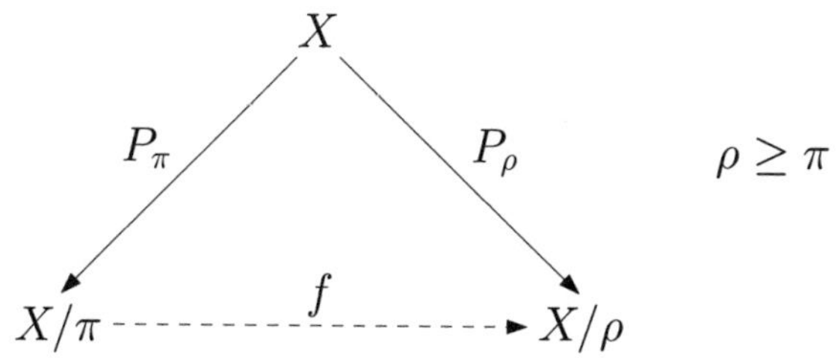

Exercise 5: Prove Propositions 2–4. In Proposition 4, by drawing a picture of X interpret $\ker f \in \mathcal{E}(X/\pi)$ in terms of the partitions π, ρ of X.

Exercise 6: **Parable of the toolmaker**

To illustrate the benefits of functional factorization, consider a large field of $N \gg 1$ stones. Each stone is suited for making exactly one of k possible tools: hammer, arrowhead, etc. One sunny day, the toolmaker organizes his tribe to sort all the stones in the field into k piles, one for each tool. Under suitable probabilistic assumptions, and taking into account search and sort times, prove that the 'sorted architecture' is roughly k times more efficient than no architecture, given that the toolmaker will make $N(\to \infty)$ tools over his lifetime.

Example 7: **Congruence of a dynamic system**

A *dynamic system* on a set X is a map $\alpha : X \to X$ with the following interpretation. The elements $x \in X$ are the system 'states', and α is the 'state transition function'. Select $x_0 \in X$ as 'initial' state and let the system evolve successively through states $x_1 = \alpha(x_0), x_2 = \alpha(x_1), \ldots$. Write α^k for the k-fold composition of α (with α^0 = identity, $\alpha^1 = \alpha$). The sequence $\{\alpha^k(x_0) | k \in \mathbb{N}\} \in X^{\mathbb{N}}$ is the *path* of (X, α) with initial state x_0.

Let $\pi \in \mathcal{E}(X)$ with canonical projection $P_\pi : X \to \bar{X} := X/\pi$. We say that π is a *congruence* for α if there exists a map $\bar{\alpha} : \bar{X} \to \bar{X}$ such that $\bar{\alpha} \circ P_\pi = P_\pi \circ \alpha$, namely the following diagram commutes.

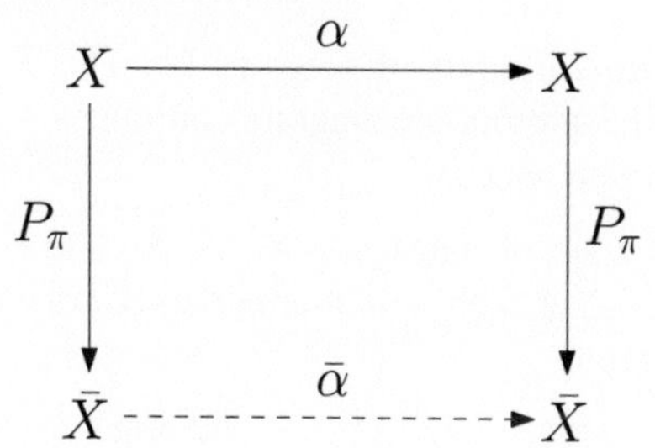

Proposition 8

π is a congruence for α iff

$$\ker P_\pi \leq \ker(P_\pi \circ \alpha)$$

namely

$$(\forall x, x')(x, x') \in \pi \Rightarrow (\alpha(x), \alpha(x')) \in \pi$$

Proof

Immediate from Proposition 2, with the identifications $(Y, f, \rho, g) = (\bar{X}, P_\pi \circ \alpha, \pi, \bar{\alpha})$. □

If π is a congruence for α, $\bar{\alpha}$ is the map *induced* by α on $\bar{X}$. The condition says that α 'respects' the partition corresponding to π, in the sense that cells are mapped under α consistently into cells. Thus the dynamic system $(\bar{X}, \bar{\alpha})$ can be regarded as a consistent aggregated (or 'high-level' or 'lumped') model of (X, α).

Exercise 9: With (X, α) fixed, let $\mathcal{C}(X) \subseteq \mathcal{E}(X)$ be the set of all congruences for α. Show that $\mathcal{C}(X)$ is a complete sublattice of $\mathcal{E}(X)$ that contains the elements $\top, \perp$ of $\mathcal{E}(X)$.

Exercise 10: Let $X = X_1 \times X_2$ and write $x = (x_1, x_2)$. Suppose

$$\alpha : X \to X : (x_1, x_2) \mapsto (\alpha_1(x_1, x_2), \alpha_2(x_2))$$

i.e. $\alpha : x \mapsto \hat{x}$ is coordinatized as

$$\hat{x}_1 = \alpha_1(x_1, x_2), \quad \hat{x}_2 = \alpha_2(x_2)$$

for suitable maps α_1, α_2. Let $(x, x') \in \pi$ iff $x_2 = x_2'$. Show that π is a congruence for α and find a coordinatization (i.e. a concrete representation) of $(\bar{X}, \bar{\alpha})$.

Exercise 11: Consider a clock with hour, minute and second hands. Identify the corresponding congruences. **Hint:** For the second hand alone, consider

$$\alpha : \mathbb{R} \to \mathbb{R} : t \mapsto t + 1$$

where t is real-valued time in units of seconds. Let $P : \mathbb{R} \to \mathbb{R}/60\mathbb{N} = \bar{\mathbb{R}}$, and identify $\bar{\alpha} : \bar{\mathbb{R}} \to \bar{\mathbb{R}}$ so the appropriate diagram commutes. Generalize the picture to include the hour and minute hands.

Exercise 12: Let A be a set of maps $\alpha : X \to X$. Show that if $\pi \in \mathcal{C}(X)$ is a congruence for every $\alpha \in A$ then π is a congruence for every composition $\alpha_k \circ \alpha_{k-1} \circ \cdots \circ \alpha_1$ of maps in A.

Exercise 13: Given a finite set X with $|X| = n$, construct $\alpha : X \to X$ 'randomly', for instance by assigning each evaluation $\alpha(x)$ independently by a uniform probability distribution over X. Discuss the probability that (X, α) admits at least one nontrivial congruence (i.e. other than $\perp, \top$), especially as $n \to \infty$. **Hint:** Show that (X, α) admits a nontrivial congruence if α is not bijective. For fixed n calculate the fractional number of such α and then use Stirling's formula.

Exercise 14: Let X be a finite-dimensional linear vector space (over $\mathbb{R}$, say) and assume $\alpha : X \to X$ is linear. Describe $\mathcal{C}(X)$ (as in Exercise 9).

Exercise 15: With X a set and $x_0 \in X$, let $\alpha : X \to Pwr(X)$. The triple (X, x_0, α) is a *nondeterministic dynamic system*. A *path* of (X, α) with initial state x_0 is either an infinite sequence $\{x_k | k \geq 0\}$ with $x_k \in \alpha(x_{k-1})$ $(k \geq 1)$ or a finite sequence $\{x_k | 0 \leq k \leq n\}$ with $x_k \in \alpha(x_{k-1})$ $(1 \leq k \leq n)$ and $\alpha(x_n) = \varnothing$. Let $\pi \in \mathcal{E}(X)$, $P_\pi : X \to \bar{X} := X/\pi$. If $S \subseteq X$ write $P_\pi S = \{P_\pi x | x \in S\} \subseteq \bar{X}$. We say that π is a *quasi-congruence* for (X, α) if

$$(\forall x, x' \in X) P_\pi x = P_\pi x' \Rightarrow P_\pi \circ \alpha(x) = P_\pi \circ \alpha(x')$$

where $P_\pi \circ \alpha(x) := P_\pi (\alpha(x))$.

(i) Show that $\bot \in \mathcal{E}(X)$ is a quasi-congruence and that the quasi-congruences for (X, x_0, α) form a complete upper semilattice under the join operation of $\mathcal{E}(X)$, namely if $\pi_\lambda \in \mathcal{E}(X)$ is a quasi-congruence for each $\lambda \in \Lambda$ (some index set) then so is

$$\sup\{\pi_\lambda | \lambda \in \Lambda\}$$

Find an example showing that if π_1, π_2 are quasi-congruences, $\pi_1 \wedge \pi_2$ need not be. Show also that $\top \in \mathcal{E}(X)$ need not be a quasi-congruence.

(ii) Show that, if π is a quasi-congruence, there exists a unique (induced) mapping $\bar{\alpha} : \bar{X} \to Pwr(\bar{X})$ such that the following diagram commutes.

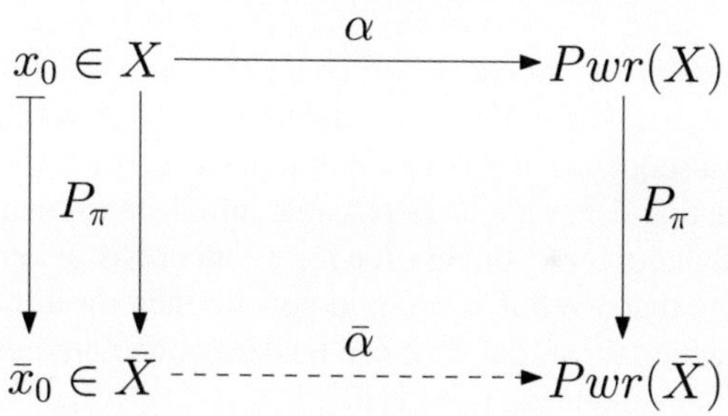

In particular let

$$\rho := \sup\{\pi | \pi \text{ is a quasi-congruence for } (X, x_0, \alpha)\}$$

with $\bar{X} := P_\rho X$ and $(\bar{X}, \bar{x}_0, \bar{\alpha})$ the corresponding nondeterministic dynamic system. Show that $\bot \in \mathcal{E}(\bar{X})$ is the only quasi-congruence for $(\bar{X}, \bar{x}_0, \bar{\alpha})$.

(iii) For $\pi \in \mathcal{E}(X)$ and $\alpha : X \to Pwr(X)$ define $\pi \cdot \alpha \in \mathcal{E}(X)$ according to

$$x \equiv x' (\text{mod } \pi \cdot \alpha) \text{ iff } P_\pi \circ \alpha(x) = P_\pi \circ \alpha(x') \text{ in } Pwr(X/\pi)$$

Let $E = \{x \in X | \alpha(x) \neq \varnothing\}$ and let $\rho_0 \in \mathcal{E}(X)$ be given by the partition $\rho_0 = \{E, X - E\}$. Now consider the sequence $\rho_n \in \mathcal{E}(X)$:

$$\rho_n = \rho_{n-1} \wedge (\rho_{n-1} \cdot \alpha), \ n \geq 1$$

Assuming X is finite, and with ρ as in (ii), show that

$$\rho = \lim \rho_n (n \to \infty)$$

the limit being achieved in finitely many steps. In particular, if $E = X$, so $\rho_0 = \top$, then also $\rho = \top$.

(iv) Let $X = \{0, 1, \ldots, 10\}$, $x_0 = 0$, and let $\alpha : X \to Pwr(X)$ be given by the following table.

x	0	1	2	3	4	5	6	7	8	9	10
$\alpha(x)$	3, 4	3	$\varnothing$	0, 1, 2, 7	1, 2, 5	0, 2, 9	1, 8, 10	1, 2, 9	$\varnothing$	0, 2	1, 8

Use the algorithm in (iii) to compute ρ, and specify $(\bar{X}, \bar{x}_0, \bar{\alpha})$.

(v) In (iv) choose arbitrarily a state-output function $\gamma : X \to \{0, 1\}$, so $\ker \gamma$ is a binary partition of X. Recompute ρ subject to the additional condition $\rho \leq \ker \gamma$, say by setting

$$\rho_0 = \{E, X - E\} \wedge \ker \gamma$$

Is it possible to choose γ in such a way that $\rho = \perp$?

(vi) In general, assume that $\rho = \perp$ and let $x', x'' \in X$ with $x' \neq x''$. Prove that there exists a sequence of possible output measurements, say $Y = \{y(0), y(1), \ldots\}$ that distinguishes x', x'' as possible initial states, namely Y is either possible for x' but not for x'' or possible for x'' but not for x'. Provide a counterexample to show that, even if $\rho = \perp$, in general there need not exist any sequence Y from which the initial state can be determined uniquely.

(vii) Consider a sequence $Y_k = \{y(0), y(1), \ldots, y(k)\}$ of $k+1$ possible consecutive output measurements corresponding to some (unknown) initial state x_0. That is,

$$Y_k = \{\gamma(x(j)) | x(j) \in \alpha^j(x_0), j = 0, 1, \ldots, k\}$$

Obtain a formula for $X_k \subseteq X$, the subset of all x_0 in X that are compatible with Y_k. Namely, X_k is 'the best information available about x_0 based on the measurements Y_k'.

(viii) Obtain a recursive formula for the best estimate of the current state $x(k)$ based on Y_k. Namely, if this is the subset $W_k \subseteq X$, show how W_{k+1} can be updated from W_k and $y(k+1)$.

Exercise 16: Let $\mathcal{L}$ be a complete lattice. A function $\psi : \mathcal{L} \to \mathcal{L}$ is *monotone* if, for all $\omega, \omega' \in \mathcal{L}$, $\omega \leq \omega'$ implies $\psi(\omega) \leq \psi(\omega')$. An element $\omega \in \mathcal{L}$ is a *fixpoint* of ψ if $\omega = \psi(\omega)$.

Show that if ψ is monotone it has at least one fixpoint. **Hint:** Let

$$\Omega^{\uparrow} := \{\omega \in \mathcal{L} | \omega \leq \psi(\omega)\}$$

Note that $\perp \in \Omega^{\uparrow}$. Define $\omega^* := \sup \Omega^{\uparrow}$, and show that ω^* is a fixpoint of ψ.

Now let Ω be the set of all fixpoints of ψ. Clearly $\Omega \subseteq \Omega^{\uparrow}$. Show that ω^* is the greatest fixpoint of ψ, namely $\omega^* \in \Omega$ and thus $\omega^* = \sup \Omega$. Dually let

$$\Omega^{\downarrow} := \{\omega \in \mathcal{L} | \omega \geq \psi(\omega)\}$$

Note that $\top \in \Omega^{\downarrow}$, and define $\omega_* := \inf \Omega^{\downarrow}$. Show that ω_* is the least fixpoint of ψ, namely $\omega_* \in \Omega$ and thus $\omega_* = \inf \Omega$.

Suppose that $\omega_1, \omega_2 \in \Omega$. Is it true that $\omega_1 \vee \omega_2, \omega_1 \wedge \omega_2 \in \Omega$? In each case either prove the positive statement or provide a counterexample.

For a sequence $\{\omega_n \in \mathcal{L} | n \in \mathbb{N}\}$ write $\omega_n \downarrow$ if the ω_n are nonincreasing, i.e. $\omega_n \geq \omega_{n+1}$ for all n. The function ψ is *downward continuous* if, whenever $\omega_n \in \mathcal{L}$ and $\omega_n \downarrow$, then

$$\psi(\inf_n \omega_n) = \inf_n \psi(\omega_n)$$

Show that if ψ is downward continuous, it is necessarily monotone. For a counterexample to the converse see Exercise 19.

In case ψ is downward continuous show that

$$\omega^* = \inf\{\psi^k(\top) | k \in \mathbb{N}\}$$

Exercise 17: Let $\mathcal{L}$ be a complete lattice. For a sequence $\{\omega_n \in \mathcal{L} | n \in \mathbb{N}\}$ write $\omega_n \uparrow$ if the ω_n are nondecreasing, i.e. $\omega_n \leq \omega_{n+1}$ for all n. Say the function $\psi : \mathcal{L} \to \mathcal{L}$ is *upward continuous* if, whenever $\omega_n \in \mathcal{L}$ and $\omega_n \uparrow$, then

$$\psi(\sup_n \omega_n) = \sup_n \psi(\omega_n)$$

Dualize the results of Exercise 16 for upward continuity.

Exercise 18: Let $\psi : \mathcal{E}(\mathbb{N}) \to \mathcal{E}(\mathbb{N})$. Show that ψ monotone and upward (resp. downward) continuous does not imply ψ downward (resp. upward) continuous.

Exercise 19: (Yingcong Guan) In $\mathcal{E}(\mathbb{N})$ define

$$\begin{aligned}
\omega_0 &= \top \\
\omega_1 &= \{(0, 1), (2, 3, \ldots)\} \\
\omega_n &= \{(0, 1), (2), \ldots, (n), (n + 1, n + 2, \ldots)\}, \quad n \geq 2
\end{aligned}$$

Write $|\pi|$ for the number of cells of $\pi \in \mathcal{E}(\mathbb{N})$. Define $\psi : \mathcal{E}(\mathbb{N}) \to \mathcal{E}(\mathbb{N})$ according to

$$\psi(\omega_n) = \omega_{n+1}, \qquad n \geq 0$$

$$\psi(\pi) = \begin{cases} \omega_n \text{ if } \pi \neq \omega_n \text{ and } |\pi| = n+1 \quad (n \geq 1) \\ \perp \text{ if } |\pi| = \infty \end{cases}$$

Show that ψ is monotone, investigate the fixpoint(s) of ψ, and calculate $\psi(\inf_n \omega_n)$ and $\inf_n \psi(\omega_n)$.

Exercise 20: Observers

Let $\alpha : X \to X$ be an arbitrary function. Recall that an element $\omega \in \mathcal{E}(X)$ is a *congruence* for α if, whenever $x \equiv x'(\text{mod}\ \omega)$ then $\alpha(x) \equiv \alpha(x')(\text{mod}\ \omega)$, i.e. α 'respects' the partition induced by ω. Define $\omega \cdot \alpha \in \mathcal{E}(X)$ according to:

$$\omega \cdot \alpha = \ker(P_\omega \circ \alpha)$$

Thus $x \equiv x'(\text{mod}\ \omega \cdot \alpha)$ iff $\alpha(x) \equiv \alpha(x')(\text{mod}\ \omega)$. Then ω is a congruence for α iff $\omega \leq \omega \cdot \alpha$.

(i) Let $\gamma : X \to Y$ be a function from X to a set Y. We define the *observer* for the triple (X, α, γ) to be the equivalence relation

$$\omega_o := \sup\{\omega \in \mathcal{E}(X) | \omega \leq (\ker \gamma) \wedge (\omega \cdot \alpha)\}$$

Thus the observer is the coarsest congruence for α that is finer than $\ker \gamma$. Define $\psi : \mathcal{E}(X) \to \mathcal{E}(X)$ according to $\psi(\omega) := (\ker \gamma) \wedge (\omega \cdot \alpha)$. Show that ψ is monotone and that ω_o is the greatest fixpoint of ψ. Prove that ψ is downward continuous, and that consequently

$$\omega_o = \inf\{(\ker \gamma) \cdot \alpha^{i-1} | i = 1, 2, \ldots\}$$

where α^j is the j-fold composition of α with itself.

(ii) We interpret the observer as follows. Consider a dynamic system with state $x \in X$ and discrete-time *transition function* α. When started in state $x(0)$ at $t = 0$, the system generates the sequence of states, or *trajectory*, given by

$$x(t+1) = \alpha\left(x(t)\right), \quad t = 0, 1, 2, \ldots$$

With γ an *output map*, the system generates the corresponding sequence of outputs, or *observations*

$$y(t) = \gamma\left(x(t)\right), \quad t = 0, 1, 2, \ldots$$

Thus ω_o represents information available about the initial state $x(0)$ after observation of the entire output sequence $\{y(t)|t = 0, 1, 2, \ldots\}$: the observations cannot resolve the uncertainty about $x(0)$ more finely than ω_o. On this basis the pair (γ, α) is said to be *observable* if $\omega_o = \perp$, namely the observation sequence determines the initial state uniquely.

(iii) Calculate ω_o when α, γ are defined as follows:

$$X = \{1, 2, 3, 4, 5, 6, 7\}$$

x	1	2	3	4	5	6	7
$\alpha(x)$	2	3	4	5	1	7	5
$\gamma(x)$	r	r	r	g	g	r	g

Here r, g stand for 'red' and 'green'. What conclusions can be drawn about the system from your result for ω_o? With reference to Example 7, take $\pi = \omega_0$, explicitly coordinatize $\bar{X} = X/\omega_0$, P_{ω_0}, and the induced transition map $\bar{\alpha}$, and verify that $\bar{\alpha} \circ P_{\omega_0} = P_{\omega_0} \circ \alpha$.

Similarly, establish the existence and uniqueness, and a coordinatization, of the induced output map $\bar{\gamma} : \bar{X} \to Y$ such that $\bar{\gamma} \circ P_{\omega_0} = \gamma$. Verify that the pair $(\bar{\gamma}, \bar{\alpha})$ is observable. These results can be displayed as shown below.

(iv) Calculate ω_o for the following. Let $X = \mathbb{N} \times \mathbb{N}$, and for $(i, j) \in X$,

$$\alpha(i, j) = (i + 1, j)$$

$$\gamma(i, j) = \begin{cases} 0 & i \leq j \\ 1 & i > j \end{cases}$$

Sketch the cells of ω_o in the $\mathbb{N}^2$ plane, and coordinatize as in (iii).

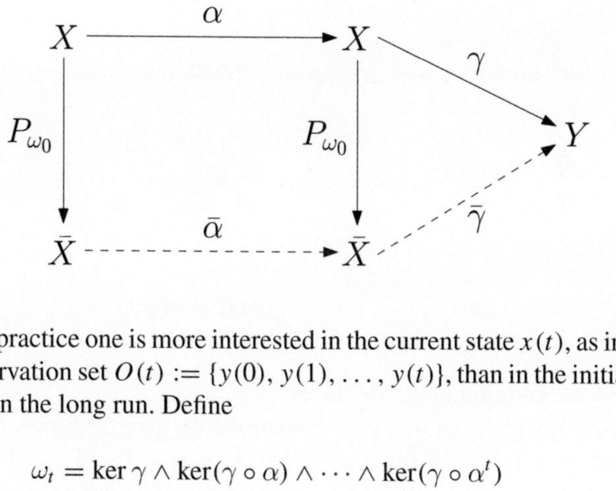

(v) Very often in practice one is more interested in the current state $x(t)$, as inferred from the observation set $O(t) := \{y(0), y(1), \ldots, y(t)\}$, than in the initial state as estimated in the long run. Define

$$\omega_t = \ker \gamma \wedge \ker(\gamma \circ \alpha) \wedge \cdots \wedge \ker(\gamma \circ \alpha^t)$$

$$\bar{X}_t = X/\omega_t$$

After observations $O(t)$, the uncertainty in $x(0)$ is represented by an element of $\bar{X}_t$ (i.e. cell of ω_t in X), inducing a map $P_t : Y^{t+1} \to \bar{X}_t$. The corresponding uncertainty in $x(t)$ is just the subset

$$\{\alpha^t(x')|x' \in P_t(O(t))\} \subseteq X$$

Discuss how to compute this subset in recursive fashion, 'online', and illustrate using the examples in (iii), (iv).

(vi) Just because a system is observable, it need not be true that much useful information can be gained about $x(t)$ from $O(t)$, even for large t. From this viewpoint investigate the 'chaotic' system defined as follows.

$$X := \{x \in \mathbb{R}|0 \leq x < 1\}$$

For $x \in \mathbb{R}$, $x \geq 0$, write $x = \text{integ}(x) + \text{fract}(x)$, where $\text{integ}(x) \in \mathbb{N}$ and $0 \leq \text{fract}(x) < 1$; and define $\alpha : X \to X$, $\gamma : X \to \mathbb{N}$ according to

$$\alpha(x) := \text{fract}(10x), \qquad \gamma(x) := \text{integ}(10x)$$

Exercise 21: Exponentiation

Let $\alpha, \beta, \gamma \in \mathcal{E}(X)$ and write $[x]$ for the cell of $x \in X$ (mod γ). Define $f_\alpha : X \to Pwr(X/\alpha)$ according to

$$f_\alpha(x) := \{P_\alpha(x')|x' \in [x]\}$$

where $P_\alpha : X \to X/\alpha$ is the canonical projection; and similarly for f_β. Write

$$\wp(\alpha) := \ker f_\alpha$$

the *exponentiation of* α *with respect to* γ. Show that $|\wp(\alpha)| \leq 2^{|\alpha|}$ and that

$$\begin{aligned} \wp(\alpha \wedge \wp(\beta)) &= \wp(\alpha) \wedge \wp(\beta) \\ &= \wp(\wp(\alpha) \wedge \beta) \end{aligned}$$

If in particular $\alpha \leq \beta$ then $\wp(\alpha) \leq \wp(\beta)$, so

$$\wp(\alpha \wedge \wp(\beta)) = \wp(\alpha)$$

Exercise 22: Independent functions

Let $\mu, \nu \in \mathcal{E}(X)$. Define the *composition* $\mu \circ \nu$ to be the binary relation $R \subseteq X \times X$ such that $(x, x') \in R$ iff for some $x'' \in X$, $x \equiv x''(\text{mod } \mu)$ and $x'' \equiv x'(\text{mod } \nu)$. Show that $\mu \circ \nu \in \mathcal{E}(X)$ iff μ and ν are *permutable*, namely $\mu \circ \nu = \nu \circ \mu$.

Now let $f : X \to Y$, $g : X \to Z$, both surjective. Say that the functions f, g are *independent* if

$$(\ker f) \circ (\ker g) = \top \in \mathcal{E}(X)$$

Show that f, g are independent iff

$$(\forall x \in X)\ g_* \circ f^{-1}(\{f(x)\}) = Z$$

or equivalently

$$(\forall x \in X)\ f_* \circ g^{-1}(\{g(x)\}) = Y$$

In that case 'knowing the result of a measurement $f(x)$ (resp. $g(x)$) provides no information about the corresponding measurement $g(x)$ (resp. $f(x)$)'.

Let $f : X \to Y$ be surjective, and $K \subseteq X$ with $f(K) = Y$. Let $\kappa = \{K, X-K\} \in \mathcal{E}(X)$. Show that $\kappa \circ \ker f = \top$, namely 'knowledge merely that $x \in K$ provides no information about $f(x) \in Y$'.

Finally, generalize to an arbitrary family of equivalences $\mu_1, \ldots, \mu_k \in \mathcal{E}(X)$ or surjections $f_i : X \to Y_i$ $(i = 1, \ldots, k)$.

1.5 Application: Internal Model Principle

In this section[2] we develop further the ideas of dynamic system (Example 1.4.7), congruence and observer (Exercise 1.4.20), to obtain a simple abstract version of the Internal Model Principle (IMP) of regulation theory. The latter states that an ideal feedback regulator will include in its internal structure a dynamic model of the external 'world' behavior that the regulator tracks or regulates against.

First we consider the restriction of an observer to an invariant subset. As in Example 1.4.7 let $\alpha : X \to X$ be a dynamic system, and let $Z \subseteq X$ be a subset of X. It is convenient to represent the subset property by means of a so-called *insertion map* as follows. Regard Z simply as a set in its own right, initially with no special relation to X, and bring in an injection

$$\mathrm{in}_Z : Z \to X : z \longmapsto \mathrm{in}_Z(z) \in X$$

Thus in_Z is a mechanism by which elements $z \in Z$ are endowed with status as elements $\mathrm{in}_Z(z) \in X$, perhaps by a transformation from 'local' (in Z) to 'global' (in X) coordinates. To highlight the distinction between Z as a 'local' object and a 'global' object embedded in X we may sometimes write

$$\mathrm{in}_Z : Z_{loc} \to X,\ Z = \mathrm{in}_Z(Z_{loc})$$

Say Z is an *α-invariant subset of* X if there exists a map $\alpha_Z : Z_{loc} \to Z_{loc}$ such that

$$\alpha \circ \mathrm{in}_Z = \mathrm{in}_Z \circ \alpha_Z$$

---

[2]Not needed in the sequel.

namely the following diagram commutes:

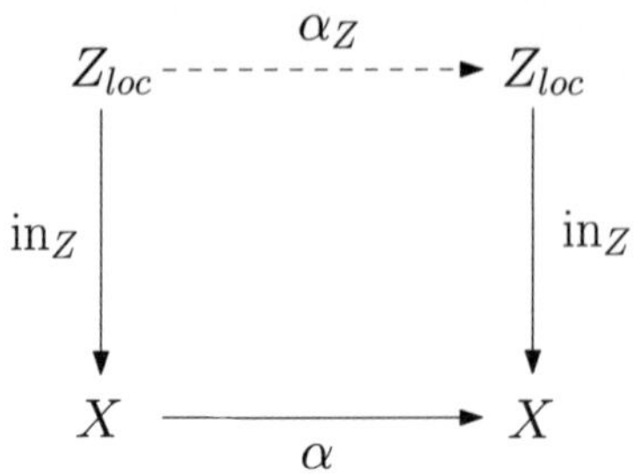

The map α_Z is the *restriction of* α *to* Z, and is defined just when Z is α-invariant. [In traditional terminology, '$Z \subseteq X$ is α-invariant if $\alpha(z) \in Z$ for every $z \in Z$'.] Often α_Z is written $\alpha|Z$.

Similarly, if $f : X \to Y$ and $Z \subseteq X$ then the *restriction of* f *to* Z, written f_Z or $f|Z$, is defined to be $f_Z := f \circ \text{in}_Z$, as in the diagram

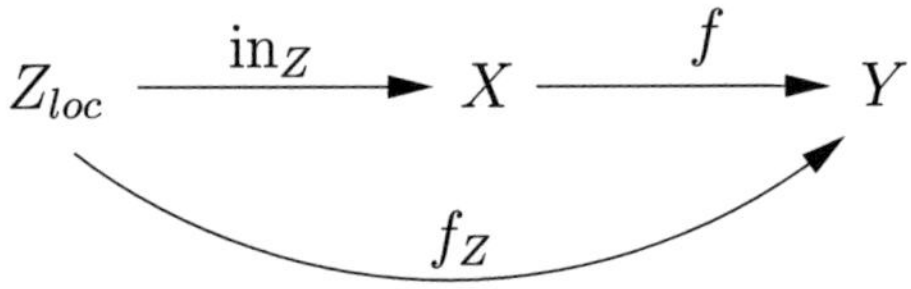

Next let $\pi \in \mathcal{E}(X)$ be an equivalence relation on X. We restrict π to $Z \subseteq X$ simply by intersecting the cells of π with Z. Namely $\pi|Z \in \mathcal{E}(Z)$ is defined by

$$z \equiv z'(\text{mod } \pi|Z) \text{ iff } \text{in}_Z(z) \equiv \text{in}_Z(z')(\text{mod } \pi) \tag{1.1}$$

for all $z, z' \in Z$. This rule inserts cells of $Z/\pi|Z$ (considered as subsets of Z) into cells of X/π (considered as subsets of X), and in turn determines an insertion

$$\text{in}_{Z/\pi|Z} : Z/\pi|Z \to X/\pi$$

Indeed, distinct cells of $\pi|Z$ are subsets (in Z) of distinct cells of π (in X). In terms of natural projections $p_{(.)}$ we now have the diagram

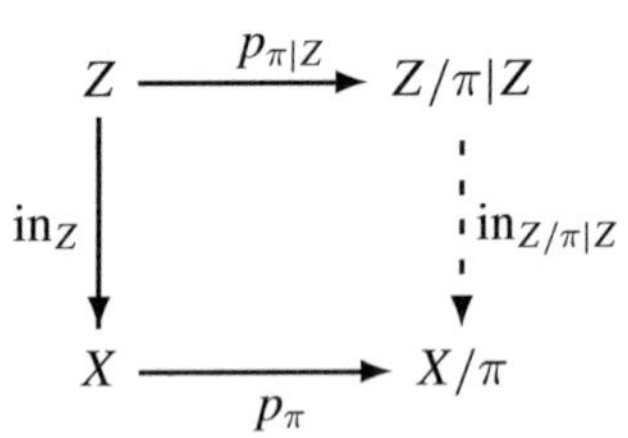

(1.2)

Exercise 1: Verify in detail that $\pi|Z$ as defined by (1.1) is an element of $\mathcal{E}(Z)$, and then that $\text{in}_{Z/\pi|Z}$ in (1.2) exists, is unique, and is injective. Alternatively, one could proceed more formally as follows. In the diagram below, let $\rho := \ker(P_\pi \circ \text{in}_Z)$ and consider the canonical factorization

$$P_\pi \circ \text{in}_Z = j \circ P_\rho \quad (j \text{ injective})$$

Then ρ is the 'restriction of π to Z' while j is the required insertion.

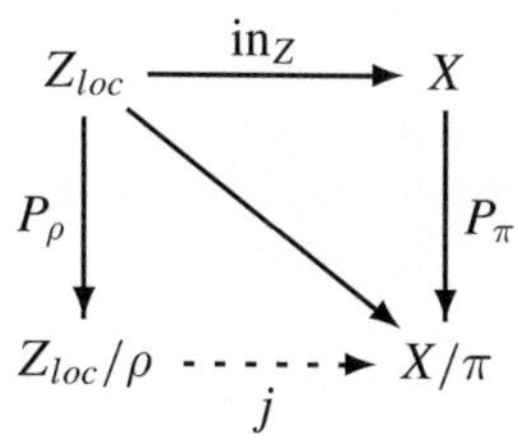

As in Exercise 1.4.20 let $\gamma : X \to Y$ denote an 'output' map for the dynamic system (X, α). If $Z \subseteq X$ and $\alpha(Z) \subseteq Z$ we could compute an 'observer on Z' in two different ways:

1. Compute the 'local observer' on Z using the restrictions $\alpha|Z$ and $\gamma|Z$. Call the result $\omega_Z \in \mathcal{E}(Z)$.
2. Compute the 'global observer' $\omega \in \mathcal{E}(X)$, and then restrict to Z to get $\omega|Z \in \mathcal{E}(Z)$.

Happily the results coincide.

Proposition 2 (Local observer = restricted global observer)

$$\omega_Z = \omega|Z$$

□

We say that $Z \subseteq X$ with $\alpha(Z) \subseteq Z$ is *detectable* with respect to (X, α, γ) if $\omega|Z = \perp$, namely Z is 'locally observable'.

Exercise 3: Prove Proposition 2.

Consider now the following standard block diagram for regulation.

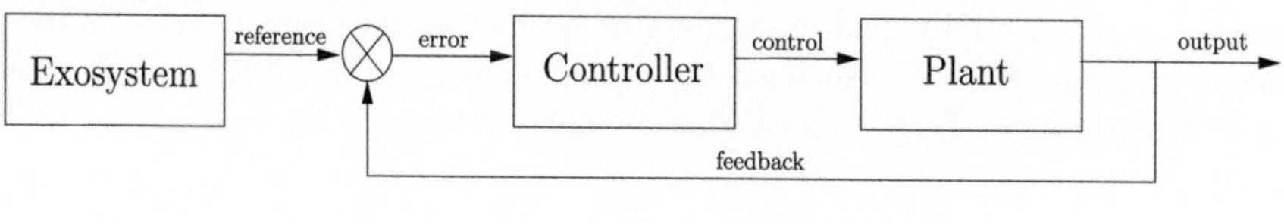

Here Exosystem (**E**), Controller (**C**) and Plant (**P**) are dynamic systems which generate, via suitable output maps, the reference, control and output signals respectively. The objective of regulation is to ensure that the output signal coincides (eventually) with the reference, namely the system 'tracks'. To this end the output is 'fed back' and compared (via $\otimes$) to the reference, and the resulting tracking error signal used to 'drive' the controller. The latter in turn controls the plant, causing its output to approach the reference, so that the tracking error eventually (perhaps as $t \to \infty$) approaches 'zero'.

Our aim is to show that this setup implies, when suitably formalized, that the controller incorporates a model of the exosystem: this statement is the Internal Model Principle.

We formalize the exosystem **E** as a dynamic system $(X_{\mathbf{E}}, \alpha_{\mathbf{E}})$, and the total system, comprising **E**, **C** and **P**, as a dynamic system (X, α). Very often X will be of simple product form $X = X_{\mathbf{E}} \times X_{\mathbf{C}} \times X_{\mathbf{P}}$ and $\alpha = \alpha_{\mathbf{E}} \times \alpha_{\mathbf{C}} \times \alpha_{\mathbf{P}} : X \to X$, where

$$\begin{aligned}
\alpha_{\mathbf{E}} &: X_{\mathbf{E}} \to X_{\mathbf{E}} \\
\alpha_{\mathbf{C}} &: X_{\mathbf{E}} \times X_{\mathbf{C}} \times X_{\mathbf{P}} \to X_{\mathbf{C}} \\
\alpha_{\mathbf{P}} &: X_{\mathbf{E}} \times X_{\mathbf{C}} \times X_{\mathbf{P}} \to X_{\mathbf{P}}
\end{aligned}$$

In fact we do not need this degree of detail here. We shall, however, assume that the feedback loop, comprising **C**, **P** interconnected as shown, enjoys a stability property with the consequence that **E** 'induces' a corresponding action on a unique α-invariant subset of X, namely there exists an injection $i_{\mathbf{E}} : X_{\mathbf{E}} \to X$ such that the following diagram commutes:

$$\begin{array}{ccc}
X_{\mathbf{E}} & \xrightarrow{\alpha_{\mathbf{E}}} & X_{\mathbf{E}} \\
{\scriptstyle i_{\mathbf{E}}}\downarrow & & \downarrow{\scriptstyle i_{\mathbf{E}}} \\
X & \xrightarrow[\alpha]{} & X
\end{array} \tag{1.3}$$

Thus $i_{\mathbf{E}}$ is assumed to be determined by

$$\alpha \circ i_{\mathbf{E}} = i_{\mathbf{E}} \circ \alpha_{\mathbf{E}} \tag{1.4}$$

together with injectivity, and uniqueness of $i_{\mathbf{E}}(X_{\mathbf{E}})$. In that case $i_{\mathbf{E}}$ itself is only unique up to an automorphism $\varphi_{\mathbf{E}} : X_{\mathbf{E}} \to X_{\mathbf{E}}$, namely $\varphi_{\mathbf{E}}$ is a bijection such that $\varphi_{\mathbf{E}} \circ \alpha_{\mathbf{E}} = \alpha_{\mathbf{E}} \circ \varphi_{\mathbf{E}}$. Then $i_{\mathbf{E}}$ can always be replaced by $\tilde{i}_{\mathbf{E}} := i_{\mathbf{E}} \circ \varphi_{\mathbf{E}}$.

Exercise 4: Suppose $X_{\mathbf{E}} = \mathbb{R}^1$, $X_{\mathbf{S}} = \mathbb{R}^1$, $X = X_{\mathbf{E}} \oplus X_{\mathbf{S}}$,

$$\alpha_{\mathbf{E}} = 2, \ \alpha = \begin{bmatrix} 2 & 0 \\ 1 & 1/3 \end{bmatrix}$$

Show that $i_{\mathbf{E}} = \begin{bmatrix} 1 \\ 3/5 \end{bmatrix}$, determined by (1.4), is the eigenvector of α with eigenvalue 2, and that $i_{\mathbf{E}}(X_{\mathbf{E}})$ is its span.

As to the controller, we assume that its state set $X_{\mathbf{C}}$ can be projected from X by a surjection $\gamma : X \to X_{\mathbf{C}}$.

The target subset for regulation (conventionally, states $x \in X$ where the tracking error is zero) will be modeled as a subset $K \subseteq X$. We assume that the condition $x \in K$ places no restriction on $\gamma(x) \in X_{\mathbf{C}}$; that is, knowledge merely that $x \in K$ yields no information about the control state $x_{\mathbf{C}} = \gamma(x) \in X_{\mathbf{C}}$; briefly

$$\gamma(K) = X_{\mathbf{C}} \tag{1.5}$$

namely any control state $x_{\mathbf{C}} \in X_{\mathbf{C}}$ corresponds to some $x \in K$.

From now on we write $i_{\mathbf{E}}(X_{\mathbf{E}}) = \tilde{X}_{\mathbf{E}} \subseteq X$. We assume that the controller is capable of identifying the state in $\tilde{X}_{\mathbf{E}}$ induced in X by the exosystem. Thus consider the time history

$$x(0) = x_0 \in X, \ x(t+1) = \alpha(x(t)), \ t \in \mathbb{N}$$

Suppose $x_0 \in \tilde{X}_{\mathbf{E}}$ and the controller records

$$x_{\mathbf{C}}(t; x_0) = \gamma(x(t)), \ t \in \mathbb{N}$$

Then we assume x_0 can be determined from the data

$$\{x_{\mathbf{C}}(t; x_0) | t \in \mathbb{N}\}$$

This means precisely that

$$\tilde{X}_{\mathbf{E}} \text{ is detectable with respect to } (X, \alpha, \gamma) \tag{1.6}$$

Next we turn to feedback structure. By this we mean that the controller is externally driven only when the system state deviates from the desired subset $K \subseteq X$ (where 'the error is zero'). Equivalently, as long as $x \in K$, the controller behaves as an autonomous dynamic system on $X_{\mathbf{C}}$. Formally, let $x \in K$. The control state is $x_{\mathbf{C}} = \gamma(x) \in X_{\mathbf{C}}$. Feedback implies that the 'next' controller state

$$x'_{\mathbf{C}} = \gamma(x') = \gamma(\alpha(x)) = \gamma \circ \alpha(x)$$

depends only on $x_{\mathbf{C}} = \gamma(x)$; i.e. for $x \in K$, $\gamma \circ \alpha(x)$ can be computed from $\gamma(x)$. Equivalently

$$\ker(\gamma|K) \le \ker(\gamma \circ \alpha|K) \tag{1.7}$$

There is no implication, of course, that $x' = \alpha(x) \in K$; in general K itself will not be α-invariant.

Finally we bring in the regulation condition, namely that the induced subset $\tilde{X}_{\mathbf{E}}$ lies in K:

$$\tilde{X}_{\mathbf{E}} \subseteq K \tag{1.8}$$

In applications, loop stability will typically ensure that $\tilde{X}_{\mathbf{E}} \subseteq X$ is a global attractor, namely for any initial condition $x(0) \in X$ the motion $x(t)$ will converge to $\tilde{X}_{\mathbf{E}}$ as $t \to \infty$; then (1.8) will ensure asymptotic regulation, as required. Note that $K \subseteq X$ can be modeled by its insertion $\kappa : K_{loc} \to X$, leading in view of (1.8) to an injection $j_{\mathbf{E}} : X_{\mathbf{E}} \to K_{loc}$ as shown below.

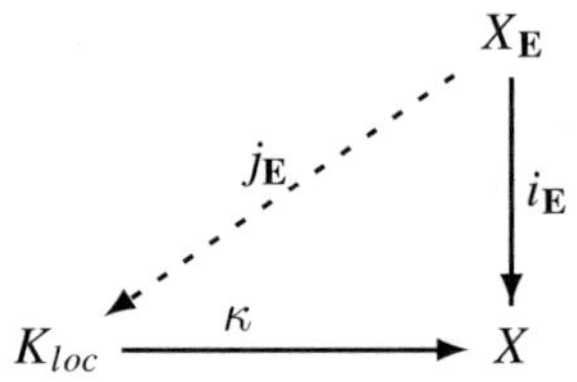

$$i_{\mathbf{E}} = \kappa \circ j_{\mathbf{E}} \tag{1.9}$$

Exercise 5: Prove that $j_{\mathbf{E}}$ in (1.9) exists, is unique, and is injective.

Now we have the sought-for result. Write

$$\tilde{\alpha}_{\mathbf{E}} = \alpha|\tilde{X}_{\mathbf{E}}, \ \tilde{\gamma}_{\mathbf{E}} = \gamma|\tilde{X}_{\mathbf{E}}$$

Theorem 6: **Internal Model Principle**

Subject to the foregoing assumptions:

(i) There exists a unique mapping $\alpha_{\mathbf{C}} : X_{\mathbf{C}} \to X_{\mathbf{C}}$ determined by

$$\alpha_{\mathbf{C}} \circ \gamma|K = \gamma \circ \alpha|K$$

(ii) $\alpha_{\mathbf{C}} \circ \tilde{\gamma}_{\mathbf{E}} = \tilde{\gamma}_{\mathbf{E}} \circ \tilde{\alpha}_{\mathbf{E}}$

(iii) $\tilde{\gamma}_{\mathbf{E}}$ is injective

Proof

(i) Let $x_{\mathbf{C}} \in X_{\mathbf{C}}$. By (1.5) there is $x \in K$ (so $x \in X$) with $\gamma(x) = x_{\mathbf{C}}$. Define $\alpha_{\mathbf{C}}(x_{\mathbf{C}}) := \gamma \circ \alpha(x)$. The definition is unambiguous, for $x' \in K$ and $\gamma(x') = x_{\mathbf{C}}$ imply $x' \equiv x(\text{mod } \ker \gamma)$, namely (as $x, x' \in K$) $x' \equiv x(\text{mod } \ker \gamma|K)$. By (1.7), $x' \equiv x(\text{mod } \ker(\gamma \circ \alpha|K))$ and so $\gamma \circ \alpha(x') = \gamma \circ \alpha(x)$. The situation is displayed below.

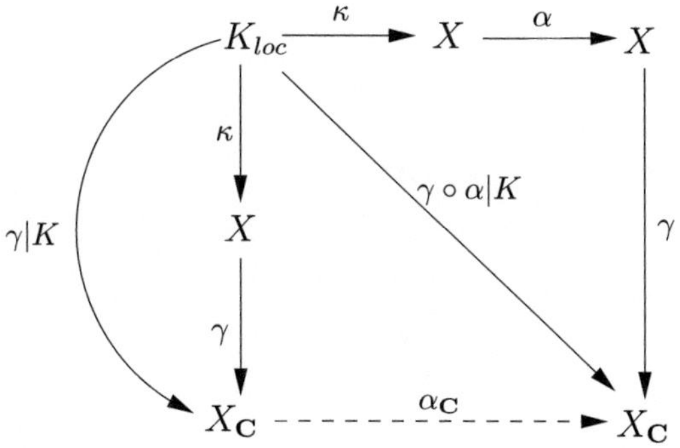

(ii) Let $x \in \tilde{X}_{\mathbf{E}}$. Since $\tilde{X}_{\mathbf{E}} \subseteq K$, from (i) we get $\alpha_{\mathbf{C}} \circ \gamma(x) = \gamma \circ \alpha(x)$. But $\alpha(x) = \tilde{\alpha}_{\mathbf{E}}(x)$ and $\alpha(x) \in \tilde{X}_{\mathbf{E}}$, so

$$\alpha_{\mathbf{C}} \circ \tilde{\gamma}_{\mathbf{E}}(x) = \tilde{\gamma}_{\mathbf{E}} \circ \tilde{\alpha}_{\mathbf{E}}(x)$$

as claimed.

(iii) Recalling Exercise 1.4.20, let $\omega \in \mathcal{E}(X)$ be the observer for (γ, α), and $\tilde{\omega}_{\mathbf{E}} := \omega|\tilde{X}_{\mathbf{E}}$ its restriction to $\tilde{X}_{\mathbf{E}}$. By observer theory

$$\begin{aligned} \tilde{\omega}_{\mathbf{E}} &= \sup\{\omega' \in \mathcal{E}(\tilde{X}_{\mathbf{E}}) | \omega' \leq \ker \tilde{\gamma}_{\mathbf{E}} \wedge (\omega' \circ \tilde{\alpha}_{\mathbf{E}})\} \\ &= \bot \end{aligned}$$

using (1.6). Also, (1.7) with restrictions to $\tilde{X}_{\mathbf{E}}$ implies

$$\ker \tilde{\gamma}_{\mathbf{E}} \leq \ker(\tilde{\gamma}_{\mathbf{E}} \circ \tilde{\alpha}_{\mathbf{E}}) = (\ker \tilde{\gamma}_{\mathbf{E}}) \circ \tilde{\alpha}_{\mathbf{E}}$$

Therefore $\ker \tilde{\gamma}_{\mathbf{E}} \leq \tilde{\omega}_{\mathbf{E}} = \bot$, namely $\tilde{\gamma}_{\mathbf{E}}$ is injective as asserted. □

Our results can be displayed in the commutative diagram:

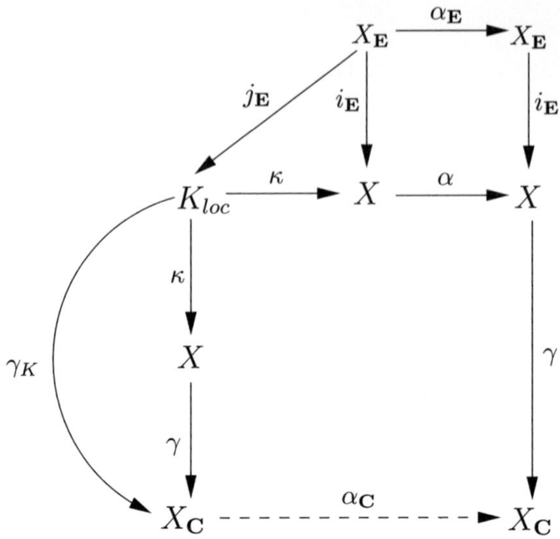

Here $\gamma_K := \gamma | K = \gamma \circ \kappa$. Thus

$$\begin{aligned}\tilde{\gamma}_{\mathbf{E}} = \gamma|\tilde{X}_{\mathbf{E}} &= \gamma \circ i_{\mathbf{E}} = \gamma \circ (\kappa \circ j_{\mathbf{E}}) \\ &= (\gamma \circ \kappa) \circ j_{\mathbf{E}} \\ &= \gamma_K \circ j_{\mathbf{E}}\end{aligned}$$

Finally, therefore, we have the diagram

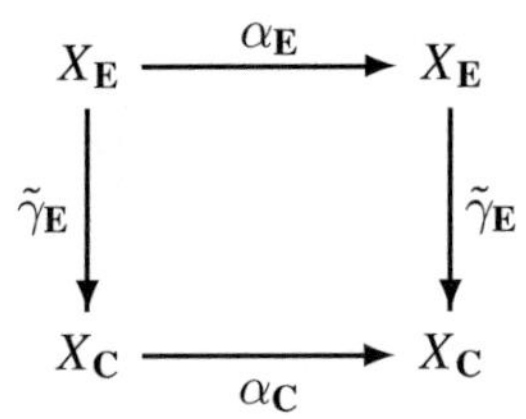

This may be paraphrased by saying that the controller dynamics faithfully simulate the dynamics of the exosystem under the condition of perfect regulation. The 'internal model' in the controller is the dynamic system $(\tilde{\gamma}_{\mathbf{E}}(X_{\mathbf{E}}), \alpha_C | \tilde{\gamma}_{\mathbf{E}}(X_{\mathbf{E}}))$.

Example 7: Let the exosystem **E** be given by

$$X_{\mathbf{E}} = \{0, 1, \ldots, N-1\}, \;\; \alpha_{\mathbf{E}}(i) = i + 1 (\mathrm{mod}\; N), \;\; i \in X_{\mathbf{E}}$$

We identify the plant **P** with the controller **C** and take

$$X_{\mathbf{C}} = \{0, 1, \ldots, N-1\} =: \mathbf{N}, \text{ say}$$

We take $X = X_{\mathbf{E}} \times X_{\mathbf{C}}$ and let $\alpha : X \to X$ be given by

$$\alpha(i, j) = \begin{cases} (i+1, j+1) & \text{if } j = i \\ (i+1, j) & \text{if } j \neq i \end{cases}$$

where evaluations of i, j are (mod N). Thus controller logic is to maintain tracking of $x_{\mathbf{C}} = j \in X_{\mathbf{C}}$ with respect to $x_{\mathbf{E}} = i \in X_{\mathbf{E}}$, if $j = i$, or else to delay $x_{\mathbf{C}}$ until tracking is achieved.

For the action induced by **E** in X we have $i_{\mathbf{E}} : X_{\mathbf{E}} \to X$ given by

$$i_{\mathbf{E}}(i) = (i, i), \; i \in \mathbf{N}$$

Clearly $\alpha \circ i_{\mathbf{E}} = i_{\mathbf{E}} \circ \alpha_{\mathbf{E}}$ as required by (1.4), and $i_{\mathbf{E}}$ is essentially unique and is injective.

Next, $\gamma : X \to X_{\mathbf{C}}$ is given by

$$\gamma(i, j) = j, \; (i, j) \in \mathbf{N} \times \mathbf{N}$$

Setting

$$\begin{aligned} K &= \{(i, j) \in X | j = i\} \\ &= \{(i, i) | i \in \mathbf{N}\} \end{aligned}$$

we have $\gamma(K) = \{i | i \in \mathbf{N}\} = X_{\mathbf{C}}$ as in (1.5).

To check detectability as in (1.6) we compute the local observer corresponding to the restriction of (γ, α) to

$$\tilde{X}_{\mathbf{E}} = i_{\mathbf{E}}(X_{\mathbf{E}}) = \{(i, i) | i \in \mathbf{N}\}$$

Then

$$\tilde{\gamma}_{\mathbf{E}}(i, i) = i, \; \tilde{\alpha}_{\mathbf{E}}(i, i) = (i+1, i+1)$$

and thus $\tilde{\omega}_{\mathbf{E}} = \bot$ as required. The regulation condition (1.8) is simply

$$\tilde{X}_{\mathbf{E}} = i_{\mathbf{E}}(X_{\mathbf{E}}) = \{(i, i) | i \in \mathbf{N}\} = K$$

For (1.9), we can model

$$K_{loc} = \mathbf{N}, \; \kappa : K_{loc} \to X : i \longmapsto (i, i)$$

while

$$j_{\mathbf{E}} : X_{\mathbf{E}} \to K_{loc} : i \longmapsto i$$

so that evidently $i_{\mathbf{E}} = \kappa \circ j_{\mathbf{E}}$ as stated.

Finally, Theorem 6 provides

$$\alpha_{\mathbf{C}} : X_{\mathbf{C}} \to X_{\mathbf{C}} : i \longmapsto i + 1$$

with

$$\alpha_{\mathbf{C}} \circ \tilde{\gamma}_{\mathbf{E}}(i, i) = \alpha_{\mathbf{C}}(i) = i + 1$$
$$\tilde{\gamma}_{\mathbf{E}} \circ \tilde{\alpha}_{\mathbf{E}}(i, i) = \tilde{\gamma}_{\mathbf{E}}(i + 1, i + 1) = i + 1$$

as the theorem asserts.

It is interesting to note that $\tilde{X}_{\mathbf{E}} \subseteq X$ plays the role of 'global attractor' in the sense that if $x(0) = (i_0, j_0)$ is arbitrary, then

$$x(t) \longrightarrow \tilde{X}_{\mathbf{E}} \text{ as } t \to \infty \tag{1.10}$$

In fact $x(t) \in \tilde{X}_{\mathbf{E}}$ for all $t \geq N-1$. With this feature we can augment the diagram (1.3) as follows. Consider the mapping

$$p_{\mathbf{E}} : X \to X_{\mathbf{E}} : (i, j) \longmapsto i$$

and notice that ker $p_{\mathbf{E}}$ is a congruence for α, namely

$$\ker p_{\mathbf{E}} \leq \ker(p_{\mathbf{E}} \circ \alpha)$$

Thus we have the diagram

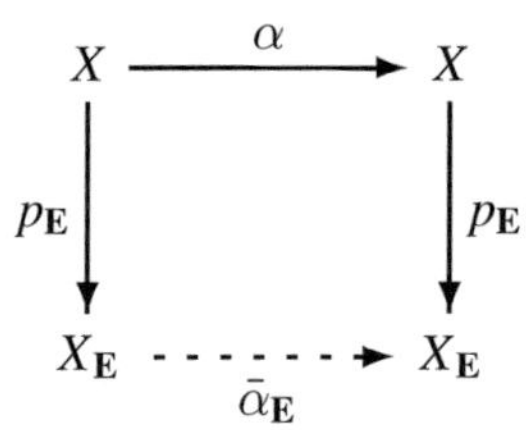

for a unique induced map $\bar{\alpha}_{\mathbf{E}}$. Since for $i \in X_{\mathbf{E}}$ we have $i = p_{\mathbf{E}}(i, j)$ for any j, there follows:

$$\begin{aligned} \bar{\alpha}_{\mathbf{E}}(i) &= \bar{\alpha}_{\mathbf{E}} \circ p_{\mathbf{E}}(i, j) \\ &= p_{\mathbf{E}} \circ \alpha(i, j) \\ &= i + 1 \\ &= \alpha_{\mathbf{E}}(i) \end{aligned}$$

and we can identify $\bar{\alpha}_{\mathbf{E}} = \alpha_{\mathbf{E}}$. Also

$$p_{\mathbf{E}} \circ i_{\mathbf{E}}(i) = p_{\mathbf{E}}(i, i) = i$$

namely $p_{\mathbf{E}} \circ i_{\mathbf{E}} = id_{\mathbf{E}}$, the identity on $X_{\mathbf{E}}$. Thus the diagram (1.3) can be augmented to

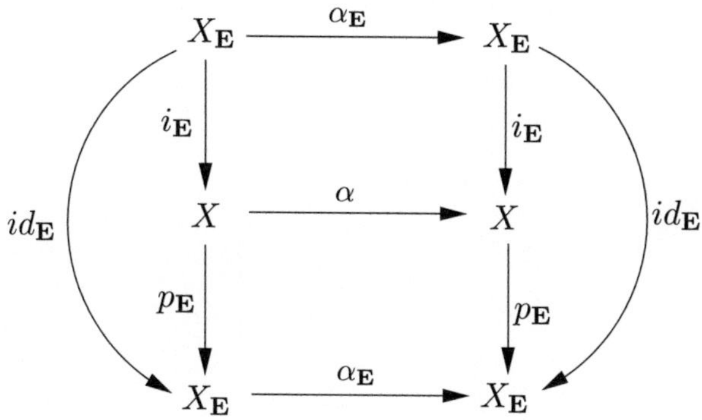

Intuitively, $p_{\mathbf{E}}$ 'projects out the stable behavior of α' to recover the 'unstable behavior' $\alpha_{\mathbf{E}}$. Notice that the augmented diagram is valid whenever X is of product form, having $X_{\mathbf{E}}$ as a factor, and $p_{\mathbf{E}} : X \to X_{\mathbf{E}}$.

Exercise 8: In a sketch of $\mathbf{N} \times \mathbf{N}$ locate $\tilde{X}_{\mathbf{E}}$ and exhibit the convergence (1.10) and the motion on $\tilde{X}_{\mathbf{E}}(= K)$. What are the cells of ker $p_{\mathbf{E}}$ and how do they move under $\bar{\alpha}_{\mathbf{E}}$?

Exercise 9: Extend Exercise 4 by taking $X_{\mathbf{P}} = X_{\mathbf{S}}$ and introducing suitable feedback control structure to ensure that $\mathbf{P}$ tracks $\mathbf{E}$ as $t \to \infty$. Identify all the elements discussed in this section.

Exercise 10: Extend Example 7 to a richer structure with physical significance.

Exercise 11: **Parametrized Internal Model**

In this exercise we propose a generalization of Theorem 6 which accommodates a family of plant models distinguished by a parameter μ with values in a set M. With $X_{\mathbf{E}}$, $\alpha_{\mathbf{E}}$ as before, now consider the diagram

$$\begin{array}{ccc} M \times X_{\mathbf{E}} & \xrightarrow{id_M \times \alpha_{\mathbf{E}}} & M \times X_{\mathbf{E}} \\ {\scriptstyle id_M \times i_{\mathbf{E}}} \downarrow & & \downarrow {\scriptstyle id_M \times i_{\mathbf{E}}} \\ M \times X & \xrightarrow[id_M \times \alpha]{} & M \times X \end{array} \qquad (1.11)$$

Here

$$(id_M \times \alpha_{\mathbf{E}})(\mu, x_{\mathbf{E}}) = (\mu, \alpha_{\mathbf{E}}(x_{\mathbf{E}})) \in M \times X_{\mathbf{E}} \tag{1.11a}$$

$$(id_M \times i_{\mathbf{E}})(\mu, x_{\mathbf{E}}) = (\mu, i_{\mathbf{E}}(\mu, x_{\mathbf{E}})) \in M \times X \tag{1.11b}$$

$$(id_M \times \alpha)(\mu, x) = (\mu, \alpha(\mu, x)) \in M \times X \tag{1.11c}$$

Thus α, but not $\alpha_{\mathbf{E}}$, will depend on μ and, as the diagram is assumed to commute, $i_{\mathbf{E}}$ will depend on μ as well. The regulation condition (1.9) generalizes to:

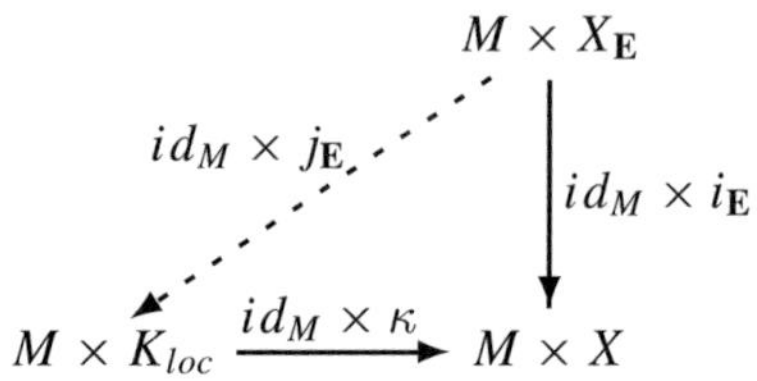

(1.12)

Here

$$(id_M \times j_{\mathbf{E}})(\mu, x_{\mathbf{E}}) = (\mu, j_{\mathbf{E}}(\mu, x_{\mathbf{E}})) \in M \times K_{loc}$$
$$(id_M \times \kappa)(\mu, \xi) = (\mu, \kappa(\xi)) \in M \times X$$

Thus $K_{loc} \stackrel{\kappa}{\longrightarrow} X$ is a fixed subset of X as before. Note that $i_{\mathbf{E}} = \kappa \circ j_{\mathbf{E}}$. The diagram (1.12) states that regulation holds for all values of $\mu \in M$.

For simplicity we assume an explicit product structure

$$\alpha = \alpha_{\mathbf{E}} \times \alpha_{\mathbf{P}} \times \alpha_{\mathbf{C}}, \; X = X_{\mathbf{E}} \times X_{\mathbf{P}} \times X_{\mathbf{C}} \tag{1.13}$$

where

$$\begin{aligned} \alpha_{\mathbf{P}} &: M \times X \longrightarrow X_{\mathbf{P}} : (\mu, x) \mapsto \alpha_{\mathbf{P}}(\mu, x) \\ \alpha_{\mathbf{C}} &: M \times X \longrightarrow X_{\mathbf{C}} : (\mu, x) \mapsto \alpha_{\mathbf{C}}(\mu, x) \end{aligned} \tag{1.14}$$

for suitable maps $\alpha_{\mathbf{P}}, \alpha_{\mathbf{C}}$. Here the formal dependence of $\alpha_{\mathbf{C}}$ on μ is really gratuitous. In practice it is usually assumed that the controller can be fixed and made independent of plant parameters; on this basis we shall write simply $\alpha_{\mathbf{C}} = \alpha_{\mathbf{C}}(x)$.

In view of (1.13), (1.14) we write

$$i_{\mathbf{E}} = id_{\mathbf{E}} \times i_{\mathbf{P}} \times i_{\mathbf{C}}$$

where

$$i_{\mathbf{P}} : M \times X_{\mathbf{E}} \longrightarrow X_{\mathbf{P}} : (\mu, x_{\mathbf{E}}) \mapsto i_{\mathbf{P}}(\mu, x_{\mathbf{E}})$$
$$i_{\mathbf{C}} : M \times X_{\mathbf{E}} \longrightarrow X_{\mathbf{C}} : (\mu, x_{\mathbf{E}}) \mapsto i_{\mathbf{C}}(\mu, x_{\mathbf{E}})$$

The diagram (1.11) is now equivalent to:

$$\alpha_{\mathbf{P}}(\mu, x_{\mathbf{E}}, i_{\mathbf{P}}(\mu, x_{\mathbf{E}}), i_{\mathbf{C}}(\mu, x_{\mathbf{E}})) = i_{\mathbf{P}}(\mu, \alpha_{\mathbf{E}}(x_{\mathbf{E}})) \tag{1.15a}$$

$$\alpha_{\mathbf{C}}(x_{\mathbf{E}}, i_{\mathbf{P}}(\mu, x_{\mathbf{E}}), i_{\mathbf{C}}(\mu, x_{\mathbf{E}})) = i_{\mathbf{C}}(\mu, \alpha_{\mathbf{E}}(x_{\mathbf{E}})) \tag{1.15b}$$

to hold for all $\mu \in M$, $x_{\mathbf{E}} \in X_{\mathbf{E}}$.

The regulation condition will be specialized to constrain the pair $(x_{\mathbf{E}}, x_{\mathbf{P}})$ but not $x_{\mathbf{C}}$. Thus the target set for regulation is

$$K = K_{\mathbf{EP}} \times X_{\mathbf{C}}$$

where $K_{\mathbf{EP}} \subseteq X_{\mathbf{E}} \times X_{\mathbf{P}}$. We therefore take

$$K_{loc} = K_{\mathbf{EP},loc} \times X_{\mathbf{C}}$$

and

$$(id_M \times \kappa)(\mu, x_{\mathbf{EP},loc}, x_{\mathbf{C}}) = (\mu, \kappa(x_{\mathbf{EP},loc}, x_{\mathbf{C}}))$$

where

$$\kappa(x_{\mathbf{EP},loc}, x_{\mathbf{C}}) = (x_{\mathbf{EP}}, x_{\mathbf{C}}) \in K_{\mathbf{EP}} \times X_{\mathbf{C}}$$

with $x_{\mathbf{EP}}$ independent of $x_{\mathbf{C}}$. Regulation requires that

$$i_{\mathbf{E}}(\mu, x_{\mathbf{E}}) = (x_{\mathbf{E}}, i_{\mathbf{P}}(\mu, x_{\mathbf{E}}), i_{\mathbf{C}}(\mu, x_{\mathbf{E}}))$$

with

$$(x_{\mathbf{E}}, i_{\mathbf{P}}(\mu, x_{\mathbf{E}})) \in K_{\mathbf{EP}}$$

By (1.12),

$$i_{\mathbf{E}}(\mu, x_{\mathbf{E}}) = \kappa \circ j_{\mathbf{E}}(\mu, x_{\mathbf{E}})$$

so that

$$(x_{\mathbf{E}}, i_{\mathbf{P}}(\mu, x_{\mathbf{E}}), i_{\mathbf{C}}(\mu, x_{\mathbf{E}})) = \kappa \circ j_{\mathbf{E}}(\mu, x_{\mathbf{E}}) \tag{1.16}$$

Let $P_{\mathbf{P}} : X \to X_{\mathbf{P}}$ be the natural projection. Then (1.16) implies

$$i_{\mathbf{P}}(\mu, x_{\mathbf{E}}) = P_{\mathbf{P}} \circ \kappa \circ j_{\mathbf{E}}(\mu, x_{\mathbf{E}}) \tag{1.17}$$

If this expression is substituted for $i_{\mathbf{P}}(\mu, x_{\mathbf{E}})$ in (1.15), we obtain equations of the form

$$\tilde{\alpha}_{\mathbf{P}}(\mu, x_{\mathbf{E}}, j_{\mathbf{E}}(\mu, x_{\mathbf{E}}), i_{\mathbf{C}}(\mu, x_{\mathbf{E}})) = P_{\mathbf{P}} \circ \kappa \circ j_{\mathbf{E}}(\mu, \alpha_{\mathbf{E}}(x_{\mathbf{E}})) \tag{1.18a}$$

$$\tilde{\alpha}_{\mathbf{C}}(x_{\mathbf{E}}, j_{\mathbf{E}}(\mu, x_{\mathbf{E}}), i_{\mathbf{C}}(\mu, x_{\mathbf{E}})) = i_{\mathbf{C}}(\mu, \alpha_{\mathbf{E}}(x_{\mathbf{E}})) \tag{1.18b}$$

to hold for all $\mu \in M$, $x_{\mathbf{E}} \in X_{\mathbf{E}}$. We make the basic assumption that $X_{\mathbf{C}}$ and $\alpha_{\mathbf{C}}$ have been chosen such that the functional equations (1.18) are solvable for functions $j_{\mathbf{E}}$ and $i_{\mathbf{C}}$, identically in μ and $x_{\mathbf{E}}$. This is the property of *structurally stable* regulation. We note from (1.18b) that if the function $\tilde{\alpha}_{\mathbf{C}}$ can be shown to be independent of its first two arguments $x_{\mathbf{E}}, j_{\mathbf{E}}$ (i.e. the controller is autonomous on K), and if $i_{\mathbf{C}}$ is injective, then error feedback structure and the internal model property of the controller will follow. The following is a simple illustration, in which we focus on regulation, for simplicity taking into account structural, but not dynamic, stability.

Example 12: **Oscillator driving first-order lag**

Let **E** be given by

$$x_1' = x_2,\ x_2' = -x_1,\ x_{\mathbf{E}} = [x_1\ x_2] \in \mathbb{R}^{2\times 1}$$

P by

$$x_3' = \mu_1 x_1 + \mu_3 x_3 + \mu_4 x_4,\ x_{\mathbf{P}} = x_3 \in \mathbb{R}^1,\ \mu = [\mu_1\ \mu_3\ \mu_4] \in \mathbb{R}^3$$

and **C** by

$$\begin{aligned} x_4' &= c_{41}x_1 + c_{43}x_3 + c_{44}x_4 + c_{45}x_5 \\ x_5' &= c_{51}x_1 + c_{53}x_3 + c_{54}x_4 + c_{55}x_5,\ c_{ij} \in \mathbb{R},\ x_{\mathbf{C}} = [x_4\ x_5] \in \mathbb{R}^{2\times 1} \end{aligned}$$

Thus in matrix form the relevant maps for this linear system are:

$$\begin{aligned} \bar{\alpha}_{\mathbf{E}} &= \begin{bmatrix} 0\ 1\ 0\ 0\ 0 \\ -1\ 0\ 0\ 0\ 0 \end{bmatrix} \in \mathbb{R}^{2\times 5} \\ \bar{\alpha}_{\mathbf{P}} &= [\mu_1\ 0\ \mu_3\ \mu_4\ 0] \in \mathbb{R}^{1\times 5} \\ \bar{\alpha}_{\mathbf{C}} &= \begin{bmatrix} c_{41}\ 0\ c_{43}\ c_{44}\ c_{45} \\ c_{51}\ 0\ c_{53}\ c_{54}\ c_{55} \end{bmatrix} \in \mathbb{R}^{2\times 5} \\ \bar{\alpha} &= \begin{bmatrix} \bar{\alpha}_{\mathbf{E}} \\ \bar{\alpha}_{\mathbf{P}} \\ \bar{\alpha}_{\mathbf{C}} \end{bmatrix} \in \mathbb{R}^{5\times 5} \end{aligned}$$

Let

$$i_{\mathbf{E}} = \begin{bmatrix} 1 & 0 \\ 0 & 1 \\ i_1 & i_2 \\ i_{11} & i_{12} \\ i_{21} & i_{22} \end{bmatrix} \in \mathbb{R}^{5\times 2}$$

Then (1.11) translates as

$$\begin{bmatrix} 0 & 1 & 0 & 0 & 0 \\ -1 & 0 & 0 & 0 & 0 \\ \mu_1 & 0 & \mu_3 & \mu_4 & 0 \\ c_{41} & 0 & c_{43} & c_{44} & c_{45} \\ c_{51} & 0 & c_{53} & c_{54} & c_{55} \end{bmatrix} \begin{bmatrix} 1 & 0 \\ 0 & 1 \\ i_1 & i_2 \\ i_{11} & i_{12} \\ i_{21} & i_{22} \end{bmatrix} = \begin{bmatrix} 1 & 0 \\ 0 & 1 \\ i_1 & i_2 \\ i_{11} & i_{12} \\ i_{21} & i_{22} \end{bmatrix} \begin{bmatrix} 0 & 1 \\ -1 & 0 \end{bmatrix} \tag{1.19}$$

For the regulation condition we assume $x_1 - x_3 = 0$, namely

$$x \in K := \ker\,[1\ 0\ -1\ 0\ 0] \subseteq X = \mathbb{R}^5$$

so that we may take

$$K_{loc} = \mathbb{R}^4 = \mathrm{im}\ I^{4\times 4}$$

and then $\kappa : K_{loc} \longrightarrow K$ is represented by

$$\kappa := \begin{bmatrix} 1 & 0 & 0 & 0 \\ 0 & 1 & 0 & 0 \\ 1 & 0 & 0 & 0 \\ 0 & 0 & 1 & 0 \\ 0 & 0 & 0 & 1 \end{bmatrix}$$

Now (1.12), or $\kappa \circ j_{\mathbf{E}} = i_{\mathbf{E}}$, becomes

$$\begin{bmatrix} 1 & 0 & 0 & 0 \\ 0 & 1 & 0 & 0 \\ 1 & 0 & 0 & 0 \\ 0 & 0 & 1 & 0 \\ 0 & 0 & 0 & 1 \end{bmatrix} \begin{bmatrix} j_{11} & j_{12} \\ j_{21} & j_{22} \\ j_{31} & j_{32} \\ j_{41} & j_{42} \end{bmatrix} = \begin{bmatrix} 1 & 0 \\ 0 & 1 \\ i_1 & i_2 \\ i_{11} & i_{12} \\ i_{21} & i_{22} \end{bmatrix}$$

which yields (uniquely)

$$j_{\mathbf{E}} = \begin{bmatrix} 1 & 0 \\ 0 & 1 \\ i_{11} & i_{12} \\ i_{21} & i_{22} \end{bmatrix}, \quad i_{\mathbf{E}} = \begin{bmatrix} 1 & 0 \\ 0 & 1 \\ 1 & 0 \\ i_{11} & i_{12} \\ i_{21} & i_{22} \end{bmatrix}$$

Substitution in (1.19) results in

$$\begin{bmatrix} 0 & 1 & 0 & 0 & 0 \\ -1 & 0 & 0 & 0 & 0 \\ \mu_1 & 0 & \mu_3 & \mu_4 & 0 \\ c_{41} & 0 & c_{43} & c_{44} & c_{45} \\ c_{51} & 0 & c_{53} & c_{54} & c_{55} \end{bmatrix} \begin{bmatrix} 1 & 0 \\ 0 & 1 \\ 1 & 0 \\ i_{11} & i_{12} \\ i_{21} & i_{22} \end{bmatrix} = \begin{bmatrix} 1 & 0 \\ 0 & 1 \\ 1 & 0 \\ i_{11} & i_{12} \\ i_{21} & i_{22} \end{bmatrix} \begin{bmatrix} 0 & 1 \\ -1 & 0 \end{bmatrix} \tag{1.20}$$

giving the 6 equations

$$\begin{aligned} &\mu_1 + \mu_3 + \mu_4 i_{11} = 0, \ \mu_4 i_{12} = 1 \\ &c_{41} + c_{43} + c_{44} i_{11} + c_{45} i_{21} = -i_{12}, \ c_{44} i_{12} + c_{45} i_{22} = i_{11} \\ &c_{51} + c_{53} + c_{54} i_{11} + c_{55} i_{21} = -i_{22}, \ c_{54} i_{12} + c_{55} i_{22} = i_{21} \end{aligned}$$

Formally these yield, in turn, the 4 evaluations

$$i_{11} = -(\mu_1 + \mu_3)/\mu_4, \quad i_{12} = 1/\mu_4$$

$$i_{21} = \frac{-1 - \mu_4(c_{41} + c_{43}) + (\mu_1 + \mu_3)c_{44}}{\mu_4 c_{45}}$$

$$i_{22} = \frac{-(\mu_1 + \mu_3) - c_{44}}{\mu_4 c_{45}}$$

and then the 2 relations

$$(\mu_1+\mu_3)(c_{44}c_{55}-c_{45}c_{54}-1)+\mu_4\left(c_{55}(c_{41}+c_{43})-c_{45}(c_{51}+c_{53})\right)-(c_{44}+c_{55})=0 \tag{1.21a}$$

$$(\mu_1 + \mu_3)(c_{44} + c_{45}) - \mu_4(c_{41} + c_{43}) + (c_{44}c_{55} - c_{45}c_{54} - 1) = 0 \tag{1.21b}$$

Now (1.21) will hold identically in μ iff the controller coefficients c_{ij} $(i = 4, 5;\ j = 1, 3, 4, 5)$ are chosen to satisfy

$$c_{41} + c_{43} = 0 \tag{1.22a}$$

$$c_{44} + c_{55} = 0 \tag{1.22b}$$

$$c_{45}(c_{51} + c_{53}) = 0 \tag{1.22c}$$

$$c_{44}c_{55} - c_{45}c_{54} = 1 \tag{1.22d}$$

These equations hold if, for instance,

$$\begin{bmatrix} c_{41} & c_{43} & c_{44} & c_{45} \\ c_{51} & c_{53} & c_{54} & c_{55} \end{bmatrix} = \begin{bmatrix} 0 & 0 & 0 & 1 \\ 1 & -1 & -1 & 0 \end{bmatrix}$$

Note that (1.22a) is just the error feedback condition that input to the controller x_4-component vanishes so long as the tracking error $x_1 - x_3$ remains zero. If in (1.22c)

$c_{45} = 0$ then by (1.22d) $c_{44}c_{55} = 1$, which is impossible by (1.22b), and therefore $c_{51} + c_{53} = 0$, the error feedback condition for the controller component x_5.

To interpret (1.22b), (1.22d) note from (1.20) that

$$CJ = JE$$

where

$$C := \begin{bmatrix} c_{44} & c_{45} \\ c_{54} & c_{55} \end{bmatrix}, \quad J := \begin{bmatrix} i_{11} & i_{12} \\ i_{21} & i_{22} \end{bmatrix}, \quad E := \begin{bmatrix} 0 & 1 \\ -1 & 0 \end{bmatrix}$$

Formally

$$\det(J) = i_{11}i_{22} - i_{12}i_{21} = \frac{1 + (\mu_1 + \mu_3)^2}{\mu_4^2 c_{45}}$$

We assume $\mu_4 \neq 0$, as otherwise **P** is decoupled from **C**; and necessarily $c_{45} \neq 0$ as noted above. Then

$$C = JEJ^{-1}$$

which states that the dynamics of controller and exosystem are isomorphic, exactly the Internal Model Principle. Equating the characteristic polynomials of C and E yields (1.22b), (1.22d).

Consistency of our results with the general case is easily verified, on noting

$$\begin{aligned} \gamma &= [0^{2\times 3} \; I^{2\times 2}], \;\; \alpha_C = C \\ \alpha_C \circ \gamma | K &= \alpha_C \circ \gamma \circ \kappa = [0^{2\times 2} \; C] \\ \gamma \circ \alpha | K &= \gamma \circ \alpha \circ \kappa = [0^{2\times 2} \; C] \\ \tilde{\gamma}_{\mathbf{E}} &= \gamma | \tilde{X}_{\mathbf{E}} = \gamma \circ i_{\mathbf{E}} = J \end{aligned}$$

We return to the general discussion and introduce a parametrization for which (1.15)–(1.18) can be analyzed further. Specifically we assume that the structural form of the system transition function α remains invariant under parameter variations, the latter inducing only changes in coordinatization of the state sets $X_{\mathbf{E}}$, $X_{\mathbf{P}}$, $X_{\mathbf{C}}$ in accordance with the diagrams below. Here $R(\mu)$, $S(\mu)$, $T(\mu)$ are bijections for each $\mu \in M$. We assume that $\alpha_{\mathbf{E}} \times \alpha_{\mathbf{P}} \times \alpha_{\mathbf{C}}$ undergoes perturbation only in the $\alpha_{\mathbf{P}}$-component, and for some 'nominal' value $\mu = \mu_0$ simply write $\alpha_{\mathbf{P}} := \alpha_{\mathbf{P}}(\mu_0)$. On the other hand, for all μ, $\alpha_{\mathbf{E}}(\mu) = \alpha_{\mathbf{E}}$ and (as assumed before) $\alpha_{\mathbf{C}}(\mu) = \alpha_{\mathbf{C}}$ remain fixed, with $R(\mu)$ and $T(\mu)$ restricted accordingly.

$$\begin{array}{ccc} X_{\mathbf{E}} & \xrightarrow{\alpha_{\mathbf{E}}} & X_{\mathbf{E}} \\ {\scriptstyle R(\mu)}\downarrow & & \downarrow{\scriptstyle R(\mu)} \\ X_{\mathbf{E}} & \xrightarrow[\alpha_{\mathbf{E}}(\mu)]{} & X_{\mathbf{E}} \end{array} \qquad \begin{array}{ccc} X & \xrightarrow{\alpha_{\mathbf{P}}} & X_{\mathbf{P}} \\ {\scriptstyle (R\times S\times T)(\mu)}\downarrow & & \downarrow{\scriptstyle S(\mu)} \\ X & \xrightarrow[\alpha_{\mathbf{P}}(\mu)]{} & X_{\mathbf{P}} \end{array} \qquad \begin{array}{ccc} X & \xrightarrow{\alpha_{\mathbf{C}}} & X_{\mathbf{C}} \\ {\scriptstyle (R\times S\times T)(\mu)}\downarrow & & \downarrow{\scriptstyle T(\mu)} \\ X & \xrightarrow[\alpha_{\mathbf{C}}(\mu)]{} & X_{\mathbf{C}} \end{array}$$

Incorporating these transformations into the system dynamics we obtain the commutative diagram below.

$$
\begin{array}{ccc}
X_{\mathbf{E}} & \xrightarrow{\alpha_{\mathbf{E}}} & X_{\mathbf{E}} \\
{\scriptstyle R(\mu)^{-1}} \downarrow & & \downarrow {\scriptstyle R(\mu)^{-1}} \\
X_{\mathbf{E}} & \xrightarrow{\alpha_{\mathbf{E}}} & X_{\mathbf{E}} \\
{\scriptstyle id_{\mathbf{E}} \times i_{\mathbf{P}} \times i_{\mathbf{C}}} \downarrow & & \downarrow {\scriptstyle id_{\mathbf{E}} \times i_{\mathbf{P}} \times i_{\mathbf{C}}} \\
X_{\mathbf{E}} \times X_{\mathbf{P}} \times X_{\mathbf{C}} & \xrightarrow{\alpha_{\mathbf{E}} \times \alpha_{\mathbf{P}} \times \alpha_{\mathbf{C}}} & X_{\mathbf{E}} \times X_{\mathbf{P}} \times X_{\mathbf{C}} \\
{\scriptstyle R(\mu) \times S(\mu) \times T(\mu)} \downarrow & & \downarrow {\scriptstyle R(\mu) \times S(\mu) \times T(\mu)} \\
X_{\mathbf{E}} \times X_{\mathbf{P}} \times X_{\mathbf{C}} & \xrightarrow{\alpha_{\mathbf{E}} \times \alpha_{\mathbf{P}}(\mu) \times \alpha_{\mathbf{C}}} & X_{\mathbf{E}} \times X_{\mathbf{P}} \times X_{\mathbf{C}}
\end{array}
$$

From the bottom square,

$$\alpha_{\mathbf{P}}(\mu)\left(R(\mu)(x_{\mathbf{E}}), S(\mu)(x_{\mathbf{P}}), T(\mu)(x_{\mathbf{C}})\right) = S(\mu) \circ \alpha_{\mathbf{P}}(x_{\mathbf{E}}, x_{\mathbf{P}}, x_{\mathbf{C}})$$

giving the plant perturbation model

$$\alpha_{\mathbf{P}}(\mu)(x_{\mathbf{E}}, x_{\mathbf{P}}, x_{\mathbf{C}}) = S(\mu) \circ \alpha_{\mathbf{P}}\left(R(\mu)^{-1}(x_{\mathbf{E}}), S(\mu)^{-1}(x_{\mathbf{P}}), T(\mu)^{-1}(x_{\mathbf{C}})\right)$$

Also, from the bottom and middle squares,

$$\alpha_{\mathbf{C}}\left(R(\mu)(x_{\mathbf{E}}), S(\mu) \circ i_{\mathbf{P}}(x_{\mathbf{E}}), T(\mu) \circ i_{\mathbf{C}}(x_{\mathbf{E}})\right) = T(\mu) \circ i_{\mathbf{C}} \circ \alpha_{\mathbf{E}}(x_{\mathbf{E}})$$

Substitution of the regulation condition (1.17) for $\mu = \mu_0$, namely

$$i_{\mathbf{P}}(x_{\mathbf{E}}) = P_{\mathbf{P}} \circ \kappa \circ j_{\mathbf{E}}(x_{\mathbf{E}})$$

yields

$$\alpha_{\mathbf{C}}\left(R(\mu)(x_{\mathbf{E}}), S(\mu) \circ P_{\mathbf{P}} \circ \kappa \circ j_{\mathbf{E}}(x_{\mathbf{E}}), T(\mu) \circ i_{\mathbf{C}}(x_{\mathbf{E}})\right) = T(\mu) \circ i_{\mathbf{C}} \circ \alpha_{\mathbf{E}}(x_{\mathbf{E}}) \tag{1.23}$$

assumed to hold for all $x_{\mathbf{E}} \in X_{\mathbf{E}}$ and $\mu \in M$.

For simplicity we take $M = M_{\mathbf{E}} \times M_{\mathbf{P}} \times M_{\mathbf{C}}$, where the component parameter sets are pairwise disjoint, and write $R(\mu) = \hat{R}(\mu_{\mathbf{E}})$, $S(\mu) = \hat{S}(\mu_{\mathbf{P}})$, $T(\mu) = \hat{T}(\mu_{\mathbf{C}})$. In addition suppose that parametrization is 'rich' in the sense that

$$(\forall x_{\mathbf{E}} \in X_{\mathbf{E}})\{\hat{R}(\mu_{\mathbf{E}})(x_{\mathbf{E}}) | \mu_{\mathbf{E}} \in M_{\mathbf{E}}\} = X_{\mathbf{E}} \tag{1.24a}$$

$$(\forall x_{\mathbf{E}} \in X_{\mathbf{E}})\{\hat{S}(\mu_{\mathbf{P}}) \circ P_{\mathbf{P}} \circ \kappa \circ j_{\mathbf{E}}(x_{\mathbf{E}}) | \mu_{\mathbf{P}} \in M_{\mathbf{P}}\} = X_{\mathbf{P}} \tag{1.24b}$$

Under these conditions (1.23) implies that, as long as

$$x \in \tilde{X}_{\mathbf{E}} := \{(x_{\mathbf{E}}, i_{\mathbf{P}}(x_{\mathbf{E}}), i_{\mathbf{C}}(x_{\mathbf{E}})) | x_{\mathbf{E}} \in X_{\mathbf{E}}\}$$

and regulation holds as in (1.17), the function $\alpha_{\mathbf{C}}(x_{\mathbf{E}}, x_{\mathbf{P}}, x_{\mathbf{C}})$ must be independent of its first two arguments $x_{\mathbf{E}}, x_{\mathbf{P}}$. But this amounts precisely to the statement that the controller is then autonomous on $\tilde{X}_{\mathbf{E}}$, namely incorporates 'error feedback structure'. Finally, setting $T(\mu) \circ i_{\mathbf{C}} = \tilde{i}_{\mathbf{C}}$ for brevity, and $\tilde{\alpha}_{\mathbf{C}} = \alpha_{\mathbf{C}} | \tilde{X}_{\mathbf{E}}$, we obtain

$$\tilde{\alpha}_{\mathbf{C}} \circ \tilde{i}_{\mathbf{C}}(x_{\mathbf{E}}) = \tilde{i}_{\mathbf{C}} \circ \alpha_{\mathbf{E}}(x_{\mathbf{E}})$$

as displayed below.

$$\begin{CD} X_{\mathbf{E}} @>\alpha_{\mathbf{E}}>> X_{\mathbf{E}} \\ @V{\tilde{i}_{\mathbf{C}}}VV @VV{\tilde{i}_{\mathbf{C}}}V \\ X_{\mathbf{C}} @>>\tilde{\alpha}_{\mathbf{C}}> X_{\mathbf{C}} \end{CD}$$

We conclude that both error feedback and an internal model are necessary for structurally stable regulation.

As an illustration consider again Example 12, with plant perturbation model

$$\mu_3 x_3' = \mu_1 x_1 + \mu_2 x_2 + \mu_3 x_3 + \mu_4 x_4 + \mu_5 x_5$$

namely $\mu = [\mu_1 \cdots \mu_5] \in \mathbb{R}^5$. This corresponds to $x_{\mathbf{E}} = [x_1 \; x_2]$, $x_{\mathbf{P}} = x_3$, $x_{\mathbf{C}} = [x_4 \; x_5]$ and

$$R(\mu) = \hat{R}(\mu_1, \mu_2) = \begin{bmatrix} \mu_1 & \mu_2 \\ -\mu_2 & \mu_1 \end{bmatrix}^{-1}, \quad S(\mu) = \hat{S}(\mu_3) = \mu_3^{-1}, \quad T(\mu) = \hat{T}(\mu_4, \mu_5)$$

With K and therefore κ, $j_{\mathbf{E}}$ as before, we obtain

$$P_{\mathbf{P}} \circ \kappa \circ j_{\mathbf{E}}(x_1, x_2) = x_1$$

and then (1.23) becomes

$$\alpha_{\mathbf{C}}\left(\hat{R}(\mu_1, \mu_2)x_{\mathbf{E}}, \mu_3^{-1}x_1, \hat{T}(\mu_4, \mu_5)i_{\mathbf{C}}(x_{\mathbf{E}})\right) = \hat{T}(\mu_4, \mu_5)i_{\mathbf{C}} \circ \alpha_{\mathbf{E}}(x_{\mathbf{E}})$$

to hold identically in μ, $x_{\mathbf{E}}$. Formally

$$\begin{aligned} \hat{R}(\mu_1, \mu_2)x_{\mathbf{E}} &= \begin{bmatrix} \mu_1 & \mu_2 \\ -\mu_2 & \mu_1 \end{bmatrix}^{-1} \begin{bmatrix} x_1 \\ x_2 \end{bmatrix} \\ &= (\mu_1^2 + \mu_2^2)^{-1} \begin{bmatrix} x_1 & -x_2 \\ x_2 & x_1 \end{bmatrix} \begin{bmatrix} \mu_1 \\ \mu_2 \end{bmatrix} = \begin{bmatrix} x_1' \\ x_2' \end{bmatrix} \end{aligned}$$

say, provided

$$(\mu_1^2 + \mu_2^2)^{-1} \begin{bmatrix} \mu_1 \\ \mu_2 \end{bmatrix} = (x_1^2 + x_2^2)^{-1} \begin{bmatrix} x_1 & x_2 \\ -x_2 & x_1 \end{bmatrix} \begin{bmatrix} x_1' \\ x_2' \end{bmatrix} = \begin{bmatrix} \xi_1 \\ \xi_2 \end{bmatrix}$$

say, which holds iff

$$\begin{bmatrix} \mu_1 \\ \mu_2 \end{bmatrix} = (\xi_1^2 + \xi_2^2)^{-1} \begin{bmatrix} \xi_1 \\ \xi_2 \end{bmatrix}$$

This verifies (1.24a), although with $X_{\mathbf{E}}$ replaced by $X_{\mathbf{E}} - \{0\}$. Similarly (1.24b) amounts to $\mu_3^{-1}x_1 = x_3'$, say, or $\mu_3 = x_1/x_3'$, which verifies (1.24b) for $X_{\mathbf{P}} - \{0\}$ and $x_1 \neq 0$. It follows that $\alpha_{\mathbf{C}}(\cdot, \cdot, \tilde{i}_{\mathbf{C}})$ is independent of its first two arguments provided $x_1 \neq 0$. By the assumed linearity (or even just continuity) of $\alpha_{\mathbf{C}}(\cdot, \cdot, \cdot)$ in $x_{\mathbf{E}} = [x_1 \; x_2]$ we conclude that $\alpha_{\mathbf{C}}(\cdot, \cdot, \tilde{i}_{\mathbf{C}})$ depends on its third argument $\tilde{i}_{\mathbf{C}}$ alone. From this the results for error feedback and the internal model follow as before.

1.6 Notes

Much of the material in this chapter is standard. For Sects. 1.1–1.4 see especially Mac Lane and Birkhoff (1993), and Davey and Priestley (1990). Section 1.5 follows Wonham (1976), which may be consulted for more detailed background; see also Wonham (1985), Chap. 8. Since 1980 the literature on the Internal Model Principle has grown considerably, but we make no attempt to review it here.

References

Davey BA, Priestley HA (1990) Introduction to lattices and order. Cambridge University Press, Cambridge

Mac Lane S, Birkhoff G (1993) Algebra, 3rd edn. Chelsea, New York

Wonham WM (1976) Towards an abstract internal model principle. IEEE Trans Syst Man Cybern SMC 6(11):735–740

Wonham WM (1985) Linear multivariable control: a geometric approach, 3rd edn. Springer, Berlin

Chapter 2
Linguistic Preliminaries

In supervisory control, formal languages provide a level of abstraction at which control concepts can be formulated in a way that is independent of any specific concrete implementation of the plant and controller. Some basic definitions are presented, leading to the Nerode equivalence relation as the bridge from a language to its dynamical state description. Emphasis is placed on regular (finite-state) languages as the class that is simplest and most directly applicable in the control context.

2.1 Languages

Let Σ be a finite set of symbols $\sigma, \tau, \ldots$. We refer to Σ as an *alphabet*. Let Σ^+ denote the set of all finite symbol sequences, of the form $\sigma_1\sigma_2\cdots\sigma_k$ where $k \geq 1$ is arbitrary and the $\sigma_i \in \Sigma$. It is convenient also to bring in the empty sequence (sequence with no symbols), denoted by the new symbol ϵ, where $\epsilon \notin \Sigma$. We then write

$$\Sigma^* := \{\epsilon\} \cup \Sigma^+$$

An element of Σ^* is a *word* or *string* over the alphabet Σ; ϵ is the *empty string*. If $\Sigma = \varnothing$ then $\Sigma^+ = \varnothing$ and $\Sigma^* = \{\epsilon\}$.

Next we define the operation of *catenation* of strings:

$$\text{cat} : \Sigma^* \times \Sigma^* \to \Sigma^*$$

according to

$$\text{cat}(\epsilon, s) = \text{cat}(s, \epsilon) = s, \qquad s \in \Sigma^*$$
$$\text{cat}(s, t) = st, \qquad s, t \in \Sigma^+$$

W. M. Wonham and K. Cai, *Supervisory Control of Discrete-Event Systems*, Communications and Control Engineering,
https://doi.org/10.1007/978-3-319-77452-7_2

Thus ϵ is the unit element of catenation. Evidently $\text{cat}(\cdot, \cdot)$ is associative, for clearly

$$\text{cat}(\text{cat}(s, t), u) = \text{cat}(s, \text{cat}(t, u))$$

if $s, t, u \in \Sigma^+$, and the other possibilities are easily checked.

With catenation as the 'product' operation, the foregoing relationships turn Σ^* into a *multiplicative monoid* (or *multiplicative semigroup with identity*).

Notice that a symbol sequence like

$$\sigma_1\sigma_2\epsilon\sigma_3\epsilon\epsilon\sigma_4 \quad (\sigma_i \in \Sigma)$$

is not (syntactically) an element of Σ^*. It will be taken as a convenient abbreviation for

$$\text{cat}(\text{cat}(\text{cat}(\text{cat}(\text{cat}(\sigma_1\sigma_2, \epsilon), \sigma_3), \epsilon), \epsilon), \sigma_4)$$

As such it evaluates, of course, to the Σ^*-element $\sigma_1\sigma_2\sigma_3\sigma_4$. Also, brackets may sometimes be inserted in a string for clarity of exposition.

The *length* $|s|$ of a string $s \in \Sigma^*$ is defined according to

$$|\epsilon| = 0; \quad |s| = k, \text{ if } s = \sigma_1 \cdots \sigma_k \in \Sigma^+$$

Thus $|\text{cat}(s, t)| = |s| + |t|$.

A *language* over Σ is any subset of Σ^*, i.e. an element of the power set $Pwr(\Sigma^*)$; thus the definition includes both the empty language $\varnothing$, and Σ^* itself. Note the distinction between $\varnothing$ (the language with no strings) and ϵ (the string with no symbols). For instance the language $\{\epsilon\}$ is nonempty, but contains only the empty string.

2.2 Nerode Equivalence and Right Congruence

Let $L \subseteq \Sigma^*$ be an arbitrary language. We would like to construct, if possible, a decision procedure to test any given string $s \in \Sigma^*$ for membership in L. We might visualize a machine into which s is fed as input and which emits a beep just in case $s \in L$. To this end we first construct a partition of Σ^* that is finer than the partition $\{L, \Sigma^* - L\}$ and which has a certain invariance property with respect to L.

The *Nerode equivalence relation on* Σ^* *with respect to* L (or mod L) is defined as follows. For $s, t \in \Sigma^*$,

$$s \equiv_L t \quad \text{or} \quad s \equiv t(\text{mod } L) \quad \text{or} \quad (s, t) \in \text{Ner}(L)$$

iff

$$(\forall u \in \Sigma^*)su \in L \quad \text{iff} \quad tu \in L$$

In other words $s \equiv_L t$ iff s and t can be continued in exactly the same ways (if at all) to form a string of L.

We write $\|L\|$ for the *index* (cardinality of the set of cells) of the Nerode equivalence relation $\equiv_L$, i.e. $\|L\| := | \equiv_L |$. Since Σ^* is countable, $\|L\|$ is at most countable infinity (cardinality of the integers). If $\|L\| < \infty$, the language L is said to be *regular*.

Let $R \in \mathcal{E}(\Sigma^*)$ be an equivalence relation on Σ^*. R is a *right congruence* on Σ^* if

$$(\forall s, t, u \in \Sigma^*) sRt \;\Rightarrow\; (su)R(tu)$$

In other words, R is a right congruence iff the cells of R are 'respected' by the operation of right catenation. As elements of $\mathcal{E}(\Sigma^*)$ the right congruences on Σ^* inherit the partial order $\leq$ on $\mathcal{E}(\Sigma^*)$.

Proposition 1

Nerode equivalence is a right congruence.

Proof

Let $s \equiv t (\text{mod } L)$ and $u \in \Sigma^*$. It must be shown that $su \equiv tu (\text{mod } L)$. Let $v \in \Sigma^*$ and $(su)v \in L$. Then $s(uv) \in L$, so $t(uv) \in L$, i.e. $(tu)v \in L$. Similarly $(tu)v \in L$ implies $(su)v \in L$, hence $su \equiv tu (\text{mod } L)$ as claimed. □

Proposition 2

Nerode equivalence is finer than the partition $\{L, \Sigma^* - L\}$.

Proof

Let $s \equiv t (\text{mod } L)$. Then $s \in L$ iff $s\epsilon \in L$, iff $t\epsilon \in L$, iff $t \in L$. □

Proposition 3

Let R be a right congruence on Σ^* such that $R \leq \{L, \Sigma^* - L\}$. Then

$$R \leq \equiv_L$$

Proof

Let sRt. We show that $s \equiv t (\text{mod } L)$. Let $su \in L$. Now sRt implies $(su)R(tu)$. Since $R \leq \{L, \Sigma^* - L\}$ and $su \in L$, it follows that $tu \in L$. Similarly if $tu \in L$ then $su \in L$. □

We summarize Propositions 1–3 as

Theorem 4

Let $L \subseteq \Sigma^*$. Nerode equivalence (mod L) is the coarsest right congruence on Σ^* that is finer than $\{L, \Sigma^* - L\}$. □

Exercise 5: Let $E, F \in \mathcal{E}(\Sigma^*)$ with $E \leq F$. Let

$$\mathcal{R} = \{R | R \text{ is a right congruence on } \Sigma^*, \text{ with } E \leq R \leq F\}$$

Assuming that $\mathcal{R} \neq \varnothing$, show that $\mathcal{R}$ is a complete sublattice of $\mathcal{E}(\Sigma^*)$. (Recall from Sect. 1.2 that a *sublattice* L of a lattice M is a subset of M that is a lattice under the operations of meet and join inherited from M.)

Exercise 6: In the notation of Exercise 5, assume that $E, F \in \mathcal{R}$. Consider the statement

$$(\forall G)G \in \mathcal{R} \Rightarrow (\exists H)H \in \mathcal{R} \quad \& \quad G \wedge H = E \quad \& \quad G \vee H = F$$

Either prove it's always true or show by counterexample that it can be false.

For $s \in \Sigma^*$ we say $t \in \Sigma^*$ is a *prefix* of s, and write $t \leq s$, if $s = tu$ for some $u \in \Sigma^*$. Thus $\epsilon \leq s$ and $s \leq s$ for all $s \in \Sigma^*$. If $L \subseteq \Sigma^*$ the (*prefix*) *closure* of L is the language $\bar{L}$ consisting of all prefixes of strings of L:

$$\bar{L} = \{t \in \Sigma^* | t \leq s \text{ for some } s \in L\}$$

Clearly $L \subseteq \bar{L}$. If $L = \varnothing$ then $\bar{L} = \varnothing$; if $L \neq \varnothing$ then $\epsilon \in \bar{L}$. For a string s we usually write $\bar{s}$ instead of $\overline{\{s\}}$ for its set of prefixes. A language L is *closed* if $L = \bar{L}$.

Exercise 7: If $A \subseteq \Sigma^*$ is closed and $B \subseteq \Sigma^*$ is arbitrary, show that $A - B\Sigma^*$ is closed.

Exercise 8: Let $L \subseteq \Sigma^*$, $L_{co} := \Sigma^* - L$. Show that $\text{Ner}(L_{co}) = \text{Ner}(L)$. If also $M \subseteq \Sigma^*$ then $\text{Ner}(L \cap M) \geq \text{Ner}(L) \wedge \text{Ner}(M)$. What can be said about $\text{Ner}(L \cup M)$?

The closure of a language L is often relevant to control problems because it embodies the 'evolutionary history' of words in L. Notice that if $s, t \in \Sigma^* - \bar{L}$ then

$$(\forall w \in \Sigma^*) sw \notin L \;\&\; tw \notin L$$

so that $s \equiv t(\text{mod } L)$. In other words the subset $\Sigma^* - \bar{L}$ of Σ^* is, if nonempty, a single Nerode cell, which we call the *dump cell*; a string that enters the dump cell can never exit from it. On the other hand if $s \in \bar{L}$ and $s \equiv t(\text{mod } L)$ then $sw \in L$ for some $w \in \Sigma^*$ and so $tw \in L$, i.e. $t \in \bar{L}$. If $s \in \bar{L} - L$ and $s \equiv t(\text{mod } L)$ then $s = s\epsilon \notin L$, so $t = t\epsilon \notin L$ but (as just proved) $t \in \bar{L}$. These remarks are summarized in

Proposition 9

Nerode equivalence $\equiv_L$ refines the partition

$$\{L, \bar{L} - L, \Sigma^* - \bar{L}\}$$

of Σ^*. □

2.3 Canonical Recognizers

The fact that Nerode equivalence is invariant under right catenation allows us to construct abstractly an automaton that tracks the Nerode cells which a string in Σ^*

visits as it evolves symbol-by-symbol. Because Nerode equivalence is as coarse as can be for this purpose, the corresponding automaton is said to be *canonical*.

Thus let $L \subseteq \Sigma^*$ and write

$$X := \Sigma^* / \equiv_L$$

with

$$P_L : \Sigma^* \to X : s \mapsto [s]$$

the canonical projection. Write

$$\text{cat} : \Sigma^* \times \Sigma \to \Sigma^* : (s, \sigma) \mapsto s\sigma$$

for catenation,

$$id_\Sigma : \Sigma \to \Sigma : \sigma \mapsto \sigma$$

for the identity on Σ, and

$$P_L \times id_\Sigma : \Sigma^* \times \Sigma \to X \times \Sigma : (s, \sigma) \mapsto ([s], \sigma)$$

Proposition 1

There exists a unique map

$$\xi : X \times \Sigma \to X$$

such that

$$\xi \circ (P_L \times id_\Sigma) = P_L \circ \text{cat}$$

namely the following diagram commutes:

$$\begin{array}{ccc} \Sigma^* \times \Sigma & \xrightarrow{\ \text{cat}\ } & \Sigma^* \\ {\scriptstyle P_L \times id_\Sigma}\downarrow & & \downarrow{\scriptstyle P_L} \\ X \times \Sigma & \xrightarrow{\ \xi\ } & X \end{array}$$

Proof

By Proposition 1.4.3, for the existence of ξ it is enough to check that

$$\ker(P_L \times id_\Sigma) \leq \ker(P_L \circ \text{cat})$$

Uniqueness will follow by the fact that $P_L \times id_\Sigma$ is surjective. Let

$$((s, \sigma), (s', \sigma')) \in \ker(P_L \times id_\Sigma)$$

namely

$$(P_L \times id_\Sigma)(s, \sigma) = (P_L \times id_\Sigma)(s', \sigma')$$

or

$$([s], \sigma) = ([s'], \sigma'),$$

that is,

$$s \equiv_L s', \qquad \sigma = \sigma'$$

Since $\equiv_L$ is a right congruence,

$$s\sigma \equiv_L s'\sigma'$$

namely

$$[s\sigma] = [s'\sigma']$$

or

$$P_L(\text{cat}(s, \sigma)) = P_L(\text{cat}(s', \sigma'))$$

so

$$(P_L \circ \text{cat})(s, \sigma) = (P_L \circ \text{cat})(s', \sigma')$$

or finally

$$((s, \sigma), (s', \sigma')) \in \ker(P_L \circ \text{cat})$$

as required. □

The elements $x \in X$ are the *states* of L; X is the *state set* of L; ξ is the *transition function* of L; and the triple (X, Σ, ξ) is the *transition structure* of L. It is convenient to extend ξ to a map

$$\hat{\xi} : X \times \Sigma^* \to X$$

as follows. Define

$$\begin{aligned} \hat{\xi}(x, \epsilon) &= x, \quad x \in X \\ \hat{\xi}(x, \sigma) &= \xi(x, \sigma) \quad x \in X,\ \sigma \in \Sigma \\ \hat{\xi}(x, s\sigma) &= \xi(\hat{\xi}(x, s), \sigma) \quad x \in X,\ s \in \Sigma^*,\ \sigma \in \Sigma \end{aligned}$$

It is easily checked (say, by induction on length of strings) that $\hat{\xi}$ is well defined. From now on we omit the ˆ and write ξ in place of $\hat{\xi}$.

If $x = [t]$ then by definition of ξ, $\xi(x, \sigma) = [t\sigma]$. Assuming inductively that $\xi(x, s) = [ts]$ we have

$$\begin{aligned} \xi(x, s\sigma) &= \xi(\xi(x, s), \sigma) && \text{by definition of } \xi \text{ (i.e. } \hat{\xi}) \\ &= \xi([ts], \sigma) && \text{by the inductive assumption} \\ &= [ts\sigma] && \text{by definition of } \xi \end{aligned}$$

so that $\xi(x, u) = [tu]$ for all $u \in \Sigma^*$. From this we get the composition property: for all $x \in X$ and $s, u \in \Sigma^*$,

$$\begin{aligned}\xi(x, su) &= [tsu] \\ &= \xi(x', u),\ x' = [ts] \\ &= \xi(\xi(x, s), u)\end{aligned}$$

We distinguish an element x_0 and a subset X_m of X as follows. Let

$$x_0 = [\epsilon], \quad X_m = \{[s] | s \in L\}$$

The state x_0 is the *initial state* of L, and if $x \in X_m$, x is a *marker state* of L. We have by definition

$$\xi(x_0, s) = [\epsilon s] = [s], \quad s \in \Sigma^*$$

In particular, if $s \in L$ then

$$\xi(x_0, s) = [s] \in X_m$$

Thus one can think of X_m as the subset of states of L that 'mark' precisely the strings of L: imagine that a beep sounds just when such a state is reached. These definitions are displayed in the diagram below.

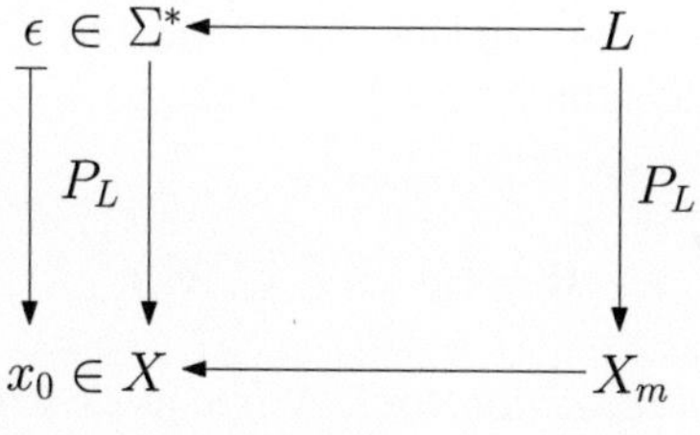

Here the horizontal arrows are the natural subset injections.

Finally the *dump state* of L is the (single) state corresponding to the dump cell $\Sigma^* - \bar{L}$, when the latter is nonempty.

In general, then, a language L can be visualized as shown in Fig. 2.1. The shaded divisions demarcate cells of $\equiv_L$. The marker states are the cells contained in L. The initial state (cell) x_0 belongs to L if the empty string $\epsilon \in L$; otherwise (provided $L \neq \varnothing$) x_0 belongs to $\bar{L}$. In case $L = \varnothing$ then $\bar{L} = \varnothing$, so

$$X = \{\Sigma^*\}, \quad x_0 = \Sigma^*, \quad X_m = \varnothing$$

On the other hand if $L = \Sigma^*$ then $\bar{L} = \Sigma^*$ and

$$X = \{\Sigma^*\}, \quad x_0 = \Sigma^*, \quad X_m = \{\Sigma^*\}$$

In general if L is closed then

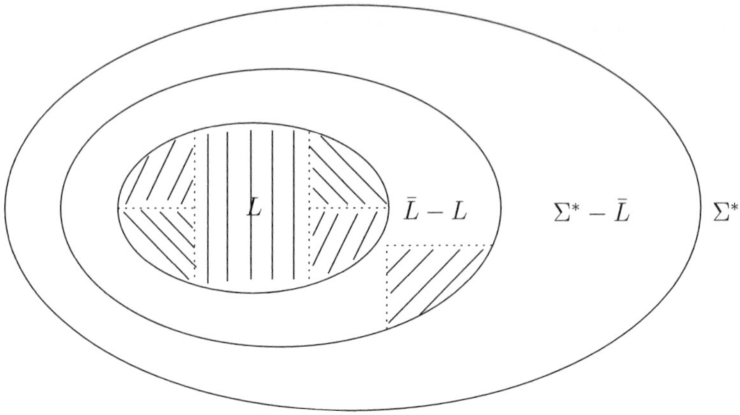

Fig. 2.1 Subset diagram for a typical language $L \subseteq \Sigma^* = L \cup (\bar{L} - L) \cup (\Sigma^* - \bar{L})$

$$X_m = X - \{\Sigma^* - L\}$$

namely all states except the dump state are marked.

Figure 2.2 displays alternative 'high-level' transition graphs for L, showing in a general way how transitions may occur in the two cases where (a) the initial state is not a marker state (namely the empty string ϵ does not belong to L), and (b) the initial state is a marker state. If L were nonempty and closed then in the corresponding graph (b) the right-hand state (identified with $\bar{L} - L$) and its associated transitions would be deleted.

The 5-tuple

$$\mathbf{R} = (X, \Sigma, \xi, x_0, X_m)$$

will be called a *canonical recognizer* for L. While its existence has now been established abstractly there is, of course, no implication in general that $\mathbf{R}$ can actually be implemented by some constructive procedure. Naturally this issue is fundamental in the applications.

In general, a recognizer for L (see below, Sect. 2.4) will be called *canonical* if its state set X is in bijective correspondence with the cells of $\equiv_L$; thus $|X| = |\equiv_L| = \|L\|$. Subject to this requirement, X may be chosen according to convenience, for example as a subset of the integers. In another common representation, X is identified with the nodes of a directed graph $\mathcal{G}$ whose edges are labeled with symbols $\sigma \in \Sigma$; namely $[x, \sigma, x']$ is a labeled edge

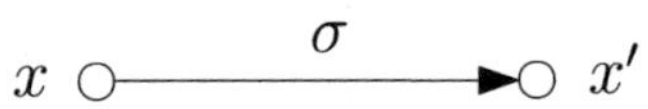

of $\mathcal{G}$ iff $\xi(x, \sigma) = x'$. Such a graph $\mathcal{G}$ is a *state transition graph* for $\mathbf{R}$ (or L). In $\mathcal{G}$ we attach to x_0 an entering arrow, as in

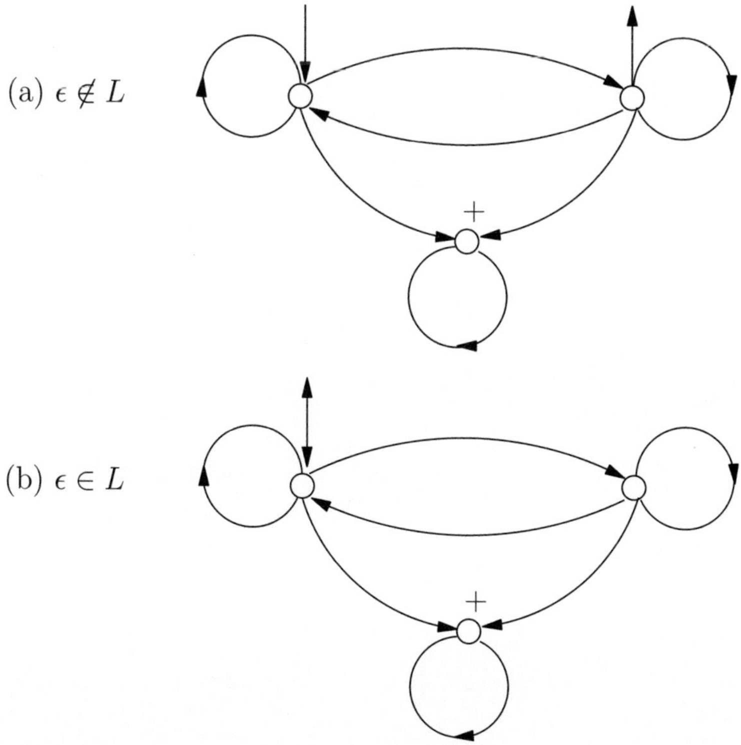

Fig. 2.2 'High-level' state transition graphs for a language L
Case **a**: $\epsilon \in \bar{L} - L$; $x_0 \notin X_m$
Case **b**: $\epsilon \in L$; $x_0 \in X_m$

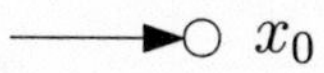

and to state $x \in X_m$ an exiting arrow, as in

If x_0 happens also to be a marker state we may attach a double arrow, as in

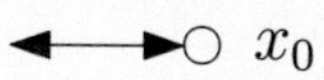

The dump state will be labeled $+$.

Example 2: **Canonical recognizers**

Let $\Sigma = \{\alpha, \beta\}$.

(i) $L = \varnothing$

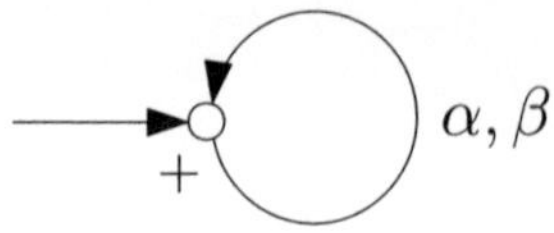

(ii) $L = \{\epsilon\}$

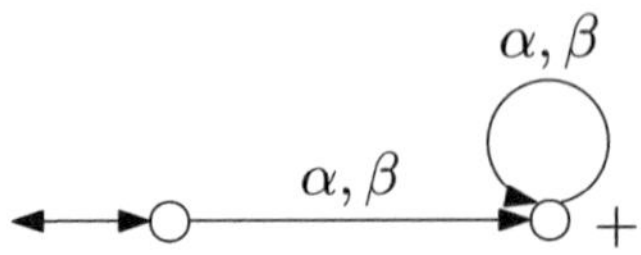

(iii) $L = \Sigma^*$

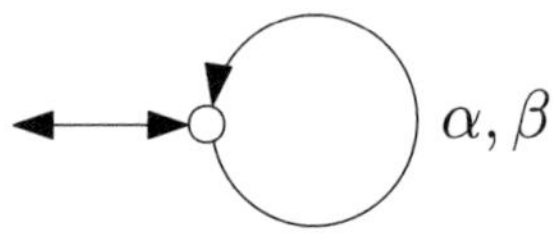

(iv) $L = \{\alpha^n | n = 0, 1, 2, \ldots\}$

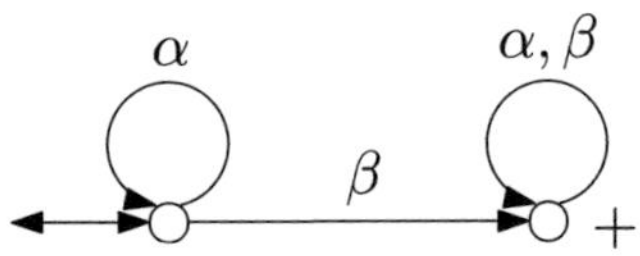

(v) $L = \{\alpha^n \beta^n | n = 0, 1, 2, \ldots\}$

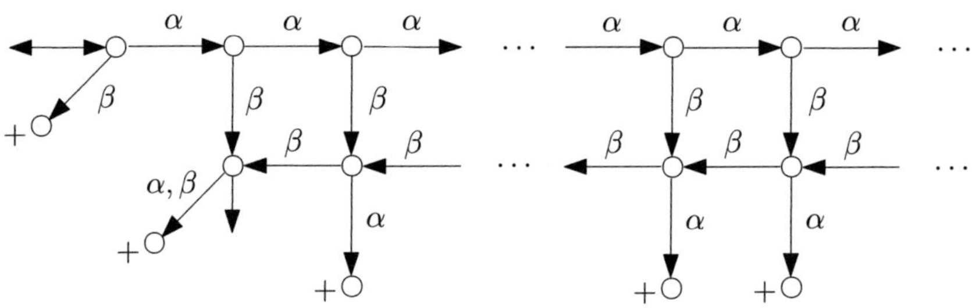

In the above transition graph, the nodes labeled + should be merged to a single 'dump' node self-looped with $\{\alpha, \beta\}$.

(vi) $L = \{s | \#\alpha(s) = \#\beta(s)\}$, where $0 \leq \#\sigma(s)$ = number of σ's in the string s.

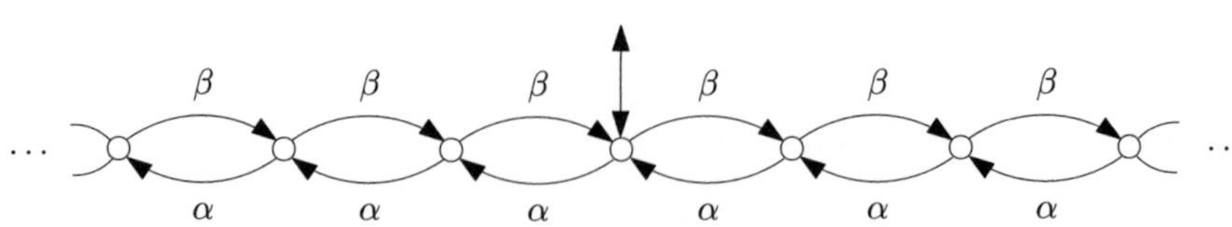

We conclude this section with a useful extension of the concept of canonical recognizer, obtained by starting with an arbitrary partition θ of Σ^* in place of the binary partition $\{L, \Sigma^* - L\}$. Let T be a new alphabet, not necessarily disjoint from Σ, with $|T| = |\theta|$ (here we allow the possibility that T is countably infinite); and label the cells of θ in bijective correspondence with T. Thus $T \simeq \Sigma^*/\theta$. There is then a well-defined projection, say $P_\theta : \Sigma^* \to T$, taking strings $s \in \Sigma^*$ to their labels in T, as shown below. An element $\tau \in T$ can be thought of as signaling the presence of a string $s \in \Sigma^*$ in the corresponding cell of θ (or ker P_θ); for this reason T will be called the *output* alphabet.

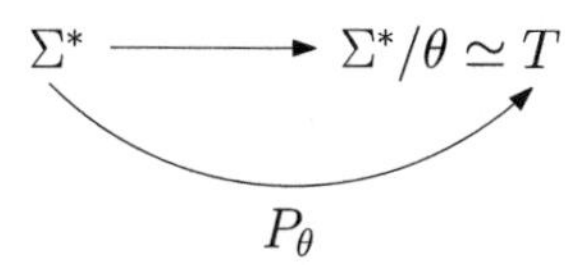

It is straightforward to construct (abstractly) an automaton A that maps any string

$$s = \sigma_1\sigma_2 \cdots \sigma_k \in \Sigma^*$$

into the corresponding sequence of output elements

$$t = P_\theta(\epsilon)P_\theta(\sigma_1)P_\theta(\sigma_1\sigma_2) \cdots P_\theta(\sigma_1\sigma_2 \cdots \sigma_k) \in T^*$$

For the state space X it suffices to take the set of cells of the coarsest right congruence ω on Σ^* that is finer than θ. There is then a unique output map $\lambda : X \to T$ such that $\lambda(x[s]) = P_\theta(s)$, where $x[s]$ is the cell of s (mod ω). The transition function $\xi : X \times \Sigma \to X$ is defined as before, together with the initial state $x_0 = x[\epsilon]$. The 6-tuple

$$\mathbf{A} = (X, \Sigma, \xi, x_0, T, \lambda)$$

is sometimes called a *Moore automaton*. Evidently the previous construction of a canonical recognizer for L is recovered on taking $\theta = \{L, \Sigma^* - L\}$, $T = \{0, 1\}$, $P_\theta(s) = 1$ iff $s \in L$, and $X_m = \{x|\lambda(x) = 1\}$.

Exercise 3: Let $\bar{K} = L$ and let κ, λ be the Nerode right congruences for K, L respectively. Show that $\kappa \leq \lambda$. In other words, closing a language coarsens its right congruence.

Exercise 4: Verify in detail that transition graphs (i)–(vi) in Example 2 above indeed display the canonical recognizers for the indicated languages. **Hint**: For each node n of the given transition graph, let $\Sigma^*(n) \subseteq \Sigma^*$ be the subset of strings that correspond to paths through the graph from the initial node n_0 to n. Show that the $\Sigma^*(n)$ are precisely the Nerode cells for the given language L: namely every string in Σ^* belongs to $\Sigma^*(n)$ for some n, for every n any pair of strings in $\Sigma^*(n)$ are Nerode equivalent, and no two strings in distinct subsets $\Sigma^*(n)$, $\Sigma^*(n')$ are equivalent.

Exercise 5: As in the construction of a Moore automaton, let θ be a given partition of Σ^* and let T label the cells of θ. Construct a recognizer that generates a new output symbol from T only when the θ-cell membership of the input string in Σ^* actually changes, as the string evolves symbol-by-symbol. Show that the number of states of this recognizer need be no more than twice that of the original Moore automaton. **Hint**: Start by augmenting T with a 'silent symbol' $\tau_o \notin T$, corresponding to 'no change in θ-cell membership'. Let $T_o = T \cup \{\tau_o\}$. Examine the relationship between Nerode cells when the output alphabet is T and when it is T_o. Provide a concrete example displaying both the original and the new Moore automata.

Exercise 6: Let Σ, T be alphabets, let $L \subseteq \Sigma^*$, and let $P : \Sigma^* \to T^*$ be a map with the properties

$$P(\epsilon) = \epsilon$$
$$P(s\sigma) = \text{either } P(s) \text{ or } P(s)\tau, \ \text{ some } \tau \in T$$

Notice that P is *prefix-preserving* in the sense that

$$(\forall s, s' \in \Sigma^*) s \le s' \Rightarrow P(s) \le P(s')$$

With τ_o a 'silent symbol' as in Exercise 5, define $T_o = T \cup \{\tau_o\}$, and then $Q : \Sigma^* \to T_o$ according to

$$Q(\epsilon) = \tau_o$$
$$Q(s\sigma) = \begin{cases} \tau_o \text{ if } P(s\sigma) = P(s) \\ \tau \ \text{ if } P(s\sigma) = P(s)\tau \end{cases}$$

Evidently Q maps a string $s \in \Sigma^*$ either into τ_o, or into the last symbol of $P(s)$ in T upon its fresh occurrence. Let $\nu \in \mathcal{E}(\Sigma^*)$ be the equivalence relation defined by $s \equiv s' (\text{mod } \nu)$ iff $Q(s) = Q(s')$ and

$$(\forall u \in \Sigma^*)(\forall t \in T^*) P(su) = P(s)t \Leftrightarrow P(s'u) = P(s')t$$

Show that ν is the coarsest right congruence that is finer than $\ker Q$. Then show how to construct (abstractly) a Moore automaton **A** that recognizes the language L and, for each string $s \in \bar{L}$, produces the output $Q(s)$ in T_o. **A** is both a recognizer for L and a *realization* of the restriction of P to $\bar{L}$. Create a simple but nontrivial example for which your construction can be carried out explicitly.

Exercise 7: Let $K \subseteq L \subseteq \Sigma^*$ and suppose the canonical recognizers for K, L have the same numbers (assumed finite) of states, marker states, and transitions. Investigate whether or not $K = L$, i.e. either prove equality or find an example with $K \subsetneq L$.

Exercise 8: Let $\Sigma = \{0\}$. Construct $L \subseteq \Sigma^*$ such that $\text{Ner}(L) = \perp$.

2.4 Automata

Let

$$\mathbf{A} = (Y, \Sigma, \eta, y_0, Y_m)$$

be a 5-tuple with Σ as before, Y a nonempty set, $y_0 \in Y$, $Y_m \subseteq Y$, and

$$\eta : Y \times \Sigma \to Y$$

a function. $\mathbf{A}$ is an *automaton over the alphabet* Σ. As before, η is the *state transition function*, y_0 is the *initial state* and Y_m is the subset of *marker states*; again we extend η to a function

$$\eta : Y \times \Sigma^* \to Y$$

by induction on length of strings.

The language $L \subseteq \Sigma^*$ *recognized* by $\mathbf{A}$ is

$$L := \{s \in \Sigma^* | \eta(y_0, s) \in Y_m\}$$

$\mathbf{A}$ is said to be a *recognizer* for L.

A state $y \in Y$ is *reachable* if $y = \eta(y_0, s)$ for some $s \in \Sigma^*$; and $\mathbf{A}$ is *reachable* if y is reachable for all $y \in Y$. Evidently a state that is not reachable can play no role in the recognition process. If $Y_{rch} \subseteq Y$ is the subset of reachable states then the *reachable subautomaton* $\mathbf{A_{rch}}$ of $\mathbf{A}$ is defined as

$$\mathbf{A_{rch}} = (Y_{rch}, \Sigma, \eta_{rch}, y_0, Y_{m,rch})$$

where

$$\eta_{rch} = \eta | Y_{rch} \times \Sigma, \quad Y_{m,rch} = Y_m \cap Y_{rch}$$

Clearly $\mathbf{A_{rch}}$ recognizes $L \subseteq \Sigma^*$ iff $\mathbf{A}$ does.

Define an equivalence relation λ on Y according to

$$y_1 \equiv y_2 (\text{mod } \lambda)$$

iff

$$(\forall s \in \Sigma^*) \eta(y_1, s) \in Y_m \Leftrightarrow \eta(y_2, s) \in Y_m$$

That is, two states of $\mathbf{A}$ are λ-equivalent if the same input strings map each of them into the subset of marker states of $\mathbf{A}$. As usual we write $s \equiv s' (\text{mod } L)$ for Nerode equivalence of strings with respect to L. Now we can state

Proposition 1

(i) $(\forall t, t' \in \Sigma^*) \eta(y_0, t) \equiv \eta(y_0, t') (\text{mod } \lambda) \Leftrightarrow t \equiv t' (\text{mod } L)$

(ii) $(\forall y, y' \in Y) y \equiv y' (\text{mod } \lambda) \Leftrightarrow (\forall s \in \Sigma^*) \eta(y, s) \equiv \eta(y', s) (\text{mod } \lambda)$
(iii) $(\forall y, y' \in Y) y \in Y_m \ \& \ y' \equiv y (\text{mod } \lambda) \Rightarrow y' \in Y_m$ □

Here (iii) states that λ refines the partition $\{Y_m, Y - Y_m\}$.

Define $X := Y/\lambda$ and let $P : Y \to X$ be the canonical projection. Let $X_m := PY_m$ and $x_0 := Py_0$. For $x = Py$ define

$$\xi(x, \sigma) = P\eta(y, \sigma)$$

Then ξ is well defined (Exercise 5 below) and extends inductively to $X \times \Sigma^*$. Properties of the extended map are summarized in

Proposition 2

(i) $(\forall s \in \Sigma^*) \xi(x, s) = P\eta(y, s)$ for $x = Py$
(ii) $(\forall s \in \Sigma^*) s \in L \Leftrightarrow \xi(x_0, s) \in X_m$
(iii) $(\forall s, s' \in \Sigma^*) \xi(x_0, s) = \xi(x_0, s') \Leftrightarrow s \equiv s' (\text{mod } L)$ □

Thus the 5-tuple

$$\mathbf{B} = (X, \Sigma, \xi, x_0, X_m)$$

is an automaton over Σ. The foregoing definitions and relationships are displayed in the commutative diagrams below.

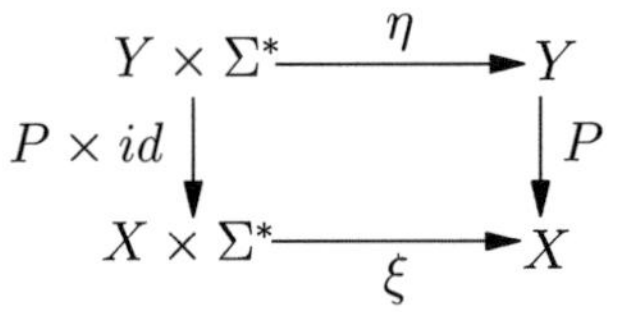

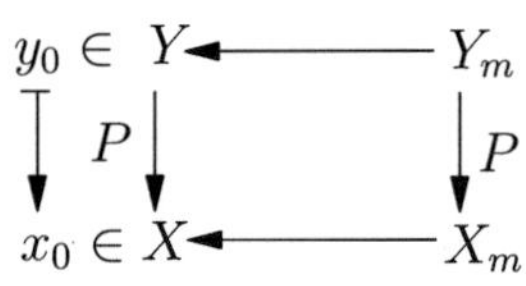

Informally, 'P projects **A** onto **B**'. The automaton **B** is reachable if **A** is reachable. By Proposition 2(iii) the reachable states of **B** can be identified with the cells of the Nerode equivalence relation on Σ^* with respect to L. We therefore have the following.

Theorem 3

If $\mathbf{B} = (X, \Sigma, \xi, x_0, X_m)$ is reachable then **B** is a canonical recognizer for L. □

Let $\mathbf{A} = (Y, \Sigma, \eta, y_0, Y_m)$ as before. Define the *complementary automaton*

$$\mathbf{A_{co}} = (Y, \Sigma, \eta, y_0, Y - Y_m)$$

Clearly **A** recognizes $L \subseteq \Sigma^*$ iff $\mathbf{A_{co}}$ recognizes the complementary language $L_{co} := \Sigma^* - L$. It is easy to see that $s \equiv s' (\text{mod } L)$ iff $s \equiv s' (\text{mod } L_{co})$, and thus $\|L_{co}\| = \|L\|$.

Similarly if $\mathbf{A_1}$, $\mathbf{A_2}$ are automata over Σ then in obvious notation the *product automaton* $\mathbf{A_1} \times \mathbf{A_2}$ is defined to be

$$\mathbf{A_1} \times \mathbf{A_2} = (Y_1 \times Y_2, \Sigma, \eta_1 \times \eta_2, (y_{10}, y_{20}), Y_{1m} \times Y_{2m})$$

where $\eta_1 \times \eta_2 : Y_1 \times Y_2 \times \Sigma \to Y_1 \times Y_2$ is given by

$$(\eta_1 \times \eta_2)((y_1, y_2), \sigma) = (\eta_1(y_1, \sigma), \eta_2(y_2, \sigma))$$

If $\mathbf{A_i}$ recognizes $L_i \subseteq \Sigma^*$ then it is easily seen that $\mathbf{A_1} \times \mathbf{A_2}$ recognizes $L_1 \cap L_2$.

Exercise 4: Prove Proposition 1.

Exercise 5: Show that ξ as described above exists and is unique by first verifying $\ker(P \times id) \leq \ker(P \circ \eta)$ on $Y \times \Sigma$. Then extend ξ to $X \times \Sigma^*$ and prove Proposition 2.

Exercise 6: Consider the automaton

$$\mathbf{A} = (Y, \Sigma, \eta, y_0, Y_m)$$

with

$$\begin{aligned} Y &= \{0, 1, 2, 3, 4\} \\ \Sigma &= \{\alpha, \beta\} \\ y_0 &= 0 \\ Y_m &= \{0, 1, 2\} \end{aligned}$$

and transitions

$$\begin{array}{l} [0, \alpha, 1]\ [2, \beta, 3] \\ [0, \beta, 4]\ [3, \alpha, 3] \\ [1, \alpha, 2]\ [3, \beta, 3] \\ [1, \beta, 4]\ [4, \alpha, 4] \\ [2, \alpha, 2]\ [4, \beta, 4] \end{array}$$

Construct the automaton **B** as above, and tabulate P. Use *TCT* **minstate** to check your result.[1]

Exercise 7: Given recognizers for L_1 and $L_2 \subseteq \Sigma^*$, construct a recognizer for $L_1 \cup L_2$. **Hint**: use

$$L_1 \cup L_2 = [\,(L_1)_{co} \cap (L_2)_{co}\,]_{co}$$

Exercise 8: Let $L_1, L_2 \subseteq \Sigma^*$ and let \$ be a symbol not in Σ. Let $L \subseteq (\Sigma \cup \{\$\})^*$ be the language $L_1\$L_2$, consisting of strings $s_1\$s_2$ with $s_1 \in L_1$ and $s_2 \in L_2$. Given recognizers for L_1 and L_2, construct a recognizer for L.

Exercise 9: In the regular case where $\|L_1\| = n_1$ and $\|L_2\| = n_2$ are both finite, derive tight upper bounds on $\|L_1 \cap L_2\|$, $\|L_1 \cup L_2\|$ and $\|L_1\$L_2\|$: that is, show by examples that your bounds cannot be improved for any (n_1, n_2).

---

[1]The software package *TCT* will be introduced in Example 3.2.1.

Exercise 10: Let $\Sigma = \{0\}$. Construct a language L with $\mathrm{Ner}(L)$ equal to the bottom element $\bot \in \mathcal{E}(\Sigma^*)$, and specify the canonical recognizer.

2.5 Generators

In Sect. 2.4 we saw that a language can be represented concretely by specifying a corresponding recognizer. For many purposes, a similar but more flexible and economical representation is provided by a *generator*, namely a transition structure in which, in general, only a proper subset of the totality of events can occur at each state. For example, a generator might simply be a recognizer from which the dump state (if any) and all transitions to it have been dropped. Let

$$\mathbf{G} = (Y, \Sigma, \eta, y_0, Y_m)$$

In the present case the transition function $\eta : Y \times \Sigma \rightarrow Y$ is defined at each $y \in Y$ only for a subset of the elements $\sigma \in \Sigma$, namely η is a *partial function* (pfn) and we write

$$\eta : Y \times \Sigma \rightarrow Y \quad \text{(pfn)}$$

The notation $\eta(y, \sigma)!$ will mean that $\eta(y, \sigma)$ is defined.[2]

Much as before, η is extended to a partial function $\eta : Y \times \Sigma^* \rightarrow Y$ by the rules

$$\eta(y, \epsilon) = y$$
$$\eta(y, s\sigma) = \eta(\eta(y, s), \sigma)$$

provided $y' := \eta(y, s)!$ and $\eta(y', \sigma)!$.

Example 1: Generator for a 'machine'

The state transition graph of a simple generator is displayed below. The generator represents a 'machine' with possible states (I)dle, (W)orking, (D)own and (S)crapped. Starting in I, the machine may take a workpiece (event α), thereby moving to W. From W the machine may either complete its work cycle, returning to I (event β), or else break down (event λ), moving to D. It remains at D until it is either repaired (events μ, ν) or scrapped (event κ). Repair event μ corresponds to loss of workpiece and return to I, completing a cycle of working, breakdown and repair, while ν corresponds to saving the workpiece to continue work at W. The initial state is I and the marker states both I and S. This process may be thought of as repeating an arbitrary finite number of times.

[2] If $\eta(y, \sigma)!$ for all $(y, \sigma) \in Y \times \Sigma$, η may be called a *total function* (η is then an 'ordinary' function, as in Sect. 2.4).

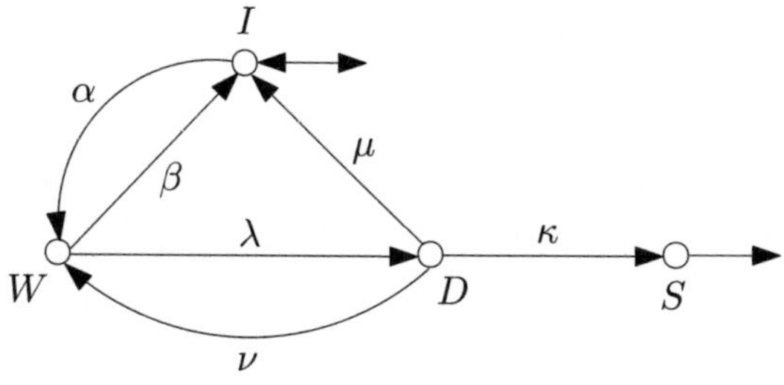

Exercise 2: Let $\Sigma = \{\alpha, \beta\}$, $L = \Sigma^*\alpha\beta\alpha\Sigma^*$. Thus L consists of all finite strings of the form $s_1\alpha\beta\alpha s_2$, where s_1 and s_2 are arbitrary strings over Σ.

(i) Give an alternative verbal description of L, regarding the occurrence of $\alpha\beta\alpha$ as the signal of an 'emergency'.
(ii) Design a (deterministic) finite-state recognizer for L. **Hint**: This can be done using just 4 states.

Exercise 3: Develop a finite-state operational (not electrical) model of an ordinary household telephone, as seen by a single subscriber able to place or receive a call. **Note**: this can become surprisingly complicated, so start with something simple and then refine it in stages, up to a model on the order of 10 to 20 states.

In general one may think of **G** as a device that 'generates' strings by starting at the initial state y_0 and executing only transitions for which its transition function η is defined; if more than one transition is defined to exit from a given state y, the device may be supposed to choose just one of these possibilities, on any particular occasion, by some quasi-random internal mechanism that is unmodeled by the system analyst. In this sense the generating action is 'possibilistic'. It may be thought of as carried out in repeated 'trials', each trial generating just one of the possible strings s for which $\eta(y_0, s)$ is defined. In this account, 'choose' needn't always be interpreted literally: most machines do not choose to start work autonomously, but are forced by some external agent. The generation model is independent of (but consistent with) causative factors, which should be examined in context.

The set of strings $s \in \Sigma^*$ such that $\eta(y_0, s)!$ is the *closed behavior* of **G**, denoted by $L(\mathbf{G})$, while the subset of strings $s \in L(\mathbf{G})$ such that $\eta(y_0, s) \in Y_m$ is the *marked behavior* of **G**, denoted by $L_m(\mathbf{G})$. Clearly $L(\mathbf{G})$ is closed and contains $L_m(\mathbf{G})$. A generator **G** is to be thought of as representing both its closed and marked behaviors; we refer to these two languages as the *semantics* of **G**. Frequently only the semantics of **G** will be of interest in our control theory, **G** itself being simply a modeling entity or computational device. As in the case of an automaton, a state $y \in Y$ is *reachable* if there is a string $s \in \Sigma^*$ with $\eta(y_0, s)!$ and $\eta(y_0, s) = y$; **G** itself is *reachable* if y is reachable for all $y \in Y$. A state $y \in Y$ is *coreachable* if there is $s \in \Sigma^*$ such that $\eta(y, s) \in Y_m$; and **G** is *coreachable* if y is coreachable for every $y \in Y$. **G** is *nonblocking* if every reachable state is coreachable, or equivalently $L(\mathbf{G}) = \overline{L_m(\mathbf{G})}$; the latter condition says that any string that can be generated by **G** is a prefix of (i.e. can always be completed to) a marked string of **G**. Finally **G** is *trim* if it is both reachable and coreachable. Of course **G** trim implies **G** nonblocking, but the converse is false: a nonblocking generator might have nonreachable states (that might or might not be coreachable). The 'machine' of Example 1 is trim; if the state S were not marked, the resulting generator would still be reachable, but not coreachable. In practice one usually models a generator **G** to be reachable; however, it may be quite realistic for **G** not to be coreachable, for instance to have a 'dump' or 'dead-end' state from which no marker state is accessible. As a rule it is therefore advisable not to overlook non-coreachable states, or inadvertently remove them from the model.

As an aid to intuition the following interpretation of the behavior of **G** as a stochastic system may be helpful. Assume that the state set Y and the alphabet Σ are both finite, the usual case in applications. At each state $y \in Y$ assign a discrete probability distribution $p(y, \sigma)$ over the events $\sigma \in \Sigma$ for which $\eta(y, \sigma)!$; here the $p(y, \sigma) > 0$ and at each state y, sum to 1. If there is no such σ, i.e. the system halts at y, attach (optionally) an artificial selfloop (labeled τ, say) with $p(y, \tau) = 1$. Imagine that **G** moves by randomly selecting σ at y according to $p(y, \sigma)$, with successive choices statistically independent. This rule assigns to **G** the structure of a discrete Markov chain; if preferred, a stationary continuous-time Poisson-Markov structure could be assigned in similar fashion. Probability theory now provides an appealing interpretation for the behavior of **G** 'in the long run'. For instance, at any state y which **G** visits 'infinitely often', every defined event σ will be chosen infinitely often. And if **G** is nonblocking, i.e. some definite marker state always remains reachable, then with probability 1 it will indeed eventually be reached. Standard calculations can confirm intuition. For instance in the DES **G** shown below, suppose event α is chosen at $y = 0$ with probability p $(0 < p < 1)$, event β with probability $q = 1 - p$. Then

$$L_m(\mathbf{G}) = \{\alpha^n \beta | n \geq 0\}$$

the nth string having probability of execution $p^n q$; while the mean number of transitions to reach the marker state 1 will be

$$\sum_{n=0}^{\infty} (n + 1) p^n q = 1/q$$

G

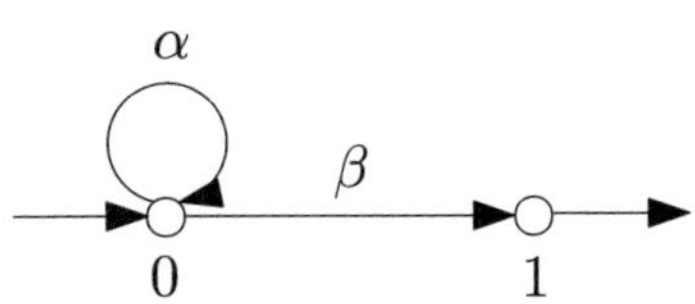

While often plausible, this Markov chain picture of **G** should nevertheless be invoked with caution: it might not be appropriate in situations involving several 'agents' in a competitive or antagonistic relationship, where a stronger notion of nonblocking (or liveness), perhaps requiring use of a queue, should be enforced (cf. Exercises 4.3.3, 4.9.6).

Example 4: **Generators**

The following trim generators correspond to the languages of Example 2.3.2, over $\Sigma = \{\alpha, \beta\}$. Here L_m denotes the language represented, while the closed behavior $L = \bar{L}_m$.

(i) $L_m = \varnothing$ **EMPTY** (having empty state set)

(ii) $L_m = \{\epsilon\}$

(iii) $L_m = \Sigma^*$

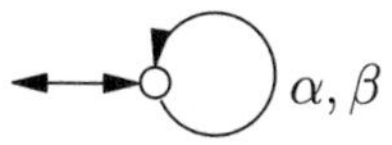

(iv) $L_m = \{\alpha^n | n = 0, 1, 2, \ldots\}$

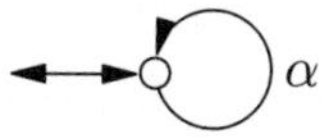

(v) $L_m = \{\alpha^n \beta^n | n = 0, 1, 2, \ldots\}$

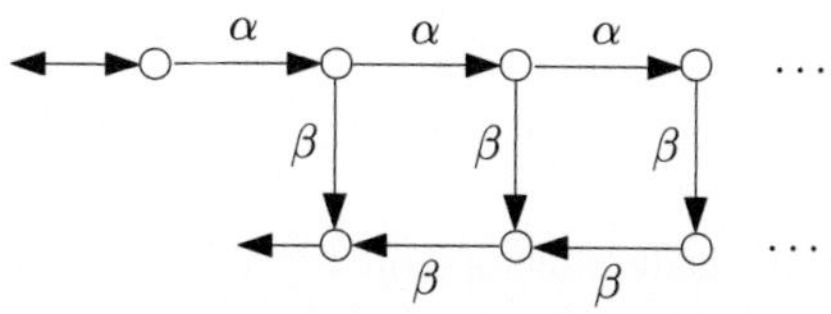

(vi) $L_m = \{s | \#\alpha(s) = \#\beta(s)\}$

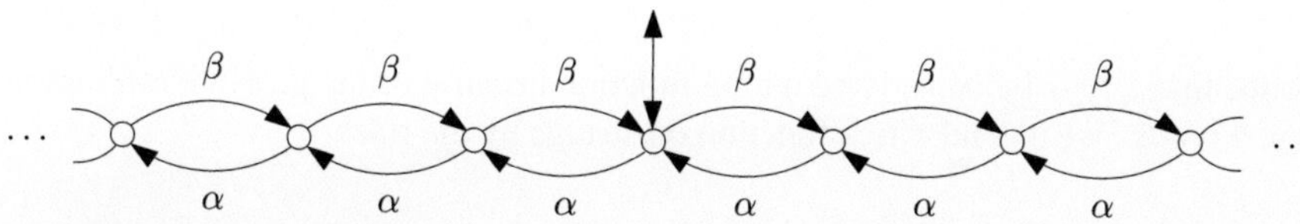

The following counterpart of Proposition 2.4.1 can be used to reduce a (reachable) generator **G** to a minimal-state version having the same closed and marked behaviors. This time we need Nerode equivalence relations on Σ^* for both $L(\mathbf{G})$ ($\equiv_c$, say) and $L_m(\mathbf{G})$ ($\equiv_m$). Define $\lambda \in \mathcal{E}(Y)$ according to $y \equiv y'(\text{mod } \lambda)$ provided

(i) $(\forall s \in \Sigma^*)\eta(y, s)! \Leftrightarrow \eta(y', s)!$

(ii) $(\forall s \in \Sigma^*)\eta(y, s)! \;\&\; \eta(y, s) \in Y_m \Leftrightarrow \eta(y', s)! \;\&\; \eta(y', s) \in Y_m$

Proposition 5

(i) $(\forall s, s' \in \Sigma^*)\eta(y_0, s) \equiv \eta(y_0, s')(\text{mod } \lambda) \Leftrightarrow s \equiv_c s' \;\&\; s \equiv_m s'$

(ii) $(\forall y, y' \in Y) y \equiv y'(\text{mod } \lambda) \Leftrightarrow (\forall s \in \Sigma^*)\eta(y, s) \equiv \eta(y', s)(\text{mod } \lambda)$

(iii) $(\forall y, y' \in Y) y \in Y_m \;\&\; y' \equiv y(\text{mod } \lambda) \Rightarrow y' \in Y_m$

Exercise 6: Prove Proposition 5 and provide a nontrivial application.

Reduction of a generator **G** to a minimal (reachable) version by projection (mod λ) is implemented in *TCT* as the procedure **minstate** (see footnote 1).

Exercise 7: Let $L \subseteq \Sigma^*$ and **G** be its minimal-state recognizer. Show how to construct a recognizer **H** whose current state encodes both the current state of **G** and the last previous state of **G**. Generalize your result to encode the list of n most recent states of **G** (in temporal order). Alternatively, construct a recognizer **K** whose current state encodes the list of n most recent events of **G**, in order of occurrence. Illustrate your results using **G** as shown below.

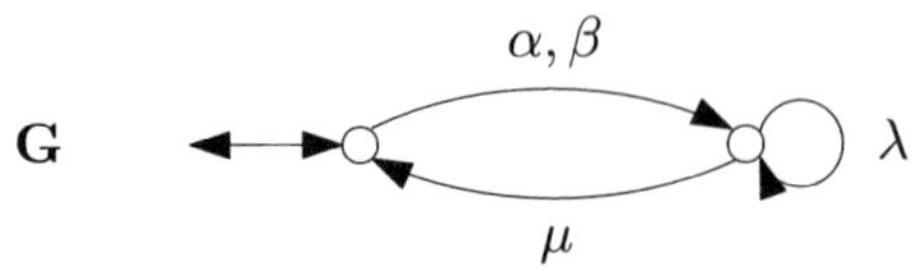

Exercise 8: Let $\Sigma = \{0, 1\}$ and let $L \subseteq \Sigma^*$ be those strings in which a 1 occurs in the third-to-last place. Design a recognizer for L.

Occasionally it will be useful to bring in a nondeterministic generator, namely one in which more than one transition defined at a given exit state may carry the same label $\sigma \in \Sigma$. Formally a *nondeterministic generator* is a 5-tuple

$$\mathbf{T} = (Y, \Sigma, \tau, y_0, Y_m)$$

as before, but with the difference that the transition function τ now maps pairs (y, σ) into subsets of Y:

$$\tau : Y \times \Sigma \rightarrow Pwr(Y)$$

Notice that τ may be considered a total function because of the possible evaluation $\tau(y, \sigma) = \varnothing$. We extend τ to a function on strings by the rules

$$\tau(y, \epsilon) = \{y\}$$
$$\tau(y, s\sigma) = \bigcup\{\tau(y', \sigma) | y' \in \tau(y, s)\}$$

We define the *closed behavior* $L(\mathbf{T})$ of the nondeterministic generator $\mathbf{T}$ to be the set of all strings $s \in \Sigma^*$ for which $\tau(y_0, s) \neq \varnothing$. The *marked behavior* $L_m(\mathbf{T})$ of $\mathbf{T}$ is the set of all strings in $L(\mathbf{T})$ for which at least one particular realization (path through the transition graph) corresponds to a sequence of states starting at y_0 and ending in Y_m:

$$L_m(\mathbf{T}) = \{s \in \Sigma^* | \tau(y_0, s) \cap Y_m \neq \varnothing\}$$

If the nondeterministic generator $\mathbf{T}$ is given, then a deterministic generator $\mathbf{T_{det}}$ that generates the same languages $L(\mathbf{T})$ and $L_m(\mathbf{T})$ can be constructed by taking as the states of $\mathbf{T_{det}}$ the nonempty subsets of Y, i.e. $Y_{det} = Pwr(Y) - \{\varnothing\}$; usually only a small fraction of the new states turn out to be reachable and therefore worthy of retention. This process of converting a nondeterministic to a deterministic generator is known as the *subset construction*. While a nondeterministic generator can always be converted in this way to a deterministic one, in some applications the use of a nondeterministic generator may result in greater convenience or economy of description.

A word of warning. Conversion by the subset construction may obliterate blocking situations in the nondeterministic model. For instance

T

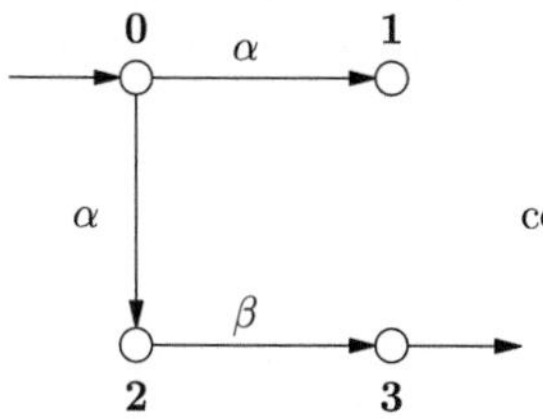

converts to $\mathbf{T}_{\mathbf{det}}$

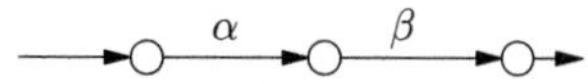

obscuring the fact that **T** can block at state 1. A solution is first to enhance **T** by selflooping all non-coreachable states with a new event label $\kappa \notin \Sigma$. Thus

T′

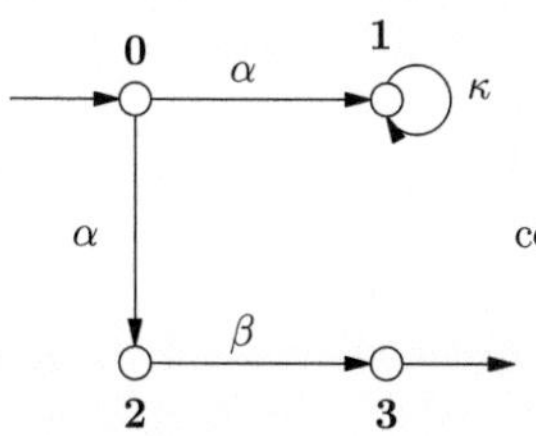

converts to $\mathbf{T}'_{\mathbf{det}}$

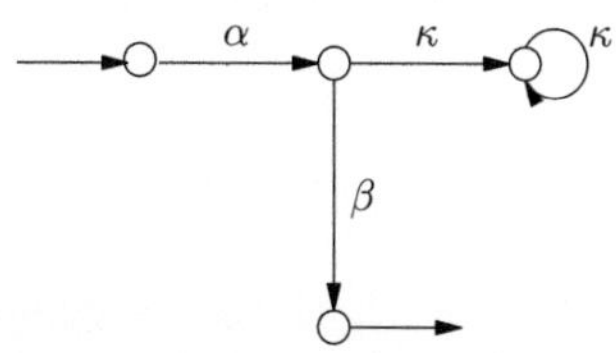

Exercise 9: **Subset construction**

Supply the details of the subset construction. Namely let

$$\mathbf{H} = (Y, \Sigma, \tau, y_0, Y_m)$$

be a nondeterministic generator, and let

$$\mathbf{G} = (X, \Sigma, \xi, x_0, X_m)$$

be the deterministic generator defined by

$$X = Pwr(Y) - \{\varnothing\},\ \xi(x, \sigma) = \bigcup\{\tau(y, \sigma)|y \in x\}$$
$$x_0 = \{y_0\}, \qquad X_m = \{x|x \cap Y_m \neq \varnothing\}$$

Here ξ is a partial function on $X \times \Sigma$ with $\xi(x, \sigma)!$ iff the defining evaluation is nonempty. Show that $L(\mathbf{G}) = L(\mathbf{H})$ and $L_m(\mathbf{G}) = L_m(\mathbf{H})$. Check the two examples above in detail.

Exercise 10: Implementation of the subset construction is unattractive in that it may require exponential computational effort in the state size of **T**. Why? Can you exhibit a 'worst case'?

2.6 Regular Expressions

In discussing small examples we may use, in addition to state transition graphs, a representation for regular languages known as regular expressions. These may be combined by *regular algebra* to represent complex languages in terms of simpler ones.

If $s \in \Sigma^*$ we may write s for the language $\{s\} \subseteq \Sigma^*$. Let $L, M \subseteq \Sigma^*$. New languages $L + M$, LM and L^* are defined as follows.

$$L + M := \{s | s \in L \text{ or } s \in M\}$$
$$LM := \{st | s \in L \text{ and } t \in M\}$$
$$L^* := \epsilon + \bigcup_{k=1}^{\infty} \{s_1 \cdots s_k | s_1, \ldots, s_k \in L\}$$

Thus

$$L^* = \epsilon + L + L^2 + \cdots = \bigcup_{k=0}^{\infty} L^k, \quad L^0 := \epsilon$$

If $\Sigma = \{\alpha, \beta\}$ we may sometimes write $\Sigma = \alpha + \beta$. In accordance with the definition of catenation and the properties of ϵ we have

$$\epsilon s = s\epsilon = s, \quad \epsilon\epsilon = \epsilon, \quad \epsilon^* = \epsilon$$

and for the empty language $\varnothing$,

$$\varnothing + L = L, \quad \varnothing L = L\varnothing = \varnothing, \quad \varnothing^* = \epsilon$$

A *regular expression over* Σ is a formal expression obtained by a finite number of applications of the operations listed above, to elements in the list: elements of Σ, ϵ, $\varnothing$, and all expressions so obtained. Since Σ, $\{\epsilon\}$, $\varnothing$ are subsets of Σ^*, a regular expression represents a subset of Σ^*. It is shown in the literature (Kleene's Theorem) that the subsets represented by the regular expressions over Σ are exactly the regular sublanguages of Σ^*, namely the sublanguages whose canonical recognizers have a finite state set.

Regular algebra admits numerous identities that are useful for simplifying regular expressions. They may be proved by comparing the corresponding subsets of Σ^*. While we are not likely to undertake complicated manipulations of regular expressions, the following catalog of identities is provided for reference. Here $L, M, N \subseteq \Sigma^*$ are arbitrary languages over Σ.

$$L\epsilon = \epsilon L = L, \quad L + L = L, \quad L + M = M + L$$

$$(L + M) + N = L + (M + N) \quad \text{(so we write } L + M + N\text{)}$$

$$(LM)N = L(MN) \quad \text{(so we write } LMN\text{)}, \quad (L+M)N = LN + MN$$

$$L^* = \epsilon + LL^*, \quad L^*L^* = L^*, \quad (L^*)^* = L^*, \quad LL^* = L^*L$$

$$(L^* + M^*)^* = (L^*M^*)^* = (L^*M)^*L^* = (L+M)^*$$

$$(LM)^*L = L(ML)^*, \quad (L^*M)^* = \epsilon + (L+M)^*M$$

The following result is fundamental for the solution of systems of equations.

Proposition 1

(i) If $L = M^*N$ then $L = ML + N$
(ii) If $\epsilon \notin M$ then $L = ML + N$ implies $L = M^*N$ □

Part (ii) is known as Arden's rule. Taken with (i) it says that if $\epsilon \notin M$ then $L = M^*N$ is the unique solution of $L = ML + N$; in particular if $L = ML$ (with $\epsilon \notin M$) then $L = \emptyset$.

Exercise 2: Show by counterexample that the restriction $\epsilon \notin M$ in Arden's rule cannot be dropped.

Exercise 3: Prove Arden's rule. **Hint**: If $L = ML + N$ then for every $k \geq 0$

$$L = M^{k+1}L + (M^k + M^{k-1} + \cdots + M + \epsilon)N$$

As an application of Arden's rule it will be shown how to find a regular expression for the language generated by a finite nondeterministic transition structure. In the example displayed below we write by definition

$$L_m(\mathbf{G}) = \{s \in \Sigma^* | \tau(x_0, s) \cap X_m \neq \emptyset\}$$

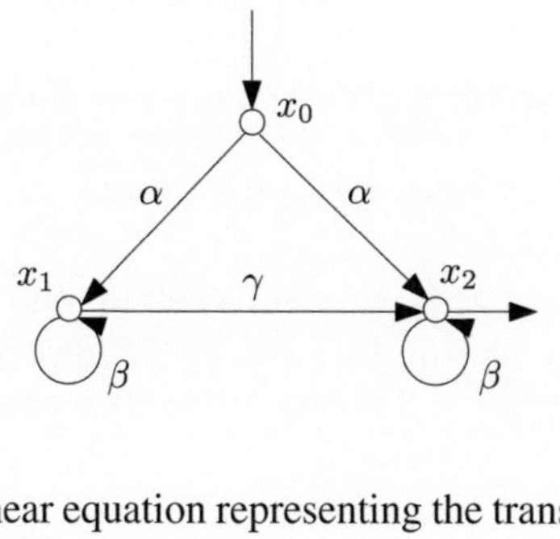

Step 1. Write a formal linear equation representing the transitions at each state of **G**:

$$x_0 = \alpha x_1 + \alpha x_2$$
$$x_1 = \beta x_1 + \gamma x_2$$
$$x_2 = \beta x_2 + \epsilon$$

Note that the 'forcing' term ϵ is added on the right side if $x \in X_m$.
Step 2. Consider the states x_i as tokens for the 'unknown' regular languages

$$X_i = \{s \in \Sigma^* | \tau(x_i, s) \cap X_m \neq \varnothing\}$$

Thus it is easy to see that the X_i satisfy exactly the regular-algebraic equations just written. Solve these equations using Arden's rule:

$$\begin{aligned} X_2 &= \beta^* \\ X_1 &= \beta^* \gamma X_2 = \beta^* \gamma \beta^* \\ X_0 &= \alpha\beta^* \gamma \beta^* + \alpha\beta^* \\ &= \alpha\beta^*(\epsilon + \gamma\beta^*) \end{aligned}$$

Step 3. Since x_0 is the initial state we obtain

$$L_m(\mathbf{G}) = X_0 = \alpha\beta^*(\epsilon + \gamma\beta^*)$$

As a second example consider the transition graph below, with states labeled a, b, c. We have

$$a = \alpha b + \epsilon, \quad b = \beta a + \gamma c, \quad c = \gamma c$$

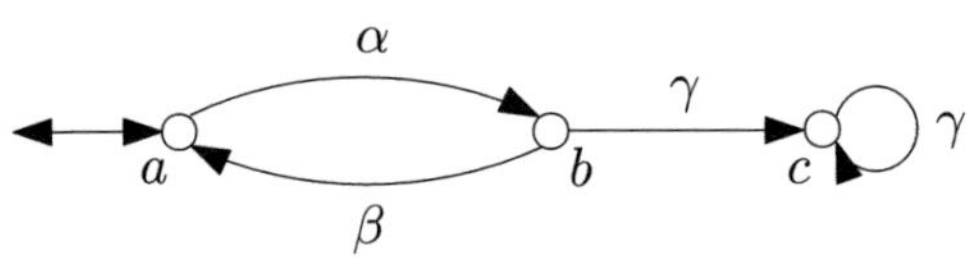

These equations give

$$c = \varnothing, \quad b = \beta a, \quad a = (\alpha\beta)^*$$

Exercise 4: Consider the problem of designing a recognizer for the language

$$L = \Sigma^* 0\,1\,0\,0\,1\,0 \Sigma^*$$

where $\Sigma = \{0, 1\}$. In other words, we want a bell to beep (and go on beeping with each new input symbol) as soon as the string indicated occurs for the first time in an arbitrary sequence of 0's and 1's. It is easy to specify a nondeterministic recognizer, as shown below.

NDR

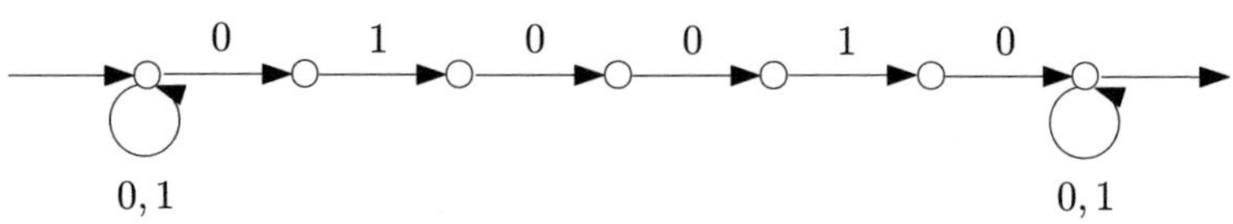

By contrast, it's somewhat tricky to design a deterministic recognizer 'by hand': try it and see. To do this using *TCT*,[1] modify the initial selfloop to obtain

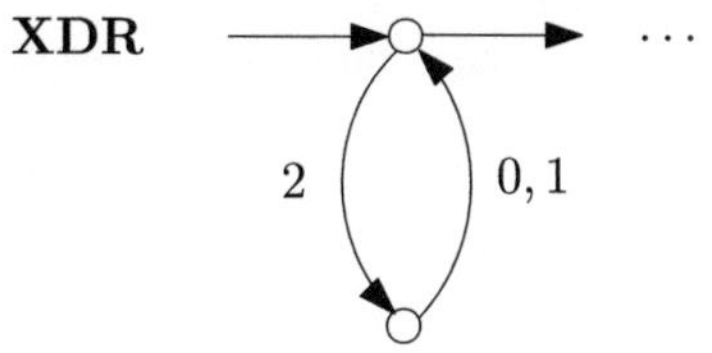

where 2 is a dummy symbol to make **XDR** deterministic. Compute

$$\mathbf{DR} = \mathbf{project}(\mathbf{XDR}, \mathbf{null}[2])$$

Draw the state diagram and convince yourself that **DR** does the job.

Exercise 5: Consider the generators

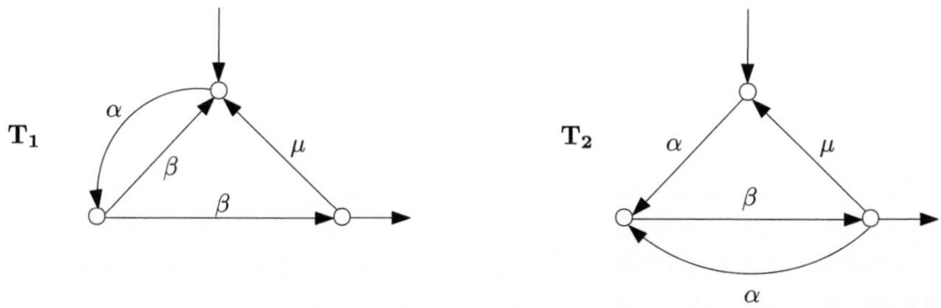

By application of the subset construction to $\mathbf{T_1}$, show that $\mathbf{T_1}$ and $\mathbf{T_2}$ determine the same regular language. Using Arden's rule, obtain corresponding regular expressions

$$L_1 = (\alpha\beta(\epsilon + \mu))^*\alpha\beta$$
$$L_2 = (\alpha(\beta\alpha)^*\beta\mu)^*\alpha(\beta\alpha)^*\beta$$

and prove by regular algebra that indeed $L_1 = L_2$. **Hint**: First prove the identity

$$(L^*M)^*L^* = (L + M)^*$$

and then reduce L_2 to L_1.

2.7 Causal Output Mapping and Hierarchical Aggregation

In this section[3] we develop in greater detail some of the ideas in Sects. 2.2, 2.3 and 2.5. For easier reading, some definitions are repeated.

[3]Not needed in the sequel.

Let $\mathbf{G} = (Q, \Sigma, \delta, q_0)$ be a *generator*: namely Q is a nonempty *state set*, $q_0 \in Q$ is the *initial state*, Σ is a nonempty set, the *alphabet of transition labels*, and $\delta : Q \times \Sigma \to Q$ is the (partial) *transition function*. The action of δ may sometimes be written by juxtaposition: $\delta(q, \sigma) = q\sigma$; and if this convention is understood then $\mathbf{G}$ may be written $(Q, \Sigma, \_, q_0)$. Also δ is extended by iteration in the natural way to a partial function on the set Σ^* of all finite strings s of elements in Σ. We write $\delta(q, s)!$ or $qs!$ to mean that the action of δ is defined at $(q, s) \in Q \times \Sigma^*$. If $\epsilon \in \Sigma^*$ is the *empty string* then $q\epsilon := \delta(q, \epsilon) := q$. The *closed behavior* of $\mathbf{G}$ is the subset of strings

$$L(\mathbf{G}) = \{s \in \Sigma^* | \delta(q_0, s)!\}$$

In this section marked behavior plays no role.

For brevity write $L(\mathbf{G}) =: L$. Note that L contains ϵ, and that L is *prefix-closed*, namely if $s \in L$ and $s' \leq s$ (s' is a prefix of s) then $s' \in L$ as well. Now let T be a second alphabet and suppose that $P : L \to T^*$ is a (total) map with the properties

$$P(\epsilon) = \epsilon$$

$$P(s\sigma) = \begin{cases} \text{either } P(s) \\ \\ \text{or } P(s)\tau, \ \text{some } \tau \in T \end{cases} \qquad s \in \Sigma^*, \quad \sigma \in \Sigma$$

We call P the *reporter* of $\mathbf{G}$. P is *causal* (or *nonanticipative*); in particular it is prefix preserving: if $s \leq s'$ then $P(s) \leq P(s')$. The pair $(\mathbf{G}, P)$ may be called a *generator with reporter*.

The simplest way to visualize the behavior of $(\mathbf{G}, P)$ is via the *reachability tree* of $L(\mathbf{G})$, a tree in which each node n is identified with a string s of L by a bijection $n : L \to$ Nodes: the root node is $n(\epsilon)$, and for each $s \in L$ the children of $n(s)$ are exactly the nodes $\{n(s\sigma) | s\sigma \in L\}$. Notice that it is possible to drop the distinction between nodes and strings, taking as one particular version of $\mathbf{G}$ the 4-tuple

$$\mathcal{T}(L) := \{L, \Sigma, \_, \epsilon\}$$

In the reachability tree the action of P can be indicated as follows. Bring in an element $\tau_o \notin T$ and write $T_o = T \cup \{\tau_o\}$. Define the *tail map* $\omega_o : L \to T_o$ according to

$$\omega_o(\epsilon) = \tau_o$$

$$\omega_o(s\sigma) = \begin{cases} \tau_o \text{ if } P(s\sigma) = P(s) \\ \tau \ \text{ if } P(s\sigma) = P(s)\tau \end{cases}$$

Thus ω_o identifies the last output symbol 'reported' by P, with τ_o interpreted as the 'silent output symbol'. Using $\mathcal{T}(L)$ to represent $\mathbf{G}$, we can now represent $(\mathbf{G}, P)$ by the 6-tuple

$$\mathcal{T}(L, P) := \{L, \Sigma, \_, \epsilon, T_o, \omega_o\}$$

In graphical representation the edges of the tree (transitions $s \xrightarrow{\sigma} s\sigma$ of L) are labeled by the corresponding element σ of Σ, while the nodes are labeled by the corresponding output in T_o.

Define the *output language* of $(\mathbf{G}, P)$ to be $PL \subseteq T^*$, and let $\mathcal{T}(PL)$ be the reachability tree of PL. Write ϵ for the empty string in T^* as well. Clearly PL contains ϵ and is prefix-closed. Again, ignoring the distinction between nodes of the tree and strings of PL, we have that P induces a surjection $\hat{P}$ from $\mathcal{T}(L, P)$ to $\mathcal{T}(PL)$:

$$\hat{P} : L \to PL : s \mapsto Ps$$

It is convenient to introduce the modified tail map

$$\hat{\omega}_o : L \times \Sigma \to T_o : (s, \sigma) \mapsto \omega_o(s\sigma)$$

and define the product map

$$\hat{P} \times \hat{\omega}_o : L \times \Sigma \to PL \times T_o : (s, \sigma) \mapsto (\hat{P}(s), \hat{\omega}_o(s, \sigma))$$

Then we have the obvious commutative diagram

$$\begin{array}{ccc} L \times \Sigma & \longrightarrow & L \\ \Big\downarrow \hat{P} \times \hat{\omega}_o & & \Big\downarrow \hat{P} \\ PL \times T_o & \longrightarrow & PL \end{array}$$

Here the horizontal arrows represent the transition action, with the extended definition for the bottom arrow that $(t, \tau_o) \mapsto t$. One can think of the diagram as a display of how transitions in the tree (L) are 'tracked' by transitions in $\mathcal{T}(PL)$. Notice that by composition of transition functions the diagram can be iterated arbitrarily far to the right:

$$\begin{array}{ccc|ccc|ccc} L \times \Sigma & \longrightarrow & L & \times\Sigma & \longrightarrow & L & \times\Sigma & \longrightarrow & \cdots \\ \Big\downarrow & & & \Big\downarrow & & & \Big\downarrow & & \\ PL \times T_o & \longrightarrow & PL & \times T_o & \longrightarrow & PL & \times T_o & \longrightarrow & \cdots \end{array}$$

thus extending the tracking feature to strings of L.

While the reachability trees of L and PL are useful for purposes of visualization, more efficient representations are available in principle, which are often more convenient in practical applications. These are obtained by aggregating nodes of the tree (i.e. strings of the language) by means of suitable equivalence relations. For any language $K \subseteq A^*$ over an alphabet A, write

$$K/s := \{u \in A^* | su \in K\}, \quad s \in A^*$$

The *Nerode equivalence relation* Ner(K) on A^* is defined by

$$s \equiv s'(\text{mod Ner}(K)) \text{ iff } K/s = K/s'$$

For brevity write (mod K) for (mod Ner(K)). In more detail the definition can be stated:

$$s \equiv s'(\text{mod } K) \ \text{ iff } \ (\forall u \in A^*) su \in K \ \Leftrightarrow \ s'u \in K$$

Ner(K) is a *right congruence* on A^*, namely

$$(\forall u \in A^*) s \equiv s'(\text{mod } K) \ \Rightarrow \ su \equiv s'u(\text{mod } K)$$

as is quickly seen from the identity

$$K/(su) = (K/s)/u$$

If $[s]$ is the cell of s (mod K) it then makes sense to define

$$[s]\alpha = [s\alpha] \quad s \in A^*, \quad \alpha \in A$$

Setting $Z = \{[s] | s \in L\}$, $z_0 = [\epsilon]$ and with transitions (as just described) written by juxtaposition, we obtain the generator

$$\mathbf{N} = (Z, A, \_\_, z_0)$$

with $L(\mathbf{N}) = K$. We refer to $\mathbf{N}$ as the *Nerode generator* of K. Its states are the cells (equivalence classes) of Ner(K). It can easily be shown that the Nerode generator of K is 'universal' in the sense that any other generator for K (over A), say

$$\tilde{\mathbf{N}} = (\tilde{Z}, A, \_\_, \tilde{z}_0)$$

can be mapped onto $\mathbf{N}$ in accordance with the commutative diagram

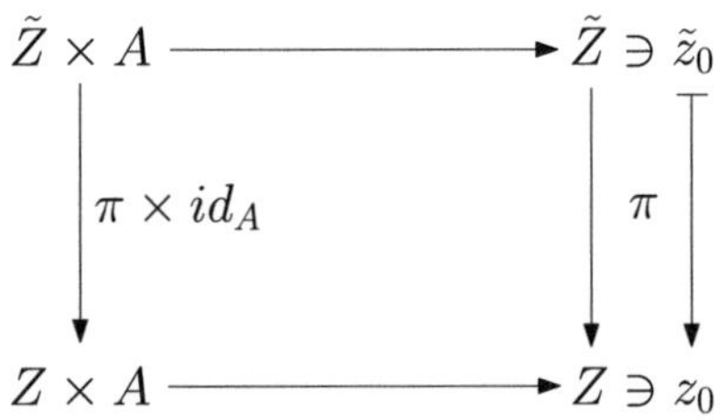

where $\pi : \tilde{Z} \rightarrow Z$ is a suitable surjection and id_A is the identity map on A.

Let $\mathcal{N}(PL) = (X, T, \_\_, x_0)$ be the Nerode generator for the language PL discussed above.

In order to find a suitably economical representation of the pair (L, P) we must incorporate into the new state structure both the information required for the generation of L and the additional information required to specify P. To this end, define an equivalence relation $\mathrm{Fut}(P)$ on L as follows:

$$s \equiv s' (\mathrm{mod}\ \mathrm{Fut}(P)) \text{ iff } (\forall u \in \Sigma^*) P(su)/Ps = P(s'u)/Ps'$$

or in more detail

$$s \equiv s' (\mathrm{mod}\ \mathrm{Fut}(P)) \text{ iff}$$
$$(\forall u \in \Sigma^*)(\forall w \in T^*) P(su) = (Ps)w \Leftrightarrow P(s'u) = P(s')w$$

Thus $\mathrm{Fut}(P)$ aggregates strings whose corresponding outputs share a common future. It is well to note that equivalence $\mathrm{Fut}(P)$ does not imply that corresponding outputs share a common present, namely that the output map ω_o determined by P takes the same value on equivalent strings. In the poset of equivalence relations on L the equivalence kernel of ω_o is in general not comparable with $\mathrm{Fut}(P)$.

The identity

$$P(suv)/P(su) = (P(suv)/Ps) \,/\, (P(su)/Ps)$$

shows that $\mathrm{Fut}(P)$ is actually a right congruence on L. Since the meet of right congruences is again a right congruence, we may define the right congruence

$$\mathrm{Mealy}(L, P) := \mathrm{Ner}(L) \wedge \mathrm{Fut}(P)$$

Now let $\mathbf{G} = (Q, \Sigma, \_\_, q_0)$ be specifically the generator of $L = L(\mathbf{G})$ based on $\mathrm{Mealy}(L, P)$, i.e. $q \in Q$ stands for a cell (mod $\mathrm{Mealy}(L, P)$) in L. We define the *Mealy output map* λ of $\mathbf{G}$ according to

$$\lambda : Q \times \Sigma \to T_o : (q, \sigma) \mapsto \omega_o(s\sigma)$$

for all $s \in q$. From the definitions it easily follows that λ is well defined. The 5-tuple

$$\mathcal{M}(L, P) := (Q, \Sigma, \_\_, q_0, \lambda)$$

will be called the *Mealy* generator for (L, P). One can verify that any other generator for (L, P) of the same type, say

$$\tilde{\mathcal{M}}(L, P) = (\tilde{Q}, \Sigma, \_\_, \tilde{q}_0, \tilde{\lambda})$$

maps onto $\mathcal{M}(L, P)$ in the sense that for a suitable surjection π the following diagram commutes:

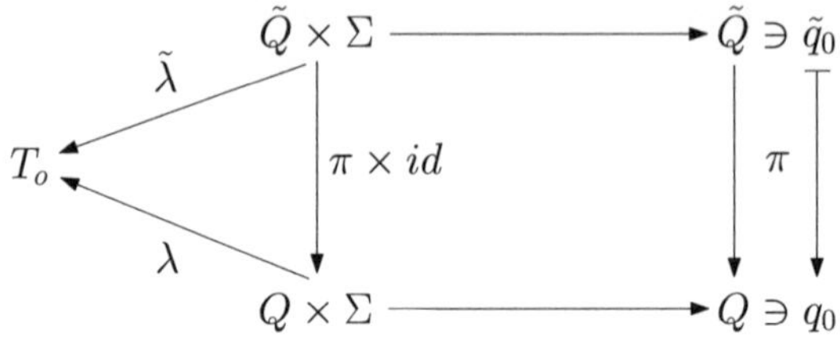

In particular such a diagram exists with $\tilde{\mathcal{M}}$ taken to be the Mealy description

$$(L, \Sigma, \_\_, \epsilon, \hat{\omega}_o)$$

The situation now is that we have obtained two 'economical' descriptions: the Mealy generator $\mathcal{M}(L, P)$ of (L, P), and the Nerode generator $\mathcal{N}(PL)$ of PL. However, while the items P and PL can certainly be recovered from $\mathcal{M}(L, P)$, the state set of $\mathcal{M}(L, P)$ is a little too coarse to allow tracking in $\mathcal{N}(PL)$ of transitions in $\mathcal{M}(L, P)$. The problem is just that $s \equiv s'(\text{mod Mealy}(L, P))$ does not imply $Ps \equiv Ps'(\text{mod } PL)$. The cure is to refine equivalence (mod Mealy(L, P)) as follows. Define

$$\text{Hier}(L, P) := \text{Mealy}(L, P) \wedge \text{Ner}(PL) \circ P$$

where $s \equiv s'(\text{mod Ner}(PL) \circ P)$ is defined to mean $Ps \equiv Ps'(\text{mod } PL)$.

Proposition 1

Hier(L, P) is a right congruence on L.

Proof

In the proof write Hier, Mealy, and Norode for brevity. Suppose $s \equiv s'(\text{mod Hier})$, let $u \in \Sigma^*$, and let $P(su) = (Ps)w$. Since $s \equiv s'(\text{mod Mealy})$, it follows both that $su \equiv s'u(\text{mod Mealy})$ and that $P(s'u) = (Ps')w$. Since $Ps \equiv Ps'(\text{mod } PL)$ we therefore have $P(su) \equiv P(s'u)(\text{mod } PL)$ and thus Nerode $\circ\, P(su) \equiv$ Nerode $\circ\, P(s'u)$. □

With Hier(L, P) as the basis of our new description of (L, P), let the corresponding generator of Mealy type be

$$\mathcal{H}(L, P) = (Y, \Sigma, \_\_, y_0, \eta)$$

Just as above we have that $(L, \Sigma, \_\_, \epsilon, \hat{\omega}_o)$ projects onto $\mathcal{H}$ according to the diagram

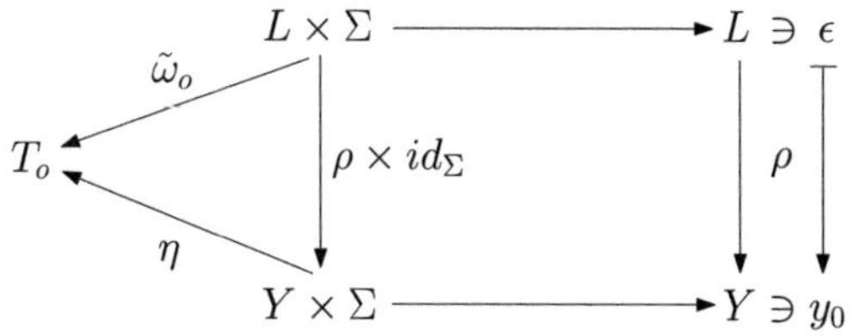

It will be seen that the output map η can be identified with the map g of the following proposition, which states that there is a natural connection between $\mathcal{H}(L, P)$ and $\mathcal{N}(PL)$ that admits step-by-step transition tracking.

Proposition 2

There exist a surjection $f : Y \to X$ and a map $g : Y \times \Sigma \to T_o$ such that the following diagram commutes, where

$$f \times g : Y \times \Sigma \to X \times T_o : (y, \sigma) \longmapsto (f(y), g(y, \sigma))$$

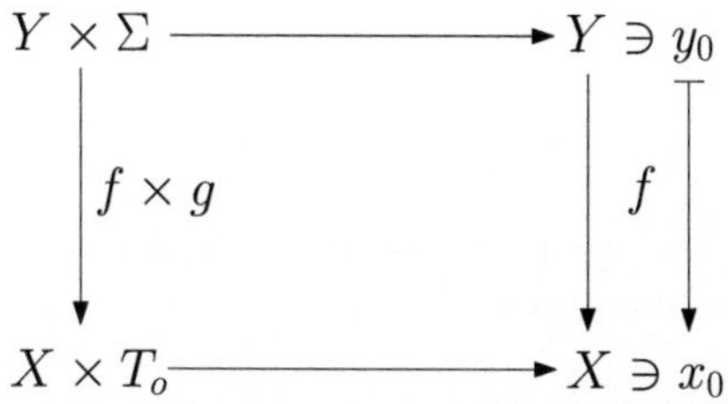

For the bottom arrow we recall the extended definition $(x, \tau_o) \mapsto x$.

Proof

Let $\rho : L \to Y$ be the natural projection (mod Hier), $\nu : PL \to X$ the natural projection (mod PL), and consider the diagram

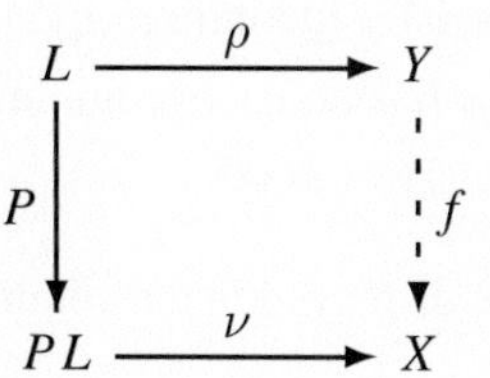

(2.1)

By definition of Hier, $\rho(s) = \rho(s')$ implies $\nu \circ P(s) = \nu \circ P(s')$, namely $\ker \rho \leq \ker(\nu \circ P)$, which shows that f exists as displayed. That f is surjective follows because P and ν are surjective, hence $\nu \circ P$ is surjective. Furthermore $\nu \circ P(\epsilon) = \nu(\epsilon) = x_0$ and $\rho(\epsilon) = y_0$ by definition of generator, so $f(y_0) = x_0$.

To complete the proof we need another version of the tail map, namely $\omega : L \times \Sigma \to T^*$, defined according to

$$\omega(s,\sigma) = \begin{cases} \epsilon \text{ if } \hat{\omega}_o(s,\sigma) = \tau_o \\ \tau \text{ if } \hat{\omega}_o(s,\sigma) = \tau \in T \end{cases}$$

With the usual definition $t = t\epsilon$ for catenation by ϵ, we have the identity

$$P(s\sigma) = P(s)\omega(s,\sigma) \tag{2.2}$$

Next consider the diagram

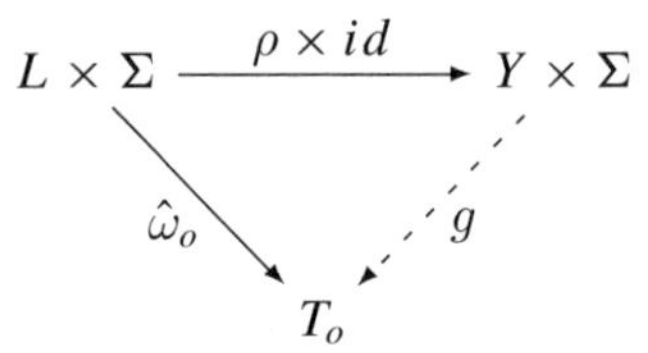

(2.3)

where $id := id_\Sigma$. We have $(\rho \times id)(s,\sigma) = (\rho \times id)(s',\sigma')$ iff $\rho(s) = \rho(s')$ and $\sigma = \sigma'$, which implies $s \equiv s'(\text{mod Fut}(P))$. But then $P(s\sigma) = P(s)\omega(s,\sigma)$ iff $P(s'\sigma) = P(s')\omega(s,\sigma)$, so that $\omega_o(s\sigma) = \omega_o(s'\sigma) \in T_o$, namely $\hat{\omega}_o(s,\sigma) = \hat{\omega}_o(s',\sigma)$. This shows that

$$\ker(\rho \times id) \leq \ker \hat{\omega}_o$$

proving the existence of g as displayed. To check that $f(y\sigma) = f(y)g(y,\sigma)$ we assume that $y = \rho(s)$ and compute

$$\begin{aligned} f(y\sigma) &= f(\rho(s)\sigma) \\ &= f \circ \rho(s\sigma) \quad [\ker \rho := \text{ Hier is a right congruence}] \\ &= \nu \circ P(s\sigma) \quad [\text{commutativity of (2.1)}] \\ &= \nu\,(P(s)\omega(s,\sigma)) \quad [\text{identity (2.2) for } \omega] \\ &= \nu\,(P(s))\,\omega(s,\sigma) \quad [\ker \nu := \text{ Nerode is a right congruence}] \\ &= f\,(\rho(s))\,\omega(s,\sigma) \quad [\text{commutativity of (2.1)}] \\ &= f\,(\rho(s))\,\hat{\omega}_o(s,\sigma) \quad [\text{definitions of } \omega,\ \text{transition function}] \\ &= f(y)(g \circ (\rho \times id))(s,\sigma) \quad [\text{commutativity of (2.3)}] \\ &= f(y)g(y,\sigma) \quad [y = \rho(s)] \end{aligned}$$

□

The results so far can all be displayed in the commutative cube below.

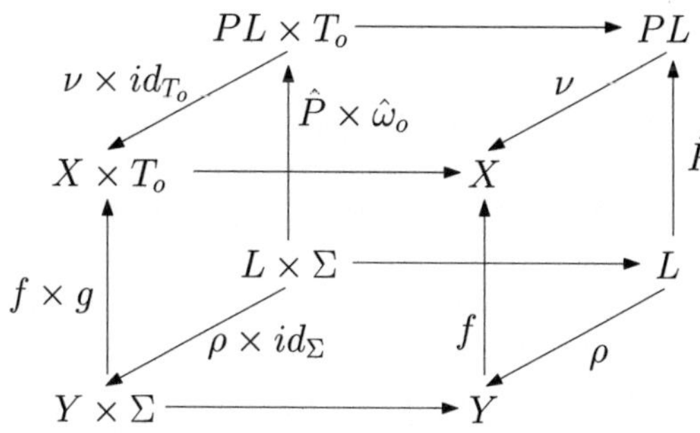

In the cube, unlabeled arrows represent transition action. The bottom face can be thought of as representing 'fast' dynamics, originating with the generating action $L \times \Sigma \rightarrow L$, while the top face represents the 'slow' dynamics that result from hierarchical aggregation. The rear face of the cube represents fine-grained behavioral descriptions in terms of strings, while the front face carries the corresponding more economical state descriptions. The scheme is summarized below.

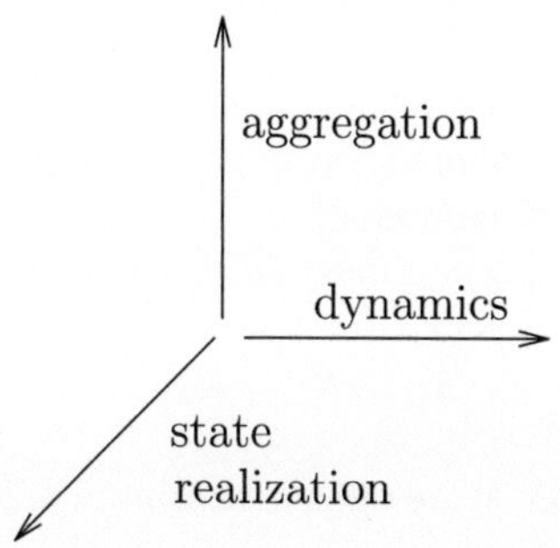

As it stands, the scheme is purely passive (it is nothing more than a 'clock with two hands'); the dynamic action is purely deterministic, and there is no way for an 'agent' to intervene. However, the scheme admits an interesting elaboration that incorporates the action of a controller: this is the subject of *hierarchical control theory*, to be considered in Chap. 5.

To conclude this section we note for completeness' sake a slightly more fine-grained state realization of (L, P) in which the next output symbol corresponding to a state transition $q \xrightarrow{\sigma} q'$ in $\mathcal{M}(L, P)$ or in $\mathcal{H}(L, P)$ becomes a function of the entrance state q' alone (as distinct from being a function of the pair (q, σ) – of course q' may be the entrance state for several other transitions too). Such a representation is more convenient than the Mealy description for graphical representation and certain data processing operations. For this, refine the equivalence Fut(P) to include the present:

$$\text{Pfut}(P) := \text{Fut}(P) \wedge \ker \omega_o$$

In detail,

$$s \equiv s'(\text{mod Pfut}(P)) \text{ iff } s \equiv s'(\text{mod Fut}(P)) \ \& \ \omega_o(s) = \omega_o(s')$$

It is easy to see that Pfut(P) is a right congruence on L, so we may define the right congruence

$$\text{Moore}(L, P) := \text{Ner}(L) \wedge \text{Pfut}(P)$$

All the previous considerations now apply with Mealy(L, P) replaced by Moore (L, P). It can be checked that the finer granularity of state description for $\mathcal{H}(L, P)$ is reflected in the property that the output map g of Proposition 2 now factors through the transition function (say $\delta : Y \times \Sigma \to Y$) of $\mathcal{H}(L, P)$: namely ker $g \geq$ ker δ, hence there exists a map $\phi : Q \to T_o$ such that $g = \phi \circ \delta$.

Exercise 3: Show how to include marked behavior in the foregoing discussion by 'marking' states of **G** with an auxiliary selfloop.

2.8 Chains of Regular Languages

In this section we consider descending chains of regular languages and their finite convergence. The situation commonly arises when computing languages with special properties by successive approximation.

Let $K_i \subseteq \Sigma^*$ $(i = 0, 1, \ldots)$ be a descending chain, namely

$$K_0 \supseteq K_1 \supseteq K_2 \ldots \tag{2.4}$$

where the languages K_i over Σ are otherwise arbitrary. The chain (or set-theoretic) limit K always exists, and is

$$K = \bigcap_{i=0}^{\infty} K_i$$

Let $\mathcal{P}$ be some 'property' common to the K_i (e.g. regularity) that is closed under finite intersection, so for every n, $\cap_{i=0}^{n} K_i$ has property $\mathcal{P}$ as well. In general there is no guarantee that K also has $\mathcal{P}$. For instance with $\Sigma = \{\alpha, \beta\}$, the languages

$$K_i := \Sigma^* - \{\alpha^j \beta^j | j = 0, 1, \ldots, i\}$$

are regular, but

$$K = \Sigma^* - \{\alpha^j \beta^j | j = 0, 1, \ldots\}$$

is not, as it is the complement of a non-regular language (Example 2.3.2(v)).

The simplest condition under which K will have $\mathcal{P}$ is evidently that the chain (2.4) terminate finitely, namely for some n,

$$K_i = K_n, \; i \geqslant n \tag{2.5}$$

To obtain a useful criterion for (2.5), consider the Nerode equivalence $\equiv_L$ of a language $L \subseteq \Sigma^*$ (Sect. 2.2), and write $\|L\|$ for its index. We recall that $\|L\|$ is the state size of a canonical recognizer for L.

Proposition 1

Fix the alphabet Σ, with $|\Sigma| = M < \infty$. For every $R = 1, 2, \ldots$ the number of distinct languages $L \subseteq \Sigma^*$ with $\|L\| \leq R$ is finite.

Proof

Let $\mathbf{R} = (X, \Sigma, \xi, x_0, X_m)$ be a canonical recognizer for L. Then $|X| \leq R$, the number of choices for $X_m \subseteq X$ is no more than 2^R, and the number of choices for $\xi : X \times \Sigma \to X$ no more than $|X|^{|X \times \Sigma|} \leq R^{RM}$. □

To establish (2.5) we now have the following.

Proposition 2

For fixed Σ with $|\Sigma| < \infty$, let $K_i \subseteq \Sigma^*$ be a descending chain, as in (2.4). Assume that for some R and all i, $\|K_i\| \leq R$. Then for some n, $K_i = K_n$ for all $i \geqslant n$.

Proof

If the conclusion fails then for every n there is $m > n$ with $K_n \supset K_m$, hence there is an infinite subsequence

$$K_{i(1)} \supset K_{i(2)} \supset K_{i(3)} \supset \cdots$$

where the inclusions are strict. By Proposition 1 the $K_{i(j)}$ belong to a common finite collection of sublanguages, and therefore

$$K_{i(r)} = K_{i(r+s)}$$

for some r, s. From this contradiction the result follows. □

In the remainder of this section we consider a condition on regular sublanguages of Σ^* by which a state size bound as in Proposition 2 can be established. Note first that if, for languages $K, L \subseteq \Sigma^*$, it happens to be true that

$$\equiv_L \leq \equiv_K \tag{2.6}$$

for their Nerode equivalences, then obviously $\|K\| \leq \|L\|$. However the refinement condition (2.6), equivalent to the existence of a morphism P as in Proposition 2.4.2, is rarely applicable in the present context, and we introduce a weaker condition that is more relevant.

Let $K, L \subseteq \Sigma^*$ be arbitrary, with prefix-closures $\bar{K}, \bar{L}$. We say that K is *weakly supported on* L, and write[4] $K \vdash L$, if

---

[4]The symbol $\vdash$ may be pronounced 'bu'.

$$\{\bar{K}, \Sigma^* - \bar{K}\} \wedge \equiv_L \ \leq \ \equiv_K \tag{2.7}$$

Here $\{\cdot, \cdot\}$ is the binary partition separating $\bar{K}$ from its complement. Explicitly, (2.7) states

$$(\forall s_1, s_2 \in \Sigma^*) s_1 \equiv s_2 (\text{mod } \{\bar{K}, \Sigma^* - \bar{K}\}) \ \& \ s_1 \equiv_L s_2 \ \Rightarrow \ s_1 \equiv_K s_2 \tag{2.8}$$

In fact if $s_1, s_2 \in \Sigma^* - \bar{K}$ then, as the latter is a single cell of $\equiv_K$ (see Sect. 2.2), (2.8) is automatic; only the condition $s_1, s_2 \in \bar{K}$ plays a role. Thus (2.8) is equivalent to

$$(\forall s_1, s_2 \in \Sigma^*) s_1, s_2 \in \bar{K} \ \& \ s_1 \equiv_L s_2 \ \Rightarrow \ \big((\forall w \in \Sigma^*) s_1 w \in K \Leftrightarrow s_2 w \in K\big) \tag{2.9}$$

Example 3: Consider $\varnothing \subseteq \{\epsilon\} \subseteq \Sigma^*$ and write $\text{Ner}(\cdot)$ for $\equiv$. We have

$$\text{Ner}(\varnothing) = \text{Ner}(\Sigma^*) = \{\Sigma^*\}$$

and

$$\text{Ner}(\{\epsilon\}) = \{\{\epsilon\}, \Sigma^* - \{\epsilon\}\}$$

Thus for any $L \subseteq \Sigma^*$,

$$\varnothing \vdash L, \{\epsilon\} \vdash L, \ \text{and} \ \Sigma^* \vdash L$$

In general $\bar{L} \vdash L$, but as $\text{Ner}(L) \leq \text{Ner}(\bar{L})$, usually $L \nvdash \bar{L}$.

Example 4: Given $\varnothing \neq M \subseteq \Sigma^*$ closed, select $t \in M$, and let $L = \{s \in M | s \equiv_M t\}$. Thus L is the sublanguage of M consisting of strings that terminate in the cell $[t]_M$. In general L is not closed; in any case

$$\bar{L} = \{s \in \Sigma^* | (\exists w \in \Sigma^*) sw \equiv_M t\} \subseteq M$$

One can think of $[t]$ as a 'marker state' of M, and of L as the 'trim' of M with respect to it.

We show that $L \vdash M$. For this let $s_1, s_2 \in \bar{L}$, $s_1 \equiv_M s_2$, and $s_1 w \in L$. Then $s_1 w \equiv_M t$, $s_2 w \equiv_M s_1 w$, so $s_2 w \in L$, as required.

One easily verifies the corresponding more general result, based on selecting $t_1, \ldots, t_k \in M$ and defining

$$L := \bigcup_{j=1}^{k} \{s \in M | s \equiv_M t_j\}$$

Example 5: Fix $t \in \Sigma^*$, $L \subseteq \Sigma^*$, and let

$$K := L - \{s \in \Sigma^* | (\exists s' \leq s) s' \equiv_L t\}$$

Setting $\lambda(t) := \{s' \in \Sigma^* | s' \equiv_L t\}$, the Nerode cell of t (mod L), show that

$$K = \sup\{J \subseteq L | \bar{J} \cap \lambda(t) = \varnothing\}$$

We use (2.9) to verify that $K \vdash L$. Let $s_1, s_2 \in \bar{K}$, $s_1 \equiv_L s_2$, and $s_1 w \in K$. It must be shown that $s_2 w \in K$. Now $s_1 w \in K \subseteq L$ along with $s_2 \equiv_L s_1$ implies $s_2 w \in L$. Next, $s_2 \in \bar{K}$ implies $s_2 x \in K$ for some $x \in \Sigma^*$, hence for all $s' \leq s_2 x$ there holds $s' \not\equiv_L t$, in particular for all $s' \leq s_2$. Let $y \leq w$. Then $s_1 w \in K$ implies $s_1 y \not\equiv_L t$, and then $s_2 y \equiv_L s_1 y$ yields $s_2 y \not\equiv_L t$, so that now $s' \not\equiv_L t$ for all $s' \leq s_2 w$. There follows $s_2 w \in K$, as required.

We next investigate the Nerode equivalence relation of K and its relation to the cells of $\equiv_L$. Consider the family of subsets of Σ^* given by

$$\Phi := \{[s]_L \cap \bar{K} | s \in \bar{K}\} \cup \{\Sigma^* - \bar{K}\}$$

Proposition 6

The family Φ (restricted to its nonempty members) is a partition of Σ^*, determining (say) an equivalence relation $\varphi \in \mathcal{E}(\Sigma^*)$. If $K \vdash L$ then φ is a right congruence, with $\varphi \leq \equiv_K$. In that case

$$\|K\| \leq |\varphi| \leq \|L\| + 1$$

If, furthermore, $\bar{K} \subseteq \bar{L} \subsetneqq \Sigma^*$ then $|\varphi| \leq \|L\|$.

Proof

Since $\cup\{[s]_L | s \in \bar{K}\} \supseteq \bar{K}$, it is clear that Φ covers Σ^*. Suppose

$$t \in ([s_1]_L \cap \bar{K}) \cap ([s_2]_L \cap \bar{K})$$

Then $[s_1]_L \cap [s_2]_L \neq \varnothing$, so $[s_1]_L = [s_2]_L$, and Φ is a partition. Now assume $K \vdash L$ and let $s_1 \equiv s_2 (\text{mod } \varphi)$. If $s_1, s_2 \in \Sigma^* - \bar{K}$, clearly $s_1 w, s_2 w \in \Sigma^* - \bar{K}$ for all $w \in \Sigma^*$, hence $s_1 w \equiv s_2 w (\text{mod } \varphi)$. Suppose $s_1, s_2 \in \bar{K}$, so that $s_1, s_2 \in [s]_L \cap \bar{K}$ for some $s \in \bar{K}$. Thus $s_1 \equiv_L s_2$. As $K \vdash L$, we know

$$\{\bar{K}, \Sigma^* - \bar{K}\} \wedge \equiv_L \leq \equiv_K$$

Therefore $s_1 \equiv_K s_2$ (and we have now proved $\varphi \leq \equiv_K$). Thus if, for some w, $s_1 w \in \Sigma^* - \bar{K}$, then also $s_2 w \in \Sigma^* - \bar{K}$. If instead $s_1 w \in \bar{K}$ then $s_1 w \in [s_1 w]_L \cap \bar{K}$, and as $s_2 w \equiv_L s_1 w$ we have $s_2 w \in [s_1 w]_L$. But $s_1 \equiv_K s_2$ and $s_1 w \in \bar{K}$ implies $s_2 w \in \bar{K}$ (otherwise $s_2 wx \notin K$ for all $x \in \Sigma^*$, so $s_1 wx \notin K$ for all x, contradicting $s_1 w \in \bar{K}$); therefore $s_2 w \in [s_1 w]_L \cap \bar{K}$. It follows that, for all $w \in \Sigma^*$, $s_1 \equiv s_2 (\text{mod } \varphi)$ implies

$s_1 w \equiv s_2 w (\mathrm{mod}\ \varphi)$, namely φ is a right congruence on Σ^*. Next, from $\varphi \le \equiv_K$ the inequality $\|K\| \le |\varphi|$ follows at once; and obviously $|\varphi| \le \|L\| + 1$. The latter inequality can be improved if some cell of $\equiv_L$ does not meet $\bar{K}$. Thus if $\bar{L} \neq \Sigma^*$ then $\Sigma^* - \bar{L} \neq \varnothing$, and $\bar{K} \subseteq \bar{L}$ implies $\Sigma^* - \bar{L} \subseteq \Sigma^* - \bar{K}$. Thus $t \in \Sigma^* - \bar{L}$ yields $[t]_L = \Sigma^* - \bar{L}$, with $[t]_L \cap \bar{K} = \varnothing$. In that case, $|\varphi| \le \|L\|$. □

Proposition 6 suggests a useful strengthening of the 'weak support' relation $\vdash$. For $K, L \subseteq \Sigma^*$ we say that K is *supported on* L, and write[5] $K \underline{\vdash} L$, if both $\bar{K} \subseteq \bar{L}$ and $K \vdash L$. Then we have

Corollary 7

If $K \underline{\vdash} L$ and $\bar{L} \subsetneqq \Sigma^*$ then $\|K\| \le \|L\|$. □

The condition $\bar{L} \neq \Sigma^*$ in this corollary cannot be dropped altogether, as shown by the example $K = \{\epsilon\}$, $L = \Sigma^*$, where $\|K\| = 2,\ \|L\| = 1$.

We remark that it need not be true that $\varphi \ge \equiv_L$. For instance if $\Sigma = \{\alpha, \beta\}$, $L = \alpha^*$, $K = \{\epsilon\}$, then $K \underline{\vdash} L$, and $\mathrm{Ner}(K) = \varphi = \{\{\epsilon\}, \Sigma^* - \{\epsilon\}\}$, $\mathrm{Ner}(L) = \{[\epsilon], [\beta]\}$. Thus $\epsilon \equiv_L \alpha$ but $\epsilon \not\equiv \alpha (\mathrm{mod}\ \varphi)$.

Our next result is the useful fact that $\underline{\vdash}$ is transitive: if K is supported on L, and L on M, then so is K on M.

Proposition 8

The binary relation $\underline{\vdash}$ on $Pwr(\Sigma^*)$ is transitive.

Proof

Suppose $K \underline{\vdash} L$ and $L \underline{\vdash} M$. To show $K \underline{\vdash} M$, we have by definition

$$\bar{K} \subseteq \bar{L},\ \{\bar{K}, \Sigma^* - \bar{K}\} \wedge \equiv_L \le \equiv_K$$
$$\bar{L} \subseteq \bar{M},\ \{\bar{L}, \Sigma^* - \bar{L}\} \wedge \equiv_M \le \equiv_L$$

By transitivity of $\le$ on $\mathcal{E}(\Sigma^*)$,

$$\{\bar{K}, \Sigma^* - \bar{K}\} \wedge \{\bar{L}, \Sigma^* - \bar{L}\} \wedge \equiv_M \le \equiv_K$$

Now if $s_1, s_2 \in \bar{K}$ then $s_1, s_2 \in \bar{L}$, so

$$(s_1, s_2) \in \{\bar{K}, \Sigma^* - \bar{K}\} \wedge \{\bar{L}, \Sigma^* - \bar{L}\}$$

Then $s_1 \equiv_M s_2$ yields $s_1 \equiv_K s_2$. This together with $\bar{K} \subseteq \bar{M}$ gives $K \underline{\vdash} M$, as required. □

For application to chain convergence, it is convenient to generalize slightly the relations $\vdash$ and $\underline{\vdash}$ as follows. Let ρ be an arbitrary right congruence on Σ^* (not necessarily the Nerode equivalence of a sublanguage). For $K, L \subseteq \Sigma^*$ we say that K is *weakly ρ-supported on* L, and write $K \vdash_\rho L$, if

[5] The symbol $\underline{\vdash}$ may be pronounced 'shang'.

$$\{\bar{K}, \Sigma^* - \bar{K}\} \wedge \rho \wedge \equiv_L \ \leq \ \equiv_K$$

If in addition $\bar{K} \subseteq \bar{L}$, we say that K is *ρ-supported on L*, written $K \underset{\rho}{\vDash} L$.

Setting $\pi := \rho \wedge \equiv_L$, consider the family of subsets of Σ^* given by

$$\Psi := \{[s]_\pi \cap \bar{K} | s \in \bar{K}\} \cup \{\Sigma^* - \bar{K}\}$$

It is straightforward to prove the following generalizations of Propositions 6 and 8.

Proposition 9

The family Ψ is a partition, determining (say) an equivalence relation ψ, on Σ^*. If $K \vdash_\rho L$ then ψ is a right congruence, with $\psi \leq \equiv_K$. In that case

$$\|K\| \leq |\psi| \leq |\pi| + 1 \leq |\rho| \cdot \|L\| + 1$$

If, furthermore, $\bar{K} \subseteq \bar{L} \subsetneqq \Sigma^*$ then

$$\|K\| \leq |\rho| \cdot \|L\|$$

Proof

The proof closely follows that of Proposition 8 and may be left to the reader. For the third statement, note that

$$[s]_\pi \cap \bar{K} = \varnothing$$

for some s provided

$$[s]_L \cap \bar{K} = \varnothing$$

By use of the obvious estimate $|\pi| \leq |\rho| \cdot \|L\|$, the argument is completed as before. □

Proposition 10

The binary relation $\underset{\rho}{\vDash}$ on $Pwr(\Sigma^*)$ is transitive: if $K \underset{\rho}{\vDash} L$ and $L \underset{\rho}{\vDash} M$ then $K \underset{\rho}{\vDash} M$. □

Our main result is now the following.

Theorem 11

For a finite alphabet Σ, let $K_i \subseteq \Sigma^*$ $(i = 0, 1, 2, \ldots)$ be a descending chain, with K_0 regular. Assume that, for some fixed right congruence ρ on Σ^* with $|\rho| < \infty$,

$$K_i \underset{\rho}{\vDash} K_{i-1}, \ i = 1, 2, \ldots$$

Then each K_i is regular, and the chain is finitely convergent to a sublanguage K. Furthermore, $K \underset{\rho}{\vDash} K_0$ and

$$\|K\| \leq |\rho| \cdot \|K_0\| + 1$$

Proof

By transitivity of $\underset{\rho}{\vdash}$, Proposition 10, $K_i \underset{\rho}{\vdash} K_0$ for all $i \geqslant 1$, and therefore

$$\|K_i\| \leq |\rho| \cdot \|K_0\| + 1, \ i \geqslant 1$$

by Proposition 9. Now finite convergence follows by Proposition 2. □

Applications of Theorem 11 will be given in Sects. 3.5 and 6.1.

Exercise 12: $K \vdash L$ does not imply $K \subseteq L$. **Hint**: $K = \{\epsilon\}$.

Exercise 13: Let **G**, **K** be two generators over the alphabet Σ. Suppose $L_m(\mathbf{K})$ is supported on $L_m(\mathbf{G})$. Show by example that it needn't be true that $L(\mathbf{K})$ is supported on $L(\mathbf{G})$.

The following definition will be useful in Chap. 6. Let **G**, **K** be two generators over the alphabet Σ. Say **K** is *supported* on **G** if $L_m(\mathbf{K})$ is supported on $L_m(\mathbf{G})$ and $L(\mathbf{K})$ is supported on $L(\mathbf{G})$.

2.9 Notes

Except for Sect. 2.8, which originates here, most of the material in this chapter is standard. For Sects. 2.1–2.6 see especially Hopcroft and Ullman (1979). Exercise 2.6.4 is adapted from Carroll and Long (1989), p. 123. Our distinction between 'automaton' and 'generator' is perhaps non-standard, but is helpful in control theory. Section 2.7 also originates here, but is not used in the sequel.

For an early appearance of (robotic) automata in Western literature, see Homer, *Iliad*, V (749–751) and XVIII (373–377).

References

Carroll J, Long D (1989) Theory of finite automata. Prentice-Hall, Englewood Cliffs

Hopcroft JE, Ullman JD (1979) Introduction to automata theory, languages and computation. Addison-Wesley, Boston

Chapter 3
Supervision of Discrete-Event Systems: Basics

We introduce the concept of controlled discrete-event system, by adjoining to the structure of a language generator a control technology. This amounts to partitioning the set of events into controllable events and uncontrollable, the former being amenable to disablement by an external controller or supervisor. Starting from the fundamental definition of a controllable language, it is shown how to formulate and solve a basic problem of optimal supervision. The formulation is extended to treat event forcing, reconfiguration, mutual state exclusion, and forbidden state subsets. Computation is illustrated using the software package *TCT*.

3.1 Introduction

Discrete-event systems encompass a wide variety of physical systems that arise in technology. These include manufacturing systems, traffic systems, logistic systems (for the distribution and storage of goods, or the delivery of services), database management systems, communication protocols, and data communication networks. Typically the processes associated with these systems may be thought of as discrete (in time and state space), asynchronous (event-driven rather than clock-driven), and in some sense generative (or nondeterministic). The underlying primitive concepts include events, conditions and signals.

Our approach in this monograph will be to regard the discrete-event system to be controlled, i.e. the 'plant' in traditional control terminology, as the generator of a formal language. By adjoining control structure, it will be possible to vary the

W. M. Wonham and K. Cai, *Supervisory Control of Discrete-Event Systems*, Communications and Control Engineering,
https://doi.org/10.1007/978-3-319-77452-7_3

language generated by the system within certain limits. The desired performance of such a controlled generator will be specified by stating that its generated language must be contained in some specification language. It is often possible to meet this specification in an 'optimal', that is, minimally restrictive, fashion. The control problem will be considered fully solved when a controller that forces the specification to be met has been shown to exist and to be constructible. In accordance with widely accepted control methodology, we take the state description of a system (and, in this case, a language) to be fundamental.

In parallel with the formal theory, we shall provide a guide to the software package *TCT*, which can be used for developing small-scale examples on a personal computer.

3.2 Representation of Controlled Discrete-Event Systems

The formal structure of a DES to be controlled is that of a *generator* in the sense of Sect. 2.5. As usual, let

$$\mathbf{G} = (Q, \Sigma, \delta, q_0, Q_m)$$

Here Σ is a finite alphabet of symbols that we refer to as *event labels*, Q is the *state set* (at most countable), $\delta : Q \times \Sigma \to Q$ is the (partial) *transition function*, q_0 is the *initial state*, and $Q_m \subseteq Q$ is the subset of *marker states*. The *transition graph* shown below represents a primitive 'machine' named **MACH**, with 3 states, labeled I, W, D for 'idle', 'working' and 'broken down'. Its state set is thus $Q = \{I, W, D\}$ and alphabet is $\Sigma = \{\alpha, \beta, \lambda, \mu\}$, while $q_0 = I$ and $Q_m = \{I\}$.

In a transition graph the initial state is labeled with an entering arrow (⟶○), while a state labeled with an exiting arrow (○⟶) will denote a

MACH

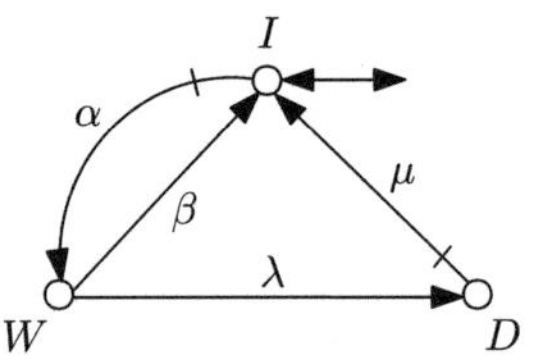

marker state. If the initial state is also a marker state, it may be labeled with a double arrow (⟷○). Formally a transition or *event* of **G** is a triple of the form $[q, \sigma, q']$ where $\delta(q, \sigma) = q'$. Here $q, q' \in Q$ are respectively the *exit state* and the *entrance state*, while $\sigma \in \Sigma$ is the *event label*. The *transition set* of **G** is the set of all such triples.

The mode of operation of a DES, of which **MACH** is typical, may be pictured as follows. Starting from state I, **MACH** executes a sequence of events in accordance with its transition graph. Each event is instantaneous in time. The events occur at quasi-random (unpredictable) time instants. Upon occurrence of an event, the event label is 'signaled' to some external agent. In this way **MACH** generates a string of event labels over the alphabet Σ. At a state such as W from which more than one event may occur, **MACH** will be considered to select just one of the possibilities, in accordance with some mechanism that is hidden from the system analyst and is therefore unmodeled. Such a mechanism could be 'forcing' by an external agent. In this sense the operation of **MACH** is nondeterministic. However, it will be assumed that the labeling of events is 'deterministic' in the sense that distinct events exiting from a given state always carry distinct labels. In general it may happen that two or more events exiting from distinct states may carry the same label.

The marker states serve to distinguish those strings that have some special significance, for instance represent a completed 'task'. In many applications a task is cyclic, for example a production cycle, and its completion will be represented by the return of one or more system components to their initial state, which is therefore marked, as shown for **MACH**. If all states of a component have equal 'task-completion' status (in other words the modeler is indifferent as to which state of this component the process might terminate in, or just reach) then *all* its states should be marked.

For the alphabet Σ we have the partition (disjoint union)

$$\Sigma = \Sigma_c \dot{\cup} \Sigma_u$$

where the disjoint subsets Σ_c and Σ_u comprise respectively the *controllable* events and the *uncontrollable* events. In a transition graph a controllable event may be indicated by an optional tick on its transition arrow (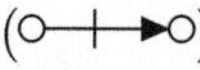). For **MACH**, $\Sigma_c = \{\alpha, \mu\}$ and $\Sigma_u = \{\beta, \lambda\}$.

The controllable event labels, in this case $\{\alpha, \mu\}$, label transitions that may be enabled or disabled by an external agent. A controllable event can occur only if it is enabled. Thus if the event (labeled) α is enabled, but not otherwise, **MACH** can execute the transition $[I, \alpha, W]$ to W from I; if **MACH** is at D, enablement of μ may be interpreted as the condition that **MACH** is under repair, and so may (eventually) execute the transition $[D, \mu, I]$. For brevity we shall often refer to 'the event σ', meaning any or all events (transitions) that happen to be labeled by σ.

By contrast, the uncontrollable event labels, in this case $\{\beta, \lambda\}$, label transitions over which (as expected) an external agent has no control: an uncontrollable event simply 'happens' in accordance (let us say) with internal machine volition, as when **MACH** 'chooses', at state W, to execute either β or λ. In a DES the assignment of controllability status (controllable or uncontrollable) to an event is entirely up to the modeler.

The *TCT* procedure **create** allows the user to create and file a new DES. In response to the prompt, the user enters the DES name, number of states or *size*, the list of marker states and list of transitions (event triples). The *TCT* standard state set is the integer set {0, 1, ..., size − 1}, with 0 as the initial state. Event labels must be entered as integers between 0 and 999, where controllable events are odd and uncontrollable events are even. For instance **MACH** could be created as displayed below.

Example 1: *TCT* procedure create

Name? MACH
States? 3
% *TCT* selects standard state set {0,1,2}
Marker state(s)? 0
% User selects event labels {0,1,2,3}:
% events labeled 1 or 3 are controllable

Transitions?
[0,1,1]
[1,0,0]
[1,2,2]
[2,3,0]

The *TCT* procedure **SE** displays an existing DES in approximately the format indicated above.

We recall from Chap. 2 that the languages associated with a DES **G** are the *closed behavior*

$$L(\mathbf{G}) = \{s \in \Sigma^* | \delta(q_0, s)!\}$$

and the *marked behavior*

$$L_m(\mathbf{G}) = \{s \in L(\mathbf{G}) | \delta(q_0, s) \in Q_m\}$$

These two languages are the *semantics* of **G**. Note that $\varnothing \subseteq L_m(\mathbf{G}) \subseteq L(\mathbf{G})$, and always $\epsilon \in L(\mathbf{G})$ (provided $\mathbf{G} \neq \mathbf{EMPTY}$, the DES with empty state set). The *reachable (state) subset* of **G** is

$$Q_r = \{q \in Q | (\exists s \in \Sigma^*)\, \delta(q_0, s) = q\}$$

G is *reachable* if $Q_r = Q$. The *coreachable subset* is

$$Q_{cr} = \{q \in Q | (\exists s \in \Sigma^*)\, \delta(q, s) \in Q_m\}$$

G is *coreachable* if $Q_{cr} = Q$. **G** is *trim* if it is both reachable and coreachable.

The *TCT* procedure **trim** returns the trimmed version of its argument:

$$\mathbf{TDES} = \mathbf{trim(DES)}$$

possibly after state recoding, as illustrated below.

Example 2: *TCT* procedure trim

TDES = trim(DES)

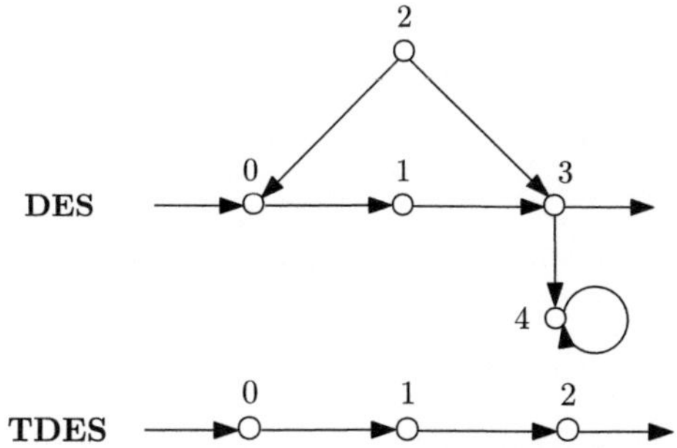

$$Q_r = \{0, 1, 3, 4\}, \quad Q_{cr} = \{0, 1, 2, 3\}, \quad Q_{new} = Q_r \cap Q_{cr} = \{0, 1, 3\}$$

Note that state 3 in Q_{new} has been recoded as 2.

The DES **G** is *nonblocking* if every reachable state is coreachable, namely $Q_r \subseteq Q_{cr}$. In linguistic terms,

$$\overline{L_m(\mathbf{G})} = L(\mathbf{G})$$

In particular **G** is nonblocking if it is trim. If $K \subseteq \Sigma^*$ then **G** *represents* K if **G** is nonblocking and $L_m(\mathbf{G}) = K$. Then $L(\mathbf{G}) = \bar{K}$, although **G** might possibly be non-coreachable. Normally, if **G** is intended to represent K, it is taken to be both reachable and coreachable (i.e.trim).

3.3 Synchronous Product, Shuffle, and Meet

In this section we describe a way of combining several DES into a single, more complex DES. The technique will be standard for the specification of control problems involving the coordination or synchronization of several DES together. We define the operations required on languages, and then the counterpart *TCT* operations on their generators.

Let $L_1 \subseteq \Sigma_1^*$, $L_2 \subseteq \Sigma_2^*$, where it is allowed that $\Sigma_1 \cap \Sigma_2 \neq \varnothing$. Let $\Sigma = \Sigma_1 \cup \Sigma_2$. Define

$$P_i : \Sigma^* \to \Sigma_i^* \quad (i = 1, 2)$$

according to

$$P_i(\epsilon) = \epsilon$$

$$P_i(\sigma) = \begin{cases} \epsilon \text{ if } \sigma \notin \Sigma_i \\ \sigma \text{ if } \sigma \in \Sigma_i \end{cases}$$

$$P_i(s\sigma) = P_i(s)P_i(\sigma) \quad s \in \Sigma^*, \sigma \in \Sigma$$

Clearly $P_i(st) = P_i(s)P_i(t)$, i.e. P_i is *catenative*. The action of P_i on a string s is just to erase all occurrences of σ in s such that $\sigma \notin \Sigma_i$. P_i is the *natural projection* of Σ^* onto Σ_i^*. Let

$$P_i^{-1} : Pwr(\Sigma_i^*) \rightarrow Pwr(\Sigma^*)$$

be the inverse image function of P_i (cf. Sect. 1.4), namely for $H \subseteq \Sigma_i^*$,

$$P_i^{-1}(H) := \{s \in \Sigma^* | P_i(s) \in H\}$$

For $L_1 \subseteq \Sigma_1^*$, $L_2 \subseteq \Sigma_2^*$ we define the *synchronous product* $L_1 \| L_2 \subseteq \Sigma^*$ according to

$$L_1 \| L_2 := P_1^{-1} L_1 \cap P_2^{-1} L_2$$

Thus $s \in L_1 \| L_2$ iff $P_1(s) \in L_1$ and $P_2(s) \in L_2$. If $L_1 = L_m(\mathbf{G_1})$ and $L_2 = L_m(\mathbf{G_2})$, one can think of $\mathbf{G_1}$ and $\mathbf{G_2}$ as generating $L_1 \| L_2$ 'cooperatively' by agreeing to synchronize those events with labels σ which they possess in common.

The *TCT* procedure **sync** returns $\mathbf{G} = \mathbf{sync}(\mathbf{G_1}, \mathbf{G_2})$ where

$$L_m(\mathbf{G}) = L_m(\mathbf{G_1}) \| L_m(\mathbf{G_2}), \quad L(\mathbf{G}) = L(\mathbf{G_1}) \| L(\mathbf{G_2})$$

Example 1: Synchronous product

$\Sigma_1 = \{\alpha, \beta\}$ $\mathbf{G_1}$

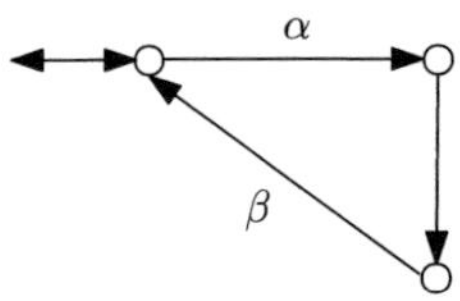

$\Sigma_2 = \{\beta, \gamma\}$ $\mathbf{G_2}$

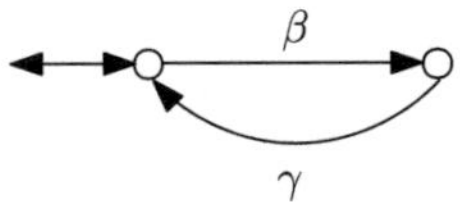

$\mathbf{sync}(\mathbf{G_1}, \mathbf{G_2})$

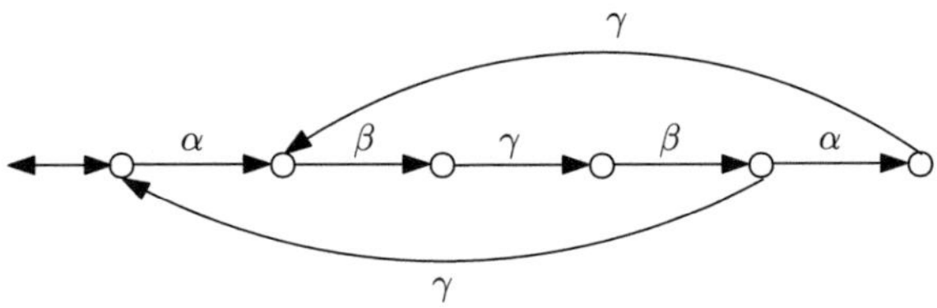

Exercise 2: Show that, in general,

$$L_1 \| L_2 = (L_1 \| (\Sigma_2 - \Sigma_1)^*) \cap (L_2 \| (\Sigma_1 - \Sigma_2)^*)$$

Here the alphabet for $(\Sigma_i - \Sigma_j)^*$ should be taken as $\Sigma_i - \Sigma_j$. Discuss the boundary cases $\Sigma_1 = \Sigma_2$ and $\Sigma_1 \cap \Sigma_2 = \varnothing$.

Exercise 3: With L_i, P_i, Σ_i $(i = 1, 2)$ as above, let

$$P_{i0} : \Sigma_i^* \to (\Sigma_1 \cap \Sigma_2)^* \quad i = 1, 2$$

be the natural projections. Show that

$$P_1(L_1\|L_2) = L_1 \cap P_{10}^{-1}(P_{20}L_2)$$
$$P_2(L_1\|L_2) = L_2 \cap P_{20}^{-1}(P_{10}L_1)$$

This result describes how L_1 (resp. L_2) is constrained when synchronized with L_2 (resp. L_1). Interpret the boundary cases when $P_i(L_1\|L_2) = L_i$ or $P_i(L_1\|L_2) = \emptyset$, provide examples, and check with *TCT*.

Notation: If μ, ν are binary relations on a set X, then $\mu \circ \nu$ is the relation given by

$$x(\mu \circ \nu)x' \ \text{ iff } \ (\exists x'')x\mu x'' \ \& \ x''\nu x'$$

Exercise 4: With $\Sigma_1 \cup \Sigma_2 \subseteq \Sigma$ and $P_i : \Sigma^* \to \Sigma_i^*$ $(i = 1, 2)$ natural projections, show that $P_1 P_2 = P_2 P_1$, and that

$$\begin{aligned} \ker P_1 \vee \ker P_2 &= \ker(P_1 P_2) \\ &= (\ker P_1) \circ (\ker P_2) \end{aligned}$$

Exercise 5: Show that synchronous product is associative, namely

$$(L_1\|L_2)\|L_3 = L_1\|(L_2\|L_3)$$

where L_i is defined over Σ_i and the Σ_i bear no special relationship to one another.

It is well to note a subtlety in the definition of synchronous product. While it must be true that Σ_i includes all the event labels that explicitly appear in $\mathbf{G_i}$, it may be true that some label in Σ_i does not appear in $\mathbf{G_i}$ at all. If $\sigma \in \Sigma_1 \cap \Sigma_2$ does not actually appear in $\mathbf{G_2}$ but may appear in $\mathbf{G_1}$, then **sync** will cause $\mathbf{G_2}$ to block σ from appearing anywhere in $\mathbf{G}$. Thus if, in Example 1, Σ_2 is redefined as $\{\alpha, \beta, \gamma\}$, then α is blocked, with the result that now

$$L_m(\mathbf{G_1})\|L_m(\mathbf{G_2}) = \{\epsilon\}$$

Thus in general $L_1\|L_2$ depends critically on the specification of Σ_1, Σ_2. Currently *TCT* implements **sync** by taking for Σ_i exactly the events that appear in $\mathbf{G_i}$.

Exercise 6: Nonassociativity of *TCT* sync

The *TCT* implementation **sync** of synchronous product need not respect associativity, since the events appearing in **sync**$(\mathbf{G_1}, \mathbf{G_2})$ may form a proper subset of $\Sigma_1 \cup \Sigma_2$. Consider

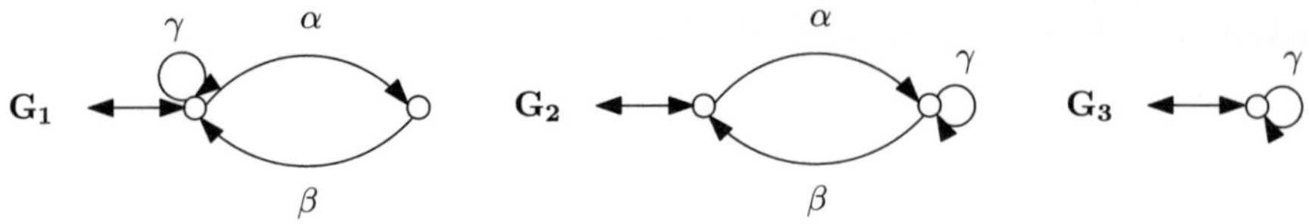

Check that

$$\mathbf{sync}(\mathbf{sync}(\mathbf{G_1}, \mathbf{G_2}), \mathbf{G_3})$$

$$\neq \mathbf{sync}(\mathbf{G_1}, \mathbf{sync}(\mathbf{G_2}, \mathbf{G_3}))$$

and explain why.

TCT will warn the user if some event in $\Sigma_1 \cup \Sigma_2$ fails to appear in the synchronous product: such an event is 'blocked'. This remedy was deemed preferable to maintaining a separate event list for a DES throughout its history. For more on this issue see Exercise 12 below.

Exercise 7: With Σ_i, P_i, L_i $(i = 1, 2)$ as before, take strings $s_i \in L_i$.
(i) Show that $\{s_1\}||\{s_2\} \neq \varnothing$ iff $P_2(s_1) = P_1(s_2)$.
(ii) Calculate $\{\epsilon\}||\{s_2\}$, where ϵ is the empty string (considered as an element of Σ_1^*). Be careful: there are two relevant cases for s_2.
(iii) Illustrate your results with *TCT* examples. **Hint**: Remember that *TCT* takes for Σ_i exactly the events that occur in $\mathbf{G_i}$, so to model $\{\epsilon\}$ over Σ_1^* use, for example

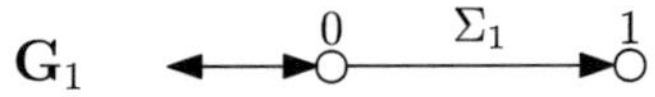

Exercise 8: For alphabets Σ_0, Σ_1, Σ_2 with $\Sigma_0 \subseteq \Sigma_1 \cup \Sigma_2$, let

$$L_1 \subseteq \Sigma_1^*, \quad L_2 \subseteq \Sigma_2^*$$

and let

$$P_0 : (\Sigma_1 \cup \Sigma_2)^* \to \Sigma_0^*$$

be the natural projection. Show that

$$P_0(L_1 \| L_2) \subseteq (P_0 L_1) \| (P_0 L_2)$$

always, and that equality holds provided

$$\Sigma_1 \cap \Sigma_2 \subseteq \Sigma_0$$

namely 'every shared event is observable under P_0'. Here consider $P_0 L_i \subseteq (\Sigma_0 \cap \Sigma_i)^*$. Now extend this result to languages $L_i \subseteq \Sigma_i^*$ $(i = 1, ..., n)$.

Let $\mathbf{G}$ be a DES over alphabet Σ, and Σ_{loop} be an arbitrary event list with no particular relation to Σ. As its name suggests, the *TCT* procedure **selfloop** forms

$$\mathbf{SLG} = \mathbf{selfloop}(\mathbf{G}, \Sigma_{loop})$$

by attaching a transition $[q, \sigma, q]$ for each $\sigma \in \Sigma_{loop}$, at each state q in $\mathbf{G}$ where a transition $[q, \sigma, q']$ is not already defined.

Assume temporarily that $\Sigma_1 \cap \Sigma_2 = \emptyset$. With L_1, L_2 as before the *shuffle product* $L_1 \| L_2$ is defined to be the synchronous product for this special case. Thus the shuffle product of two languages L_1, L_2 over disjoint alphabets is the language consisting of all possible interleavings ('shuffles') of strings of L_1 with strings of L_2. One can think of $\mathbf{G_1}$ and $\mathbf{G_2}$ as generating $L_1 \| L_2$ by independent and asynchronous generation of L_1 and L_2, respectively.

In this monograph we may write **shuffle** to stand for the *TCT* procedure **sync** when the component alphabets Σ_1 and Σ_2 are disjoint.

Reverting to the case of general Σ_1, Σ_2 and $\mathbf{G_1}, \mathbf{G_2}$, the *TCT* procedure **meet** returns a reachable DES $\mathbf{G} = \mathbf{meet}(\mathbf{G_1}, \mathbf{G_2})$ such that

$$L_m(\mathbf{G}) = L_m(\mathbf{G_1}) \cap L_m(\mathbf{G_2}), \quad L(\mathbf{G}) = L(\mathbf{G_1}) \cap L(\mathbf{G_2})$$

Thus **meet** is really the special case of **sync** corresponding to the assumption of a common alphabet $\Sigma_1 \cup \Sigma_2$; namely all events are considered shared and synchronization is total. In particular, **meet** will block any event whose label does not occur in *both* $\mathbf{G_1}$ and $\mathbf{G_2}$. Note also that $\overline{L_m(\mathbf{G})}$ may be a proper sublanguage of $L(\mathbf{G})$, even when each of $\mathbf{G_1}$ and $\mathbf{G_2}$ is trim.

Example 9: Meet

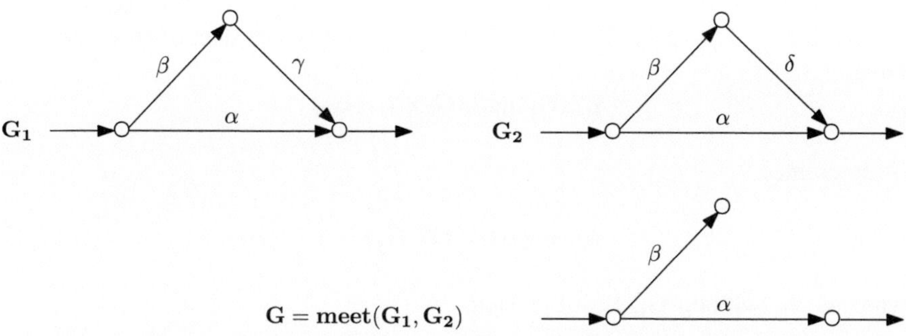

While $L_m(\mathbf{G}) = \{\alpha\} = L_m(\mathbf{G_1}) \cap L_m(\mathbf{G_2})$ and $L(\mathbf{G}) = \{\epsilon, \alpha, \beta\} = L(\mathbf{G_1}) \cap L(\mathbf{G_2})$, nevertheless $\overline{L_m(\mathbf{G})} \subsetneqq L(\mathbf{G})$ even though each of $\mathbf{G_1}$ and $\mathbf{G_2}$ is trim.

Exercise 10: The *TCT* procedure **meet** is implemented for DES $\mathbf{G_1}$ and $\mathbf{G_2}$ as follows. Let $\mathbf{G_i} = (Q_i, \Sigma, \delta_i, q_{0i}, Q_{mi})$, $i = 1, 2$. First define the product $\mathbf{G_1} \times \mathbf{G_2} = (Q, \Sigma, \delta, q_0, Q_m)$, where $Q = Q_1 \times Q_2$, $\delta = \delta_1 \times \delta_2$, $q_0 = (q_{01}, q_{02})$, and $Q_m = Q_{m1} \times Q_{m2}$, with

$$(\delta_1 \times \delta_2)((q_1, q_2), \sigma) := (\delta_1(q_1, \sigma), \delta_2(q_2, \sigma))$$

whenever $\delta_1(q_1, \sigma)!$ and $\delta_2(q_2, \sigma)!$. In other words the product is defined like the product of two automata, due account being taken of the fact that the component transition functions are partial functions, hence the product transition function is a partial function as well. *TCT* now returns $\mathbf{G} = \mathbf{meet}(\mathbf{G_1}, \mathbf{G_2})$ as the reachable sub-DES of $\mathbf{G_1} \times \mathbf{G_2}$, and will number the states from 0 to size−1 (in some arbitrary fashion) as usual. Note that one can think of $\mathbf{G}$ as a structure that is capable of tracking strings that can be generated by $\mathbf{G_1}$ and $\mathbf{G_2}$ in common, when each starts at its initial state. Calculate by hand the meet of two DES and check your result using *TCT*. Using Exercise 2, explain how **meet** can be extended to obtain **sync**.

Exercise 11: Let $L \subseteq \Sigma^*$. Writing

$$L = \left[L \cup (\Sigma^* - \bar{L})\right] \cap \bar{L}$$

show that the two languages intersected on the right are represented by two generators in which, respectively, all events are enabled (i.e. can occur) at each state, and every state is marked. The former places no constraints on local choice but does distinguish 'successful' (marked) strings from others, while the latter declares every string to be 'successful' but constrains event choice. For $\Sigma = \{\alpha, \beta\}$, $L = \alpha(\beta\alpha)^*$, illustrate by drawing the state diagrams and check using **meet**.

Exercise 12: With reference to the $\mathbf{G_i}$ of Exercise 6, adopt the *TCT* specification

$$\Sigma_1 = \Sigma_2 = \{\alpha, \beta, \gamma\}, \;\; \Sigma_3 = \{\gamma\}$$

and write $\Sigma = \Sigma_1 \cup \Sigma_2 \cup \Sigma_3$. Show that the synchronous product of languages,

$$L_m(\mathbf{G_1})||L_m(\mathbf{G_2})||L_m(\mathbf{G_3})$$

is represented by

$$\mathbf{meet}(\mathbf{G_1'}, \mathbf{G_2'}, \mathbf{G_3'}) \tag{3.1}$$

where $\mathbf{G_i'} = \mathbf{selfloop}(\mathbf{G_i}, \Sigma - \Sigma_i)$. Here

$$\mathbf{meet}(\mathbf{F}, \mathbf{G}, \mathbf{H}) := \mathbf{meet}(\mathbf{meet}(\mathbf{F}, \mathbf{G}), \mathbf{H})$$

is always independent of the order of arguments. Thus (3.1), and its generalization to k arguments, provides a correct, order-independent (hence, associative) implementation of synchronous product, as long as *all* the relevant $\mathbf{G_i}$ are specified in advance.

Which of the two results in Exercise 6 agrees with (3.1)? Explain.

Exercise 13: Let $I = \{1, ..., k\}$, and take Σ_i, $L_i \subseteq \Sigma_i^*$, $\Sigma = \cup\{\Sigma_i | i \in I\}$ and $P_i : \Sigma^* \to \Sigma_i^*$ $(i \in I)$. Define

$$L := ||\{L_i | i \in I\} := \bigcap\{P_i^{-1} L_i | i \in I\}$$

Show that $P_i L \subseteq L_i$ and that

$$L = ||\{P_i L | i \in I\}$$

Thus if arbitrary $L_i' \subseteq \Sigma_i^*$ happen to satisfy $L = ||\{L_i' | i \in I\}$, then $P_i L \subseteq L_i'$ $(i \in I)$. With reference to Exercise 12, illustrate your results with *TCT* examples.

For use in the following we introduce *TCT* **project**. For a DES **G** over Σ, let $\Sigma_o \subseteq \Sigma$, $\Sigma_{null} := \Sigma - \Sigma_o$, and $P : \Sigma^* \to \Sigma_o^*$ the natural projection. Then **project**(**G**, $[\Sigma_{null}/\Sigma_o]$) returns a (state-minimal) DES **PG** over Σ_o such that

$$L_m(\mathbf{PG}) = PL_m(\mathbf{G}), \quad L(\mathbf{PG}) = PL(\mathbf{G})$$

Exercise 14: Consider a buffer having capacity n, with input event 0 and output event n:

$$\mathbf{B} = (Q, \Sigma_o, \delta, q_0, Q_m)$$

where

$$Q = \{0, 1, ..., n\}, \ \Sigma_o = \{0, n\}$$

$$q_0 = 0, \ Q_m = \{0\}$$

and δ determines transitions

$$[i - 1, 0, i], \ [i, n, i - 1], \ i = 1, ..., n$$

For $i = 1, ..., n$ let $\mathbf{B_i}$ be the 1-slot buffer with

$$Q_i = \{0, 1\}, \ \Sigma_i = \{i - 1, i\},$$

$$q_{0i} = 0, \ Q_{m,i} = \{0\}$$

and transitions $[0, i - 1, 1]$, $[1, i, 0]$. Let

$$\Sigma = \{0, 1, ..., n\}, \ \Sigma_o = \{0, n\}, P_o : \Sigma^* \to \Sigma_o^*$$

Show that

$$L_m(\mathbf{B}) = P_o[L_m(\mathbf{B_1}) \parallel \cdots \parallel L_m(\mathbf{B_n})]$$

Check your result for $n = 4$ using *TCT* and provide a physical interpretation.

Exercise 15: The *prioritized synchronous product* of languages $L_1 \subseteq \Sigma_1^*$ and $L_2 \subseteq \Sigma_2^*$ can be defined informally as follows. Let L_i be represented by generator $\mathbf{G_i} = (Q_i, \Sigma_i, \delta_i, \_, \_)$. Let $\alpha \in \Sigma_1 \cap \Sigma_2$. To assign $\mathbf{G_1}$ 'priority' over $\mathbf{G_2}$ with respect to α declare that, in $\mathbf{G_1} \times \mathbf{G_2}$,

$$\delta((q_1,q_2),\alpha) = \begin{cases} (\delta_1(q_1,\alpha),\delta_2(q_2,\alpha)) & \text{if } \delta_1(q_1,\alpha)!\ \&\ \delta_2(q_2,\alpha)! \\ (\delta(q_1,\alpha),q_2) & \text{if } \delta_1(q_1,\alpha)!\ \&\ \textit{not}\ \delta_2(q_2,\alpha)! \\ \text{undefined} & \text{otherwise} \end{cases}$$

In other words, $\mathbf{G_1}$ may execute α whenever it can; $\mathbf{G_2}$ synchronizes with $\mathbf{G_1}$ if it can, but otherwise exercises no blocking action and makes no state change. The definition of δ may be completed in the evident way, corresponding to a threefold partition of $\Sigma_1 \cap \Sigma_2$ into events prioritized for $\mathbf{G_1}$ (resp. $\mathbf{G_2}$), or non-prioritized (as in ordinary synchronous product). For simplicity assume that only $\alpha \in \Sigma_1 \cap \Sigma_2$ is prioritized, say for $\mathbf{G_1}$. Denote the required product by **psync**$(\mathbf{G_1}, \mathbf{G_2})$. To implement this in *TCT*, extend $\mathbf{G_2}$ to $\mathbf{G_2'}$ according to

$$\delta_2'(q_2,\alpha) = \begin{cases} \delta_2(q_2,\alpha) & \text{if } \delta_2(q_2,\alpha)! \\ q_2 & \text{if } \textit{not}\ \delta_2(q_2,\alpha)! \end{cases}$$

In *TCT*, $\mathbf{G_2'} =$ **selfloop**$(\mathbf{G_2}, [\alpha])$. Then

$$\textbf{psync}(\mathbf{G_1}, \mathbf{G_2}) = \textbf{sync}(\mathbf{G_1}, \mathbf{G_2'})$$

Illustrate this construction using Example 1, assigning $\mathbf{G_1}$ priority with respect to β, and compare the result with that of **sync**$(\mathbf{G_1}, \mathbf{G_2})$.

Remark 16: Flags

Since *TCT* recodes the states of a product structure generated by **sync** or **meet** sequentially as natural numbers, information about component states is discarded. To retain such information one can use auxiliary selfloops as 'flags'. For instance, to display which states of $\mathbf{G_3} =$ **sync**$(\mathbf{G_1}, \mathbf{G_2})$ correspond to states (1,2) or (6,2) of $\mathbf{G_1} \times \mathbf{G_2}$, first modify $\mathbf{G_1}$ and $\mathbf{G_2}$ by self-looping states 1 and 6 of $\mathbf{G_1}$ and state 2 of $\mathbf{G_2}$ with a new flag event σ_{flag}. After recoding, the selected product states will appear self-looped in $\mathbf{G_3}$.

Example 17: KANBAN

This example illustrates the usefulness of synchronous product (*TCT* **sync**) in building up complex systems. **KANBAN** is an instance of a kanban production system. We consider just two workcells, say **CELL1** and **CELL2**, indexing from output (right-hand) end of the system to input (left-hand) end. **CELLi** consists of an output hopper **HOPi**, input bulletin board **BBi** for kanbans (cards to signal request for input), a feeder queue **Qi** for processor machine **Mi**, and **Mi** itself. Information (via kanbans) circulates in the same order. We model each storage item **HOPi**, **BBi**, **Qi** as a 2-slot buffer. **CELLi** = **sync(HOPi,BBi,Qi,Mi)** (9,14) and **KANBAN** = **sync(CELL1,CELL2)** (81,196), where integer pairs (n, m) denote the number n of states and m of transitions in the DES.

Customer requests for product are sensed at **CELL1** and modeled by event 100. Requests are either blocked, if **HOP1** is empty, or satisfied instantaneously by decrementing **HOP1**. When **HOP1** is decremented, a card is transferred to **BB1**, incrementing **BB1** (doubly synchronized event 100). A card in **BB1** represents a signal to **CELL2** that an input part is required for processing. If **HOP2** is empty, nothing happens. Otherwise, **HOP2** and **BB1** are decremented and **Q1** is incremented, by the 4-tuple synchronized event 12. Parts are taken from **Q1** by **M1** for processing (event 11) and the finished parts deposited in **HOP1** (event 10). In **CELL2** the action is similar: event 12 increments **BB2**; if **BB2** is nonempty a raw part can be taken in by **CELL2**, causing **BB2** to be decremented and **Q2** incremented (doubly synchronized event 22). **M2** deposits its output part in **HOP2** (doubly synchronized event 20).

To display the overall input–output structure, let

PKANBAN = **project**(**KANBAN**,**null**[10,11,12,20,21]) (5,8)

The result is just a 4-slot buffer that is incremented by event 22 (raw part input) and decremented by event 100 (customer request filled). Notice that no more than 4 parts will be in progress in the system (WIP ≤ 4) at one time. If the system is initialized with **BB1**, **BB2** both full and **HOP1**, **Q1**, **HOP2**, **Q2** all empty, then initially requests are blocked and production must begin at the input end.

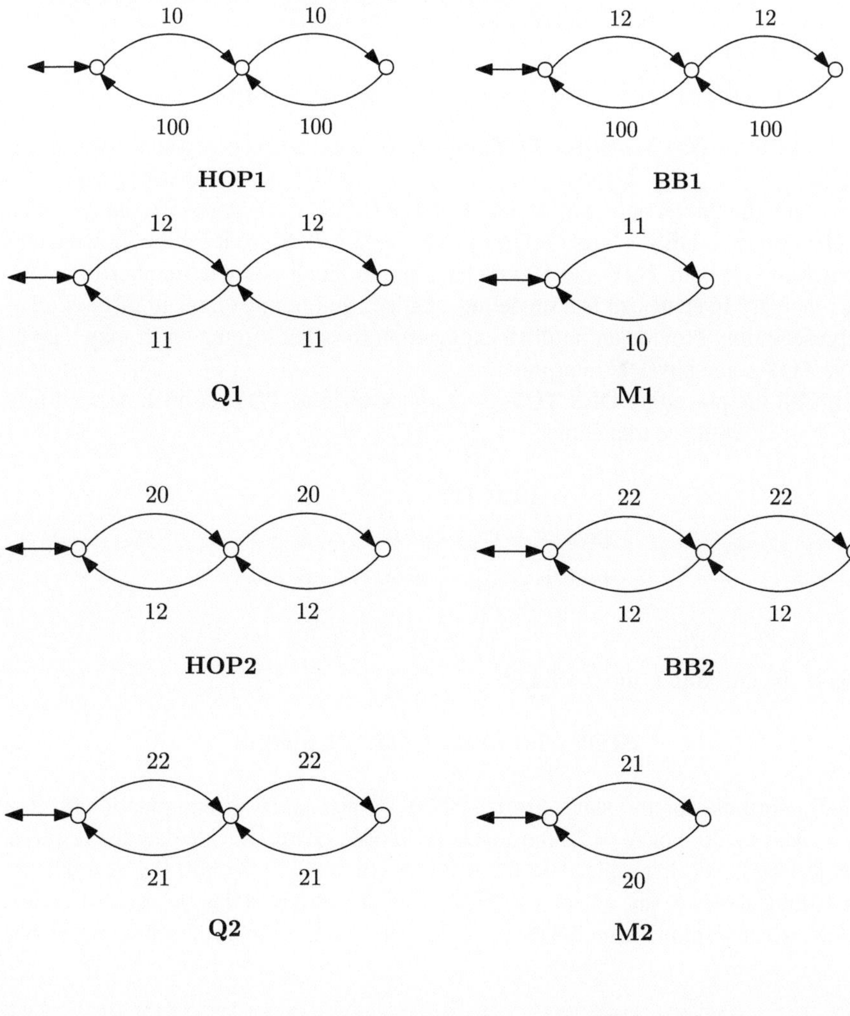

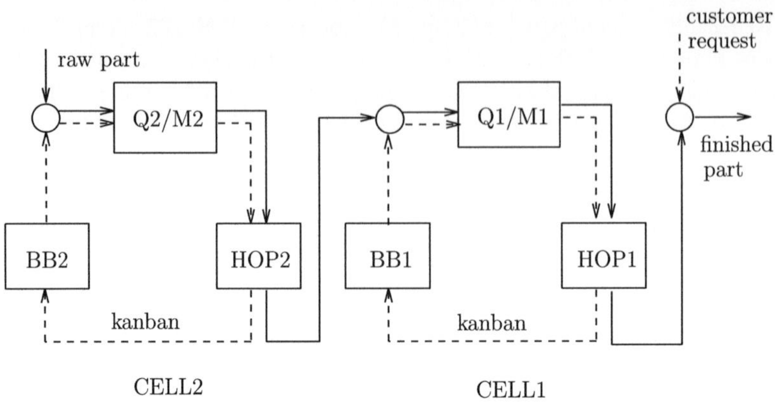

KANBAN

Exercise 18: Consider the DES **TOY** defined as follows.
#states: 3; state set: 0,1,2; initial state: 0; marker state: 0
transition table:

[0,0,1], [0,1,1], [0,2,2], [1,1,2], [1,2,0], [1,3,1], [2,1,0], [2,2,1], [2,3,2]

(i) Construct **PTOY** = **project**(**TOY**,**null**[0]) by hand and check your answer using *TCT*.
(ii) If you did things right, you should find that **PTOY** has 5 states, i.e. the state size of the projected DES is larger (cf. also Exercise 2.5.9). Since it can be shown using **minstate** that both **TOY** and **PTOY** have the minimal possible number of states, this increase in state size is a modeling 'reality', and not just a result of inefficient representation. Provide an intuitive explanation for what's going on. It may help to give **TOY** some physical interpretation.
(iii) Call the preceding DES **TOY_3** and generalize to **TOY_N** on state set {0, 1, ..., $N-1$}, with $3N$ transitions:

$$\begin{aligned} &[0, 0, 1] \\ &[k, 1, k+1], \quad k = 0, 1, ..., N-1 \\ &[k+1, 2, k], \quad k = 0, 1, ..., N-1 \\ &[k, 3, k], \quad k = 1, 2, ..., N-1 \end{aligned}$$

where the indexing is mod(N). Let

$$\mathbf{PTOY\_N} = \mathbf{project}(\mathbf{TOY\_N}, \mathbf{null}[0])$$

Find a formula for the state size of **PTOY_N** and verify it computationally for $N = 3, 4, ..., 20$. For $N = 20$ the answer is 786431. **Hint**: To calculate the result of **project** by hand, first replace each transition with label in the **null** list by a 'silent' transition labeled λ, say, where $\lambda \notin \Sigma$. Next apply a variant of the subset construction to obtain a deterministic model that is λ-free: the initial 'state' is the subset reachable from 0 on paths labeled by strings in λ^* only; the next 'state', following σ, say (with σ *not* in the **null** list), is the subset reachable on paths of form $\lambda^*\sigma\lambda^*$; and so on.

For later reference we define the *TCT* procedure **complement**. Assume that the DES **G** contains in its description exactly the event labels of some alphabet Σ, and let $T \supseteq \Sigma$. Then $\mathbf{G_{co}} = \mathbf{complement}(\mathbf{G}, T - \Sigma)$ has the semantics

$$L_m(\mathbf{G_{co}}) = T^* - L_m(\mathbf{G}), \quad L(\mathbf{G_{co}}) = T^*$$

If $T = \Sigma$ we write simply $\mathbf{G_{co}} = \mathbf{complement}(\mathbf{G}, \_)$. In terms of transition structures, **complement** forms $\mathbf{G_{co}}$ by adjoining a (non-marker) dump state q_+ to the state set of **G**, and complementary transitions from each state of $\mathbf{G_{co}}$ to q_+ as required to render the new transition function a total function; after that, the subsets of marker and non-marker states are interchanged.

Example 19: Small Factory

To conclude this section we show how the foregoing procedures can be used to build up the specifications for a supervisory control problem, to be known as Small Factory. We bring in two 'machines' **MACH1**, **MACH2** as shown.

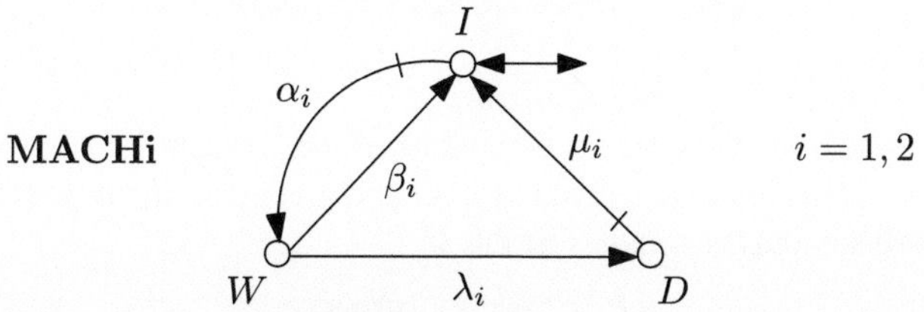

Define **FACT** = **shuffle**(**MACH1**, **MACH2**). Small Factory consists of the arrangement shown below, where **BUF** denotes a buffer with one slot.

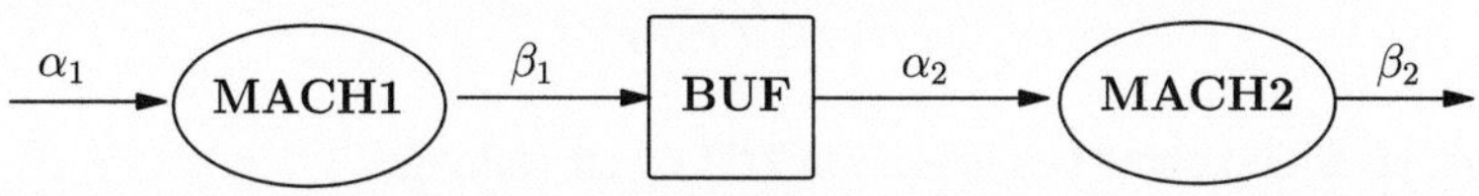

Small Factory operates as follows. Initially the buffer is empty. With the event α_1, **MACH1** takes a workpiece from an infinite input bin and enters W. Subsequently **MACH1** either breaks down and enters D (event λ_1), or successfully completes its work cycle, deposits the workpiece in the buffer, and returns to I (event β_1). **MACH2** operates similarly, but takes its workpiece from the buffer and deposits it when finished in an infinite output bin. If a machine breaks down, then its workpiece is discarded and on repair it returns to I (event μ).

The informal specifications for admissible operation are the following:

1. The buffer must not overflow or underflow.
2. If both machines are broken down, then **MACH2** must be repaired before **MACH1**.

To formalize these specifications we bring in two language generators as the DES **BUFSPEC** and **BRSPEC**, as shown below.

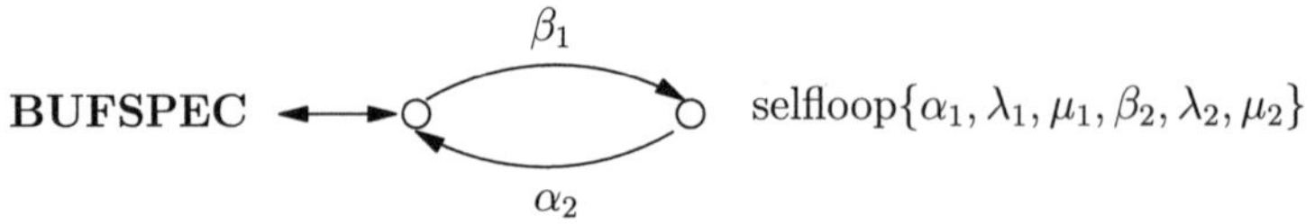

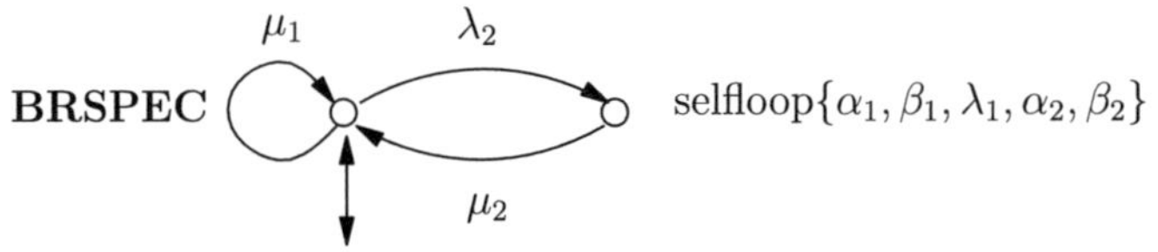

L(**BUFSPEC**) expresses the requirement that β_1 and α_2 must occur alternately, with β_1 occurring first, while L(**BRSPEC**) requires that if λ_2 occurs then μ_1 may not occur (again) until μ_2 has occurred. The assignment of the initial state as a marker state in each of these DES is largely a matter of convenience. In each case selfloops must be adjoined to account for all events that are irrelevant to the specification but which may be executed in the plant. For the combined specification we form

$$\mathbf{SPEC} = \mathbf{meet}(\mathbf{BUFSPEC}, \mathbf{BRSPEC})$$

By inspection it is clear that **SPEC** is trim.

Temporarily denote by **G** the (as yet unknown) DES that would represent '**FACT** under control'. In general, for a given DES **G** and given specification DES **SPEC** as above, we shall say that **G** *satisfies* **SPEC** if

$$L_m(\mathbf{G}) \subseteq L_m(\mathbf{SPEC})$$

Typically **G** and **SPEC** will both be trim, and then it follows on taking closures,

$$L(\mathbf{G}) \subseteq L(\mathbf{SPEC})$$

The first condition could be checked in *TCT* by computing

$$\mathbf{COSPEC} = \mathbf{complement}(\mathbf{SPEC}, \_\_)$$

and then verifying that **trim**(**meet**(**G**, **COSPEC**)) = **EMPTY**, where **EMPTY** is the DES with empty state set (it suffices to check that **meet**(**G**, **COSPEC**) has empty marker set). The results for Small Factory will be presented in Sect. 3.7 below.

3.4 Controllability and Supervision

Let

$$\mathbf{G} = (Q, \Sigma, \delta, q_0, Q_m)$$

be a (nonempty) controlled DES, with $\Sigma = \Sigma_c \dot{\cup} \Sigma_u$ as in Sect. 3.2. A particular subset of events to be enabled can be selected by specifying a subset of controllable events. It is convenient to adjoin with this all the uncontrollable events as these are automatically enabled. Each such subset of events is a *control pattern*, and we introduce the set of all control patterns

$$\Gamma = \{\gamma \in Pwr(\Sigma) | \gamma \supseteq \Sigma_u\}$$

A *supervisory control for* **G** is any map $V : L(\mathbf{G}) \to \Gamma$. The pair $(\mathbf{G}, V)$ will be written $V/\mathbf{G}$, to suggest '**G** under the supervision of V'. The *closed behavior* of $V/\mathbf{G}$ is defined to be the language $L(V/\mathbf{G}) \subseteq L(\mathbf{G})$ described as follows.

(i) $\epsilon \in L(V/\mathbf{G})$
(ii) If $s \in L(V/\mathbf{G})$, $\sigma \in V(s)$, and $s\sigma \in L(\mathbf{G})$ then $s\sigma \in L(V/\mathbf{G})$
(iii) No other strings belong to $L(V/\mathbf{G})$

We always have $\{\epsilon\} \subseteq L(V/\mathbf{G}) \subseteq L(\mathbf{G})$, with either bound a possibility depending on V and **G**. Clearly $L(V/\mathbf{G})$ is nonempty and closed.

The *marked behavior* of $V/\mathbf{G}$ is

$$L_m(V/\mathbf{G}) = L(V/\mathbf{G}) \cap L_m(\mathbf{G})$$

Thus the marked behavior of $V/\mathbf{G}$ consists exactly of the strings of $L_m(\mathbf{G})$ that 'survive' under supervision by V. We always have $\varnothing \subseteq L_m(V/\mathbf{G}) \subseteq L_m(\mathbf{G})$.

We say that V is *nonblocking* (for **G**) if

$$\overline{L_m(V/\mathbf{G})} = L(V/\mathbf{G})$$

Our main objective is to characterize those languages that qualify as the marked behavior of some supervisory control V. To this end we define a language $K \subseteq \Sigma^*$ to be *controllable* (*with respect to* **G**) if

$$(\forall s, \sigma) s \in \bar{K} \ \& \ \sigma \in \Sigma_u \ \& \ s\sigma \in L(\mathbf{G}) \Rightarrow s\sigma \in \bar{K}$$

In other words, K is controllable iff no $L(\mathbf{G})$-string that is already a prefix of K, when followed by an uncontrollable event in **G**, thereby exits from the prefixes of K: the prefix closure $\bar{K}$ is invariant under the occurrence in **G** of uncontrollable events.

For a more concise statement, we use the following notation. If $S \subseteq \Sigma^*$ and $\Sigma_o \subseteq \Sigma$, let $S\Sigma_o$ denote the set of strings of form $s\sigma$ with $s \in S$ and $\sigma \in \Sigma_o$. Then K is controllable iff

$$\bar{K}\Sigma_u \cap L(\mathbf{G}) \subseteq \bar{K}$$

It is clear that $\varnothing$, $L(\mathbf{G})$ and Σ^* are always controllable with respect to **G**.

Note that the controllability condition constrains only $\bar{K} \cap L(\mathbf{G})$, since if $s \notin L(\mathbf{G})$ then $s\sigma \notin L(\mathbf{G})$, i.e. the condition $s \in \bar{K} - L(\mathbf{G})$ & $s\sigma \in L(\mathbf{G})$ is always false.

Example 1: Controllable and uncontrollable languages

MACH

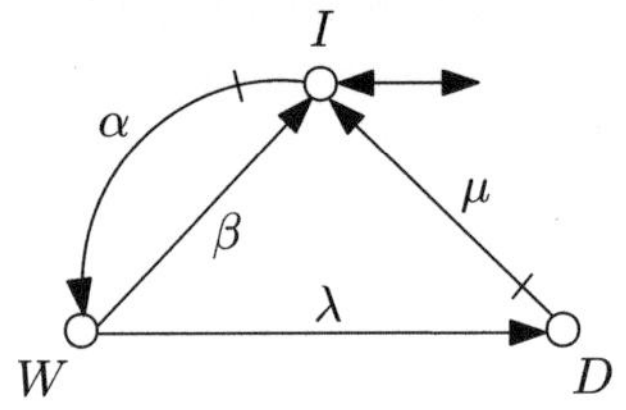

$$\Sigma_c = \{\alpha, \mu\}, \quad \Sigma_u = \{\beta, \lambda\}$$

With respect to **MACH**, $L' = \{\alpha\lambda\mu\}$ is not controllable, since $\alpha\beta$ consists of a prefix α of L' followed by an uncontrollable event β such that $\alpha\beta$ belongs to $L(\mathbf{MACH})$ but not to $\bar{L}'$. On the other hand $L'' = \{\alpha\beta, \alpha\lambda\}$ is controllable, since none of its prefixes $s \in \bar{L}'' = \{\epsilon, \alpha, \alpha\beta, \alpha\lambda\}$ can be followed by an uncontrollable event σ such that $s\sigma$ belongs to $L(\mathbf{MACH}) - \bar{L}''$.

Exercise 2: Assume that a language K is controllable with respect to a DES $\mathbf{G}$. Show that if $s \in \bar{K}$, $w \in \Sigma_u^*$, and $sw \in L(\mathbf{G})$, then $sw \in \bar{K}$. **Hint**: Use structural induction on w.

Let $K \subseteq L \subseteq \Sigma^*$. The language K is *(relatively) L-closed* if $K = \bar{K} \cap L$. Thus K is L-closed provided it contains every one of its prefixes that belongs to L.

We now present our first main result.

Theorem 3

Let $K \subseteq L_m(\mathbf{G})$, $K \neq \varnothing$. There exists a nonblocking supervisory control V for $\mathbf{G}$ such that $L_m(V/\mathbf{G}) = K$ iff

(i) K is controllable with respect to $\mathbf{G}$, and
(ii) K is $L_m(\mathbf{G})$-closed.

Proof

(If) We have $\bar{K}\Sigma_u \cap L(\mathbf{G}) \subseteq \bar{K}$ together with $\bar{K} \subseteq \overline{L_m(\mathbf{G})} \subseteq L(\mathbf{G})$. Furthermore $\epsilon \in \bar{K}$ since $K \neq \varnothing$. For $s \in \bar{K}$ define $V(s) \in \Gamma$ according to

$$V(s) = \Sigma_u \cup \{\sigma \in \Sigma_c | s\sigma \in \bar{K}\}$$

We claim that $L(V/\mathbf{G}) = \bar{K}$. First we show that $L(V/\mathbf{G}) \subseteq \bar{K}$. Suppose $s\sigma \in L(V/\mathbf{G})$, i.e. $s \in L(V/\mathbf{G})$, $\sigma \in V(s)$, and $s\sigma \in L(\mathbf{G})$. Assuming inductively that $s \in \bar{K}$ we have that $\sigma \in \Sigma_u$ implies $s\sigma \in \bar{K}\Sigma_u \cap L(\mathbf{G})$, so that $s\sigma \in \bar{K}$ (by controllability), whereas $\sigma \in \Sigma_c$ implies that $s\sigma \in \bar{K}$ by definition of $V(s)$. For the reverse inclusion, suppose $s\sigma \in \bar{K}$; thus $s\sigma \in L(\mathbf{G})$. Assuming inductively that $s \in L(V/\mathbf{G})$ we have that $\sigma \in \Sigma_u$ automatically implies that $\sigma \in V(s)$, so that $s\sigma \in L(V/\mathbf{G})$; while $\sigma \in \Sigma_c$ and $s\sigma \in \bar{K}$ imply that $\sigma \in V(s)$, so $s\sigma \in L(V/\mathbf{G})$. The claim is proved. Finally

$$\begin{aligned} L_m(V/\mathbf{G}) &= L(V/\mathbf{G}) \cap L_m(\mathbf{G}) \quad \text{(by definition)} \\ &= \bar{K} \cap L_m(\mathbf{G}) \\ &= K \quad (\text{since } K \text{ is } L_m(\mathbf{G}) - \text{closed}) \end{aligned}$$

and $\overline{L_m(V/\mathbf{G})} = \bar{K} = L(V/\mathbf{G})$, so $V/\mathbf{G}$ is nonblocking for $\mathbf{G}$.

(Only if) Let V be a supervisory control for $\mathbf{G}$ with $L_m(V/\mathbf{G}) = K$. Assuming that V is nonblocking for $\mathbf{G}$ we have $L(V/\mathbf{G}) = \bar{K}$, so

$$K = L(V/\mathbf{G}) \cap L_m(\mathbf{G}) = \bar{K} \cap L_m(\mathbf{G})$$

i.e. K is $L_m(\mathbf{G})$-closed. To show that K is controllable let $s \in \bar{K}, \sigma \in \Sigma_u, s\sigma \in L(\mathbf{G})$. Then $s \in L(V/\mathbf{G})$ and $\sigma \in V(s)$. So $s\sigma \in L(V/\mathbf{G}) = \bar{K}$, i.e.

$$\bar{K}\Sigma_u \cap L(\mathbf{G}) \subseteq \bar{K}$$

as required. □

Corollary 4

Let $K \subseteq L(\mathbf{G})$ be nonempty and closed. There exists a supervisory control V for $\mathbf{G}$ such that $L(V/\mathbf{G}) = K$ iff K is controllable with respect to $\mathbf{G}$. □

For brevity we refer to a nonblocking supervisory control (for $\mathbf{G}$, understood) as an NSC. It is useful to introduce a slight generalization of NSC in which the supervisory action includes marking as well as control. For this, let $M \subseteq L_m(\mathbf{G})$. Define a *marking nonblocking supervisory control for the pair* $(M, \mathbf{G})$, or *MNSC*, as a map $V : L(\mathbf{G}) \to \Gamma$ exactly as before, but now for the marked behavior of $V/\mathbf{G}$ we define

$$L_m(V/\mathbf{G}) = L(V/\mathbf{G}) \cap M$$

One may think of the marking action of the MNSC V as carried out by a recognizer for M that monitors the closed behavior of $V/\mathbf{G}$, sounding a beep exactly when a string in M has just been generated. As a sublanguage of $L_m(\mathbf{G})$, these strings could be thought of as representing a subset of the 'tasks' that $\mathbf{G}$ (or its underlying physical referent) is supposed to accomplish. For instance in Small Factory, one might define a 'batch' to consist of 10 fully processed workpieces. M might then be taken as the set of strings that represent the successful processing of N integral batches, $N \geq 0$, with both machines returned to the I(dle) state and the buffer empty.

The counterpart result to Theorem 3 actually represents a simplification, as the condition of $L_m(\mathbf{G})$-closedness can now be dropped.

Theorem 5

Let $K \subseteq L_m(\mathbf{G})$, $K \neq \varnothing$. There exists an MNSC V for $(K, \mathbf{G})$ such that

$$L_m(V/\mathbf{G}) = K$$

iff K is controllable with respect to $\mathbf{G}$.

Proof

(If) With V defined as in the proof of Theorem 3, it may be shown as before that $L(V/\mathbf{G}) = \bar{K}$. Then

$$L_m(V/\mathbf{G}) = L(V/\mathbf{G}) \cap K = \bar{K} \cap K = K$$

so that $\overline{L_m(V/\mathbf{G})} = \bar{K} = L(V/\mathbf{G})$; namely V is nonblocking for $\mathbf{G}$.

(Only if) We have $\bar{K} = \overline{L_m(V/\mathbf{G})} = L(V/\mathbf{G})$. Then the proof that K is controllable is unchanged from that of Theorem 3. □

3.5 Supremal Controllable Sublanguages and Optimal Supervision

Let $\mathbf{G} = (\_, \Sigma, \_, \_, \_)$ be a controlled DES with $\Sigma = \Sigma_c \dot{\cup} \Sigma_u$; the items $(\_)$ of $\mathbf{G}$ will be immaterial to the discussion of this section. Let $E \subseteq \Sigma^*$; later E will play the role of a specification language for the supervisory control of $\mathbf{G}$. We introduce the set of all sublanguages of E that are controllable with respect to $\mathbf{G}$:

$$\mathcal{C}(E) = \{K \subseteq E | K \text{ is controllable with respect to } \mathbf{G}\}$$

As a subset of the sublanguages of E, $\mathcal{C}(E)$ is a poset with respect to inclusion. It will be shown that the supremum in this poset always exists in $\mathcal{C}(E)$.

Proposition 1

$\mathcal{C}(E)$ is nonempty and is closed under arbitrary unions. In particular, $\mathcal{C}(E)$ contains a (unique) supremal element [which we denote by $\sup \mathcal{C}(E)$].

Proof

Since the empty language is controllable, it is a member of $\mathcal{C}(E)$. Let $K_\alpha \in \mathcal{C}(E)$ for all α in some index set A, and let $K = \cup\{K_\alpha | \alpha \in A\}$. Then $K \subseteq E$. Furthermore, $\bar{K} = \cup\{\bar{K}_\alpha | \alpha \in A\}$ and $\bar{K}\Sigma_u = \cup\{\bar{K}_\alpha \Sigma_u | \alpha \in A\}$. Therefore

$$\begin{aligned}
\bar{K}\Sigma_u \cap L(\mathbf{G}) &= \left[\bigcup (\bar{K}_\alpha \Sigma_u)\right] \cap L(\mathbf{G}) \\
&= \bigcup \left[\bar{K}_\alpha \Sigma_u \cap L(\mathbf{G})\right] \\
&\subseteq \bigcup \bar{K}_\alpha \\
&= \bar{K}
\end{aligned}$$

Finally we have for the supremal element

$$\sup \mathcal{C}(E) = \bigcup\{K | K \in \mathcal{C}(E)\}$$

□

It may be helpful to keep in mind the following Hasse diagram, where $K_{sup} = \sup \mathcal{C}(E)$.

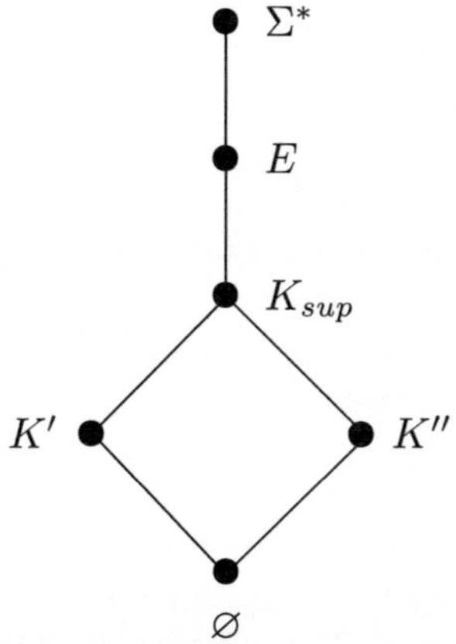

We remark that $\mathcal{C}(E)$ is not generally closed under intersection, so it is not a sublattice of the lattice of sublanguages of E. To see what goes wrong, let $K_1, K_2 \in \mathcal{C}(E)$. We must determine whether or not

$$\overline{K_1 \cap K_2}\Sigma_u \cap L(\mathbf{G}) \subseteq \overline{K_1 \cap K_2} \quad (?)$$

But $\overline{K_1 \cap K_2} \subseteq \bar{K}_1 \cap \bar{K}_2$ always, and quite possibly with strict inclusion. It follows that the left side of (?) is included in

$$\begin{aligned}(\bar{K}_1 \cap \bar{K}_2)\Sigma_u \cap L(\mathbf{G}) &= (\bar{K}_1\Sigma_u \cap L(\mathbf{G})) \cap (\bar{K}_2\Sigma_u \cap L(\mathbf{G})) \\ &\subseteq \bar{K}_1 \cap \bar{K}_2 \\ &\neq \overline{K_1 \cap K_2}\end{aligned}$$

in general. The situation may be described by saying that $\mathcal{C}(E)$ is only a complete *upper semilattice* with join operation (union) that of the lattice of sublanguages of E.

Example 2: Controllability need not be preserved by intersection

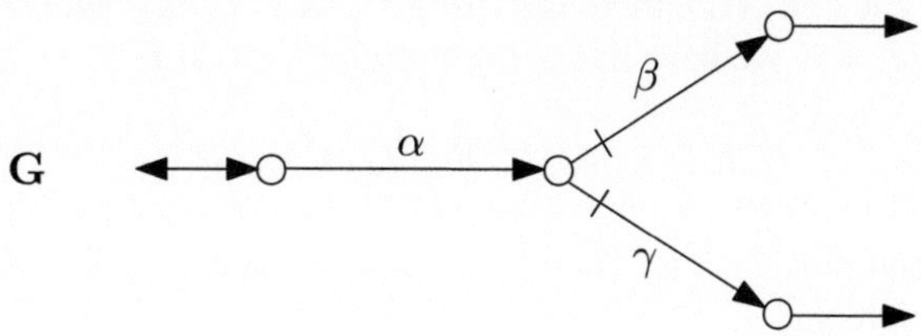

Here $\Sigma = \{\alpha, \beta, \gamma\}$, $\Sigma_c = \{\beta, \gamma\}$. The languages $K_1 = \{\epsilon, \alpha\beta\}$, $K_2 = \{\epsilon, \alpha\gamma\}$ are controllable, but $K_1 \cap K_2 = \{\epsilon\}$ is not controllable, since the event α is uncontrollable and $\alpha \notin \overline{K_1 \cap K_2}$. On the other hand, $\bar{K}_1 \cap \bar{K}_2 = \{\epsilon, \alpha\}$ is controllable.

It is easy to see that the intersection of an arbitrary collection of closed controllable languages is always closed and controllable. Proof of the following observation may now be left to the reader.

Proposition 3

With respect to a fixed controlled DES $\mathbf{G}$ with alphabet Σ, the closed controllable sublanguages of an arbitrary language $E \subseteq \Sigma^*$ form a complete sublattice of the lattice of sublanguages of E. □

To summarize, if K_α ($\alpha \in A$) are controllable then $\cup K_\alpha$ and $\cap \bar{K}_\alpha$ are controllable, but generally $\cap K_\alpha$ is not.

Together with $E \subseteq \Sigma^*$ now fix $L \subseteq \Sigma^*$ arbitrarily. Consider the collection of all sublanguages of E that are L-closed:

$$\mathcal{F}(E) = \{F \subseteq E | F = \bar{F} \cap L\}$$

It is straightforward to verify that $\mathcal{F}(E)$ is nonempty ($\varnothing$ belongs) and is closed under arbitrary unions and intersections. Thus we have the following.

Proposition 4

$\mathcal{F}(E)$ is a complete sublattice of the lattice of sublanguages of E. □

Again let $E, L \subseteq \Sigma^*$. We say that E is *L-marked* if $E \supseteq \bar{E} \cap L$, namely any prefix of E that belongs to L must also belong to E.

Proposition 5

Let $E \subseteq \Sigma^*$ be $L_m(\mathbf{G})$-marked. Then $\sup \mathcal{C}(E \cap L_m(\mathbf{G}))$ is $L_m(\mathbf{G})$-closed.

Proof

We have $E \supseteq \bar{E} \cap L_m(\mathbf{G})$, from which there follows in turn

$$\begin{aligned}
\bar{E} \cap L_m(\mathbf{G}) &\subseteq E \cap L_m(\mathbf{G}) \\
\bar{E} \cap \overline{L_m(\mathbf{G})} \cap L_m(\mathbf{G}) &\subseteq E \cap L_m(\mathbf{G}) \\
\overline{E \cap L_m(\mathbf{G})} \cap L_m(\mathbf{G}) &\subseteq E \cap L_m(\mathbf{G})
\end{aligned}$$

so that $F := E \cap L_m(\mathbf{G})$ is $L_m(\mathbf{G})$-closed. Let $K = \sup \mathcal{C}(F)$. If K is not $L_m(\mathbf{G})$-closed, i.e. $K \subsetneqq \bar{K} \cap L_m(\mathbf{G})$, there is a string $s \in \bar{K} \cap L_m(\mathbf{G})$ with $s \notin K$. Let $J = K \cup \{s\}$. Since $\bar{J} = \bar{K}$ we have that J is controllable. Also $K \subseteq F$ implies that

$$\bar{K} \cap L_m(\mathbf{G}) \subseteq \bar{F} \cap L_m(\mathbf{G}) = F$$

so that $s \in F$ and thus $J \subseteq F$. Therefore $J \in \mathcal{C}(F)$ and $J \supsetneqq K$, contradicting the fact that K is supremal. □

Now we can present the main result of this section.

Theorem 6

Let $E \subseteq \Sigma^*$ be $L_m(\mathbf{G})$-marked, and let $K = \sup \mathcal{C}(E \cap L_m(\mathbf{G}))$. If $K \neq \varnothing$, there exists a nonblocking supervisory control (NSC) V for $\mathbf{G}$ such that $L_m(V/\mathbf{G}) = K$.

Proof

K is controllable and, by Proposition 5, $L_m(\mathbf{G})$-closed. The result follows by Theorem 3.4.3. □

The result may be paraphrased by saying that K is (if nonempty) the *minimally restrictive*, or *maximally permissive*, solution of the problem of supervising $\mathbf{G}$ in such a way that its behavior belongs to E and control is nonblocking. In this sense the supervisory control provided by Theorem 6 is optimal.

Example 7: Let $\mathbf{G}$ be the controlled DES displayed below:

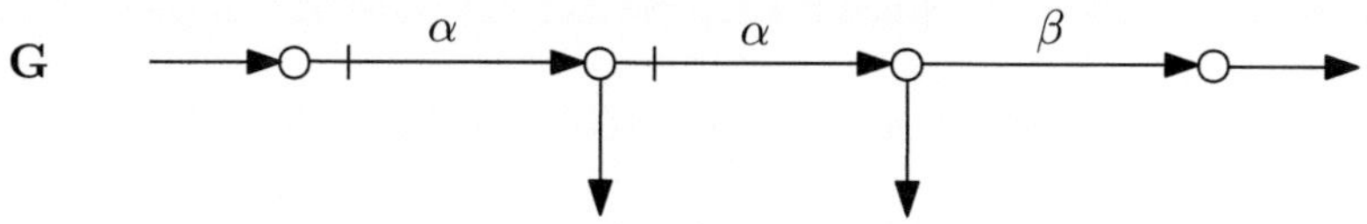

Here

$$\Sigma = \{\alpha, \beta\}, \quad \Sigma_c = \{\alpha\}, \quad L(\mathbf{G}) = \{\epsilon, \alpha, \alpha^2, \alpha^2\beta\},$$

$$L_m(\mathbf{G}) = \{\alpha, \alpha^2, \alpha^2\beta\}$$

For the specification language we take $E = \{\alpha, \alpha^2, \beta\}$. Then

$$E \cap L_m(\mathbf{G}) = \{\alpha, \alpha^2\}, \quad \bar{E} = \{\epsilon, \alpha, \alpha^2, \beta\}$$

$$\bar{E} \cap L_m(\mathbf{G}) = \{\alpha, \alpha^2\}, \quad \sup \mathcal{C}(E \cap L_m(\mathbf{G})) = \{\alpha\}$$

$$\overline{\{\alpha\}} = \{\epsilon, \alpha\}, \quad \overline{\{\alpha\}} \cap L_m(\mathbf{G}) = \{\alpha\}$$

From these results we see that E is $L_m(\mathbf{G})$-marked, and that indeed $\sup \mathcal{C}(E \cap L_m(\mathbf{G}))$ is $L_m(\mathbf{G})$-closed as asserted by Proposition 5. For the supervisory control we may take $V(\epsilon) = \{\alpha, \beta\}$, $V(\alpha) = \{\beta\}$, and $V(s) = \{\beta\}$ otherwise. Then it is clear that

$$L(V/\mathbf{G}) = \{\epsilon, \alpha\}, \quad L_m(V/\mathbf{G}) := L(V/\mathbf{G}) \cap L_m(\mathbf{G}) = \{\alpha\}$$

namely V is nonblocking for $\mathbf{G}$, as expected.

As might be expected, if we place part of the burden of 'marking action' on the supervisory control itself we may relax the prior requirement on E. By an application of Theorem 3.4.5 the reader may easily obtain the following.

Theorem 8

Let $E \subseteq \Sigma^*$ and let $K = \sup \mathcal{C}(E \cap L_m(\mathbf{G}))$. If $K \neq \varnothing$ there exists a marking nonblocking supervisory control (MNSC) V for $(K, \mathbf{G})$ such that $L_m(V/\mathbf{G}) = K$. □

Exercise 9: Let $E \subseteq L(\mathbf{G})$. Show that

$$\sup \mathcal{C}(E) = E \cap \overline{\sup \mathcal{C}(E)}$$

that is, $\sup \mathcal{C}(E)$ is E-closed. Show next that $\sup \mathcal{C}(\bar{E})$ is prefix-closed, but in general

$$\overline{\sup \mathcal{C}(E)} \subsetneqq \sup \mathcal{C}(\bar{E})$$

and

$$\sup \mathcal{C}(E) \subsetneqq E \cap \sup \mathcal{C}(\bar{E})$$

Exercise 10: For **MACH** defined in Example 3.4.1, consider the languages

$$E_1 := \{s \in L_m(\mathbf{MACH}) | \#\beta(s) \geq 5\}$$

'production runs with at least 5 items produced'; and

$$E_2 := \{s \in L_m(\mathbf{MACH}) | \#\lambda(s) \leq 10\}$$

'runs with at most 10 breakdowns'; and

$$E_3 := E_1 \cap E_2$$

In each case calculate the supremal controllable sublanguage and describe the corresponding control action.

Exercise 11: Let $E \subseteq \Sigma^*$ be arbitrary, and let $K := \sup \mathcal{C}(E \cap L_m(\mathbf{G}))$ with respect to a controlled DES $\mathbf{G}$ over $\Sigma = \Sigma_c \dot{\cup} \Sigma_u$. That is,

$$K \subseteq E \cap L_m(\mathbf{G}), \quad \bar{K}\Sigma_u \cap L(\mathbf{G}) \subseteq \bar{K}$$

and K is supremal with respect to these two properties. Set $K_0 := E \cap L_m(\mathbf{G})$, write $H_0 := \bar{K}_0$, and define the chain

$$H_j := \{s \in H_{j-1} | \overline{\{s\}}\Sigma_u \cap L(\mathbf{G}) \subseteq H_{j-1}\}, \ j = 1, 2, \ldots$$

Verify that H_j is closed. For the chain limit write

$$H_\infty := \lim_{j \to \infty} H_j = \bigcap_{j=0}^{\infty} H_j$$

Prove that

$$H_\infty = \sup \mathcal{C}(H_0)$$

Write λ for the Nerode equivalence relation of $L_m(\mathbf{G})$. With reference to Sect. 2.8, prove that

$$H_j \precsim_\lambda H_{j-1}, \quad j = 1, 2, ...$$

Assuming as usual that $|\Sigma| < \infty$, and $E, L_m(\mathbf{G})$ are both regular, apply Theorem 2.8.11 to conclude that the chain H_j is finitely convergent, and $H_\infty \precsim_\lambda H_0$. Finally, prove from the latter result that $H_\infty \precsim_\lambda K_0$.

Next define

$$K_1 := H_\infty \cap E \cap L_m(\mathbf{G})$$

and show that $K_1 \precsim_\lambda K_0$.

The foregoing definitions of K_0 and K_1, via the construction of H_∞, determine a mapping $T : Pwr(\Sigma^*) \to Pwr(\Sigma^*)$ given by

$$T(J) := J \cap \sup \mathcal{C}(\bar{J}), \quad J \subseteq \Sigma^*$$

where $T(J) \precsim_\lambda J$. Setting $K_0 = E \cap L_m(\mathbf{G})$ as above, we obtain the chain

$$K_j = T(K_{j-1}), \quad j = 1, 2, ...$$

which is finitely convergent by virtue of the transitivity of $\precsim_\lambda$ and the fact that $\|K_0\| < \infty$.

Writing K_∞ for the chain limit, show that $K_\infty = K$ (as defined above), $K \precsim_\lambda E$, and thus

$$\|K\| \leq \|E\| \cdot \|L_m(\mathbf{G})\| + 1$$

3.6 Implementation of Supervisory Controls by Automata

While theoretically convenient, the abstract definition of a supervisory control as a map $L(\mathbf{G}) \to \Gamma$ does not in itself provide a concrete representation for practical implementation. We now show how such a representation may be derived in the form of an automaton.[1] Let V be a marking nonblocking supervisory control (MNSC) for the controlled DES $\mathbf{G} = (Q, \Sigma, \delta, q_0, Q_m)$, with

$$L_m(V/\mathbf{G}) = K, \quad L(V/\mathbf{G}) = \bar{K} \tag{3.1}$$

[1] In this section we prefer the term 'automaton' for this representation, rather than 'generator', but allow the transition function to be a partial function.

In case V is nonmarking but is nonblocking, i.e. V is an NSC, the discussion will specialize in an evident way. Now let **KDES** be a reachable automaton over Σ that represents K, namely

$$L_m(\mathbf{KDES}) = K, \quad L(\mathbf{KDES}) = \bar{K}$$

Obviously

$$K = L_m(\mathbf{KDES}) \cap L_m(\mathbf{G}), \quad \bar{K} = L(\mathbf{KDES}) \cap L(\mathbf{G})$$

With K as in (3.1), now let **SDES** be any DES such that

$$K = L_m(\mathbf{SDES}) \cap L_m(\mathbf{G}), \quad \bar{K} = L(\mathbf{SDES}) \cap L(\mathbf{G}) \tag{3.2}$$

If (3.2) holds we say that **SDES** *implements* V. Notice that, in general, **SDES** need not represent $L_m(V/\mathbf{G})$ in order to implement V: the conditions (3.2) allow **SDES** to represent a superlanguage of $L_m(V/\mathbf{G})$ that may be simpler (admit a smaller state description) than $L_m(V/\mathbf{G})$ itself. This flexibility is due to the fact that closed-loop behavior is a consequence of constraints imposed by the plant **G** (i.e. the structure of $L(\mathbf{G})$) as well as by the supervisory control V.

In any case we have

Proposition 1

Let $E \subseteq \Sigma^*$ and let

$$K := \sup \mathcal{C}(E \cap L_m(\mathbf{G})) \neq \varnothing$$

Let V be an MNSC such that $L_m(V/\mathbf{G}) = K$ (which exists by Theorem 3.5.8). Let **KDES** represent K. Then **KDES** implements V. □

For the converse, let $S \subseteq \Sigma^*$ be an arbitrary language over Σ such that

$$S \text{ is controllable with respect to } \mathbf{G} \tag{3.3a}$$

$$S \cap L_m(\mathbf{G}) \neq \varnothing \tag{3.3b}$$

$$\overline{S \cap L_m(\mathbf{G})} = \bar{S} \cap L(\mathbf{G}) \tag{3.3c}$$

We note that $K := S \cap L_m(\mathbf{G})$ is controllable with respect to **G**; in fact

$$\begin{aligned}
\bar{K}\Sigma_u \cap L(\mathbf{G}) &= (\overline{S \cap L_m(\mathbf{G})})\Sigma_u \cap L(\mathbf{G}) \\
&= (\bar{S} \cap L(\mathbf{G}))\Sigma_u \cap L(\mathbf{G}) \\
&\subseteq \bar{S}\Sigma_u \cap L(\mathbf{G}) \\
&\subseteq \bar{S} \cap L(\mathbf{G}) \qquad \text{(by (3.3a))} \\
&= \bar{K} \qquad \text{(by (3.3c))}
\end{aligned}$$

Let V be an MNSC such that $L_m(V/\mathbf{G}) = K$ (which exists by Theorem 3.5.8), and let **SDES** represent S. It is easy to see that **SDES** implements V; in fact

$$L_m(\mathbf{SDES}) \cap L_m(\mathbf{G}) = S \cap L_m(\mathbf{G}) = K$$
$$L(\mathbf{SDES}) \cap L(\mathbf{G}) = \bar{S} \cap L(\mathbf{G}) = \bar{K}$$

Thus we have

Proposition 2

Let **SDES** be any nonblocking DES over Σ such that $S := L_m(\mathbf{SDES})$ satisfies conditions (3.3a) and (3.3c). Let $\varnothing \neq K := S \cap L_m(\mathbf{G})$ and let V be an MNSC such that $L_m(V/\mathbf{G}) = K$. Then **SDES** implements V. In particular

$$L_m(V/\mathbf{G}) = L_m(\mathbf{G}) \cap L_m(\mathbf{SDES}), \quad L(V/\mathbf{G}) = L(\mathbf{G}) \cap L(\mathbf{SDES}) \quad \square$$

If (3.3a) and (3.3c) hold and **SDES** represents S we say that **SDES** is a *supervisor* for **G**. It is thus convenient to include under 'supervisor' the trivial case where $L_m(\mathbf{G}) \cap L_m(\mathbf{SDES}) = \varnothing$. If, in addition, it is trim, **SDES** will be called a *proper* supervisor for **G**. It is also convenient to extend the usage of 'controllability' to DES: thus if **SDES** represents the language S and S is controllable with respect to **G**, we shall say that **SDES** is *controllable* with respect to **G**. To summarize, **SDES** is declared to be a *proper supervisor* for **G** if

(i) **SDES** is trim (reachable and coreachable)
(ii) **SDES** is controllable with respect to **G**
(iii) $\overline{L_m(\mathbf{SDES}) \cap L_m(\mathbf{G})} = L(\mathbf{SDES}) \cap L(\mathbf{G})$

As an illustration of Proposition 2, let **EDES** be an arbitrary DES over Σ and let

$$K := \sup \mathcal{C}(L_m(\mathbf{G}) \cap L_m(\mathbf{EDES}))$$

where the right side may possibly be empty. Let **KDES** represent K. Then **KDES** is a proper supervisor for **G**.

The relationships discussed above are displayed in Fig. 3.1. The *TCT* procedures **supcon** and **condat** are introduced in the following section. In general a 'simplified' supervisor language S, or its generator **SDES**, can be obtained only by intelligent guesswork, or a heuristic reduction procedure like **supreduce** (Sect. 3.12).

Let $\mathbf{S} = (X, \Sigma, \xi, x_0, X_m)$ implement V. We may interpret **S** as a state machine that accepts as 'forcing inputs' the sequence of symbols of Σ output by **G** and executes corresponding state transitions in accordance with its transition function ξ. In this interpretation, control action is exercised by **S** on **G** implicitly, by way of a state-output function

$$\psi : X \to Pwr(\Sigma)$$

defined according to

$$\psi(x) := \{\sigma \in \Sigma | \xi(x, \sigma)!\}$$

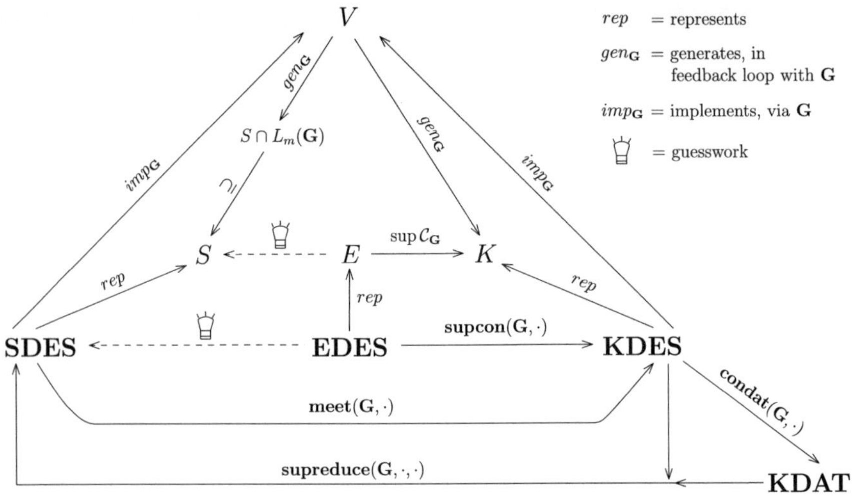

Fig. 3.1 Scheme for supervisory control implementation

The control action of **S** may be visualized as follows. Immediately upon entering the state $x \in X$, and while resident there, **S** disables in **G** just those (controllable) events $\sigma \in \Sigma_c$ such that $\sigma \notin \psi(x)$. In other words the next possible event that can be generated by **G** is any event, but only events, in the set

$$\{\sigma \in \Sigma | \xi(x, \sigma)! \,\&\, \delta(q, \sigma)!\}$$

where $q \in Q$ is the current state of **G**. The actual mechanism of disablement, which would involve instantaneous transfer of information from **S** to **G**, will be left unspecified in our interpretation. As a metaphor, one might consider the switching of signals between red and green (no amber!) in an idealized road traffic network.

To formalize the closed-loop supervisory control system that results from this construction, we denote by $\mathbf{S}/\mathbf{G}$ the product generator (cf. Sect. 2.4)

$$\mathbf{S}/\mathbf{G} = (X \times Q, \Sigma, \xi \times \delta, (x_0, q_0), X_m \times Q_m)$$

where

$$\xi \times \delta : X \times Q \times \Sigma \to X \times Q : (x, q, \sigma) \mapsto (\xi(x, \sigma), \delta(q, \sigma))$$

provided $\xi(x, \sigma)!$ and $\delta(q, \sigma)!$.

Exercise 3: From the foregoing discussion verify that

$$L_m(\mathbf{S}/\mathbf{G}) = L_m(V/\mathbf{G}), \quad L(\mathbf{S}/\mathbf{G}) = L(V/\mathbf{G})$$

In this sense our use of the term ‘implements’ is justified.

Example 4: Supervisor

Referring to Example 3.5.7 and applying the foregoing result for MNSC, we may take for **S** the recognizer for sup $\mathcal{C}(E \cap L_m(\mathbf{G})) = \{\alpha\}$ with (partial) transition function displayed below:

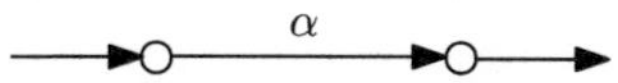

Alternatively, since in this example the specification language E is $L_m(\mathbf{G})$-marked, all marking action for sup $\mathcal{C}(E \cap L_m(\mathbf{G}))$ could be left to **G** itself, namely we could take for **S** the recognizer with $X_m = X$ corresponding to the closed language $\{\epsilon, \alpha\}$:

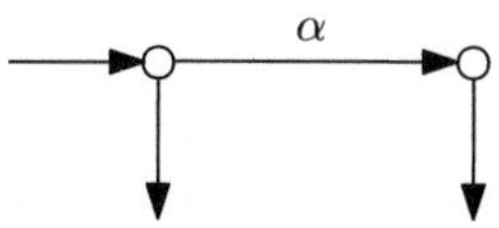

Which style is selected will depend on computational convenience. Currently *TCT* runs more efficiently the smaller the subset of marker states, so the first approach would be preferred.

The condition (3.3c) provides the basis for a useful definition. In general let K, L be arbitrary sublanguages of Σ^*. Then K, L are *nonconflicting* if

$$\overline{K \cap L} = \bar{K} \cap \bar{L}$$

Thus K and L are nonconflicting just in case every string that is both a prefix of K and a prefix of L can be extended to a string belonging to K and L in common. In *TCT* the boolean function **nonconflict(G₁,G₂)** = *true* just in case every reachable state of the product structure **meet(G₁,G₂)** is coreachable, namely $\overline{L_m(\mathbf{G_1}) \cap L_m(\mathbf{G_2})} = L(\mathbf{G_1}) \cap L(\mathbf{G_2})$. Thus to check whether two languages L_1 and L_2 are nonconflicting it is equivalent to check that $\mathbf{G_1}$ and $\mathbf{G_2}$ satisfy **nonconflict**, where $L_i = L_m(\mathbf{G_i})$ and $\bar{L}_i = L(\mathbf{G_i})$; namely $\mathbf{G_1}, \mathbf{G_2}$ represent L_1, L_2 respectively.

The next result follows immediately from the definitions.

Proposition 5

Let **SUP** be an arbitrary DES over Σ and assume that $\overline{L_m(\mathbf{G})} = L(\mathbf{G})$. Then **SUP** is a proper supervisor for **G** iff (i) $L_m(\mathbf{SUP})$ is controllable with respect to **G**, (ii) **SUP** is trim, and (iii) $L_m(\mathbf{SUP})$ and $L_m(\mathbf{G})$ are nonconflicting. □

In case $L_m(\mathbf{SUP}) \subseteq L_m(\mathbf{G})$ the 'nonconflicting' condition is automatic.

Notice that condition (iii) for **SUP** is a nontrivial property that may require separate verification whenever **SUP** is not, or is not already known (perhaps by construction) to be, the trim (or at least nonblocking) generator of a controllable sublanguage of $L_m(\mathbf{G})$.

The following example illustrates how the conclusion of Proposition 5 fails when $L_m(\mathbf{SUP})$ and $L_m(\mathbf{G})$ conflict.

Example 6: Conflict

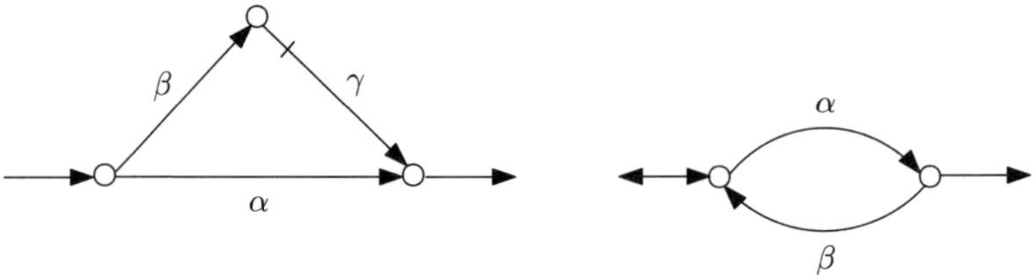

G **SUP**

Here $\Sigma = \{\alpha, \beta, \gamma\}$ with $\Sigma_c = \{\gamma\}$. We have $L_m(\mathbf{SUP}) \cap L_m(\mathbf{G}) = \{\alpha\}$, whereas $L(\mathbf{SUP}) \cap L(\mathbf{G}) = \{\epsilon, \alpha, \beta\}$.

Taken literally, the interpretation of supervisory control action proposed earlier in this section assigns to the supervisor a role that is largely passive, consisting only in the disablement of events whose occurrence or nonoccurrence is otherwise actively 'decided' by the plant. While often physically plausible, this is not the only interpretation that is possible or even desirable. It should be borne in mind that, from a formal point of view, the theory treats nothing but event synchronization among transition structures: the issue as to which (if any) among two or more synchronized transition structures actively 'causes' a given shared event to occur is not formally addressed at all. The system modeler is free to ascribe causal action as he sees fit: a machine transition from Idle to Working is (in the theory) nothing but that; it is consistent with the theory to suppose that the transition is 'caused' by spontaneous internal machine volition, or by internal volition on the part of the supervisor, or indeed by some external agent that may or may not be explicitly modeled, say a human operator or aliens on Mars. This feature provides the modeler with considerable flexibility. In the exercises of later sections, the reader is invited to test the model against whatever interpretation he deems appropriate; of course the desired interpretation may very well guide the modeling process. See, for instance, Sect. 3.8, which touches on the related issue of forced events.

3.7 Design of Supervisors Using *TCT*

We now indicate how the results of Sects. 3.5 and 3.6 can be applied to supervisor design. Let the controlled DES $\mathbf{G}$ be given, along with an upper bound $E \subseteq \Sigma^*$ on admissible marked behavior. As before we refer to E as the specification language. It will be assumed that $E = L_m(\mathbf{E})$, where the DES $\mathbf{E}$ along with $\mathbf{G}$ has been created by *TCT*. Our objective is to design an optimal (i.e. minimally restrictive) proper supervisor $\mathbf{S}$ for $\mathbf{G}$ subject to $L_m(\mathbf{S}/\mathbf{G}) \subseteq E$. In accordance with Theorem 3.5.8 and the discussion in Sect. 3.6 the most direct method is to compute a trim generator (used as recognizer) for the language $K := \sup \mathcal{C}(E \cap L_m(\mathbf{G}))$. The *TCT* procedure **supcon** computes a trim representation **KDES** of K according to

$$\mathbf{KDES} = \mathbf{supcon}(\mathbf{G}, \mathbf{E})$$

To complete the description of **S**, the *TCT* procedure **condat** returns the control pattern (specifically, the minimal set of controllable events that must be disabled) at each state of **S**:

$$\mathbf{KDAT} = \mathbf{condat}(\mathbf{G}, \mathbf{KDES})$$

In outline the procedure **supcon** works as follows. Let $Pwr(\Sigma^*)$ be the power set of Σ^*, i.e. the set of all sublanguages of Σ^*. Define the operator

$$\Omega : Pwr(\Sigma^*) \to Pwr(\Sigma^*)$$

according to

$$\Omega(Z) = E \cap L_m(\mathbf{G}) \cap \sup\{T \subseteq \Sigma^* | T = \bar{T},\ T\Sigma_u \cap L(\mathbf{G}) \subseteq \bar{Z}\}$$

With K as defined above, it can be shown that K is the largest fixpoint of Ω. In the present regular (finite-state) case, this fixpoint can be computed by successive approximation. Let

$$K_0 = E \cap L_m(\mathbf{G}), \quad K_{j+1} = \Omega(K_j) \quad (j = 0, 1, 2, ...)$$

It can be shown that

$$K = \lim K_j \quad (j \to \infty)$$

Furthermore the limit is attained after a finite number of steps that is of worst case order $\|L_m(\mathbf{G})\| \cdot \|E\|$. In *TCT* the operator Ω is implemented by a simple backtracking operation on the product transition structure **meet**(**G**, **E**).

As an example, we consider Small Factory, as described in Example 3.3.19. The result for

$$\mathbf{FACTSUP} = \mathbf{supcon}(\mathbf{FACT}, \mathbf{SPEC})$$

is displayed in Fig. 3.2 (for coding details see Appendix 3.1). By tracing through the transition graph the reader may convince himself that the specifications are satisfied; and the theory guarantees that **FACTSUP** represents the freest possible behavior of **FACT** under the stated constraints. We also tabulate the control patterns as displayed by[2]

$$\mathbf{FACTSUP} = \mathbf{condat}(\mathbf{FACT}, \mathbf{FACTSUP})$$

Only controllable events that are strictly required to be disabled appear in the table.

In practice it is rarely necessary to implement an optimal supervisor by explicit representation of the language $\sup \mathcal{C}(L_m(\mathbf{G}) \cap E)$. Often common sense and intuition

[2] *TCT* stores the result of **condat** as a .DAT file, whereas the result of **supcon** is a .DES file.

FACTSUP

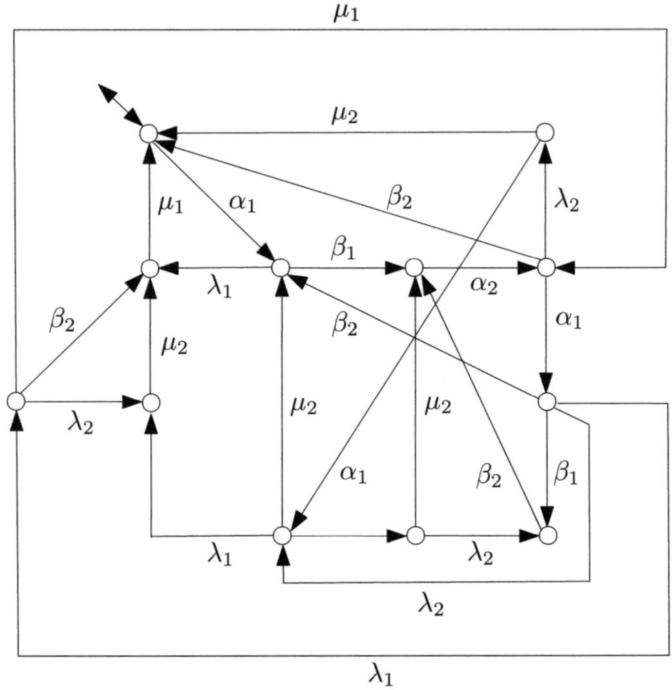

Fig. 3.2 Supremal Controllable Sublanguage L_m(**FACTSUP**) for Small Factory

will lead to an optimal supervisor with a much smaller transition structure.[3] For justification of such a proposed supervisor, we may apply the *TCT* analog of Proposition 3.6.5.

Proposition 1

Let **SUP** be a DES over Σ, such that

(i) **condat(G,SUP)** lists only controllable (i.e. odd-numbered) events as requiring disablement
(ii) **SUP = trim(SUP)**
(iii) **nonconflict(G,SUP)** = *true*

Then **SUP** is a proper supervisor for **G**. □

In analogy to the notation $V/\mathbf{G}$, denote by **SUP/G** the closed-loop controlled DES obtained by forming the **meet** of **SUP** with **G**. Then, with **SUP** a proper supervisor for **G**, we have

$$L_m(\mathbf{SUP/G}) = L_m(\mathbf{meet}(\mathbf{G}, \mathbf{SUP})), \quad L(\mathbf{SUP/G}) = L(\mathbf{meet}(\mathbf{G}, \mathbf{SUP}))$$

[3]See also Sect. 3.12.

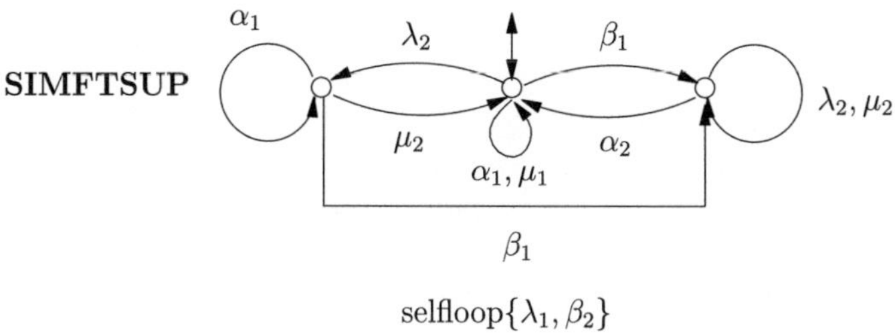

Fig. 3.3 Reduced Supervisor for Small Factory

along with the guaranteed nonblocking property

$$\overline{L_m(\mathbf{SUP/G})} = L(\mathbf{SUP/G})$$

For Small Factory we construct the candidate supervisor **SIMFTSUP** directly as shown in Fig. 3.3, where the natural decomposition of the control problem into the regimes of 'normal operation' and 'breakdown and repair' is clearly manifest. Evidently **SIMFTSUP** is trim. Controllability of the language L_m(**SIMFTSUP**) is easily checked from the table for

$$\mathbf{SIMFTSUP} = \mathbf{condat}(\mathbf{FACT}, \mathbf{SIMFTSUP})$$

inasmuch as only controllable events are required to be disabled. To test whether **SIMFTSUP** is nonblocking we apply the *TCT* procedure **nonconflict**. In the present case we find that **nonconflict(FACT,SIMFTSUP)** is *true*, and so conclude finally that **SIMFTSUP** really is a proper supervisor for **FACT**, as expected.

As yet there is no guarantee that **SIMFTSUP** is optimal. To verify that it is, we must first compute the closed-loop language

$$L_m(\mathbf{SIMFTSUP/FACT}) = L_m(\mathbf{SIMFTSUP}) \cap L_m(\mathbf{FACT})$$

as represented by, say,

$$\mathbf{TEST} = \mathbf{meet}(\mathbf{SIMFTSUP}, \mathbf{FACT})$$

and then check that $L_m(\mathbf{TEST}) = L_m(\mathbf{FACTSUP})$.

In general, suppose $M_1 = L_m(\mathbf{G_1})$ and $M_2 = L_m(\mathbf{G_2})$, with $\mathbf{G_1}, \mathbf{G_2}$ both trim. *TCT* offers two ways of investigating equality of M_1 and M_2. A general method is to check the inclusions $M_1 \subseteq M_2$ and $M_2 \subseteq M_1$ according to

$$\mathbf{trim}(\mathbf{meet}(\mathbf{G_1}, \mathbf{complement}(\mathbf{G_2}, \_))) = \mathbf{EMPTY}$$
$$\mathbf{trim}(\mathbf{meet}(\mathbf{G_2}, \mathbf{complement}(\mathbf{G_1}, \_))) = \mathbf{EMPTY}$$

Alternatively, if $\mathbf{G_1}$ and $\mathbf{G_2}$ are seen to have the same state size, number of marker states and number of transitions, then it may already be plausible that $M_1 = M_2$, and it is sufficient to check that the *TCT* procedure

$$\textbf{isomorph}(\mathbf{G_1}, \mathbf{G_2})$$

returns *true*. In fact, this method becomes definitive provided $\mathbf{G_1}$, $\mathbf{G_2}$ are both state-minimal; namely $M_1 = M_2$ iff

$$\textbf{isomorph}(\mathbf{MG_1}, \mathbf{MG_2}) = true$$

where $\mathbf{MG_i}$ = **minstate**($\mathbf{G_i}$) ($i = 1, 2$).

In the present example, one can verify by either method that L_m(**TEST**) = L_m(**FACTSUP**), as hoped. The design and justification of **SIMFTSUP** are now complete.

Exercise 2: A workcell consists of two machines **M1**, **M2** and an automated guided vehicle **AGV** as shown.

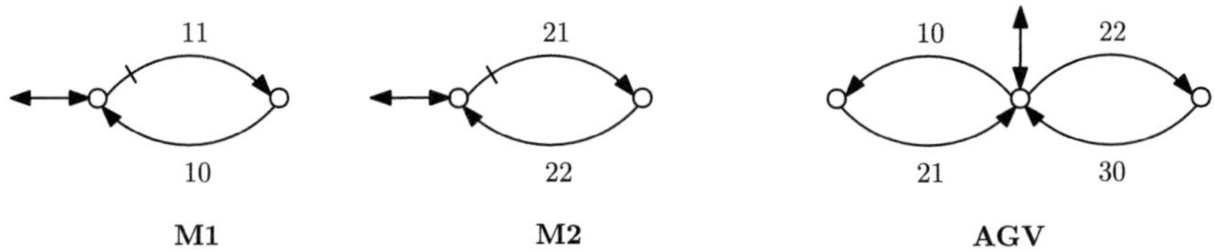

AGV can be loaded with a workpiece either from **M1** (event 10) or from **M2** (event 22), which it transfers respectively to **M2** (event 21) or to an output conveyor (event 30). Let **CELL = sync(M1,M2,AGV)**. By displaying an appropriate event sequence show that **CELL** can deadlock, i.e. reach a state from which no further transitions are possible. Interpret physically. To prevent deadlock, define the legal language **TCELL = trim(CELL)**, then **SUPER = supcon(CELL,TCELL)**. Explain how **SUPER** prevents deadlocking event sequences. Also explain why **TCELL** itself cannot serve directly as a (proper) supervisor.

Exercise 3: A transmitter is modeled by the 3-state generator **T0** tabulated, where the event α denotes arrival of a message to be transmitted, σ denotes the start of a message transmission, τ denotes timeout in case the transmitted message is not acknowledged, and ρ denotes reset for transmission of a subsequent message. An unacknowledged message is retransmitted after timeout. New messages that arrive while a previous message is being processed are ignored.

The system **T** to be controlled consists of transmitters **T1** and **T2**, where **Ti** is modeled over the alphabet $\{\alpha_i, \sigma_i, \tau_i, \rho_i\}$; thus **T = shuffle(T1,T2)**. Only the events σ_1, σ_2 are controllable. It is required that the channel be utilized by at most one transmitter at a time.

A first trial solution implements the supervisory control V as the 2-state device **C**, which can be thought of as a model of the channel. **C** ensures that if, for instance,

σ_1 occurs, then σ_2 cannot occur until ρ_1 occurs, namely **T1** has finished processing a message.

Verify that **T** and **C** conflict, compute **TC** := **meet**(**T**, **C**), and find a string that leads to a state of **TC** that is non-coreachable. Explain why **C** is not a suitable controller.

Try the new channel model **NC**, verify that **T** and **NC** are nonconflicting, and that **NC** controls **T1** and **T2** according to specification. Provide *TCT* printouts and state transition graphs as appropriate.

T0 # states: 3; marker state: 0
transitions:
$[0, \alpha, 1],\ [1, \alpha, 1],\ [1, \sigma, 2],\ [2, \rho, 0],\ [2, \tau, 1],\ [2, \alpha, 2]$

C # states: 2; marker state: 0
transitions:
$[0, \sigma_1, 1],\ [0, \sigma_2, 1],\ [1, \rho_1, 0],\ [1, \rho_2, 0]$
[adjoin selfloops with events $\alpha_i,\ \tau_i, i = 1, 2$]

NC # states: 3; marker state: 0
transitions:
$[0, \sigma_1, 1],\ [1, \sigma_1, 1],\ [1, \rho_1, 0],\ [0, \sigma_2, 2],\ [2, \sigma_2, 2],\ [2, \rho_2, 0]$
[adjoin selfloops with events $\alpha_i, \tau_i, i = 1, 2$]

Exercise 4: WORKCELL is the synchronous product of **ROBOT**, **LATHE** and **FEEDER**. The latter is a mechanism that imports new parts for **WORKCELL** to process. There is a 2-slot input buffer **INBUF** to store new parts as they are imported, and a 1-slot buffer **SBBUF** associated with **LATHE**, to hold parts on standby. **ROBOT** transfers new parts from **INBUF** to **LATHE**. If **LATHE** is idle, **ROBOT** loads the new part; if busy, **ROBOT** places the new part in **SBBUF**; in each case, **ROBOT** then returns to idle. If **ROBOT** is idle and **LATHE** is idle and there is a part in **SBBUF**, then **ROBOT** can load it. There are other tasks unrelated to **LATHE**, which **ROBOT** can initiate and return from.

Specifications are the following. **SPEC1** says that **LATHE** can be loaded only if it is idle. **SPEC2** says that if a part is on standby (i.e. **SBBUF** is not empty) then **ROBOT** cannot transfer a new part from **INBUF**. **SPEC3** says that **LATHE** can move from idle to busy only after being loaded. **SPEC4** says that a part can be put on standby only if **LATHE** is busy. **SPEC5** says that **ROBOT** must give **LATHE** priority over its other tasks: namely **ROBOT** can initiate other tasks only when: either **LATHE** is busy, or both **INBUF** and **SBBUF** are empty. To set up **SPEC5**, compute **sync(LATHE,INBUF,SBBUF)**, then selfloop with **ROBOT**'s 'initiate_unrelated_task' event at just the appropriate states (recall the method of 'flags', Remark 3.3.16). **SPEC5** automatically incorporates the usual overflow/underflow constraints on the buffers. Finally **SPEC** is the synchronous product of **SPEC1**,...,**SPEC5**, self-looped with any **WORKCELL** events not already included.

Create *TCT* models for the items described above, making your own detailed choices for controllable/uncontrollable events. Then compute **SUPER** = **supcon**

(**WORKCELL,SPEC**), as well as **SUPER(.DAT)** = **condat(WORKCELL, SUPER)**. Discuss any points of interest.

To examine the controlled behavior when **ROBOT**'s extraneous tasks are hidden from view, compute

$$\textbf{PSUPER} = \textbf{project}(\textbf{SUPER}, \textbf{null/image}[\text{appropriate\_event\_list}])$$

and discuss.

Exercise 5: Use *TCT* to re-solve Exercise 3.5.10, making sure your results are consistent.

Exercise 6: In Small Factory, compute

$$\textbf{PFACTSUP} = \textbf{project}(\textbf{FACTSUP}, \textbf{null}[10, 12, 13, 21, 22, 23])$$

and interpret the result.

Exercise 7: Using *TCT*, redo Small Factory using a buffer of capacity 2. Also design the corresponding simplified supervisor. Generalize your results to a buffer of arbitrary size N.

Exercise 8: Three Cooks

Three cooks share a common store of 5 pots. For his favorite dish, COOK1 needs 2 pots, COOK2 4 pots, and COOK3 all 5 pots. The cooks may take pots from the store individually and independently, but only one pot at a time; a cook returns all his pots to the store simultaneously, but only when he has acquired and used his full complement. Design a supervisor that is maximally permissive and guarantees nonblocking. Assume that 'take-pot' events are controllable and 'return-pot' events uncontrollable. Explain the supervisor's control action in terms of disabling entrance by the cooks to 'forbidden states'.

Exercise 9: In the context of a DES problem where the alphabet is Σ, define the self-looped DES

$$\textbf{ALL} = (\{0\}, \Sigma, \{\text{transitions } [0, \sigma, 0] | \sigma \in \Sigma\}, 0, \{0\})$$

Thus the closed and marked behaviors of **ALL** are both Σ^*. As usual let **PLANT** and **SPEC** be two DES over Σ. The corresponding supervisory control problem has solution **SUP** = **supcon(PLANT,SPEC)**. Show that this problem can always be replaced by an equivalent problem where the specification is **ALL**. **Hint**: First replace **SPEC** by **NEWSPEC**, where **NEWSPEC** merely adjoins a 'dump state' to **SPEC**, if one is needed. This makes the closed behavior of **NEWSPEC** equal to Σ^*, while the marked behavior is that of **SPEC**. In *TCT*,

$$\textbf{NEWSPEC} = \textbf{complement}(\textbf{complement}(\textbf{SPEC})) \text{ (why?)}$$

Now set

$$\textbf{NEWPLANT} = \textbf{meet}(\textbf{PLANT}, \textbf{NEWSPEC})$$
$$\textbf{NEWSUP} = \textbf{supcon}(\textbf{NEWPLANT}, \textbf{ALL})$$

Show that **NEWSUP** and **SUP** define exactly the same languages (in particular, perhaps after application of **minstate**, **NEWSUP** and **SUP** are isomorphic).

While this maneuver offers no computational advantage, it can simplify theoretical discussion, as the specification **ALL** requires only that the closed-loop language be nonblocking.

If Σ is exactly the set of events appearing in a DES **G**, then **ALL** in Exercise 9 is computed in *TCT* as

$$\textbf{ALL} = \textbf{allevents}(\textbf{G})$$

Exercise 10: Let **G** be a DES defined over the alphabet Σ. With **ALL** defined as in Exercise 9, show that **G** is nonblocking iff **nonconflict(ALL,G)** = *true*. Using *TCT* check this result against examples of your own invention.

Exercise 11: Let $\mathbf{G} = (Q, \Sigma, \delta, q_0, Q_m)$ and let $K \subseteq \Sigma^*$. Suppose K is *represented* by a DES $\mathbf{KDES} = (X, \Sigma, \xi, x_0, X_m)$, in the sense that $K = L_m(\mathbf{KDES})$, $\bar{K} = L(\mathbf{KDES})$, i.e. the marked and closed behaviors of **KDES** are K and $\bar{K}$ respectively. Let $\mathbf{PROD} = \mathbf{G} \times \mathbf{KDES}$ as defined in Sect. 2.4, and let **RPROD** be the reachable sub-DES of **PROD**.

Show that K is controllable with respect to **G** iff, at each state (q, x) of **RPROD**,

$$\{\sigma \in \Sigma_u | \delta(q, \sigma)!\} \subseteq \{\sigma \in \Sigma_u | \xi(x, \sigma)!\}$$

namely any uncontrollable event that is state-enabled ('physically executable') by **G** is also control-enabled ('legally admissible') by **KDES**.

Illustrate the foregoing result with two simple examples, for the cases K controllable and uncontrollable respectively.

Exercise 12: The result of Exercise 11 is the basis of the *TCT* procedure **condat**. The result of **condat**, say **KDAT = condat(G,KDES)**, is a table of the states x of **KDES** along with all the events which must be disabled in **G** (by a supervisory control) when **RPROD** is at (q, x) in $\mathbf{G} \times \mathbf{KDES}$ for some $q \in Q$, in order to force the inclusion

$$\{\sigma \in \Sigma | \delta(q, \sigma)!, \ \sigma \text{ control-enabled}\} \subseteq \{\sigma \in \Sigma | \xi(x, \sigma)!\}$$

Thus the set of disabled events tabulated by **condat** at x is

$$\{\sigma \in \Sigma | [(\exists q \in Q)(q, x) \text{ in } \mathbf{RPROD} \ \& \ \delta(q, \sigma)!] \ \& \ \textit{not} \ \xi(x, \sigma)!\}$$

For K to be controllable, this set must contain only events that are controllable (so they can be disabled): in other words, in the *TCT* event encoding, only events with odd-numbered labels. So to check controllability of K it is sufficient to scan the **condat** table: if only odd events are listed, K is controllable; if an even event occurs anywhere, K is uncontrollable.

Illustrate this remark by testing your two examples from Exercise 11 and supply the *TCT* printouts for **condat**.

Exercise 13: Message passing

Investigate how to implement message passing in our setup. For instance, suppose a supervisor **M0** wishes to enable an event 11 'remotely' in controlled module **M1** by sending a 'message' 0. **M1** should not execute 11 before receiving 0, but **M0** need not wait for **M1** to execute 11 before completing other tasks (say, event 2). Also, if **M1** has just executed 11, it must not do so again until it has executed task 10 (say) and once again received a 0. For this, define

$$\begin{aligned}\mathbf{M0} &= (Q, \Sigma, \delta, q_0, Q_m)\\ &= (\{0,1,2\}, \{0,1,2\}, \{[0,1,1],[1,0,2],[2,2,0]\}, 0, \{0\})\\ \mathbf{M1} &= (\{0,1\}, \{10,11\}, \{[0,11,1],[1,10,0]\}, 0, \{0\})\end{aligned}$$

To couple **M0** and **M1** as described define a 'mailbox'

$$\mathbf{MB} = (\{0,1\}, \{0,11\}, \{[0,0,1],[1,0,1],[1,11,0]\}, 0, \{0\})$$

Check that the synchronous product of **M0**, **M1**, and **MB** displays the required behavior. Show that the approach can be extended to two or more controlled modules **M1**, **M2**, ... For instance, create **M2** by relabeling events 10,11 in **M1** as 20,21; rename **MB** as **MB1**; create **MB2** by relabeling 11 in **MB1** as 21; and consider the new controlled module **M** = **sync(M1,M2)** and mailbox **MB** = **sync(MB1,MB2)**. Notice that the message 0 can always be projected out of the final structure **sync(M0,M,MB)** if it is of no external interest.

Investigate this model, with particular attention to the growth in complexity as measured by the size of the final state set.

Other message-passing semantics are possible. For instance, suppose **M0** should not progress past state 2 until both enabled events 11 and 21 have occurred—**M0** waits for its message to be acted on. For this, remodel **M0** as

$$\begin{aligned}\mathbf{N0} = (&\{0,1,2,3,4\}, \{1,2,11,21\},\\ &\{[0,1,1],[1,11,2],[1,21,3],[2,21,4],[3,11,4],[4,2,0]\}, 0, \{0\})\end{aligned}$$

Because **N0** waits, there is logically no need for a mailbox at all. Check that the result has 20 states and 36 transitions, still complicated but much less so than before.

Now explore other variations on the theme.

Exercise 14: Show that

$$\mathbf{supcon}(\mathbf{G}, \mathbf{meet}(\mathbf{E_1}, \mathbf{E_2})) = \mathbf{supcon}(\mathbf{supcon}(\mathbf{G}, \mathbf{E_1}), \mathbf{E_2})$$

and interpret.

Exercise 15: Interrupts

Let **G** be a controlled DES over alphabet $\Sigma = \Sigma_c \dot{\cup} \Sigma_u$, and let events $\alpha, \beta \notin \Sigma$. Interpret α as an 'interrupt', namely an event that causes all controllable events in **G** to be instantly disabled, pending the occurrence of event β as a signal that the interrupt has been serviced and normal activity of **G** may be resumed. Explain how **G** with the interrupt feature can be modeled as **sync(G,H)**, with **H** as below.

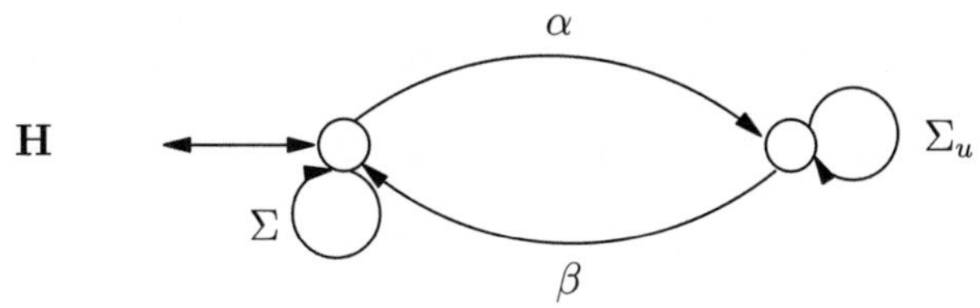

Exercise 16: Uncontrollable subsystems

It is important to be aware that in system modeling, the 'plant' may certainly include subsystem components all of whose events are uncontrollable, and that such inclusion may actually be necessary for the computed result to be meaningful. As shown below, suppose a **PUMP** increments (event 10) a **TANK**, modeled as a buffer with capacity 2, and that **TANK** is uncontrollably drained (event 12). If the plant is taken to be **PUMP** alone (but with event 12 self-looped) and **TANK** as the specification (with 11 self-looped), verify that the supremal controllable sublanguage is empty, and explain why. Alternatively, if the plant is taken to be the synchronous product of **PUMP** and **TANK**, and the specification taken as **ALL** (as in Exercise 9) with events 10,11,12, then no control action is called for by the supervisor (explain why), but this result is clearly not the one intended. A correct modeling approach is to consider **TANK** as embedded in an environment, say **POOL**, with infinite capacity; take **PLANT** to be the synchronous product of **PUMP** with **POOL**; and let **TANK** be the specification, in effect limiting the content of **POOL** as required. Here it suffices to take as 'infinity' anything larger than **TANK**, say a buffer with capacity 3. Carry through the supervisor design as suggested, including the simplified supervisor obtained by **supreduce** (see Sect. 3.12). Extend the approach to a chain of tanks downstream from **PUMP**, having possibly different capacities, each draining uncontrollably into the next.

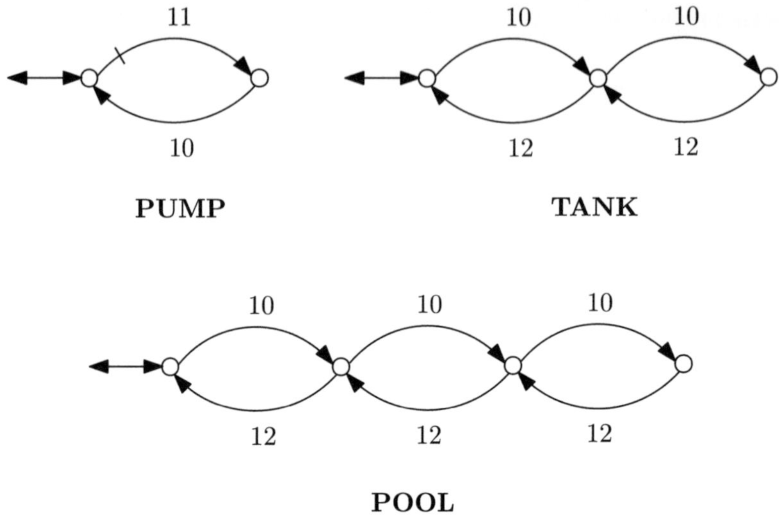

Exercise 17: Carry through an original supervisor design problem of your own, along the lines of this section. If feasible, draw the transition graph of your supremal controllable sublanguage and discuss any features of special interest.

Exercise 18: Let **G** be a possibly blocking DES, i.e. with $\overline{L_m(\mathbf{G})} \subsetneqq L(\mathbf{G})$. Let **K** be a trim 'coordinator', i.e. trim supervisor which controllably implements the sole specification of nonblocking, namely

$$L_m(\mathbf{K}) := \sup \mathcal{C}(L_m(\mathbf{G}) \cap L_m(\mathbf{ALL})) = \sup \mathcal{C}(L_m(\mathbf{G}))$$

Show that $L_m(\mathbf{K})$ is $L_m(\mathbf{G})$-closed, i.e. $L_m(\mathbf{K}) = L(\mathbf{K}) \cap L_m(\mathbf{G})$, where as usual $L(\mathbf{K})$ is the closed behavior of **K**. **Hint**: See Exercise 3.5.9, with $E = L_m(\mathbf{G})$.

Exercise 19: Let **K** be as in Exercise 18. Recalling the definition of 'support' in Sect. 2.8, show that **K** is supported on **G**. **Hint**: Exploit Exercise 3.5.11.

Exercise 20: Variation on Small Factory

With reference to Small Factory, redefine **MACH1** and **MACH2** according to

$$\begin{aligned}\mathbf{MACHi} = (&\{0, 1, 2, 3\}, \{\text{i0}, \text{i1}, \text{i2}, \text{i3}, \text{i4}\}, \{[0, \text{i1}, 1], [1, \text{i0}, 0], [1, \text{i2}, 2],\\ &[2, \text{i4}, 3], [3, \text{i3}, 0]\}, 0, \{0\})\end{aligned}$$

where the event i4 stands for 'request for repair', and the remaining events are as in Small Factory. Create a specification **TECHN** called 'technician' with transitions

$$\{[0, 14, 1], [0, 24, 2], [1, 13, 0], [1, 24, 3], [2, 14, 4], [2, 23, 0], [3, 13, 2], [4, 23, 1]\}$$

together with appropriate selfloops. Explain in words the behavior that is enforced. After adjoining a buffer specification as before, compute the optimal supervisor and its reduced version (in the sense of Sect. 3.12 below).

3.8 Forced Events

In practice one often thinks of control as 'forcing' the occurrence of some desirable result. In the asynchronous world of discrete events, forcing amounts to timely preemption: a tap or valve is closed in time to prevent overflow, a stirrer is switched on to prevent gelatification in a tank of fluid, a drainage pump is started when water in a mine reaches a defined level, a car is braked to forestall an impending collision, an industrial process is begun in time to deliver the product on schedule (before a deadline).

The crucial feature common to these examples is that the controlling agent denies permission for the occurrence of undesirable competing events; namely (directly or indirectly) such events are disabled. Enforcing the familiar specification that a buffer must not overflow or underflow is achieved by disabling the appropriate 'upstream' events in the causal (or just behavioral) sequence; meeting a deadline is achieved by 'disabling' the tick of a clock to ensure the occurrence of a desired event on schedule—how this can be modeled without violence to the physical requirement that 'time goes on' regardless of technology is explained in Chap. 9.

While in general 'forcing' is probably best placed in a temporal context (cf. Chap. 9) simple preemption can often capture the required action in the untimed framework considered so far. As a primitive example, suppose a tank T is filled by fluid flow through a valve V, which must be turned off to prevent overflow when the tank is full. V can be modeled as a one-state DES

$$\mathbf{V} = (Q, \Sigma, \delta, q_0, Q_m) = (\{0\}, \{\sigma\}, \{[0, \sigma, 0]\}, 0, \{0\})$$

with σ controllable. The event σ is interpreted as the delivery of a defined unit of fluid to the tank. The tank itself is modeled like a buffer, with its content incremented by one unit when σ occurs. If the tank capacity is N units then the transition structure could be

$$\begin{aligned}\mathbf{T} = (&\{0, 1, ..., N+1\}, \{\sigma\}, \{[0, \sigma, 1], [1, \sigma, 2], ...,\\ &[N, \sigma, N+1], [N+1, \sigma, N+1]\}, 0, \{0, 1, ..., N+1\})\end{aligned}$$

where the state $N + 1$ represents an overflow condition. To prevent overflow, let

$$\begin{aligned}\mathbf{TSPEC} = (&\{0, 1, ..., N\}, \{\sigma\}, \{[0, \sigma, 1], [1, \sigma, 2], ...,\\ &[N-1, \sigma, N]\}, 0, \{N\})\end{aligned}$$

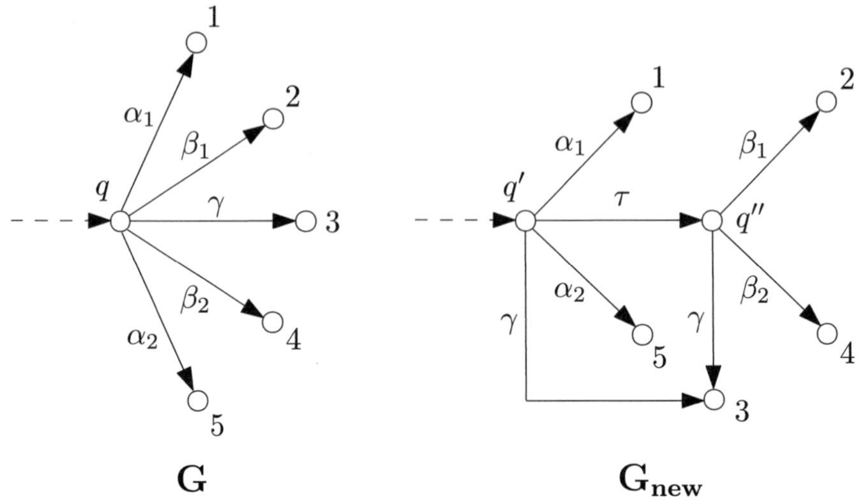

Fig. 3.4 Modeling forcing by α_1 or α_2 to preempt β_1 and β_2

thus disabling σ at state N. The closed behavior with respect to **TSPEC** is then simply $\overline{\sigma^N}$, as required. Notice that the model is consistent with the physical picture of (temporally) continuous flow through the valve, as there is no inconsistency in supposing that σ occurs one second after it is initially enabled, or re-enabled after a subsequent occurrence. As soon as σ is disabled, flow stops. However, there is no logical necessity that σ be tied to a fixed interval of time or a unit flow. The situation is much like filling up the fuel tank of a car using a hose with a spring-loaded trigger valve: when the tank is full, the trigger is released 'automatically' (or by the user) and the valve closes.

More generally, the notion of forcing as timely preemption can be formalized as follows. Define a new subset $\Sigma_f \subseteq \Sigma$ of *forcible* events, and a subset $\Sigma_p \subseteq \Sigma$ of *preemptible* events, with $\Sigma_f \cap \Sigma_p = \varnothing$. Bring in a new controllable event $\tau \notin \Sigma$ which may be thought of as a 'timeout' event. Assume that a plant model **G** has been created as usual over Σ, and we wish to adjoin the feature that any event in Σ_f can be 'forced' to preempt any event in Σ_p. For this, examine each state in **G** where some event $\alpha \in \Sigma_f$ and some event $\beta \in \Sigma_p$ are both enabled, e.g. the state q as displayed in Fig. 3.4. Notice that there may exist events γ defined at q that are neither forcible nor preemptible. Also, we impose no constraint as to whether an event in either Σ_f or Σ_p is controllable or not, although normally events in Σ_p will be uncontrollable. Now modify **G** (or in *TCT*, **edit(G)**) at q as shown: split q into q' and q'', with a transition $[q', \tau, q'']$. If, say, $\alpha_1, \alpha_2, \beta_1, \beta_2, \gamma$ are the events defined at q in **G**, then define $\alpha_1, \alpha_2, \gamma$ at q' and β_1, β_2, γ at q''. Selfloops should be treated as follows. If α was self-looped at q it should be self-looped at q'; a selfloop β at q is replaced by a transition $[q'', \beta, q']$; while a selfloop γ at q is replaced by selfloops γ at both q' and q''. In the new DES $\mathbf{G_{new}}$, say, the effect of disabling τ is to 'force' one of the events

$\alpha_1, \alpha_2, \gamma$ to preempt β_1, β_2. Observe that, unless γ is included in Σ_p, it could also preempt the other events α_1, α_2 defined at q'. Having modified **G** to $\mathbf{G_{new}}$, modify the specification DES **E** say, to $\mathbf{E_{new}}$, by self-looping each state of **E** with τ. We now have, in $(\mathbf{G_{new}}, \mathbf{E_{new}})$ a supervisory control problem of standard type, and proceed as usual to compute $\mathbf{supcon}(\mathbf{G_{new}}, \mathbf{E_{new}})$. This standard solution will 'decide' exactly when forcing (i.e. disablement of τ) is appropriate.

The foregoing procedure is implemented in *TCT* as **force**; in detail

$$\mathbf{G_{new}} = \mathbf{force}(\mathbf{G}, [\alpha_1, \alpha_2], [\beta_1, \beta_2], \tau)$$

After the design has been completed, the event τ can be hidden by being projected out. Thus all the details except initial selection of the subsets Σ_f, Σ_p can be rendered invisible to the user if desired.

Finally, it is advisable always to check that the result for forcing action is physically 'reasonable'. This is not always guaranteed, as for instance when an event chosen as forcible happens to occur as a selfloop at a state where forcing action is expected. In this case 'forcing' may be formally 'preemptive' but results in no change of state, a situation that could be physically unrealistic, depending on how the model is interpreted.

Example 1: Forcing

Consider the two machines **M1**, **M2** and the 1-slot buffer **B** in Fig. 3.5, with *TCT* encoding of events. For the plant take **M** = **sync(M1,M2)** and for the specification **E** take **B** self-looped with {11, 20}.

The solution **SUP** = **supcon(M,E)** is displayed in Fig. 3.6. Now suppose that event 21 ('**M2** starts work') is forcible with respect to event 10 ('**M1** completes work') as preemptible. Construct **MNEW** by modifying the structure of **M** as shown in Fig. 3.5, at the one state q (in this case) where events 21 and 10 are both defined. The new controllable 'timeout' event 23 can be thought of as inserting a time delay invoked by disablement of this event, thus providing event 21 with the opportunity to preempt event 10. In *TCT*, **MNEW** = **force**(**M**,[21],[10],23). Construct **ENEW** = **selfloop**(**E**,[23]), and compute **SUPNEW** = **supcon(MNEW,ENEW)**. Finally, hide the auxiliary event 23 to obtain the solution **PSUPNEW** = **project**(**SUPNEW**, **null**[23]), as displayed in Fig. 3.6. Notice that **PSUPNEW** generates a superlanguage of that of **SUP**; in general, controlled behavior with forcing will be less conservative than it is with the disablement feature alone.

To summarize, forcing is really an issue not of synthesis but of modeling; more precisely, by declaring forcibility as a modeling assumption, we eliminate forcing as a synthesis issue, and the standard framework can be utilized without further change or the addition of any new mechanism. Nevertheless it is well to note that, once an instance of a *controllable* event is designated to be forced (e.g. event 21 at state x in **PSUPNEW**, Fig. 3.6), it is no longer available for disablement in any subsequent refinement of the control design. For instance, 21 at x could be relabeled as 22 (i.e. redefined as uncontrollable) as a safeguard against inadvertent disablement in a subsequent application of *TCT*.

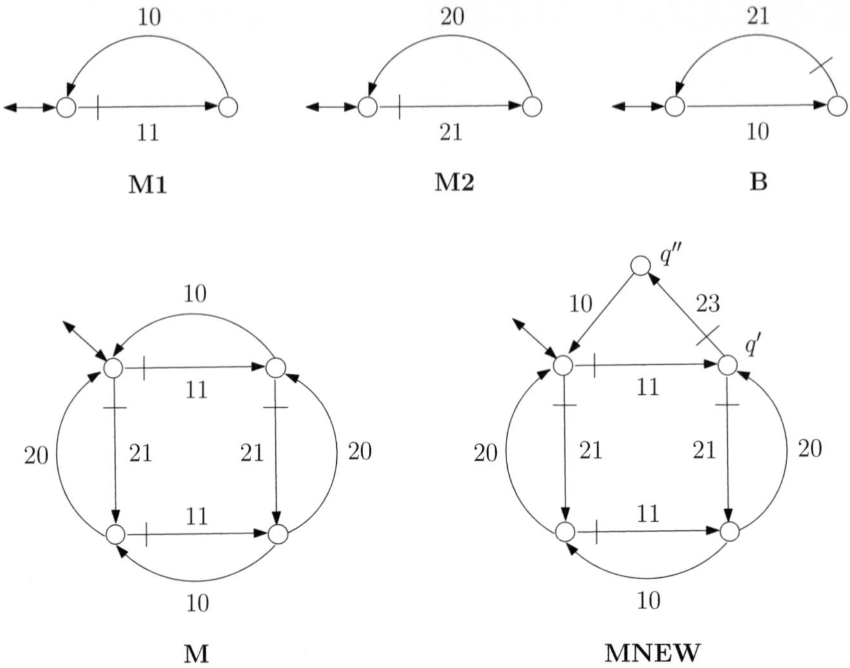

Fig. 3.5 Modeling forcing by event 21 to preempt event 10

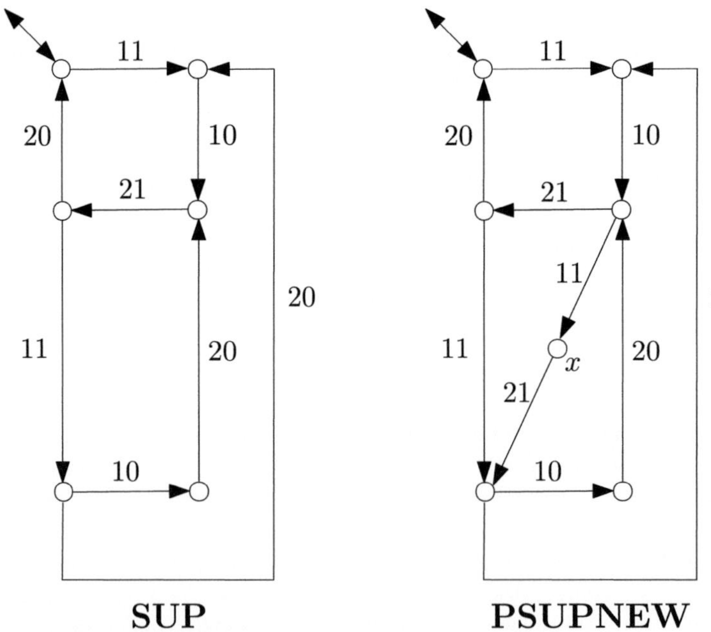

Fig. 3.6 In **PSUPNEW** event 21 preempts 10 at x

Exercise 2: Forcing in Small Factory

In Small Factory, assume that events α_2, μ_2 can be forced to preempt events α_1, β_1 with a new timeout event τ. Use *TCT* **force** to create the corresponding plant model, say **FFACT**. For the same specifications as before (now additionally self-looped with τ) compute the supervisor, say **FFACTSUP**, then project out τ to obtain, say, **NEWFACTSUP**. Show that

$$L_m(\mathbf{FACTSUP}) \subsetneqq L_m(\mathbf{NEWFACTSUP})$$

and illustrate with sample strings.

Exercise 3: Forced events

Provide examples of modeling intuitively 'forced' events as just described, carrying through a complete control design. For instance, consider a water supply tank for a country cottage, which is emptied incrementally by random household events, and filled by a pump. The pump is to be switched on when the water drops below a defined lower level and switched off when the water rises to a defined upper level. Ideally a good design would ensure that the tank is never emptied by normal household usage.

Example 4: Guaranteed reachability

Consider the DES $\mathbf{G} = (Q, \Sigma, \delta, q_0, Q_m)$, as usual with Q finite and $\Sigma = \Sigma_c \dot{\cup} \Sigma_u$. Let $q\# \in Q$ be a designated *current* state (not necessarily q_0) and $q*$ be a designated *target* state (not necessarily in Q_m). We wish to determine whether or not $q*$ is *guaranteed reachable* from $q\#$ in some upper-bounded number of steps (written $GR(q\#, q*)$). In other words, starting from $q\#$, $\mathbf{G}$ can be controlled in such a way that for some N, $\mathbf{G}$ will necessarily visit $q*$ in no more than N steps, namely execute a string s with $|s| \leq N$, such that $\delta(q\#, s) = q*$. To formalize this requirement we use a standard dynamic programming construction as follows.

For simplicity assume first that the only means employed for control is the usual SCT 'technology' of disablement of controllable events.

For $n = 0, 1, 2, ...$ define the state subset

$$Q_n := \{q \in Q | q* \text{ is } GR \text{ from } q \text{ in } n \text{ or fewer transitions}\}$$

By definition $Q_0 := \{q*\}$. Then also $q* \in Q_1$. If $q \neq q*$, clearly $q \in Q_1$ iff there exists $\sigma \in \Sigma$ such that $\delta(q, \sigma) = q*$ and, for all $\sigma' \in \Sigma$ such that $\delta(q, \sigma')!$ and $\delta(q, \sigma') \neq q*$, we have that σ' can be disabled, i.e. $\sigma' \in \Sigma_c$. Inductively,

$$\begin{aligned} Q_n = Q_{n-1} \cup \{q \in Q | q \notin Q_{n-1} \ \& \ (\exists \sigma \in \Sigma)\delta(q, \sigma) \in Q_{n-1} \\ \& \ (\forall \sigma' \in \Sigma) \big(\delta(q, \sigma')! \ \& \ \delta(q, \sigma') \notin Q_{n-1} \Rightarrow \sigma' \in \Sigma_c\big)\} \end{aligned}$$

Define $Q_{GR} := \cup\{Q_n | n \geq 0\}$, where of course the union is finitely convergent at $n = |Q|$. Obviously $q*$ is guaranteed-reachable from $q\#$, i.e. $GR(q\#, q*)$, iff $q\# \in Q_{GR}$, the required upper bound being $N = |Q|$.

Denote this computation by the proposition $GR(\mathbf{G}, q\#, q*) = true/false$, with evaluation *true* iff $GR(q\#, q*)$ for the DES $\mathbf{G}$.

A stronger result can be obtained if, in addition to controllable event disablement, event forcing is optionally employed as well. For this we need only define

$$Q_n = Q_{n-1} \cup \{q \in Q | q \notin Q_{n-1} \;\; \& \;\; (\exists \sigma \in \Sigma)\delta(q, \sigma) \in Q_{n-1} \\ \& \;\; \sigma \text{ is locally forcible at } q\}$$

In this definition, we tacitly 'localize' the original definition of 'forcible event' to allow the property of forcibility of a given event to depend both on the state at which that event is considered, and also on the subset of competing events defined there. In fact, it is reasonable to suppose that the 'technology' available to 'force' a given event may vary from one state to another and may depend on which competing events at that (source) state need to be preempted. It is then convenient to include in the definition of 'locally forcible' the simpler case first considered, when all competing events σ', such that $\delta(q, \sigma') \notin Q_{n-1}$, happen to be controllable.

As an example, consider the following.

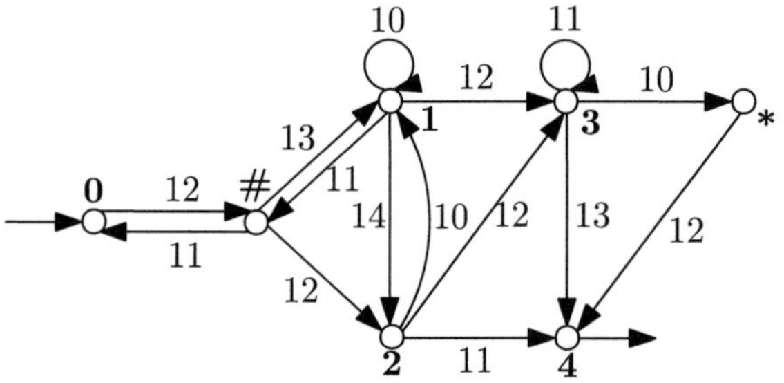

G

If standard disablement of controllable events is the only control technology available, the algorithm yields $Q_0 = \{*\}$, $Q_1 = \{*, 3\}$, $Q_{GR} = \{*, 3\}$, and therefore $GR(\mathbf{G}, \#, *) = false$. If, however, event 12 is locally forcible at state 1 (to preempt events 10 and 14) then $Q_2 = \{*, 3, 1\}$, $Q_3 = \{*, 3, 1, 2\}$, $Q_4 = \{*, 3, 1, 2, \#\}$, and $GR(\mathbf{G}, \#, *) = true$.

As a simple exercise, it will be clear how to extend the GR algorithm to compute, say, $GR(\mathbf{G}, \#, Q*)$ for an arbitrary target subset $Q* \subseteq Q$.

3.9 Supervisory Control for Reconfiguration

In the operation of any system it may be necessary, from time to time, to switch from the current 'configuration' or operating mode to some other, either because of a breakdown or emergency, or simply to replace a faulty or obsolete component. Naturally it is desirable to do so in as economical and least disruptive a manner as possible. In this section we formalize this objective as a supervisory control problem for DES. While our approach is somewhat conceptual, it may be useful as a schematic guide to design in practice.

Broadly speaking, distinct operating modes may differ according to the components which make up the plant to be controlled, or to the specifications on plant behavior which the supervisor is required to enforce, or to a combination of these features. Thus in practice reconfiguration is highly case-dependent. For convenience we shall separate the two cases mentioned, leaving the reader free to construct his own scenario according to need.

For the first case, assume that the totality of available plant components is known and fixed in advance, say $\mathbf{M}_1, \mathbf{M}_2, \ldots, \mathbf{M}_n$, where the $\mathbf{M}_i$ might be thought of as individual 'machines'. We take a plant configuration to consist of a specified subset of these components combined by synchronous product to form the ith 'mode' $Mode_i$, represented by the DES $\mathbf{G}_i$, with $i = 1, \ldots, p$:

$$\mathbf{G}_i = \mathbf{M}_{i_1}||\mathbf{M}_{i_2}|| \cdots ||\mathbf{M}_{i_{l_i}}$$

where $\{i_1, \ldots, i_{l_i}\} \subseteq \{1, \ldots, n\}$. Combining the various modes $\mathbf{G}_i$ we take for the total plant DES to be controlled

$$\mathbf{G} := \mathbf{G}_1|| \cdots ||\mathbf{G}_p$$

At any given time the system is to operate in just one plant mode, corresponding, say, to the component $\mathbf{G}_i$ of $\mathbf{G}$. A reconfiguration amounts to a switch from $\mathbf{G}_i$ to some $\mathbf{G}_j$, say by a new 'instantaneous' controllable event we call the *reconfiguration event* (RE) σ_{ij}. If Σ_i is the alphabet of events appearing in $\mathbf{G}_i$ then we may bring in the *reconfiguration specification* (RS) over the alphabet

$$\Sigma_{ij} := \Sigma_i \cup \Sigma_j \cup \{\sigma_{ij}\}$$

represented by the DES

$$\mathbf{RECON}_{ij} = (\{0, 1\}, \Sigma_{ij}, \delta_{ij}, 0, \{0, 1\})$$

$$\delta_{ij}(0, \sigma) : = \begin{cases} 0 & \text{if } \sigma \in \Sigma_i \\ 1 & \text{if } \sigma = \sigma_{ij} \end{cases}$$

$$\delta_{ij}(1, \sigma) : = 1, \quad \sigma \in \Sigma_j$$

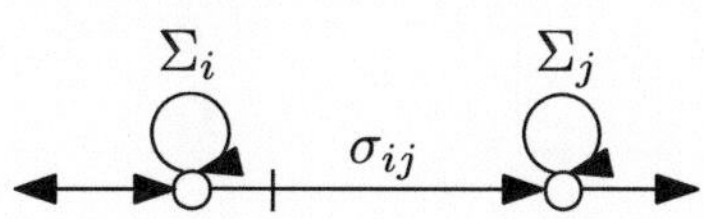

$\mathbf{RECON}_{ij}$

Note that, as $\mathbf{G}_i$ and $\mathbf{G}_j$ may include components $\mathbf{M}(\cdot)$ in common, the two alphabets Σ_i and Σ_j need not be disjoint, but σ_{ij} does not belong to either.

To reduce clutter we assume henceforth that $p = 2$ and consider switching via σ_{12} from $Mode_1$ to $Mode_2$ represented by $\mathbf{G}_1, \mathbf{G}_2$, respectively. For this we take the *reconfiguration plant* (**RG**) to be

$$\mathbf{RG} := \mathbf{G}_1||\mathbf{G}_2||\mathbf{RECON}_{12}$$

In the present case we assume that both modes satisfy the same specification as represented, say, by the DES **RSPEC**.

We may now compute the optimal reconfiguration supervisor

$$\mathbf{RSUPER}_{ij} = \mathbf{RSUPER}_{12} =: \mathbf{RSUPER}$$

given by

$$\mathbf{RSUPER} := \mathbf{supcon}(\mathbf{RG}, \mathbf{RSPEC})$$

RSUPER represents system operation initially in $Mode_1$, then optionally switching into $Mode_2$, where operation will remain. The states of **RSUPER** at which the mode switch is possible, say the subset X_{12} of the state set X of **RSUPER**, are those at which the controllable event σ_{12} is enabled. In an application, reconfiguration might be demanded at any reachable state $q\#$ of $\mathbf{G}_1$, depending on the need of the system manager. It is then required to construct a path in $Mode_1$ from $q\#$ to a state (perhaps the 'nearest') in X_{12}, a path that is feasible in terms of the disablement and forcing technology available to the manager. This is simply the *guaranteed reachability* (GR) problem already considered in Example 3.8.4.

Example 1: Reconfiguration by change of plant component

Consider a version of Small Factory (Example 3.3.19), say **FACT1**, consisting of an 'advanced' machine **MACH1** and a 'primitive' machine **MACH2**, as shown below.

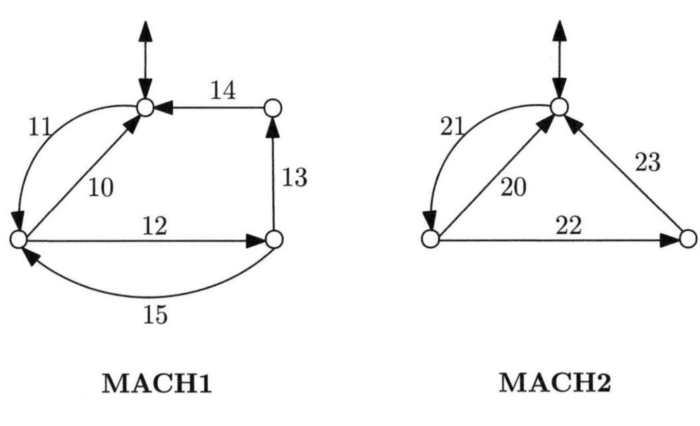

Thus **FACT1 = sync(MACH1,MACH2)** (12,34). Now consider reconfiguration of **FACT1** to **FACT2**, which differs from **FACT1** only in that the primitive machine **MACH2** is replaced by its advanced version **MACH3** similar to **MACH1**, as shown below.

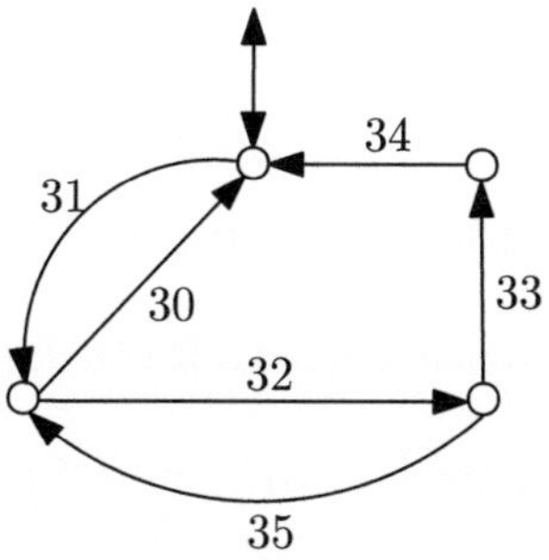

MACH3

Thus **FACT2 = sync(MACH1,MACH3)** (16,48). Let the RE $\sigma_{12} =:<91>$;[4] the RS, say **RECON**, is then as displayed below.

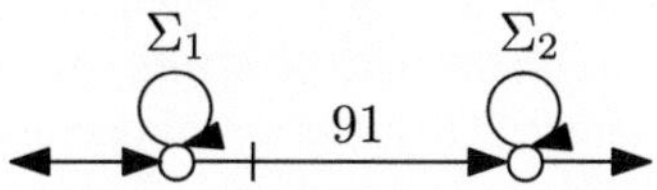

RECON

We now compute the reconfiguration plant

$$\begin{aligned}\mathbf{RG} &= \mathbf{sync(FACT1, FACT2, RECON)} \\ &= \mathbf{sync(MACH1, MACH2, MACH3, RECON)}\ (60, 190)\end{aligned}$$

Thus **RG** incorporates all available plant components plus a description of the specific reconfiguration transition. The latter places no restriction on how the transition may take place; this will be determined later. We need **ALLRG = allevents(RG)** to represent the corresponding full alphabet.

In this example the specifications are independent of configuration. We take a buffer with 3 slots (**RBUF**), and (as in Example 3.3.19) priority logic (**RBR**) for breakdown and repair as displayed below. Note that these must be compatible with both the initial and the final configuration.

Now we compute

$$\mathbf{RSPEC} = \mathbf{sync(RBUF, RBR, ALLRG)}\ (8, 110)$$

[4]It will be convenient to designate an event by enclosure in < >.

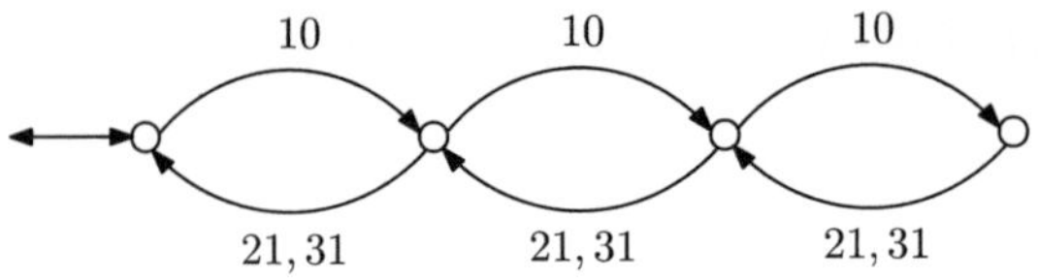

RBUF

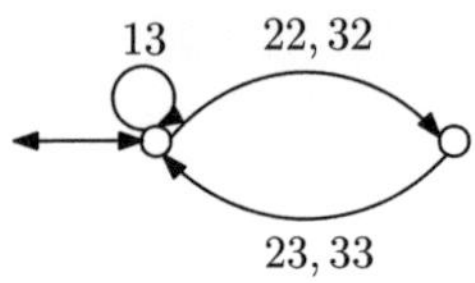

RBR

and finally the optimal reconfiguration supervisor

$$\mathbf{RSUPER} = \mathbf{supcon}(\mathbf{RG}, \mathbf{RSPEC})\ (91, 242)$$

The structure of **RSUPER** displays exactly the states at which the RE <91> is permitted to occur (when enabled by the manager). As for any result of **supcon**, **RSUPER** is controllable, nonblocking, and maximally permissive. It is essentially the nonblocking requirement that determines where the RE can be executed, in order that any plant component that is replaced (in this example, **MACH2**) is left in a marker state (for **MACH2**, its initial state). By inspection of the transition table of **RSUPER** the reader can verify that in fact <91> is executable exactly at the 15 states where **MACH2** is idle. By use only of event disablement, such a state can be reached from an arbitrary current state of $Mode_1$ by disabling <11>, allowing **MACH2** to reach its idle state [0][5] (by successful operation or via breakdown and repair) and then disabling <21>. At that point <91> can be executed to achieve nondisruptively the replacement of **MACH2** by **MACH3**.

We briefly discuss the second case distinguished above, where the plant is kept fixed but reconfiguration entails a change in operational specification. The pattern is similar to the first case, but with evident modifications. Again restricting to two modes, say $Mode_1$ and $Mode_2$, we assume that the plant **G** is fixed, but the modes are distinguished by specifications **SPEC1** and **SPEC2** respectively, over alphabets Σ_1, Σ_2. We model **G** to accommodate both **SPEC1** and **SPEC2**. From this point on the procedure is as before. The following example illustrates the steps.

Example 2: Reconfiguration under change of specification

Again we consider a version of Small Factory, with standard machines **MACH1**, **MACH2**, adapted to work with either mode specification as needed. $Mode_1$, $Mode_2$ are distinguished by buffer specifications according to the buffer capacities 3 (for **BUF1**, in $Mode_1$), or 1 (**BUF2**, in $Mode_2$). Thus reconfiguration from $Mode_1$ to $Mode_2$ must prevent both blocking (due to a workpiece left behind in **BUF1**) and buffer overflow. The plant and specification components are shown below.

[5]It will be convenient to designate a state by enclosure in [ ].

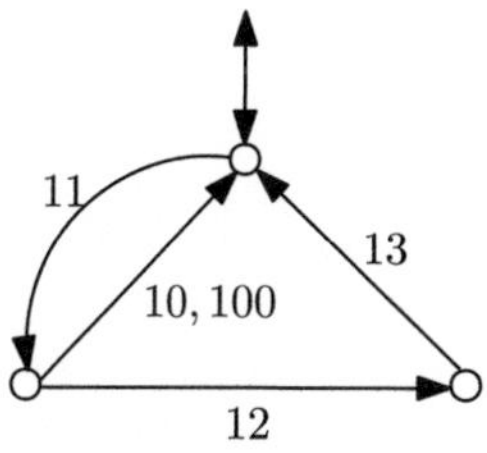

MACH1

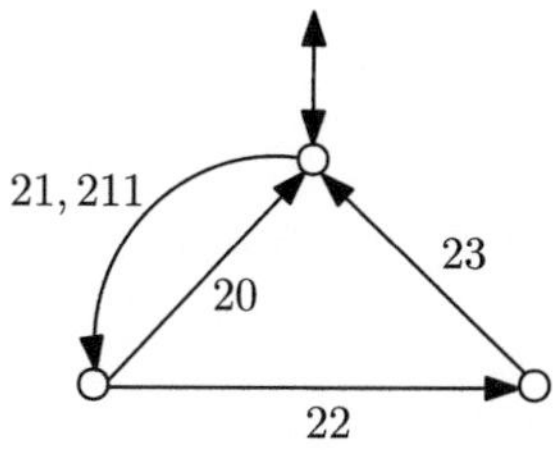

MACH2

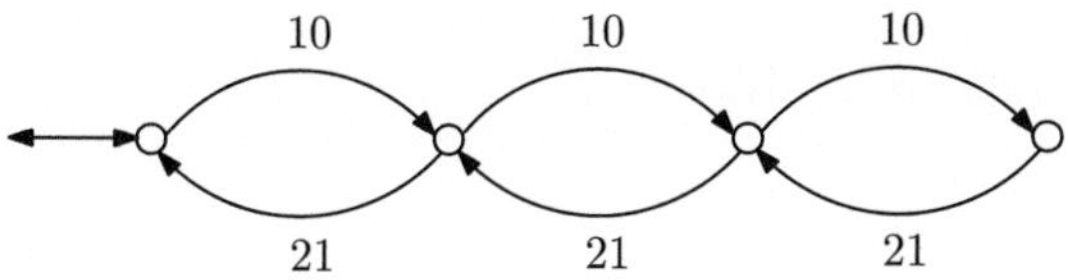

BUF1

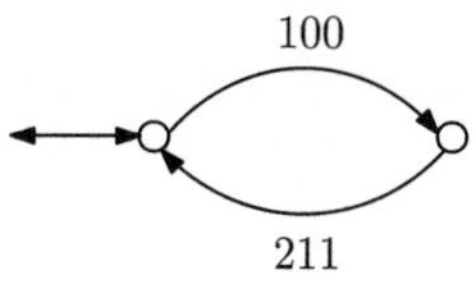

BUF2

Let **G** = **sync(MACH1,MACH2)**. As in Example 1 we model the RS **RECON** on the state set {0, 1} with both states marked. The alphabets Σ_1, Σ_2 for $Mode_1$, $Mode_2$ respectively, are

$$\Sigma_1 = \{10, 11, 12, 13, 20, 21, 22, 23\}, \quad \Sigma_2 = \{100, 11, 12, 13, 20, 211, 22, 23\}$$

In **RECON**, state [0] is self-looped with Σ_1, [1] with Σ_2, with the RE <91> as transition between them, i.e. [0, 91, 1].

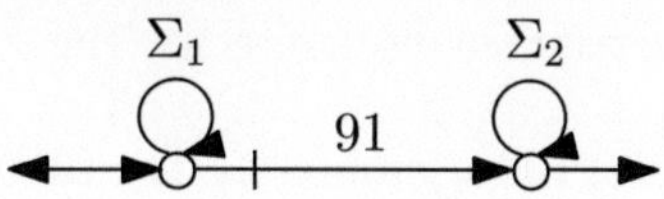

RECON

Proceeding as in Example 1,
RG = **sync(MACH1,MACH2,RECON)** (18,57)
ALLRG = **allevents(RG)** (1,10)
RSPEC = **sync(BUF1,BUF2,ALLRG)** (8,76)
RSUPER = **supcon(RG,RSPEC)** (42,107)

In **RSUPER**, <91> is enabled at exactly 8 states, namely where **BUF1** is empty, so blocking and overflow are both prevented, and the switch to $Mode_2$ is non-disruptive. Analogous to Example 1, reconfiguration from an arbitrary state of $Mode_1$ is achieved by disabling <11>, then allowing **MACH2** to empty **BUF1**, if necessary, by one or more successive cycles of successful operation or breakdown/repair, and finally executing the RE <91>. If desired, a shorter path to reconfiguration may sometimes be achieved by suitable forcing, as described in the guaranteed reachability algorithm of Example 3.8.4.

Exercise 3: Illustrate this remark.

Exercise 4: For each of Examples 1 and 2 above, investigate reconfiguration in the reverse direction, from $Mode_2$ to $Mode_1$. Explain why 'naive' modeling of **RECON** in both directions (say with two RE <91> and <93>) could lead to blocking.

Exercise 5: Explore reconfiguration for a system of your own design. An instance could be based on the AGV system of Sect. 4.7 below, for instance to achieve re-routing in the event of a transportation breakdown.

3.10 Mutual Exclusion

Assume we are given DES

$$\mathbf{G}_1 = (Q_1, \Sigma_1, \delta_1, q_{10}, Q_{1m})$$
$$\mathbf{G}_2 = (Q_2, \Sigma_2, \delta_2, q_{20}, Q_{2m})$$

with $\Sigma_1 \cap \Sigma_2 = \emptyset$. We may wish to control $\mathbf{G}_1$, $\mathbf{G}_2$ in such a way that designated state pairs $q_1 \in Q_1$, $q_2 \in Q_2$ are never occupied simultaneously. In such problems '$\mathbf{G}_i$ in q_i' typically means '$\mathbf{G}_i$ is using a single shared resource', for instance when two readers share a single textbook or two AGVs a single section of track.

Because such a constraint can be awkward to express linguistically, *TCT* provides a procedure to compute the required result directly. Thus

$$\mathbf{MXSPEC} = \mathbf{mutex}(\mathbf{G}_1, \mathbf{G}_2, \mathbf{LIST})$$

where

$$\mathbf{LIST} = [(q_1^{(1)}, q_2^{(1)}), ..., (q_1^{(k)}, q_2^{(k)})]$$

with $(q_1^{(i)}, q_2^{(i)}) \in Q_1 \times Q_2$, is the user's list of mutually exclusive state pairs. **MXSPEC** is reachable and controllable with respect to $\mathbf{G} = \mathbf{shuffle}(\mathbf{G}_1, \mathbf{G}_2)$ but need not be coreachable. If not, it may serve as a new specification for the plant **G**, and the final result computed using **supcon** in the usual way.

Exercise 1: In **FACT** (Example 3.3.19) suppose power is limited, so at most one of **MACH1**, **MACH2** may be working at once. Compute a suitable supervisor. Repeat the exercise using the constraint that at most one machine at a time may be broken down.

Exercise 2: Cat and Mouse

A cat and mouse share a maze of 5 interconnected chambers. The chambers are numbered 0,1,2,3,4 for the cat, but respectively 3,2,4,1,0 for the mouse. Adjoining chambers may be connected by one-way gates, each for the exclusive use of either the cat or the mouse. An event is a transition by either the cat or the mouse from one chamber to another via an appropriate gate; the animals never execute transitions simultaneously. Some gates are always open, corresponding to uncontrollable events; while others may be opened or closed by an external supervisory control, so passage through them is a controllable event. The cat and the mouse are initially located in their 'home' chambers, numbered 0. *TCT* models for the cat and mouse are printed below.

It is required to control the roamings of cat and mouse in such a way that (i) they never occupy the same chamber simultaneously, (ii) they can always return to their respective home chambers, and (iii) subject to the latter constraints they enjoy maximal freedom of movement.

```
CAT # states: 5    state set: 0...4    initial state: 0
marker states: 0
vocal states: none
# transitions: 8
transitions:
[0,201,1] [1,205,2] [1,207,3] [2,200,3]
[2,203,0] [3,200,2] [3,211,4] [4,209,1]
CAT printed
MOUSE # states: 5   state set: 0...4   initial state: 0
marker states: 0
vocal states: none
# transitions: 6
transitions:
[0,101,1] [1,103,2] [2,105,0] [2,107,3] [3,111,4] [4,109,2]
MOUSE printed
```

Exercise 3: Forbidden states

The 'forbidden states' problem for a DES **G** with state set Q is simply to control **G** in such a way that a given state subset $F \subseteq Q$ is never entered. Show that this problem can be treated as mutual exclusion, taking

$$\mathbf{MXSPEC} = \mathbf{mutex}(\mathbf{G}, \mathbf{ALL}, \mathbf{LIST})$$

where $\mathbf{ALL} = \mathbf{allevents}(\mathbf{G})$, and $\mathbf{LIST} = [(q, 0)|q \in F]$.

3.11 Supervisory Control by Relabeling

In practice the components of a plant may be divided into groups, according to different roles, functions, or capabilities. For example, machines in a manufacturing cell are grouped to process different types of workpieces; AGVs in a warehouse are grouped to transport items of distinct sizes and weights. In the same group, components often have similar or even identical state transition structures, for the purpose of productivity or increasing redundancy against failure. Exploiting the component similarity may effectively reduce the complexity of supervisory control design; for this we employ *event relabeling*.

Given a DES **G** defined on $\Sigma(=\Sigma_c\dot{\cup}\Sigma_u)$ and a new event set $T(=T_c\dot{\cup}T_u)$, let the *relabeling map* $R:\Sigma^*\to T^*$ be defined according to

$$\begin{aligned} R(\epsilon) &= \epsilon \\ R(\sigma) &= \tau \in T_c \quad \text{if } \sigma \in \Sigma_c \\ R(\sigma) &= \tau \in T_u \quad \text{if } \sigma \in \Sigma_u \\ R(s\sigma) &= R(s)R(\sigma) \quad s \in \Sigma^*, \sigma \in \Sigma \end{aligned}$$

Namely the empty string is relabeled as the empty string, a controllable (resp. uncontrollable) event as a new controllable (resp. uncontrollable) event, and the relabeling map is catenative. In *TCT*, R is implemented by **relabel**: given **G**, Σ and T,

$$\mathbf{RG} = \mathbf{relabel}(\mathbf{G}, [\{(\sigma,\tau)|\sigma\in\Sigma, \tau\in T, R(\sigma)=\tau\}])$$

is a (minimal-state) DES with

$$L_m(\mathbf{RG}) = RL_m(\mathbf{G}), \qquad L(\mathbf{RG}) = RL(\mathbf{G})$$

Now assume that the totality of available plant components is known and fixed in advance, say $\mathbf{M}_1, \mathbf{M}_2, \ldots, \mathbf{M}_n$, where the $\mathbf{M}_i$ might be thought of as individual 'machines'. These components are *partitioned* into p (disjoint) groups; the ith group, represented by the DES $\mathbf{G}_i$, with $i = 1, \ldots, p$ is given by:

$$\mathbf{G}_i = \mathbf{M}_{i_1}||\mathbf{M}_{i_2}||\cdots||\mathbf{M}_{i_{l_i}}$$

where $\{i_1, \ldots, i_{l_i}\} \subseteq \{1, \ldots, n\}$. Components in the same group have similar state transition structures; we capture this similarity using a suitable relabeling function R that maps events representing the same task to the same (new) event. For simplicity we shall consider a relabeling function R such that (i) for each group i there is a DES $\mathbf{H}_i$ such that

$$\mathbf{relabel}(\mathbf{M}_{i_1}) = \cdots = \mathbf{relabel}(\mathbf{M}_{i_{l_i}}) = \mathbf{H}_i$$

and (ii) for all $i \neq j \in \{1, \ldots, p\}$, $\mathbf{H}_i$ and $\mathbf{H}_j$ have pairwise disjoint event sets (i.e. different groups have no new event in common).

Using R we relabel each group $\mathbf{G}_i$ to obtain $\mathbf{RG}_i = \mathbf{relabel}(\mathbf{G}_i)$. Thus the total relabeled plant DES is

$$\mathbf{RG} := \mathbf{RG}_1 || \cdots || \mathbf{RG}_p$$

Let **E** be a control specification. Relabeling **E** by the same map R yields $\mathbf{RE} = \mathbf{relabel}(\mathbf{E})$, the relabeled specification. With **RG** and **RE**, we compute the optimal relabeled supervisor

$$\mathbf{RSUPER} = \mathbf{supcon}(\mathbf{RG}, \mathbf{RE})$$

RSUPER can be used to control the original plant **G**, where

$$\mathbf{G} := \mathbf{G}_1 || \cdots || \mathbf{G}_p$$

as follows. After a string s has occurred in **G**, **RSUPER** 'observes' the corresponding string $R(s)$ in **RG**. Say **RSUPER** decides to enable (resp. disable) an event τ following $R(s)$. Then in **G** following s, all events σ with $R(\sigma) = \tau$ will be enabled (resp. disabled). The criterion of validity is just that the controlled behavior **RSUPER** is 'equivalent' to the optimal supervisor for the original system

$$\mathbf{SUPER} = \mathbf{supcon}(\mathbf{G}, \mathbf{E})$$

For this we verify the following

$$\mathbf{isomorph}(\mathbf{RSUPER}, \mathbf{relabel}(\mathbf{SUPER})) = \textit{true/false}$$

Example 1: Small Factory by relabeling

Consider a version of Small Factory (Example 3.3.19) consisting of two input machines **MACH1**, **MACH2** and two output machines **MACH3**, **MACH4** as arranged below. We take a buffer with 2 slots and consider only the buffer specification **BUF**.

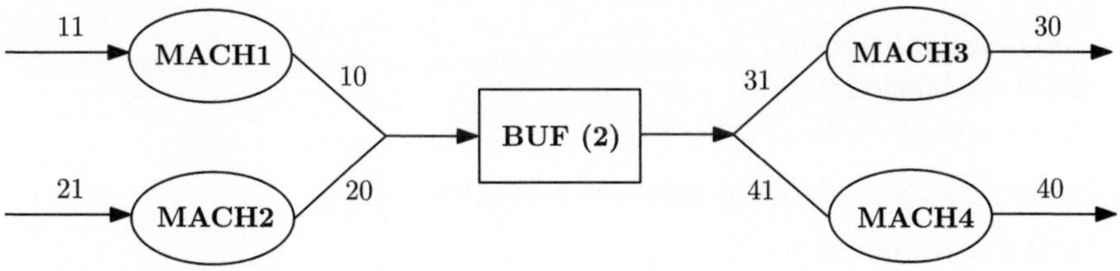

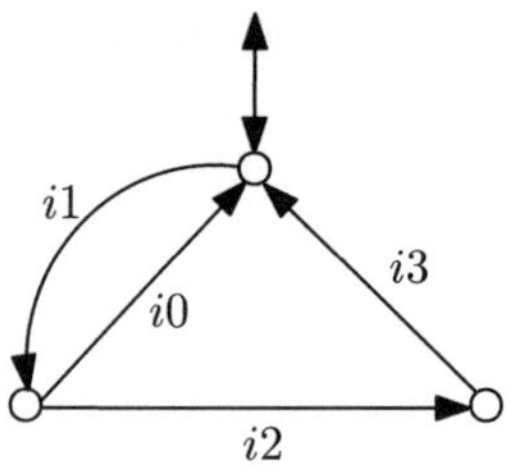

MACHi $(i = 1, 2)$

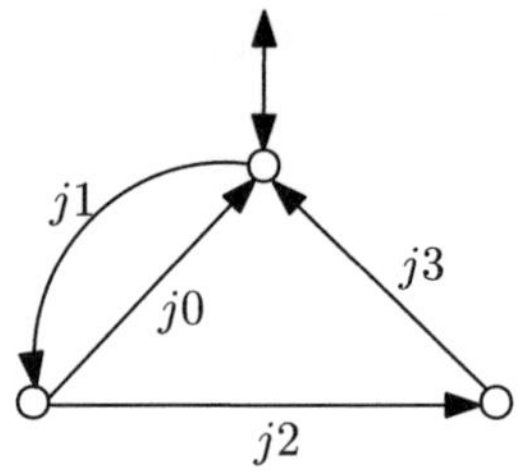

MACHj $(j = 3, 4)$

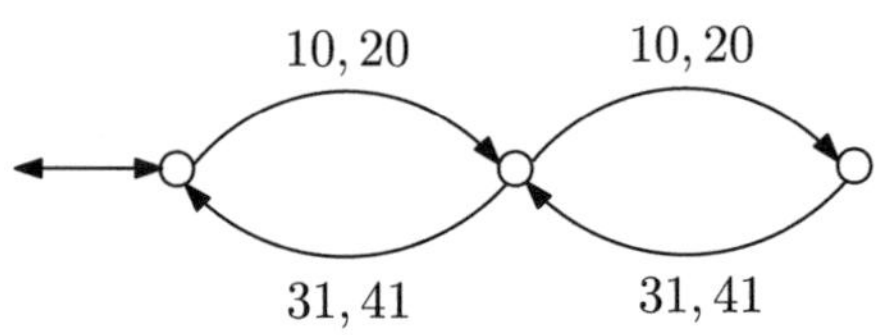

BUF

The optimal supervisor for this system is computed as follows:

G = sync(MACH1,MACH2,MACH3,MACH4) (81,432)
ALLG = allevents(G) (1,16)
SPEC = sync(BUF,ALLG) (3,44)
SUPER = supcon(G,SPEC) (180,822)

Now group **MACH1**, **MACH2** as **G1**, and **MACH3**, **MACH4** as **G2**:

G1 = sync(MACH1,MACH2) (9,24)
G2 = sync(MACH3,MACH4) (9,24)

Let the relabeling map R be given by

$$R(i0) = 0, \quad R(i1) = 1, \quad R(i2) = 2, \quad R(i3) = 3, \quad \text{for } i = 1, 2$$
$$R(j0) = 4, \quad R(j1) = 5, \quad R(j2) = 6, \quad R(j3) = 7, \quad \text{for } j = 3, 4$$

Then the relabeled plant is computed by

RG1 = relabel(G1) (6,12)
RG2 = relabel(G2) (6,12)
RG = sync(RG1,RG2) (36,144)

Also compute the relabeled specification as follows

RBUF = relabel(BUF) (3,4)
ALLRG = allevents(RG) (1,8)
RSPEC = sync(RBUF,ALLRG) (3,22)

Hence the optimal relabeled supervisor is

RSUPER = **supcon**(**RG**, **RSPEC**) (78,270)

RSUPER is defined on the event set $T := \{0, 1, 2, 3, 4, 5, 6, 7\}$. When the buffer is empty, **RSUPER** disables event 5, which corresponds to disabling events 31, 41 in the original plant **G**. Similarly when the buffer is full, **RSUPER** disables event 1 corresponding to 11, 21. The desired equivalence between **RSUPER** and the original optimal supervisor is confirmed:

isomorph(**RSUPER**, **relabel**(**SUPER**)) = *true*

We note that by use of relabeling the supervisor state size has been reduced from 180 to 78.

Exercise 2: Investigate supervisory control by relabeling for the Transfer Line system of Sect. 4.6 (below) by considering multiple machines and test units.

Exercise 3: Investigate supervisory control by relabeling for the Guideway system of Sect. 6.6 (below) by considering multiple trains.

Exercise 4: For a relabeling function $R : \Sigma^* \to T^*$, define its *inverse image function* $R^{-1} : Pwr(T^*) \to Pwr(\Sigma^*)$ by

$$R^{-1}(H) = \{s \in \Sigma^* | R(s) \in H\}, \quad H \in Pwr(T^*)$$

With the notation and assumptions in this section, show that if the specification **E** satisfies $L_m(\mathbf{E}) = R^{-1}R(L_m(\mathbf{E}))$, then

isomorph(**RSUPER**, **relabel**(**SUPER**)) = *true*

Let **XRSUPER** represent the language $R^{-1}(L_m(\mathbf{RSUPER})) \cap L_m(\mathbf{G})$. Show in addition that

isomorph(**SUPER**, **XRSUPER**) = *true*

3.12 Supervisor Reduction

As indicated in Sect. 3.7 for Small Factory, the 'standard' supervisor

SUPER = **supcon**(**PLANT**, **SPEC**)

computed by **supcon** (and representing the full optimal controlled behavior) can be much larger in state size than is actually required for the same control action. This is because the controlled behavior incorporates all the *a priori* transitional constraints embodied in the plant itself, as well as any additional constraints required by control action to enforce the specifications. The problem of finding a simplified proper supervisor, say **MINSUP**, equivalent in control action but of minimal state size, is of evident practical interest. Unfortunately, it is NP-hard. A reduction procedure called **supreduce** has been developed, based on heuristic search for a suitable congruence on the state set of **SUPER**. **Supreduce** is of polynomial complexity in the state sizes of **PLANT** and **SPEC**. While of course it cannot guarantee a simplified supervisor of minimal size, **supreduce** will often find a greatly reduced supervisor, say **SIMSUP**, and can also provide a lower bound on the size of **MINSUP**. **SIMSUP** is actually minimal if its size matches this bound. Some results found by **supreduce** are reported in Sect. 4.7 and for the examples of Appendix 8.1.

In *TCT* the syntax for computing a simplified supervisor **SIMSUP** from **SUPER** (as above) is:

$$\mathbf{SUPDAT} = \mathbf{condat}(\mathbf{PLANT}, \mathbf{SUPER})$$
$$\mathbf{SIMSUP} = \mathbf{supreduce}(\mathbf{PLANT}, \mathbf{SUPER}, \mathbf{SUPDAT})$$

Under **SIMSUP** as supervisor, the controlled behavior of **PLANT** can be represented by, say,

$$\mathbf{TEST} = \mathbf{meet}(\mathbf{PLANT}, \mathbf{SIMSUP})$$

A check on correctness is provided by

$$\mathbf{isomorph}(\mathbf{TEST}, \mathbf{SUPER}) = \mathit{true}$$

provided both **TEST** and **SUPER** are state-minimal generators, as will often be the case. Otherwise, one should verify

$$\mathbf{isomorph}(\mathbf{MTEST}, \mathbf{MSUPER}) = \mathit{true}$$

where **MTEST** = **minstate(TEST)** and **MSUPER** = **minstate(SUPER)**. When correctness is satisfied, we say that **SIMSUP** is *control-equivalent* to **SUPER** with respect to **PLANT**.

Exercise 1: With the help of **supreduce**, confirm that the simplified supervisor **SIMFTSUP** for Small Factory (Sect. 3.7) is actually minimal.

Exercise 2: Apply **supreduce** to your results (say) **SUPER** of Exercises 3.7.2, 3.7.4, 3.7.7, 3.7.8, and 3.7.15.

3.13 Notes

Supervisory control theory in the sense of this chapter originates with the doctoral thesis of P.J. Ramadge (1983) and related papers Ramadge and Wonham (1987), Wonham and Ramadge (1987), Ramadge and Wonham (1981, 1982a, b, 1984), Wonham and Ramadge (1984).

The synchronous product of languages is a convenient packaging of the more primitive language operations Meet and Shuffle. For the latter, the corresponding recognizers or generators are instances of the product construction for automata, as defined in Eilenberg (1974); from that book we also borrow the term 'trim'. A counterpart definition for the synchronous product of 'processes' was introduced by Hoare (1985) under the name 'concurrency operator' or 'parallel composition', and was further discussed by Milner (1989) as a product of 'agents' under the name 'conjunction'.

The kanban Example 3.3.17 is adapted from Viswanadham and Narahari (1992), pp. 514–524.

The extension in Exercise 3.3.8 is due to R. Su (2004, p. 16, Corollary 2.1).

Prioritized synchronous product (Exercise 3.3.15) was introduced by Heymann (1990).

Exercise 3.3.18 is based on Wong (1998) and Exercise 3.7.3 on Cassandras (1993) (Example 2.17, p. 108). Exercise 3.7.4 was suggested by Robin Qiu.

The 'guaranteed reachability' algorithm of Example 3.8.4 is a conventional application of dynamic programming, in the context of SCT. For a precursor see Brave and Heymann (1990). The latter approach was adopted in their treatment of reconfiguration by Nooruldeen and Schmidt (2015). Our version of reconfiguration in Sect. 3.9 follows similar lines, but without appeal to the terminology of stability theory.

The relabeling technique of Sect. 3.11 follows Jiao et al. (2017).

The supervisor reduction method of Sect. 3.12 was developed by Su and Wonham (2001, 2004), as an evolution of work by Vaz and Wonham (1985, 1986).

Computational methods for supervisor synthesis for DES of industrial size is an active area of current research. The reader is referred to the published theses of Germundsson (1995) and Gunnarsson (1997), as well as subsequent work by Zhang (2001), Zhang and Wonham (2002), Leduc (2002), Leduc et al. (2001a, b, c, 2005a, b), Ma (2004), and Ma and Wonham (2003, 2005a, b). An interesting large-scale industrial design is reported by Theunissen et al. (2010).

Appendix 3.1

EVENT CODING FOR SMALL FACTORY

TCT:	10	11	12	13	20	21	22	23
TEXT:	β_1	α_1	λ_1	μ_1	β_2	α_2	λ_2	μ_2

FACTSUP # states: 12 state set: 0...11 initial state: 0
marker states: 0
transitions: 24
transition table:

[0,11,1] [1,12,2] [1,10,3] [2,13,0] [3,21,4]
[4,20,0] [4,11,5] [4,22,11] [5,20,1] [5,10,6]
[5,12,8] [5,22,10] [6,20,3] [6,22,7] [7,23,3]
[8,20,2] [8,13,4] [8,22,9] [9,23,2] [10,23,1]
[10,10,7] [10,12,9] [11,23,0] [11,11,10]
FACTSUP printed.

FACTSUP
Control Data are displayed by listing the supervisor states where disabling occurs, together with the events that must be disabled there.

Control Data:
0: 21 1: 21
2: 21 3: 11
6: 11 7: 11
9: 13

FACTSUP printed.
SIMFTSUP # states: 3 state set: 0...2 initial state: 0
marker states: 0
transitions: 16
transition table:
[0,13,0] [0,11,0] [0,12,0] [0,20,0] [0,10,1]
[0,22,2] [1,21,0] [1,23,1] [1,12,1] [1,20,1]
[1,22,1] [2,23,0] [2,10,1] [2,11,2] [2,12,2] [2,20,2]
SIMFTSUP printed.

SIMFTSUP
Control Data are displayed by listing the supervisor states where disabling occurs, together with the events that must be disabled there.

Control Data:
0: 21 1: 11
2: 13
SIMFTSUP printed.

References

Brave Y, Heymann M (1990) Stabilization of discrete-event processes. Int J Control 51:1101–1117

Cassandras CG (1993) Discrete event systems. Irwin

Eilenberg S (1974) Automata, languages, and machines, vol A. Academic Press, Orlando

Germundsson R (1995) Symbolic systems - theory, computation and applications. PhD thesis no. 389, Department of Electrical Engineering, Linköping University

Gunnarsson J (1997) On modeling of discrete event dynamic systems, using symbolic algebraic methods. Thesis no. 502, Division of Automatic Control, Lund University

Heymann M (1990) Concurrency and discrete event control. IEEE Control Syst 10(4):103–112

Hoare, CAR (1985) Communicating sequential processes. Prentice-Hall

Jiao T, Gan Y, Xiao G, Wonham WM (2017) Exploiting symmetry of state tree structures for discrete-event systems with parallel components. Int J Control 90(8):1639–1651
Leduc RJ (2002) Hierarchical interface-based supervisory control. PhD thesis, Department of Electrical and Computer Engineering, University of Toronto
Leduc RJ, Brandin BA, Wonham WM, Lawford M (2001a) Hierarchical interface-based supervisory control: serial case. In: Proceedings of the fortieth IEEE conference on decision and control, Orlando, pp 4116–4121
Leduc RJ, Wonham WM, Lawford M (2001b) Hierarchical interface-based supervisory control: parallel case. In: Proceedings of the thirty-ninth annual Allerton conference on communications, control, and computing, Allerton, pp 386–395
Leduc RJ, Lawford M, Wonham WM (2001c) Hierarchical interface-based supervisory control: AIP example. In: Proceedings of the thirty-ninth annual Allerton conference on communications, control, and computing, Allerton, pp 396–405
Leduc RJ, Brandin BA, Lawford M, Wonham WM (2005a) Hierarchical interface-based supervisory control, part I: serial case. IEEE Trans Autom Control 50(9):1322–1335
Leduc RJ, Lawford M, Wonham WM (2005b) Hierarchical interface-based supervisory control, part II: parallel case. IEEE Trans Autom Control 50(9):1336–1348
Ma C (2004) Nonblocking supervisory control of state tree structures. PhD thesis, Department of Electrical and Computer Engineering. University of Toronto
Ma C, Wonham WM (2003) Control of state tree structures. In: Proceedings of the eleventh IEEE Mediterranean conference on control and automation, Rhodes, Greece, paper T4–005 (6 pp)
Ma C, Wonham WM (2005a) A symbolic approach to the supervision of state tree structures. In: Proceedings thirteenth IEEE Mediterranean conference on control and automation (MED '05). Limassol, Cyprus, pp 908–913
Ma C, Wonham WM (2005b) Nonblocking supervisory control of state tree structures. Lecture notes in control and information sciences (LNCIS), vol 317. Springer
Milner R (1989) Communication and concurrency. Prentice-Hall
Nooruldeen A, Schmidt KW (2015) State attraction under language specification for the reconfiguration of discrete event systems. IEEE Trans Autom Control 60(6):1630–1634
Ramadge PJ (1983) Control and supervision of discrete event processes. PhD thesis, Department of Electrical Engineering, University of Toronto
Ramadge PJ, Wonham WM (1981) Algebraic decomposition of controlled sequential machines. In: Eighth Triennial World Congress, International Federation Automatic Control (IFAC), Kyoto. Preprints, vol 3, pp 37–41
Ramadge PJ, Wonham WM (1982a) Supervision of discrete event processes. In: Proceedings of the twenty-first IEEE conference on decision and control, New York, pp 1228–1229
Ramadge PJ, Wonham WM (1982b) Supervisory control of discrete event processes. Joint workshop on feedback and synthesis of linear and nonlinear systems, Istituto di Automatica, University di Roma. In: Hinrichsen D, Isidori A (eds) Feedback control of linear and nonlinear systems. Lecture notes on control and information sciences, vol 39. Springer, Berlin, pp 202–214
Ramadge PJ, Wonham WM (1984) Supervisory control of a class of discrete event processes. In: Bensoussan A, Lions JL (eds) Proceedings of the sixth international conference on analysis and optimization of systems, Nice, June 1984. Analysis and optimization of systems, Lecture notes on control and information sciences, Part 2, vol 63. Springer, Berlin, pp 477–498
Ramadge PJ, Wonham WM (1987) Supervisory control of a class of discrete event processes. SIAM J Control Optim 25(1):206–230
Su R (2004) Distributed diagnosis for discrete-event systems. PhD thesis, Department of Electrical and Computer Engineering, University of Toronto
Su R, Wonham WM (2001) Supervisor reduction for discrete-event systems. In: Conference on information sciences and systems, The Johns Hopkins University, 21–23 March 2001, [6 pp]
Su R, Wonham WM (2004) Supervisor reduction for discrete-event systems. Discret Event Dyn Syst 14(1):31–53

Theunissen RJM, Petreczky M, Schiffelers RRH, van Beek DA, Rooda JE (2010) Application of supervisory control synthesis to MRI scanners: improving evolvability. SE Report: Nr. 2010-06, Eindhoven, Apr 2010

Vaz A, Wonham WM (1985) On supervisor reduction in discrete event systems. In: Proceedings of the twenty-third annual Allerton conference on communication, control and computing, University of Illinois, Urbana, pp 933–939

Vaz AF, Wonham WM (1986) On supervisor reduction in discrete-event systems. Int J Control 44(2):475–491

Viswanadham N, Narahari Y (1992) Performance modeling of automated manufacturing systems. Prentice-Hall

Wong KC (1998) On the complexity of projections of discrete-event systems. In: Proceedings of the international workshop on discrete event systems (WODES '98). Cagliari, pp 201–206

Wonham WM, Ramadge PJ (1984) On the supremal controllable sublanguage of a given language. In: Proceedings of the twenty-third IEEE conference on decision and control, New York, pp 1073–1080

Wonham WM, Ramadge PJ (1987) On the supremal controllable sublanguage of a given language. SIAM J Control Optim 25(3):637–659

Zhang ZH (2001) Smart TCT: an efficient algorithm for supervisory control design. MASc thesis, Department of Electrical and Computer Engineering, University of Toronto

Zhang ZH, Wonham WM (2001) STCT: an efficient algorithm for supervisory control design. In: Caillaud B, Xie X (eds) Proceedings of the symposium on the supervisory control of discrete event systems (SCODES '01), INRIA, Paris, pp 82–93. See also: Caillaud B et al (eds) (2002) Synthesis and control of discrete event systems, Kluwer, pp 77–100

Chapter 4
Decentralized and Distributed Supervision of Discrete-Event Systems

In applications of supervisory control, especially to large systems made up of several component subsystems and with several individual specifications, it is generally advantageous to decompose the monolithic supervisor accordingly, both for greater computational feasibility and for enhanced transparency of control action. A risk, however, is that component modular supervisors may conflict and result in blocking, leading to the often difficult requirement of effective coordination. This chapter introduces the formal issues involved.

4.1 Introduction

In this chapter we discuss a decentralized approach to the synthesis of supervisors for discrete-event systems. In this approach the overall supervisory task is divided into two or more subtasks. Each of the latter is solved using the results of Chap. 3, and the resulting individual subsupervisors are run concurrently to implement a solution of the original problem. We refer to such a construction as a *decentralized modular synthesis* and to the resultant supervisor as a *decentralized supervisor*. The control architecture is sketched below.

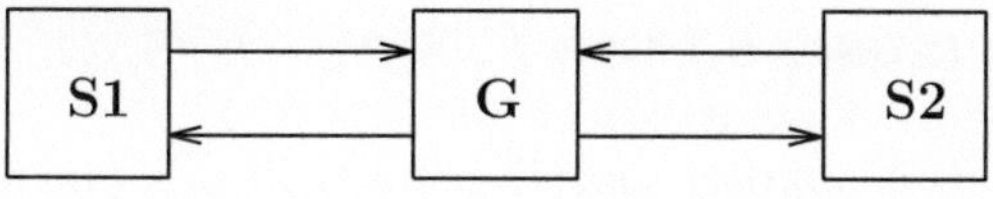

Such constructions represent a very general approach to complex problems—sometimes called 'divide and conquer'. In addition to being more easily synthesized, a decentralized supervisor should ideally be more readily modified, updated and maintained. For example, if one subtask is changed, then it should only be necessary

W. M. Wonham and K. Cai, *Supervisory Control of Discrete-Event Systems*, Communications and Control Engineering,
https://doi.org/10.1007/978-3-319-77452-7_4

to redesign the corresponding modular component supervisor: in other words, the overall decentralized supervisor should exhibit greater flexibility than its 'monolithic' counterpart.

Unfortunately, these advantages are not always to be gained without a price. The fact that the individual supervisory modules are simpler implies that their control action must be based on a partial or 'local' version of the global system state; in linguistic terms, a component supervisor processes only a 'projection' of the behavior of the DES to be controlled. A consequence of this relative insularity may be that different component supervisors, acting quasi-independently on the basis of local information, come into conflict at the 'global' level, and the overall system fails to be nonblocking. Thus a fundamental issue that almost always arises in the presence of decentralization is how to guarantee the nonblocking property of the final synthesis.

4.2 Conjunction of Supervisors

Let $\mathbf{S_1}$ and $\mathbf{S_2}$ be proper supervisors for $\mathbf{G}$: that is, each of $\mathbf{S_1}$ and $\mathbf{S_2}$ is a trim automaton,[1] is controllable with respect to $\mathbf{G}$ (equivalently, $L_m(\mathbf{S_1})$, $L_m(\mathbf{S_2})$ are controllable with respect to $\mathbf{G}$), and each of $\mathbf{S_1/G}$, $\mathbf{S_2/G}$ is nonblocking, namely

$$\overline{L_m(\mathbf{S_1/G})} = L(\mathbf{S_1/G}), \quad \overline{L_m(\mathbf{S_2/G})} = L(\mathbf{S_2/G})$$

Recalling from Sect. 2.4 the definitions of reachable subautomaton and of product automaton, we define the *conjunction* of $\mathbf{S_1}$ and $\mathbf{S_2}$, written $\mathbf{S_1} \wedge \mathbf{S_2}$, as the reachable subautomaton of the product:

$$\mathbf{S_1} \wedge \mathbf{S_2} = Rch(\mathbf{S_1} \times \mathbf{S_2}) = \mathbf{meet}(\mathbf{S_1}, \mathbf{S_2})$$

It is easily seen from the definition that the supervisory action of $\mathbf{S_1} \wedge \mathbf{S_2}$ is to enable an event σ just when σ is enabled by $\mathbf{S_1}$ and $\mathbf{S_2}$ simultaneously. To describe the action of $\mathbf{S_1} \wedge \mathbf{S_2}$ more fully we have the following.

Theorem 1

Under the foregoing conditions,

$$L_m((\mathbf{S_1} \wedge \mathbf{S_2})/\mathbf{G}) = L_m(\mathbf{S_1/G}) \cap L_m(\mathbf{S_2/G})$$

Furthermore $\mathbf{S_1} \wedge \mathbf{S_2}$ is a proper supervisor for $\mathbf{G}$ iff it is trim and the languages $L_m(\mathbf{S_1/G})$, $L_m(\mathbf{S_2/G})$ are nonconflicting.

---

[1] As in Sect. 3.6 and our usage of 'generator', 'automaton' includes the case of partial transition function.

Proof

For the first statement we have

$$\begin{aligned} L_m((\mathbf{S_1} \wedge \mathbf{S_2})/\mathbf{G}) &= L_m(\mathbf{S_1} \wedge \mathbf{S_2}) \cap L_m(\mathbf{G}) \\ &= L_m(\mathbf{S_1}) \cap L_m(\mathbf{S_2}) \cap L_m(\mathbf{G}) \\ &= L_m(\mathbf{S_1}/\mathbf{G}) \cap L_m(\mathbf{S_2}/\mathbf{G}) \end{aligned}$$

Similarly, as $L(\mathbf{S_1} \wedge \mathbf{S_2}) = L(\mathbf{S_1}) \cap L(\mathbf{S_2})$ we have

$$L((\mathbf{S_1} \wedge \mathbf{S_2})/\mathbf{G}) = L(\mathbf{S_1}/\mathbf{G}) \cap L(\mathbf{S_2}/\mathbf{G})$$

so that $(\mathbf{S_1} \wedge \mathbf{S_2})/\mathbf{G}$ is nonblocking iff $L_m(\mathbf{S_1}/\mathbf{G})$ and $L_m(\mathbf{S_2}/\mathbf{G})$ are nonconflicting. Now $\mathbf{S_1}$ and $\mathbf{S_2}$ proper implies that each is controllable, so $L(\mathbf{S_1})$ and $L(\mathbf{S_2})$ are both controllable with respect to **G**. By Proposition 3.5.3, $L(\mathbf{S_1}) \cap L(\mathbf{S_2})$ is controllable, therefore $L(\mathbf{S_1} \wedge \mathbf{S_2})$ is controllable. Thus $\mathbf{S_1} \wedge \mathbf{S_2}$ is a proper supervisor iff it satisfies the defining condition that it be trim, as claimed. □

Recall from Sect. 3.3 that in *TCT*

$$\mathbf{S}_1 \wedge \mathbf{S}_2 = \mathbf{meet}(\mathbf{S}_1, \mathbf{S}_2)$$

Obviously, if $\mathbf{S_1}$ and $\mathbf{S_2}$ satisfy all the conditions of Theorem 1 except that $\mathbf{S_1} \wedge \mathbf{S_2}$ happens not to be trim (i.e. fails to be coreachable), then $\mathbf{S_1} \wedge \mathbf{S_2}$ may be replaced by its trim version, to which the conclusions of the theorem will continue to apply. When designing with *TCT* the desirable situation is that $L_m(\mathbf{S_1})$ and $L_m(\mathbf{S_2})$ be nonconflicting (in *TCT*, $\mathbf{nonconflict}(\mathbf{S}_1, \mathbf{S}_2)$ returns *true*); then $\mathbf{S_1} \wedge \mathbf{S_2} = \mathbf{meet}(\mathbf{S}_1, \mathbf{S}_2)$ will indeed be trim.

Let the controlled DES **G** be arbitrary. The following results, which are almost immediate from the definitions, will find application when exploiting decentralization.

Proposition 2

Let $K_1, K_2 \subseteq \Sigma^*$ be controllable with respect to **G**. If K_1 and K_2 are nonconflicting then $K_1 \cap K_2$ is controllable with respect to **G**. □

Proposition 3

Let $E_1, E_2 \subseteq \Sigma^*$. If $\sup \mathcal{C}(E_1)$, $\sup \mathcal{C}(E_2)$ are nonconflicting then

$$\sup \mathcal{C}(E_1 \cap E_2) = \sup \mathcal{C}(E_1) \cap \sup \mathcal{C}(E_2)$$

□

Exercise 4: Prove Propositions 2 and 3.

To complete this section we provide a version of Theorem 1 adapted to *TCT*. For DES $\mathbf{G_1}, \mathbf{G_2}$, write $\mathbf{G_1} \approx \mathbf{G_2}$, '$\mathbf{G_1}$ and $\mathbf{G_2}$ are behaviorally equivalent', to mean

$$L_m(\mathbf{G_1}) = L_m(\mathbf{G_2}), \qquad L(\mathbf{G_1}) = L(\mathbf{G_2})$$

Theorem 5
Assume

(i) $\mathbf{S_1}, \mathbf{S_2}$ are controllable with respect to $\mathbf{G}$ [as confirmed, say, by **condat**],
(ii) $\mathbf{nonconflict}(\mathbf{S_1} \wedge \mathbf{S_2}, \mathbf{G}) = true$, and
(iii) $\mathbf{S_1} \wedge \mathbf{S_2}$ istrim

Then $\mathbf{S_1} \wedge \mathbf{S_2}$ is a proper supervisor for $\mathbf{G}$, with

$$(\mathbf{S_1} \wedge \mathbf{S_2}) \wedge \mathbf{G} \approx (\mathbf{S_1} \wedge \mathbf{G}) \wedge (\mathbf{S_2} \wedge \mathbf{G})$$

□

Notice that condition (i) holds in particular if $\mathbf{S_1}, \mathbf{S_2}$ are proper supervisors for $\mathbf{G}$. Even in that case, however, condition (ii) is not automatic and must be checked. Finally, the result is easily extended to any collection $\mathbf{S_1}, \mathbf{S_2}, \ldots, \mathbf{S_k}$.

Corollary 6
Let $\mathbf{E_1}, \mathbf{E_2}$ be arbitrary DES and

$$\mathbf{S_i} = \mathbf{supcon}(\mathbf{G}, \mathbf{E_i}), \qquad i = 1, 2$$

If $\mathbf{nonconflict}(\mathbf{S_1}, \mathbf{S_2}) = true$, then

$$\mathbf{S_1} \wedge \mathbf{S_2} \approx \mathbf{supcon}(\mathbf{G}, \mathbf{E_1} \wedge \mathbf{E_2})$$

□

Exercise 7: Prove Theorem 5 and Corollary 6.

4.3 Naive Decentralized Supervision: Deadly Embrace

Before presenting successful examples of decentralized supervision we illustrate the possibility of blocking in a simple but classical situation. Consider two users of two shared resources (e.g. two professors sharing a single pencil and notebook). To carry out his task each user needs both resources simultaneously; but the resources may be acquired in either order. We model the generators **USER1, USER2** and the legal constraint languages **RESA, RESB** in the simple manner shown. Here

$$\Sigma_c = \{\alpha_1, \beta_1, \alpha_2, \beta_2\}, \qquad \Sigma_u = \{\gamma_1, \gamma_2\}$$

The DES to be controlled is then

$$\mathbf{USER} = \mathbf{shuffle}(\mathbf{USER1}, \mathbf{USER2})$$

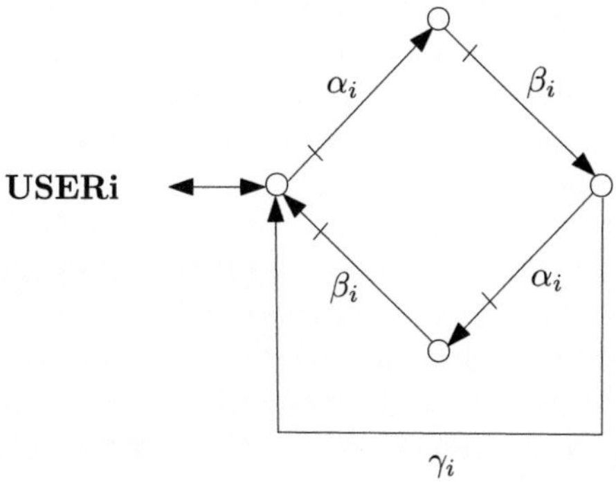

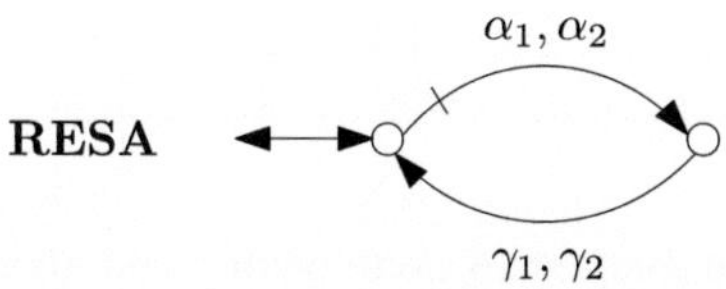

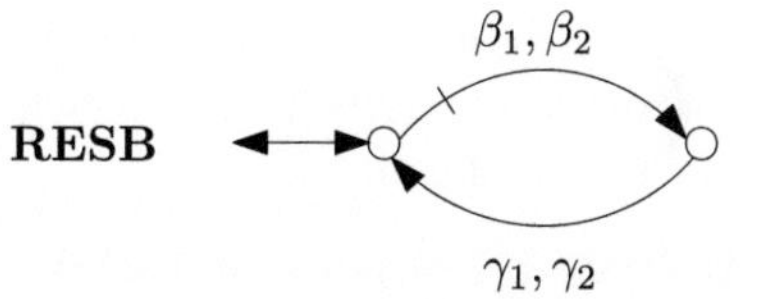

selfloop $\{\beta_1, \beta_2\}$ selfloop $\{\alpha_1, \alpha_2\}$

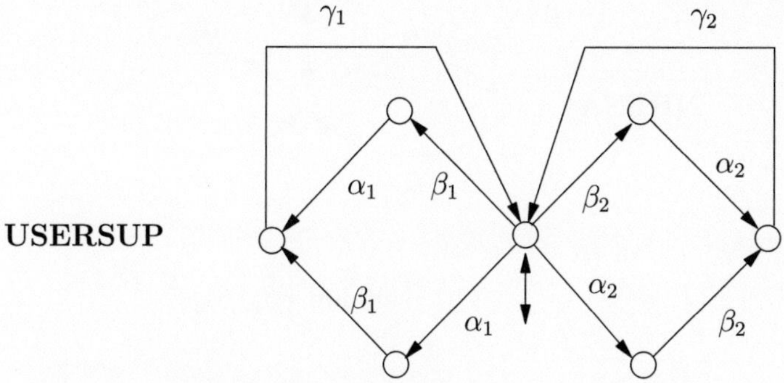

subject to the legal language

$$\mathbf{RES} = \mathbf{meet}(\mathbf{RESA}, \mathbf{RESB})$$

The optimal global supervisor is

$$\mathbf{USERSUP} = \mathbf{supcon}(\mathbf{USER}, \mathbf{RES})$$

as displayed. Initially users and resources are all idle; as soon as one user acquires one resource, **USERSUP** disables the other user from acquiring any resource until the first user has completed his task. Notice, incidentally, that the validity of this proposed control depends crucially on the assumption of the shuffle model that independent events can never occur at the same moment; if this assumption fails, the system will block if both users acquire their first resource simultaneously.

Let us now employ **RESA** and **RESB** as naive modular component supervisors. Each is controllable and nonconflicting with respect to **USER**, hence is proper. The corresponding controlled languages are

$$\mathbf{CONA} = \mathbf{meet}(\mathbf{USER}, \mathbf{RESA}), \quad \mathbf{CONB} = \mathbf{meet}(\mathbf{USER}, \mathbf{RESB})$$

However, **CONA** and **CONB** are conflicting! It is easy to see that concurrent operation of **CONA** and **CONB** could lead to blocking: because nothing prevents **USER1** from acquiring one resource (event α_1 or β_1), then **USER2** acquiring the other (resp. event β_2 or α_2), with the result that both users are blocked from further progress, a situation known as 'deadly embrace'. The example therefore illustrates the crucial role of marker states in system modeling and specification, as well as the importance of absence of conflict.

Exercise 1: Discuss control of this situation that guarantees nonblocking and also 'fairness' according to some common-sense criterion of your invention: fairness should guarantee that neither user could indefinitely shut out the other. **Hint:** Use a queue.

Exercise 2: Replace **RESA** above with the more refined model

NRESA

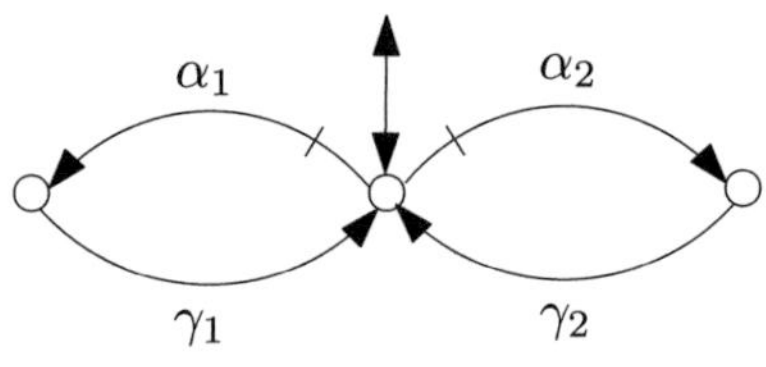

selfloop $\{\beta_1, \beta_2\}$

and similarly for **RESB**. Carry out the new design to get **NUSERSUP**. Verify that it is isomorphic to **USERSUP** and explain why this might be expected.

Exercise 3: As stated in the text, **USERSUP** depends for its validity on the assumption that events in independent agents interleave in time, i.e. never occur simultaneously. Discuss how event interleaving could be enforced, for practical purposes, by use of a queue. For this, require that **USER1**, **USER2** first request the use of a desired resource, while it is up to the supervisor to decide in what order competing requests are granted. Assume that simultaneous requests could be queued in random order.

4.4 Decentralized Supervision: Small Factory

We shall apply the results of Sect. 4.2 to the decentralized supervision of Small Factory (cf. Example 3.3.19 and Appendix 3.1). As displayed below, introduce trim automata **BUFSUP** and **BRSUP** to enforce the buffer and the breakdown/repair specifications respectively.

BUFSUP

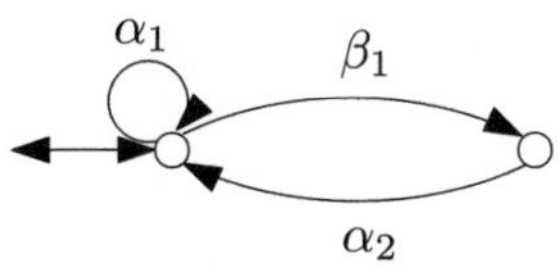

selfloop $\{\lambda_1, \mu_1, \beta_2, \lambda_2, \mu_2\}$

BRSUP

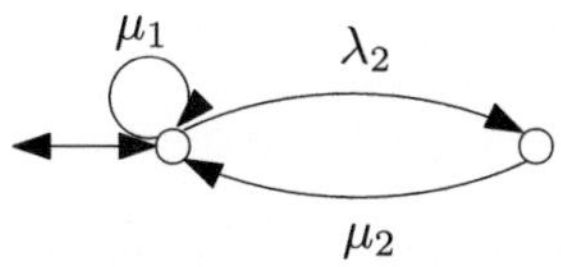

selfloop $\{\alpha_1, \beta_1, \lambda_1, \alpha_2, \beta_2\}$

By use of the *TCT* procedure **condat** it can be confirmed that **BUFSUP** and **BRSUP** are controllable with respect to **FACT**, and application of **nonconflict** to the pairs **FACT**, **BUFSUP** and **FACT**, **BRSUP** respectively, shows by Proposition 3.6.5 that **BUFSUP** and **BRSUP** are nonblocking for **FACT**; so we may conclude that each is a proper supervisor for **FACT**. For our modular decentralized supervisor we now take the conjunction

$$\mathbf{MODSUP} = \mathbf{BUFSUP} \wedge \mathbf{BRSUP}$$

It is easy to check by hand that **BUFSUP**, **BRSUP** are nonconflicting, so

$$\mathbf{MODSUP} = \mathbf{trim}(\mathbf{meet}(\mathbf{BUFSUP}, \mathbf{BRSUP}))$$

namely **MODSUP** is trim; and by application of **condat** and **nonconflict** to the pair **FACT**, **MODSUP** we now conclude by Proposition 3.7.1 that **MODSUP** is a proper supervisor for **FACT**.

We note parenthetically that, on taking **G = FACT**, Proposition 4.2.2 holds with

$$K_1 = L_m(\mathbf{BUFSUP}), \quad K_2 = L_m(\mathbf{BRSUP})$$

while Proposition 4.2.3 holds with

$$E_1 = L_m(\textbf{SPEC1}), \quad E_2 = L_m(\textbf{SPEC2})$$

Finally it may be verified that **MODSUP** is actually optimal. Various approaches are possible: perhaps the most direct is to check that

$$L_m(\textbf{FACT}) \cap L_m(\textbf{MODSUP}) = L_m(\textbf{FACTSUP})$$

via the computation

$$\textbf{isomorph}(\textbf{meet}(\textbf{FACT}, \textbf{MODSUP}), \textbf{FACTSUP}) = true$$

Another possibility, using Proposition 4.2.3, is left to the reader to develop independently.

4.5 Decentralized Supervision: Big Factory

As another example of the foregoing ideas we consider Big Factory, described below. Two machines as before operate in parallel to feed a buffer with capacity 3; a third machine empties the buffer.

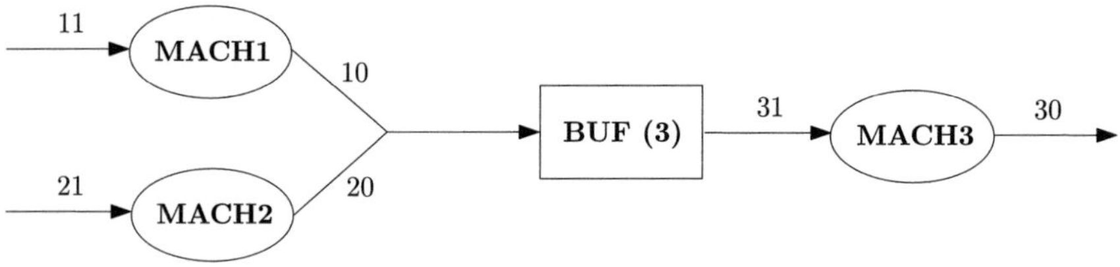

MACHi

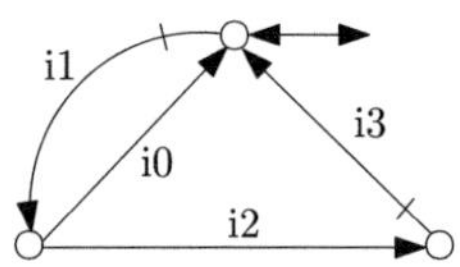

The informal specifications are:

1. Buffer must not overflow or underflow.
2. **MACH1** and **MACH2** are repaired in order of breakdown.
3. **MACH3** has priority of repair over **MACH1** and **MACH2**.

As the plant we take

$$\textbf{BFACT} = \textbf{shuffle}(\textbf{MACH1}, \textbf{MACH2}, \textbf{MACH3})$$

To formalize the specifications we construct the DES shown below:

Buffer overflow/underflow:

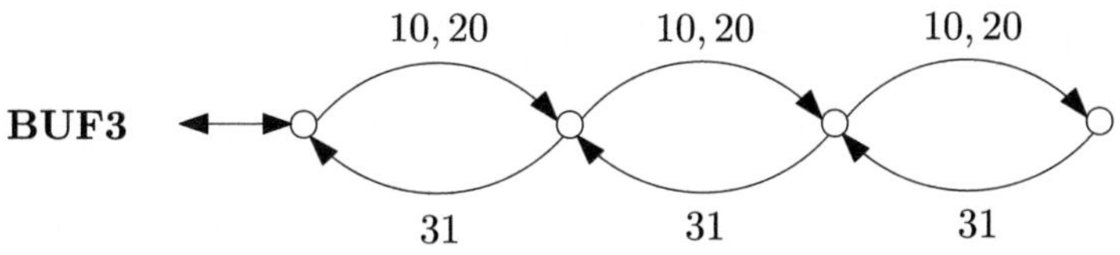

Breakdown/repair of **MACH1, MACH2**:

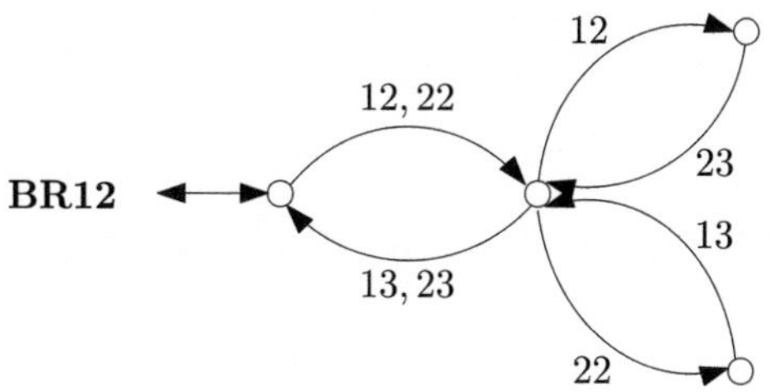

Breakdown/repair of **MACH3**:

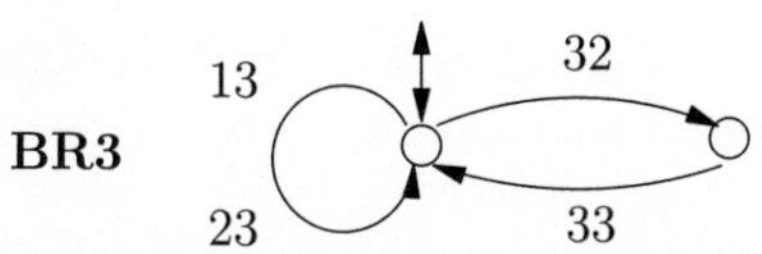

Each DES is understood to be self-looped with its complementary subalphabet.

We first consider 'monolithic' supervision. **BFACT** turns out to have 27 states and 108 transitions (written (27,108)). Combining thespecification languages into their intersection, we define

$$\mathbf{BSPEC} = \mathbf{meet}(\mathbf{BUF3}, \mathbf{BR12}, \mathbf{BR3}) \quad (32, 248)$$

For the 'monolithic'supervisor we then obtain

$$\mathbf{BFACTSUP} = \mathbf{supcon}(\mathbf{BFACT}, \mathbf{BSPEC}) \quad (96, 302)$$

By the theory, the transition structure of the DES **BFACTSUP** is that of the supremal controllable sublanguage of L_m(**BFACT**) that is contained in the specification language L_m(**BSPEC**). Thus **BFACTSUP** is guaranteed to be the optimal (i.e. minimally restrictive) proper supervisor that controls **BFACT** subject to the three legal specifications. Nevertheless, **BFACTSUP** is a rather cumbersome structure to implement directly, and it makes sense to consider a decentralized approach.

For prevention of buffer overflow alone, we compute

$$\text{status1} = \#\ \text{empty buffer slots} - \#\ \text{feeder machines at work}$$

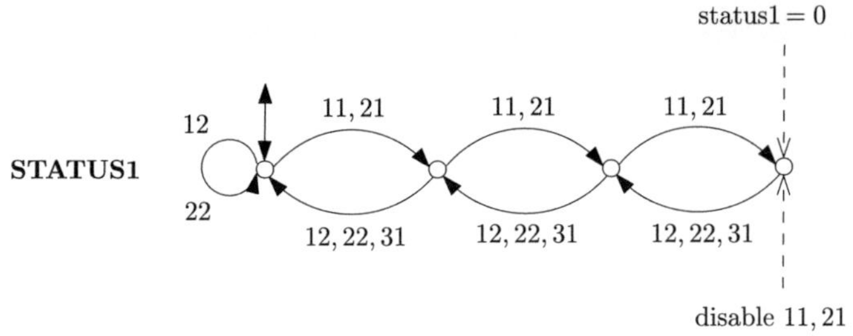

STATUS1 disables 11 and 21 when status1 = 0 and is a proper supervisor for **BFACT**. For prevention of buffer underflow alone, we compute

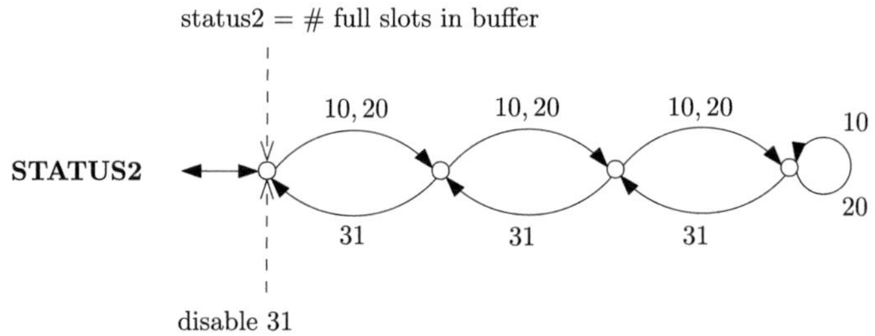

STATUS2 disables 31 when status2 = 0 and is also proper. For control of breakdown/repair, **BR12** and **BR3** are themselves proper supervisors. It can be verified that optimal (and proper) supervision of the buffer is enforced by

$$\mathbf{STATUS} = \mathbf{STATUS1} \wedge \mathbf{STATUS2}$$

while optimal (and proper) supervision of breakdown/repair is enforced by

$$\mathbf{BR} = \mathbf{BR12} \wedge \mathbf{BR3}$$

Finally, optimal (and proper) supervision with respect to all the legal specifications is enforced by

$$\mathbf{BFTMDSUP} = \mathbf{STATUS} \wedge \mathbf{BR}$$

Obviously **BFTMDSUP** is much simpler to implement than **BFACTSUP**, to which it is equivalent in supervisory action.

Exercise 1: Construct a 9-state supervisor that is equivalent in control action to **STATUS**. Check your result by *TCT* and supply the printouts.

Exercise 2: Carry out the following 'direct' approach to decentralized control of Big Factory. For each of the three specifications **BUF3, BR12, BR3** compute the 'global' controllers using **supcon**, and their corresponding reduced versions using **supreduce**, of (minimal) state size 7, 3, 2, respectively. Verify that these provide an optimal decentralized control. While no special analysis was required, supervisor reduction was essential for efficient decentralization.

4.6 Decentralized Supervision: Transfer Line

As a third example of decentralized control we consider an industrial 'Transfer Line' consisting of two machines **M1**, **M2** followed by a test unit **TU**, linked by buffers **B1** and **B2**, in the configuration shown. A workpiece tested by **TU** may be accepted or rejected; if accepted, it is released from the system; if rejected, it is returned to **B1** for reprocessing by **M2**. Thus the structure incorporates 'material feedback'. The specification is simply that **B1** and **B2** must be protected against underflow and overflow, along with the tacit requirement of nonblocking.

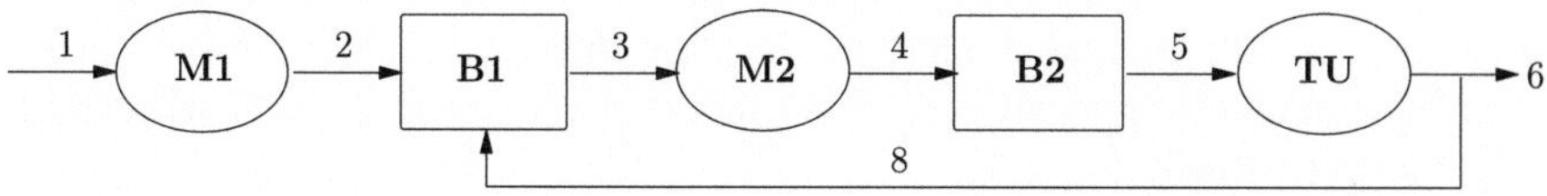

The component DES, displayed below, are taken to be as simple as possible.

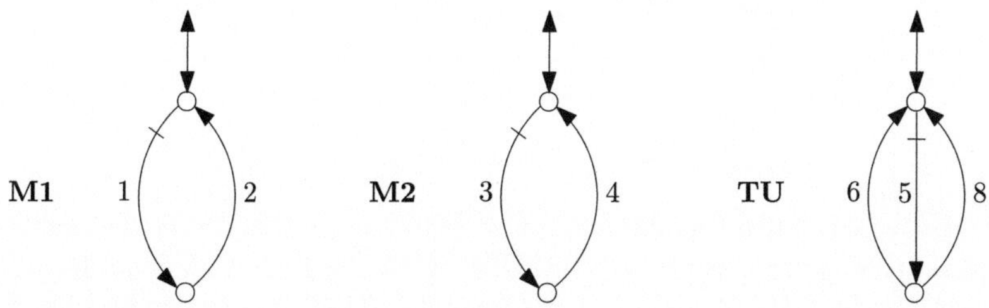

The DES representing the Transfer Line is

$$\mathbf{TL} = \mathbf{shuffle}(\mathbf{M1}, \mathbf{M2}, \mathbf{TU})$$

The capacities of **B1** and **B2** are assumed to be 3 and 1 respectively, and the specifications are modeled as **B1SP**, **B2SP** in the usual way.

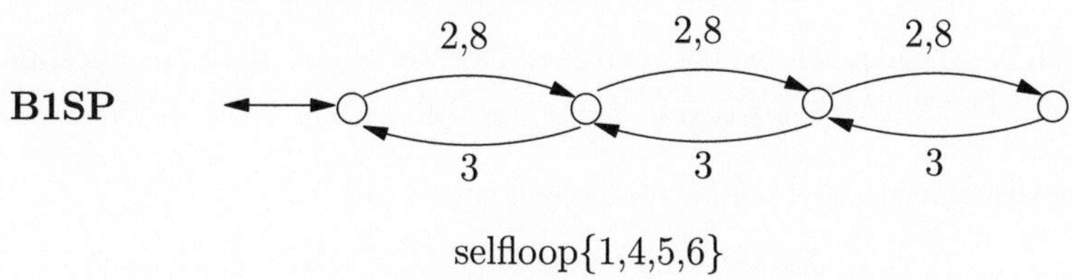

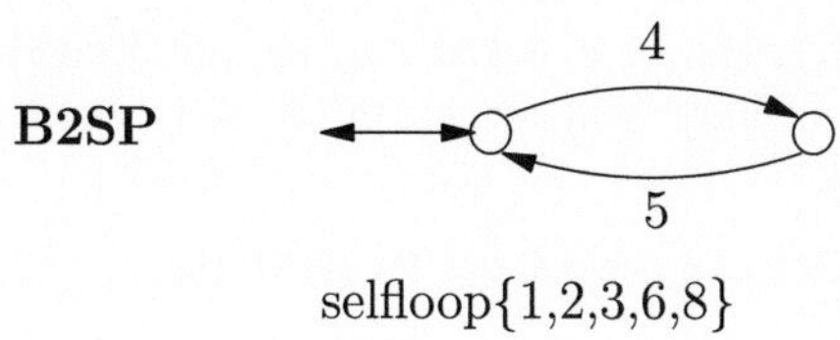

Then the total specification is

BSP = meet(B1SP,B2SP)

The centralized or monolithic supervisor is computed as

CSUP = supcon(TL,BSP) (28, 65)

and turns out to have 28 states and 65 transitions. The control data file for **CSUP** is

CSUP = condat(TL,CSUP)

For decentralized supervision we may proceed as follows. A modular component supervisor for **B2** is simple enough: we construct **B2SUP** to disable event 5 in **TU** when **B2** is empty (to prevent underflow) and to disable event 3 in **M2** when **B2** is full (to prevent overflow).

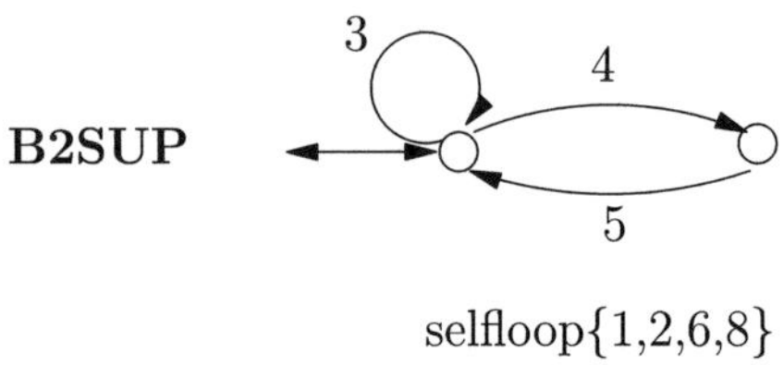

For **B1** we separate the requirements of overflow and underflow into two subtasks, assigned to component supervisors **B1SUP1**, **B1SUP2**. To prevent underflow it suffices to adopt for **B1SUP2** the specification model **B1SP**, augmented by a harmless selfloop for events 2, 8 at state 3 to render **B1SUP2** controllable. Then **B1SUP2** disables event 3 of **M2** at state 0 (where **B1** is empty), but is indifferent to possible overflow at state 3. To prevent overflow we make a first, 'naive' attempt at designing **B1SUP1**, with result **XB1SUP1**, as follows. The entities feeding **B1** (potentially causing overflow) are **M1** and **TU**: define

FB1A = shuffle(M1,TU), **FB1 = selfloop(FB1A**, [3])

FB1 will be considered to be the controlled DES for the overflow specification

FB1SP = selfloop(B1SP,[1,5,6])

leading to the proposed modular component supervisor

XB1SUP1A = supcon(FB1, FB1SP)

over the subalphabet {1,2,3,5,6,8}, and finally the global version

XB1SUP1 = selfloop(XB1SUP1A, [4]) (12,45)

over the full alphabet. It can be checked that each of **XB1SUP1** and **B1SUP2** is nonconflicting and controllable with respect to **TL**, and that **XB1SUP1** and **B1SUP2** are nonconflicting. Let

XB1SUP = meet(XB1SUP1,B1SUP2) (12, 45)

(Verify that **XB1SUP1**, **XB1SUP** are isomorphic: why is this so?) From the theory or by direct computation, **XB1SUP** is controllable and nonconflicting with respect to **TL**. It remains to combine **XB1SUP** with **B2SUP**: to our chagrin, these components turn out to be conflicting! Let

XBSUP = trim(meet(XB1SUP,B2SUP))

Because of conflict, the closed behavior of **XBSUP** (equal to the closure of its marked behavior, by definition of the operation **trim**) is a proper sublanguage of the intersection of the closed behaviors of the trim DES **XB1SUP**, **B2SUP**; and from

XBSUP = condat(TL,XBSUP)

it is seen that **XBSUP** fails to be controllable as it calls for the disablement of events 4 and 8. The concurrent operation of **XB1SUP** and **B2SUP** will certainly result in satisfaction of the specifications **B1SP** and **B2SP**. However, each of these components admits the **TL**-string

$$s = 1.2.1.2.1.2.3.4.1.2$$

which leaves **B1** and **B2** both full. Following s, **B2SUP** disables event 3 of **M2**, while **XB1SUP** disables event 1 of **M1** and event 5 of **TU**, and the system deadlocks; i.e. no further transitions are possible. The result illustrates that conflict and blocking can arise in seemingly innocent ways.[2]

A correct decentralized supervisor for overflow of **B1** can be obtained by examining the overall feedback operation of the system. It is seen that any workpiece removed from **B1** by **M2** is a candidate for eventual return to **B1** by **TU**. Thus overflow of **B1** is prevented iff the number of empty slots in **B1** is maintained at least as great as the number of workpieces being processed by **M2** and **TU** or being stored in **B2**. In terms of event counts (#event) on the current string,

$$\text{\# empty slots in } \mathbf{B1} = \text{cap}(\mathbf{B1}) + \#3 - \#2 - \#8$$

while

$$\text{\# workpieces in } \{\mathbf{M2},\mathbf{B2},\mathbf{TU}\} = \#3 - \#6 - \#8$$

To maintain the desired inequality it is therefore required to disable event 1 of **M1** iff

$$\text{\# empty slots in } \mathbf{B1} \leq \text{\# workpieces in } \{\mathbf{M2},\mathbf{B2},\mathbf{TU}\}$$

i.e. (with cap(**B1**) = 3) $3 - \#2 \leq -\#6$, or

$$\text{disable event 1 of } \mathbf{M1} \quad \text{iff} \quad \#2 - \#6 \geq 3$$

[2]This is nothing but an instance of a feedback loop going unstable if it is 'driven too hard', namely 'its gain is too high'. Here 'instability' (deadlock) is the inability of the system to return to its 'equilibrium' (initial) state.

Under these conditions event 2 can occur at most three times before an occurrence of event 6, so our new attempt at an overflow control for **B1** takes the form of **B1SUP1** as displayed. Here the harmless selfloop [0,6,0] has been adjoined to render **B1SUP1** controllable.

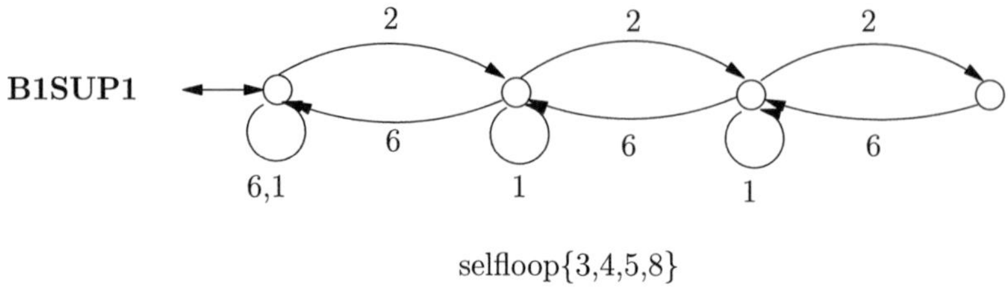

It can now be verified that **B1SUP1**, **B1SUP2** are nonconflicting, so that concurrent operation is represented by the proper supervisor

B1SUP =meet(B1SUP1,B1SUP2) (16,100)

and that **B1SUP**, **B2SUP** are nonconflicting, with their concurrent operation represented by

BSUP =meet(B1SUP,B2SUP) (32,156)

It can be checked that **BSUP** is nonconflicting and controllable with respect to **TL**. Thus the behavior of **TL** under decentralized supervision is given by

DCSUP =meet(TL,BSUP) (28,65)

Finally it can be checked that **DCSUP** is isomorphic with **CSUP**; namely the decentralized supervisor **BSUP** is optimal.

Exercise 1: Improve the recycling logic of Transfer Line as follows. A failed workpiece is sent by **TU** to a new buffer **B3** (capacity 1), and **M2** can take its workpiece from either **B1** or **B3**. Introduce a new specification that **M2** takes from **B1** only if **B3** is empty, that is, a failed workpiece has priority over a new one. Design both centralized (monolithic) and decentralized supervisors for the improved system.

Exercise 2: Carry out the following 'direct' approach to decentralized control of Transfer Line. For each of the two specifications **B1SP, B2SP** compute the 'global' controllers using **supcon**, and their corresponding versions using **supreduce**, of (minimal) state size 7, 2 respectively. Check that these do *not* provide a nonblocking decentralized control as their meet is conflicting with **TL**. Having noted as before that blocking is due to both buffers filling up, introduce a third specification which always maintains at least one buffer slot free, namely

$$\begin{aligned}\#2 - \#6 &\leq \text{(storage capacity of loop)} - 1\\ &= \text{capacity}(\mathbf{B1}) + \text{capacity}(\mathbf{B2}) - 1\\ &= 3 + 1 - 1 = 3\end{aligned}$$

Create the corresponding buffer, say **LOOPSPEC**, then the corresponding 'global' controller, and its reduced version, say **LOOPCOORD**, of (minimal) size 4. Verify that the three reduced supervisors provide an optimal decentralized control. Of course, success depended on precisely identifying the cause of blocking, a step which in many applications will be nontrivial.

Apply this method to the enhanced model of Exercise 1.

Example 3: Formal verification

We provide a formal verification of the decentralized control for Transfer Line proposed in Exercise 2. Bring in the modular specification languages

$$\begin{aligned} B_1 &= L_m(\mathbf{B1SP}) \cap L_m(\mathbf{TL}) \\ B_2 &= L_m(\mathbf{B2SP}) \cap L_m(\mathbf{TL}) \\ L_0 &= L_m(\mathbf{LOOPSPEC}) \cap L_m(\mathbf{TL}) \end{aligned}$$

The corresponding decentralized controlled behaviors are then

$$K_1 := \sup \mathcal{C}(B_1),\ \ K_2 := \sup \mathcal{C}(B_2),\ \ K_0 := \sup \mathcal{C}(L_0)$$

while the optimal monolithic controlled behavior ($L_m(\mathbf{CSUP})$ above) is

$$K := \sup \mathcal{C}(B_1 \cap B_2)$$

In these expressions $\sup \mathcal{C}(\cdot)$ is computed with respect to $\overline{L_m(\mathbf{TL})} = L(\mathbf{TL})$.

Evidently $K \subseteq K_1 \cap K_2$, but the blocking action already mentioned indicates that the inclusion is strict. We wish to show, however, that

$$K = K_1 \cap K_2 \cap K_0$$

First we check that the loop condition is necessary, namely $\bar{K} \subseteq \bar{L}_0$, i.e. if $s \in \bar{K}$ then

$$0 \leq \#2(s) - \#6(s) \leq 3$$

The first inequality is obvious for any $s \in L(\mathbf{TL})$. By the buffer constraints $s \in \bar{K}_1 \cap \bar{K}_2$, so

$$\#2(s) - \#6(s) \leq 4$$

Suppose $\#2(s) - \#6(s) = 4$. If (following s) state(**TU**) $= 1$ then by controllability of K, the string $s.8 \in \bar{K}$; similarly if state(**M2**) $= 1$ then $s.4 \in \bar{K}$; so we may assume that, following s,

$$\text{state}(\mathbf{TU}) = 0,\ \ \text{state}(\mathbf{M2}) = 0$$

Thus buffers **B1, B2** are both full; by the buffer constraints and controllability, events 3 and 5 must both be disabled (i.e. $s.3, s.5 \notin \bar{K}$) and the system blocks, hence for no string w is $sw \in K$, a contradiction to $s \in \bar{K}$, proving $\bar{K} \subseteq \bar{L}_0$.

Next, if $s \in K$ then $s \in L_m(\mathbf{TL})$, thus $0 = \#2(s) - \#6(s)$, hence $s \in L_0$, so $K \subseteq L_0$, namely $K \subseteq B_1 \cap B_2 \cap L_0$, and by controllability

$$K \subseteq \sup \mathcal{C}(B_1 \cap B_2 \cap L_0)$$

By definition of K the reverse inclusion is obvious, so

$$K = \sup \mathcal{C}(B_1 \cap B_2 \cap L_0) \tag{4.1}$$

Let

$$H_1 := \sup \mathcal{C}(B_1 \cap L_0), \quad H_2 := \sup \mathcal{C}(B_2 \cap L_0)$$

We will show that H_1, H_2 are nonconflicting. By (4.1) and Proposition 4.2.3 it will follow that

$$K = H_1 \cap H_2 \tag{4.2}$$

Let $s \in \bar{H}_1 \cap \bar{H}_2$; we must show $s \in \overline{H_1 \cap H_2}$. By controllability we may assume as above that, following s, **TU** and **M2** are both at their state 0. Since $s \in \bar{L}_0$, the possible states of **LOOPSPEC, B1, B2** are tabulated below.

LOOPSPEC	B1	B2	admissible string w
0	0	0	ϵ
1	1	0	3.4.5.6
1	0	1	5.6
2	1	1	5.6.3.4.5.6
2	2	0	3.4.5.6.3.4.5.6
3	2	1	5.6.3.4.5.6.3.4.5.6
3	3	0	3.4.5.6.3.4.5.6.3.4.5.6

It is easy to check that in each case there exists a string w (e.g. as tabulated) such that $sw \in H_1 \cap H_2$; thus $s \in \overline{H_1 \cap H_2}$ as claimed.

By an even simpler argument we obtain

$$H_1 = K_1 \cap K_0, \quad H_2 = K_2 \cap K_0 \tag{4.3}$$

From (4.2), (4.3) there results $K = K_1 \cap K_2 \cap K_0$ as was to be proved.

4.7 Decentralized Supervision: AGVs in a Manufacturing Workcell

As our fourth example of decentralized control we consider the coordination of a system of automatic guided vehicles (AGVs) serving a manufacturing workcell. As a side-benefit this will illustrate the power of supervisor reduction (Sect. 3.10, and *TCT* **supreduce**) in reducing modular components to minimal or near-minimal state size for transparency of modeling and economy in computational verification.

The system consists of two input stations IPS1, IPS2 for parts of types 1, 2; three workstations WS1, WS2, WS3; one completed parts station CPS; and five AGVs – AGV1,...,AGV5. The AGVs travel on fixed circular routes alternately loading and unloading, as shown in Fig. 4.1, or explicitly:

$$\begin{array}{l} \text{AGV1: WS2} \dashrightarrow \text{IPS1} \longrightarrow \text{WS2} \\ \text{AGV2: WS3} \dashrightarrow \text{IPS2} \longrightarrow \text{WS3} \\ \text{AGV3: WS1} \dashrightarrow \text{WS2} \longrightarrow \text{WS1} \\ \text{AGV4: WS1} \dashrightarrow \text{WS3} \longrightarrow \text{WS1} \\ \text{AGV5: CPS} \dashrightarrow \text{WS1} \longrightarrow \text{CPS} \end{array}$$

Here $\longrightarrow$ (resp. $\dashrightarrow$) indicates travel direction when loaded (resp. unloaded), with each AGV returning to its initial (parking) location just after unloading. Thus AGV1 starts unloaded from WS2, travels to IPS1 where it loads, and returns to WS2 where it unloads. Input parts of type 1 traverse the route

$$\text{IPS1} \longrightarrow (\text{AGV1}) \longrightarrow \text{WS2} \longrightarrow (\text{AGV3}) \longrightarrow \text{WS1}$$

and type 2 the route

$$\text{IPS2} \longrightarrow (\text{AGV2}) \longrightarrow \text{WS3} \longrightarrow (\text{AGV4}) \longrightarrow \text{WS1}$$

At WS1 a processed part of type 1 and one of type 2 are assembled into an output part, which is exported via the route

$$\text{WS1} \longrightarrow (\text{AGV5}) \longrightarrow \text{CPS}$$

Only one part can be processed at one time by each of WS2, WS3, while WS1 can assemble just two parts at a time into an output part.

As shown in Fig. 4.1 the AGV track systems cross one another in four shared zones, as follows:

Zone	Shared by AGVs
1	1, 2
2	2, 3
3	2, 4
4	4, 5

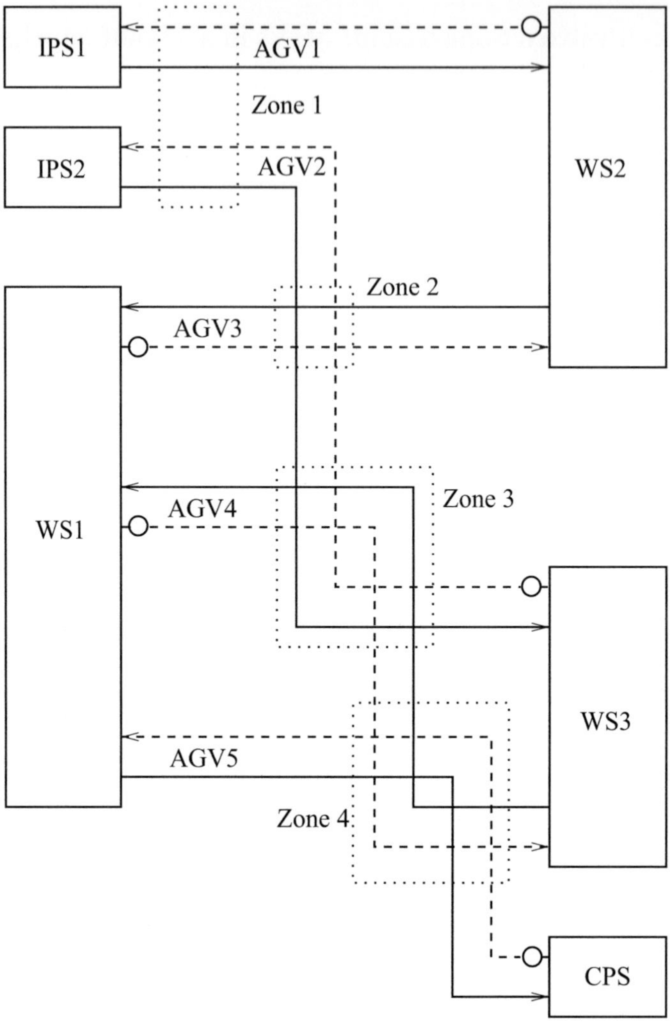

Fig. 4.1 Manufacturing workcell
AGVs: ⟶ direction loaded
⇠ direction unloaded
∘ parking location

For safety reasons each zone should be occupied by at most one AGV at a time.

Finally it will be assumed that IPS1, IPS2 share a common loading area, so only one of AGV1, AGV2 can be loaded at one time.

DES models of the AGVs are displayed in Fig. 4.2(1)–(5), while their events are listed in Table 4.1. Controllability status is denoted by odd/even labeling as usual. Our models contain no more detail than will be needed to capture the control specifications. Thus intermediate physical states implicit in (physical) event sequences like

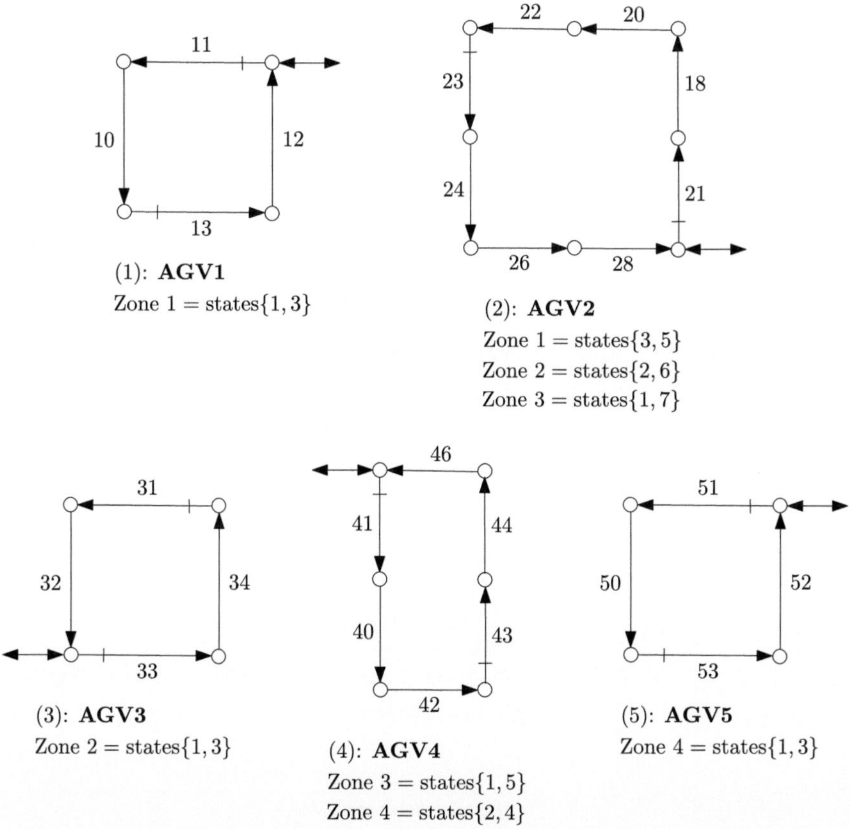

Fig. 4.2 AGVs, with zone locations. States numbered counterclockwise, from initial state 0

(interpreted) event 11 for **AGV1**, etc., are merged into one, and the corresponding sequences are replaced by single events, considered atomic.

The specifications on AGV travel are stated more formally below, and their DES models displayed in Fig. 4.3(1)–(8).

Specification 1: At most one AGV can occupy a shared zone at one time. For instance, **Z1SPEC** should prohibit **AGV1** (resp. **AGV2**) from entering Zone 1 while the latter is occupied by **AGV2** (resp. **AGV1**). We model this by saying 'if **AGV2** enters Zone 1 then **AGV1** cannot enter Zone 1 until **AGV2** exits Zone 1' and vice-versa.

Specification 2: **WS1SPEC** should enforce the cyclic sequencing allowed by WS1, namely the latter should unload **AGV3**, **AGV4** in either order, then (after the assembly process) allow **AGV5** to load the assembled output part for transport to CPS.

Specification 3: **WS2SPEC** should enforce that WS2 is alternately loaded by **AGV1** (for processing of an input part type 1) and unloaded by **AGV3** (for transport of processed part to WS1).

Specification 4: **WS3SPEC** should enforce that WS3 is alternately loaded by **AGV2** (for type 2) and unloaded by **AGV4**.

Table 4.1 Events of AGVs

Event	AGV	Interpretation
11	1	Unparks and enters Zone 1
10		Exits Zone 1 and loads from IPS1
13		Re-enters Zone 1
12		Exits Zone 1, unloads to WS2, and parks
21	2	Unparks and enters Zone 3
18*		Exits Zone 3 and enters Zone 2
20		Exits Zone 2 and enters Zone 1
22		Exits Zone 1 and loads from IPS2
23		Re-enters Zone 1
24		Exits Zone 1 and re-enters Zone 2
26		Exits Zone 2 and re-enters Zone 3
28		Exits Zone 3, unloads to WS3, and parks
33	3	Unparks and enters Zone 2
34		Exits Zone 2 and loads from WS2
31		Re-enters Zone 2
32		Exits Zone 2, unloads to WS1, and parks
41	4	Unparks and enters Zone 3
40		Exits Zone 3 and enters Zone 4
42		Exits Zone 4 and loads from WS3
43		Re-enters Zone 4
44		Exits Zone 4 and re-enters Zone 3
46		Exits Zone 3, unloads at WS1, and parks
51	5	Unparks and enters Zone 4
50		Exits Zone 4 and loads from WS1
53		Re-enters Zone 4
52		Exits Zone 4, unloads to CPS, and parks

*numbering due to need for 6 even-numbered labels

Specification 5: **IPSSPEC** should prohibit **AGV1** and **AGV2** from attempting to occupy simultaneously the assumed common work area of IPS1 and IPS2.

Our plant model will be the synchronous product of the 5 AGVs, while the full specification will be the meet of 8 components (4 for Specification 1 and one each for Specifications 2–5). System components IPS1, IPS2, CPS play no direct role in the system's control logic. While they could be created for completeness' sake, they would not be included in the computations and therefore we ignore them. Similarly, details for WS1, WS2, WS3 are not included as the workstations contribute only the sequencing specifications already provided.

After solving the monolithic centralized supervision problem we develop simple decentralized modular supervisors. For each of the 8 specifications we take for the

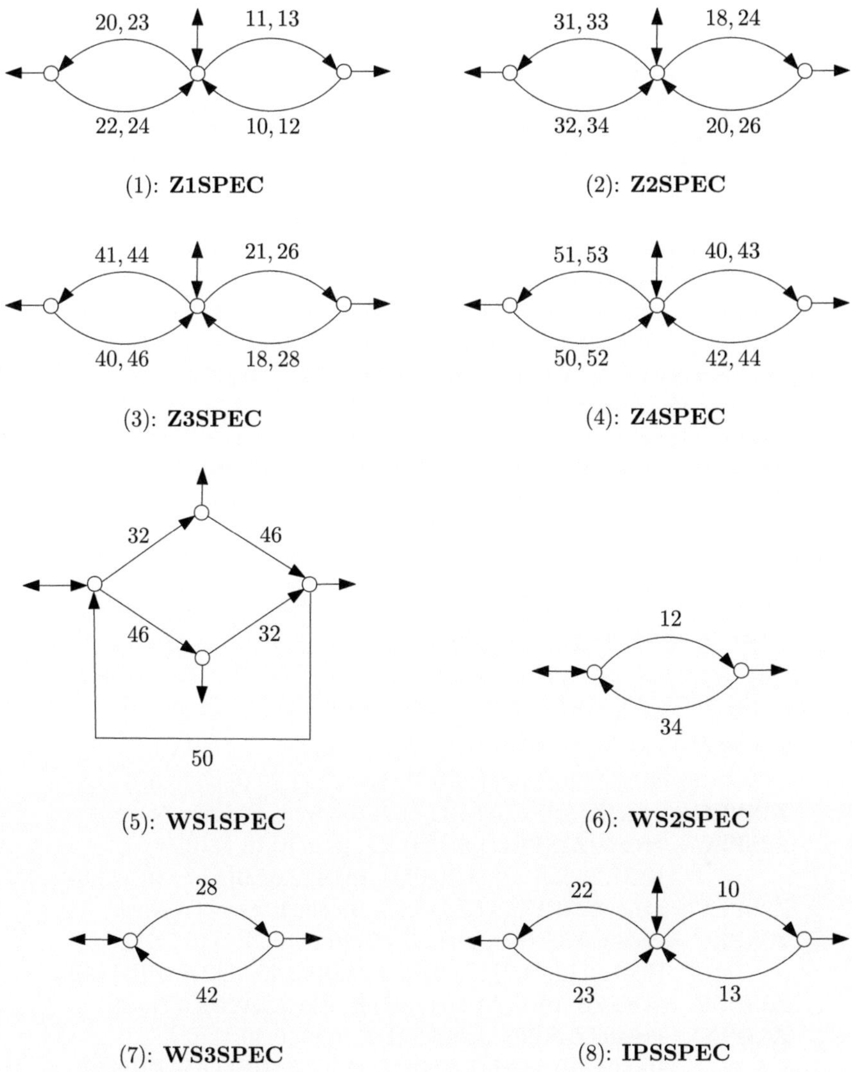

Fig. 4.3 Specifications on AGVs

'local' plant only the synchronous product of the AGVs whose events appear directly, and apply **supreduce** to the resulting local supervisor.

Finally we check for nonblocking when the modular supervisors are placed online together. As the combination fails to be nonblocking, we develop a ninth 'coordinating' modular supervisor by examining the system's overall input–output behavior. The combination of all 9 resulting modules is verified to be control-equivalent to the monolithic supervisor.

Commented MAKEIT.TXT file

Create the 5 AGVs.

AGV1 = **create**(**AGV1**,[mark 0],[tran [0,11,1],[1,10,2],[2,13,3],
[3,12,0]]) (4,4)
AGV2 = **create**(**AGV2**,[mark 0],[tran [0,21,1],[1,18,2],[2,20,3],
[3,22,4],[4,23,5],[5,24,6],[6,26,7],[7,28,0]]) (8,8)
AGV3 = **create**(**AGV3**,[mark 0],[tran [0,33,1],[1,34,2],[2,31,3],
[3,32,0]]) (4,4)
AGV4 = **create**(**AGV4**,[mark 0],[tran [0,41,1],[1,40,2],[2,42,3],
[3,43,4],[4,44,5],[5,46,0]]) (6,6)
AGV5 = **create**(**AGV5**,[mark 0],[tran [0,51,1],[1,50,2],[2,53,3],
[3,52,0]]) (4,4)
AGV = **sync**(**AGV1**,**AGV2**) (32,64) Blocked_events = None
AGV = **sync**(**AGV**,**AGV3**) (128,384) Blocked_events = None
AGV = **sync**(**AGV**,**AGV4**) (768,3072) Blocked_events = None

Compute the plant model.

AGV = **sync**(**AGV**,**AGV5**) (3072,15360) Blocked_events = None
ALL = **allevents**(**AGV**) (1,26)

Create the four zonal exclusion specifications and their meet.

Z1SPEC = **create**(**Z1SPEC**,[mark 0,1,2],[tran [0,11,1],[0,13,1],
[0,20,2],[0,23,2],[1,10,0],[1,12,0],[2,22,0],[2,24,0]]) (3,8)
Z1SPEC = **sync**(**Z1SPEC**,**ALL**) (3,62) Blocked_events = None
Z2SPEC = **create**(**Z2SPEC**,[mark 0,1,2],[tran [0,18,1],[0,24,1],
[0,31,2],[0,33,2],[1,20,0],[1,26,0],[2,32,0],[2,34,0]]) (3,8)
Z2SPEC = **sync**(**Z2SPEC**,**ALL**) (3,62) Blocked_events = None
Z3SPEC = **create**(**Z3SPEC**,[mark 0,1,2],[tran [0,21,1],[0,26,1],
[0,41,2],[0,44,2],[1,18,0],[1,28,0],[2,40,0],[2,46,0]]) (3,8)
Z3SPEC = **sync**(**Z3SPEC**,**ALL**) (3,62) Blocked_events = None
Z4SPEC = **create**(**Z4SPEC**,[mark 0,1,2],[tran [0,40,1],[0,43,1],
[0,51,2],[0,53,2],[1,42,0],[1,44,0],[2,50,0],[2,52,0]]) (3,8)
Z4SPEC = **sync**(**Z4SPEC**,**ALL**) (3,62) Blocked_events = None
ZSPEC = **meet**(**Z1SPEC**,**Z2SPEC**) (9,146)
ZSPEC = **meet**(**ZSPEC**,**Z3SPEC**) (27,318)
ZSPEC = **meet**(**ZSPEC**,**Z4SPEC**) (81,594)

Compute a modular supervisor for the combined zonal exclusion specification.

ZSUP = **supcon**(**AGV**,**ZSPEC**) (1112,3776)

Create the three workstation sequencing specifications and their meet.

WS1SPEC = **create**(**WS1SPEC**,[mark 0,1,2,3],[tran [0,32,1],
[0,46,2],[1,46,3],[2,32,3],[3,50,0]]) (4,5)
WS1SPEC = **sync**(**WS1SPEC,ALL**) (4,97)
Blocked_events = None
WS2SPEC = **create**(**WS2SPEC**,[mark 0,1],[tran [0,12,1],
[1,34,0]]) (2,2)
WS2SPEC = **sync**(**WS2SPEC,ALL**) (2,50)
Blocked_events = None
WS3SPEC = **create**(**WS3SPEC**,[mark 0,1],[tran [0,28,1],
[1,42,0]]) (2,2)
WS3SPEC = **sync**(**WS3SPEC,ALL**) (2,50)
Blocked_events = None
WSSPEC = **meet**(**WS1SPEC,WS2SPEC**) (8,186)
WSSPEC = **meet**(**WSSPEC,WS3SPEC**) (16,356)

Compute a modular supervisor for the combined workstation sequencing specification.

WSSUP = **supcon**(**AGV,WSSPEC**) (12666,47730)

Create the input parts stations exclusion specification.

IPSSPEC = **create**(**IPSSPEC**,[mark 0,1,2],[tran [0,10,1],[0,22,2],
[1,13,0],[2,23,0]]) (3,4)
IPSSPEC = **sync**(**IPSSPEC,ALL**) (3,70) Blocked_events = None

Compute a modular supervisor for the IPS specification.

IPSSUP = **supcon**(**AGV,IPSSPEC**) (2304,10944)
IPSSUP = **condat**(**AGV,IPSSUP**) Controllable.

Compute the centralized supervisor for combined zonal and workstation specifications.

ZWSSPEC = **meet(ZSPEC,WSSPEC)** (1296,7884)
ZWSSUP = **supcon(AGV,ZWSSPEC)** (6910,18226)

Compute the monolithic centralized supervisor for combined zonal, workstation and input parts stations specifications.

ZWSISPEC = **meet(ZWSSPEC,IPSSPEC)** (3888,20196)
ZWSISUP = **supcon(AGV,ZWSISPEC)** (4406,11338)
ZWSISUP = **condat(AGV,ZWSISUP)** Controllable.

We can reduce the size of this monolithic supervisor.

ZWSISMSP = **supreduce(AGV,ZWSISUP,ZWSISUP)** (494,3544;slb=169)

TEST = **meet(AGV,ZWSISMSP)** (4406,11338)
isomorph(ZWSISUP,TEST) = *true*

Modular supervisors and their reduced versions

Zone 1 modular exclusion specification

AGV12 = **sync(AGV1,AGV2)** (32,64) Blocked_events = None
ALL12 = **allevents(AGV12)** (1,12)
Z1C12 = **meet(Z1SPEC,ALL12)** (3,20)
Z1P12 = **supcon(AGV12,Z1C12)** (24,36)
Z1P12 = **condat(AGV12,Z1P12)** Controllable.
Z1R12 = **supreduce(AGV12,Z1P12,Z1P12)** (2,14;slb=2)
Z1R = **sync(Z1R12,ALL)** (2,42) Blocked_events = None

Zone 2 modular exclusion specification

AGV23 = **sync(AGV2,AGV3)** (32,64) Blocked_events = None
ALL23 = **allevents(AGV23)** (1,12)
Z2C23 = **meet(Z2SPEC,ALL23)** (3,20)
Z2P23 = **supcon(AGV23,Z2C23)** (24,36)
Z2P23 = **condat(AGV23,Z2P23)** Controllable.
Z2R23 = **supreduce(AGV23,Z2P23,Z2P23)** (2,14;slb=2)
Z2R = **sync(Z2R23,ALL)** (2,42) Blocked_events = None

Zone 3 modular exclusion specification

AGV24 = **sync(AGV2,AGV4)** (48,96) Blocked_events = None
ALL24 = **allevents(AGV24)** (1,14)
Z3C24 = **meet(Z3SPEC,ALL24)** (3,26)
Z3P24 = **supcon(AGV24,Z3C24)** (36,58)
Z3P24 = **condat(AGV24,Z3P24)** Controllable.
Z3R24 = **supreduce(AGV24,Z3P24,Z3P24)** (2,17;slb=2)
Z3R = **sync(Z3R24,ALL)** (2,41) Blocked_events = None

Zone 4 modular exclusion specification

AGV45 = **sync(AGV4,AGV5)** (24,48) Blocked_events = None
ALL45 = **allevents(AGV45)** (1,10)
Z4C45 = **meet(Z4SPEC,ALL45)** (3,14)
Z4P45 = **supcon(AGV45,Z4C45)** (18,26)
Z4P45 = **condat(AGV45,Z4P45)** Controllable.
Z4R45 = **supreduce(AGV45,Z4P45,Z4P45)** (2,11;slb=2)
Z4R = **sync(Z4R45,ALL)** (2,43) Blocked_events = None

Global combination of reduced modular supervisors for zonal exclusion specs

ZR = **meet(Z1R,Z2R)** (4,69)
ZR = **meet(ZR,Z3R)** (8,108)
ZR = **meet(ZR,Z4R)** (16,160)
nonconflict(AGV,ZR) = *true*
ZR = **condat(AGV,ZR)** Controllable.
AGVZR = **meet(AGV,ZR)** (1112,3776)
isomorph(ZSUP,AGVZR) = *true* [map omitted]

Thus the modular reduced zonal supervisor (16 states) is control-equivalent to the global zonal supervisor (1112 states) as expected.

Direct reduction of **ZSUP**

ZSUP = **condat(AGV,ZSUP)** Controllable.
ZSIM = **supreduce(AGV,ZSUP,ZSUP)** (30,270;slb=16)

Thus **ZSIM** happens to be larger than simply the meet of **Z1R**,...,**Z4R**, which has size exactly the slb.

Modular supervisors for workstation sequencing specifications

Modular supervisor for Workstation 1 sequencing

AGV345 = **sync(AGV3,AGV4)** (24,48)
Blocked_events = None
AGV345 = **sync(AGV345,AGV5)** (96,288)
Blocked_events = None
ALL345 = **allevents(AGV345)** (1,14)
WS1C345 = **meet(WS1SPEC,ALL345)** (4,49)
WS1P345 = **supcon(AGV345,WS1C345)** (222,550)
WS1P345 = **condat(AGV345,WS1P345)** Controllable.
WS1R345 = **supreduce(AGV345,WS1P345,WS1P345)**
(4,40;slb=4)
WS1R345 = **condat(AGV345,WS1R345)** Controllable.
WS1R = **sync(WS1R345,ALL)** (4,88)
Blocked_events = None
nonconflict(AGV,WS1R) = *true*
WS1R = **condat(AGV,WS1R)** Controllable.

Modular supervisor for Workstation 2 sequencing

AGV13 = **sync(AGV1,AGV3)** (16,32) Blocked_events = None
ALL13 = **allevents(AGV13)** (1,8)
WS2C13 = **meet(WS2SPEC,ALL13)** (2,14)
WS2P13 = **supcon(AGV13,WS2C13)** (24,40)
WS2P13 = **condat(AGV13,WS2P13)** Controllable.
WS2R13 = **supreduce(AGV13,WS2P13,WS2P13)** (2,12;slb=2)
WS2R = **sync(WS2R13,ALL)** (2,48) Blocked_events = None

Modular supervisor for Workstation 3 sequencing

AGV24 = **sync(AGV2,AGV4)** (48,96) Blocked_events = None
ALL24 = **allevents(AGV24)** (1,14)
WS3C24 = **meet(WS3SPEC,ALL24)** (2,26)
WS3P24 = **supcon(AGV24,WS3C24)** (62,110)
WS3P24 = **condat(AGV24,WS3P24)** Controllable.
WS3R24 = **supreduce(AGV24,WS3P24,WS3P24)** (2,21;slb=2)
WS3R = **sync(WS3R24,ALL)** (2,45) Blocked_events = None
nonconflict(AGV,WS3R) = *true*

Compute the global version of the workstation reduced supervisors.

WSR = **meet**(**WS1R,WS2R**) (8,160)
WSR = **meet**(**WSR,WS3R**) (16,264)
WSSUP = **supcon**(**AGV,WSSPEC**) (12666,47730)
AGVWSR = **meet**(**AGV,WSR**) (12666,47730)
isomorph(**WSSUP,AGVWSR**) = *true* [map omitted]

Thus **WSR** (16 states) is control-equivalent to **WSSUP** (12666 states).

Modular supervisor for IPS exclusion specification

IPSC12 = **meet**(**IPSSPEC,ALL12**) (3,28)
IPSP12 = **supcon**(**AGV12,IPSC12**) (24,42)
IPSP12 = **condat**(**AGV12,IPSP12**) Controllable.
IPSR12 = **supreduce**(**AGV12,IPSP12,IPSP12**) (2,16;slb=2)
TEST = **allevents**(**IPSR12**) (1,12)
isomorph(**ALL12,TEST**) = *true*
IPSR = **sync**(**IPSR12,ALL**) (2,44) Blocked_events = None
nonconflict(**AGV,IPSR**) = *true*
IPSR = **condat**(**AGV,IPSR**) Controllable.
TEST = **allevents**(**IPSR**) (1,26)
isomorph(**ALL,TEST**) = *true*
AGVIPSR = **meet**(**AGV,IPSR**) (2304,10944)
isomorph(**AGVIPSR,IPSSUP**) = *true* [map omitted]

Combine zonal and workstation modular supervisors into global version.

ZWS1R = **meet**(**ZR,WS1R**) (64,528)
nonconflict(**AGV,ZWS1R**) = *true*
ZWS1R = **condat**(**AGV,ZWS1R**) Controllable.
TEST = **meet**(**AGV,ZWS1R**) (3034,9220)
MTEST = **minstate**(**TEST**) (3034,9220)
nonconflict(**ZWS1R,WS2R**) = *true*
ZWS12R = **meet**(**ZWS1R,WS2R**) (128,928)
nonconflict(**ZWS12R,AGV**) = *true*
ZWS12R = **condat**(**AGV,ZWS12R**) Controllable.
ZWS123R = **meet**(**ZWS12R,WS3R**) (256,1600)

Examine **ZWS123R** as a modular supervisor for **AGV** subject to the zonal and workstation specifications.

TEST = **minstate(ZWS123R)** (256,1600)
isomorph(ZWS123R,TEST) = *true*
TZWS123R = **trim(ZWS123R)** (256,1600)
isomorph(TZWS123R,ZWS123R) = *true*
nonconflict(AGV,ZWS123R) = *true*
ZWS123R = **condat(AGV,ZWS123R)** Controllable.

Thus **ZWS123R** is a proper supervisor for **AGV**. We get the controlled behavior (for zonal and workstation specifications) as follows.

AGVZWS = **meet(AGV,ZWS123R)** (6910,18226)
nonconflict(AGVZWS,ALL) = *true*
isomorph(AGVZWS,ZWSSUP) = *true* [map omitted]

Combine zonal/workstation global modular supervisor with modular IPS supervisor. Investigate whether the result is nonblocking.

nonconflict(ZWS123R,IPSR) = *true*
ZWSIR = **meet(ZWS123R,IPSR)** (512,2608)
nonconflict(ZWSIR,ALL) = *true*

ZWSIR is state-minimized.

TEST = **minstate(ZWSIR)** (512,2608)
isomorph(TEST,ZWSIR) = *true*

---

nonconflict(ZWS123R,IPSR) = *true*
nonconflict(IPSR,ZWSSUP) = *false*

While the meet of the first 7 modular reduced supervisors (the 4 zonal and the 3 workstation sequencing modules) provides an optimal (proper) modular supervisor for **AGV**, conflict arises when the IPS specification is introduced. Conflict could be removed in the final step of the synthesis, by global supervision, but the resulting 'global' coordinator **ZWSIRSUP** merely reproduces the monolithic centralized supervisor, and is therefore unacceptable.[3]

[3]It will be shown in Sect. 6.8 how this apparently useless coordinator can be abstracted to one that is small and efficient. For now, we proceed 'naively'.

nonconflict(AGV,ZWS123R) = *true*
nonconflict(ZWS123R,IPSR) = *true*
nonconflict(AGV,ZWSIR) = *false*
ZWSIRSUP = **supcon(AGV,ZWSIR)** (4406,11338)
AGVZWSIR = **meet(AGV,ZWSIR)** (4424,11368)
TAGVZWSI = **trim(AGVZWSIR)** (4406,11338)
isomorph(ZWSIRSUP,TAGVZWSI) = *true*

Thus the final **supcon** operation simply removes inadmissible states from the meet of **AGV** with the meet of the 8 reduced modular supervisors.

It requires some trial-and-error to find a (relatively) small coordinator to achieve the same result economically: namely to remove the $4424 - 4406 = 18$ states in **AGVZWSIR** that are either not coreachable or lead to non-coreachable states along uncontrollable paths. We return to this problem after some further checks on the results so far.

Rename the final optimal nonblocking controlled behavior. We run some routine checks to confirm nonblocking and that the specifications are all enforced.

SUPER = **edit(ZWSIRSUP)** (4406,11338)
nonconflict(SUPER,ALL) = *true*
COZSPEC = **complement(ZSPEC**,[]) (82,2132)
TEST = **meet(SUPER,COZSPEC)** (4406,11338)
TTEST = **trim(TEST)** (0,0)
COWSSPEC = **complement(WSSPEC**,[]) (17,442)
TEST = **meet(SUPER,COWSSPEC)** (4406,11338)
TTEST = **trim(TEST)** (0,0)
COIPSSPE = **complement(IPSSPEC**,[]) (4,104)
TEST = **meet(SUPER,COIPSSPE)** (4406,11338)
TTEST = **trim(TEST)** (0,0)

We project to check that the AGVs are all embedded in the controlled behavior without constraint.

P1SUPER = **project(SUPER,image**[10,11,12,13]) (4,4)
isomorph(AGV1,P1SUPER) = *true*
P2SUPER = **project(SUPER,image**[18,20,21,22,23,24,26,28]) (8,8)
isomorph(AGV2,P2SUPER) = *true*
P3SUPER = **project(SUPER,image**[31,32,33,34]) (4,4)
isomorph(AGV3,P3SUPER) = *true*
P4SUPER = **project(SUPER,image**[40,41,42,43,44,46]) (6,6)
isomorph(AGV4,P4SUPER) = *true*
P5SUPER = **project(SUPER,image**[50,51,52,53]) (4,4)
isomorph(AGV5,P5SUPER) = *true*

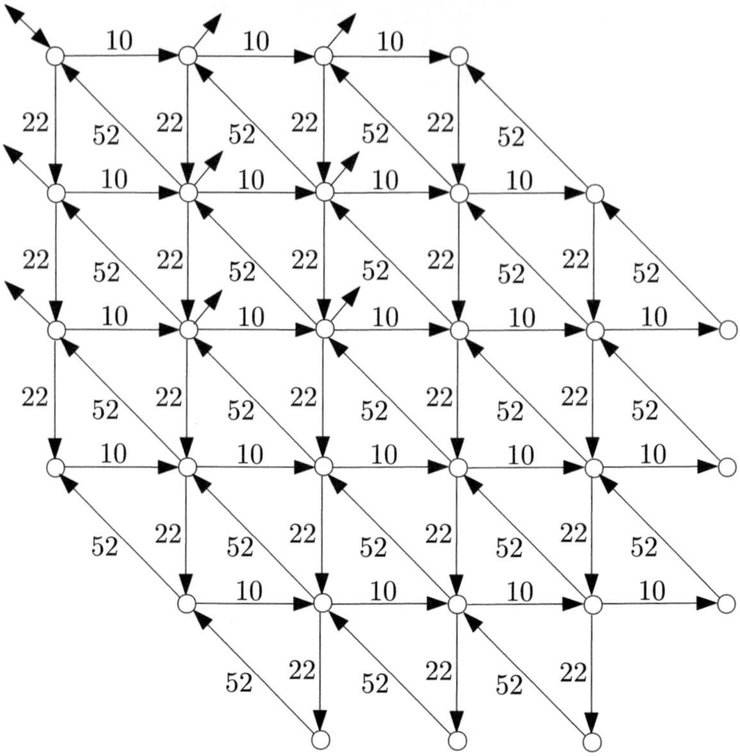

Fig. 4.4 **IOSUPER**: input–output controlled behavior

Finally we project on the input–output events (IPS to CPS) to display the input–output behavior (importing workpieces of type 1 with event <10>[4] and type 2 <22>, and exporting assembled workpiece <52>).

IOSUPER = **project(SUPER,image**[10,22,52]) (29,64)

The result is displayed in Fig. 4.4. It shows, as expected, complete symmetry between parts of type 1 and type 2, as loaded by the two IPS onto **AGV1** and **AGV2** respectively. The system can contain at most 5 parts of type 1 and 4 of type 2 (or else respectively 4, 5) before it must unload a fully processed and assembled workpiece at the CPS. The maximum work-in-progress (WIP) is therefore 9. The system can only return to its initial state after having processed an equal number of parts of each type.

Projecting **IOSUPER** onto event <10> and <52>, say, shows that, as seen by the input–output abstraction IPS1→CPS, the system behaves like a buffer of capacity 5.

P22IOSUP = **project(IOSUPER,null**[22]) (6,10)

[4]In this section, an event or string may be enclosed in brackets $< \cdot >$ for clarity.

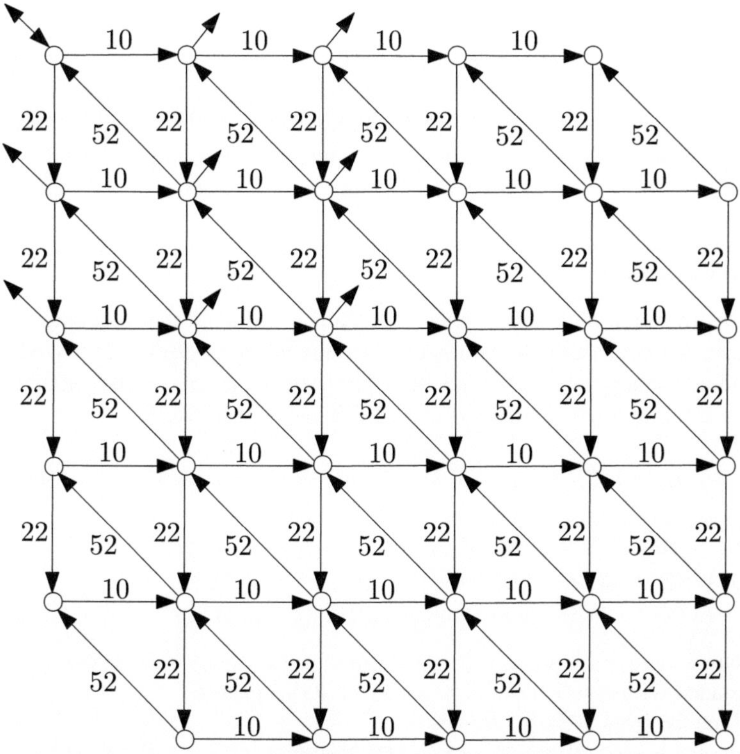

Fig. 4.5 **PAGVZWS**: input–output controlled behavior with **IPSSPEC** omitted

For comparison we project the controlled behavior under supervisor **ZWS123R** alone, i.e. subject only to the zonal exclusion and workstation sequencing constraints. The result **PAGVZWS** is more permissive, of course, than before, since the constraint **IPSSPEC** is omitted. The transition graph (Fig. 4.5) reveals that maximum WIP is now 10, namely 5 of each workpiece type can be accepted by the system before a delivery to CPS (event <52>) must be executed. Thus the '5/4' constraint mentioned above is due to **IPSSPEC**.

PAGVZWS = **project**(**AGVZWS**, **image**[10, 22, 52])(34, 81)
PAGVZWS = **BFS**(**PAGVZWS**)(34, 81)

Projecting out event <22> (importing a type 2 workpiece) yields a buffer of capacity 5 for the IPS1 $\rightarrow$ CPS abstraction, just as before.

P22AGVZW = **project**(**PAGVZWS**,**null**[22]) (6,10)
isomorph(**P22AGVZW**,**P22IOSUP**) = *true*

Conclusion so far:

Efficient modular control was achieved for the combination of zonal exclusion and workstation sequencing specifications. Although the monolithic problem with the additional input-parts-stations exclusion specification could indeed be solved, we must still find a simple modular way of doing so, which will resolve the conflict between the supervisor for this specification and the controlled behavior for the previous specifications.

Blocking analysis:

We investigate the blocking problem in detail. It turns out that blocking is due to deadlock. A search of the transition table for the blocking DES **AGVZWSIR** turns up deadlock states from which deadlocking event sequences can be extracted. One of these is the following, common to **SUPER** and **AGVZWSIR**.

```
                                       %            #         *
11 10 13 12 | 11 33 34 10 13 31 32 12 | 11 33 34 10 13 12 |   (+)
```

Problematic events in this string are the following.
%<32> AGV3 → WS1 [AGV3 unloads part type 1 to WS1]
#<34> WS2 → AGV3 [AGV3 loads part type 1 from WS2]
*<12> AGV1 → WS2 [AGV1 unloads part type 1 to WS2]

This leaves
WS1 loaded with type 1 by AGV3 %<32>, AGV3 parked
AGV3 loaded from WS2 by #<34>
WS2 loaded from AGV1 by *<12>, AGV1 parked

Thus the next events that should be enabled are
<21> AGV2 unparks
<22> IPS2 → AGV2 [type 2] <23 24 25 26>
<28> AGV2 → WS3 [type 2], AGV2 parks
<42> WS3 → AGV4 [type 2]
<46> AGV4 → WS1 [type 2]
<50> WS1 → AGV5 [output assembled part]

However, **AGZVWSIR** allows <11 10> following (+), resulting in deadlock! After <11> **AGZVWSIR** must disable <21> to prevent a Zone 1 collision via <21 18 20>. Why not allow <21> after <10>? Evidently because **IPSR** disables <21> following <11 10>. This is because of the event interpretation that AGV1 only exits from the common workspace of IPS1 and IPS2 by executing <13>. Only then, by **IPSSPEC**, can AGV2 execute <22>. The situation would not be cured by more detailed modeling of AGV1, say replacing <10> by a sequence <10 14>, with <14> the exit from workspace. However, if AGV1 executes <13> then it enters Zone 1; hence <21> still cannot be enabled lest **Z1SPEC** be violated. So AGV1 must continue on to do <12>, which is impossible because WS2 has already been loaded on the previous AGV1 cycle by *<12>, and cannot be loaded again until it has been cleared by AGV3 on event <34>. But this too is impossible, because AGV3 has already been loaded from WS2 by #<34>.

Following (+), the (nonblocking) monolithic controller **SUPER** disallows <11 10> but in effect forces <21>, so deadlock is prevented. In other words, **SUPER** prevents deadlock by forcing AGV2 into action, which eventually allows transfers <22>IPS2 → AGV2, <28>AGV2 → WS3, <42>WS3 → AGV4, <46>AGV4 → WS1. The result is to allow AGV1 to execute <11 10 13> and the transfer <12>AGV1 → WS2.

We conclude that the blocking/deadlock situation is rather complicated because of the interplay of several subsystems and their mutual constraints. Thus a modular solution (if one exists) might turn out to be complicated as well. In the present example it is now evident that deadlock is due to 'choking' on too many input parts allowed into the system at one time. With this in mind we examine the global I/O behavior **IOSUPER** (Fig. 4.4) which exhibits how inputs (or WIP) must be constrained if deadlock is to be prevented. Thus we attempt to convert **IOSUPER** into a controllable version which, as a modular supervisor, would yield the correct result for controlled behavior.

Provisional decentralized control:

We can obtain an acceptable result in two different ways. The first is simply to convert **IOSUPER** directly by moving the uncontrollable events in their respective AGVs back to their nearest controllable events upstream in the parts flow (i.e. <10>→<11>, <22>→<21> and <52>→<53>).

BUXR125 = **relabel(IOSUPER**,[[10,11],[22,21],[52,53]]) (29,64)
BUXR = **sync(BUXR125,ALL)** (29,731) Blocked_events = None
BUXR = **condat(AGV,BUXR)** Controllable.
nonconflict(AGV,BUXR) = *true*
ZWSIBR = **meet(ZWSIR,BUXR)** (14813,72482)
TEST = **meet(AGV,ZWSIBR)** (4406,11338)
isomorph(SUPER,TEST) = *true*

This confirms that **BUXR** does indeed provide the correct coordinating control action to remove the blocking. We could also have obtained **BUXR** by projecting **SUPER** onto the controllable events identified above:

CUXR125 = **project(SUPER,image**[11,21,53]) (29,64)
isomorph(CUXR125,BUXR125) = *true*

Provisional conclusion:

To identify the coordinating control action required to remove (possible) blocking (often deadlock) from the action of synchronized modular supervisors, one may need to examine the global (optimal) behavior as projected onto the relevant input–output events. This could entail trial-and-error, requiring human insight. The fact is that blocking and deadlock can arise from quite subtle interactions of the various plant components and modular controllers, and the rectifying control action may well be rather nonobvious. Deadlock in an I/O system will usually occur because the system becomes ‘choked’ with input items, and this will occur relatively far down the reachability tree of blocking behavior. This means that the required coordinating modular controller may be rather complex.

In the present example our provisional result is acceptable. Corresponding to the 8 specifications Z1, Z2, Z3, Z4, WS1, WS2, WS3, IPS we obtained 8 modular supervisors of state size 2, 2, 2, 2, 4, 2, 2, 2 having product size 512. The resulting product supervisor failed to be nonblocking for the plant. Investigation showed that to achieve nonblocking required the controlled removal of $4424 - 4406 = 18$ states. This was achieved using ‘input–output heuristics’ to obtain an auxiliary, ninth modular supervisor, or coordinator, of 29 states. The combination of all 9 modular supervisors was verified to be control-equivalent to the (optimal) monolithic centralized supervisor.

Refined decentralized control:
To refine our approach to decentralized control of the AGV system we employ a decomposition and ‘direct’ method due to L. Feng. For this we make the important preliminary observation that the zonal specifications **ZiSPEC** and their reduced modular supervisors **ZiR** ($i = 1, 2, 3, 4$) can be temporarily disregarded. In justification bring in

IPSWSSPEC = **meet(IPSSPEC, WSSPEC)** (48,940)
IPSWSSUP = **supcon(AGV, IPSWSSPEC)** (7800,26248)

for the combined IPS and WS specifications and corresponding optimal controlled behavior. Now we verify

nonconflict(ZSUP, IPSWSSUP) = *true*

namely **ZSUP** can introduce no new blocking behavior if **IPSWSSUP** is nonblocking already. By Proposition 4.2.3 we conclude

ZWSISUP = **supcon(AGV, ZWSISPEC)** (4406,11338)
$\simeq$ **meet(ZSUP, IPSWSSUP)**

where $\simeq$ denotes isomorphism. This means that global optimal supervision is achieved by online supervision of the combined zonal specifications, and the combined input-parts and workstation specifications, independently. Thus it will be sufficient to construct a refined decentralized supervisor that is control-equivalent to **IPSWSSUP**.

To this end we make the second key observation that **WS1SPEC** can be decomposed as

WS1SPEC = **sync(WS13SPEC, WS14SPEC)**

separating the couplings of **WS1** to **AGV3** and **AGV4**, as displayed below.

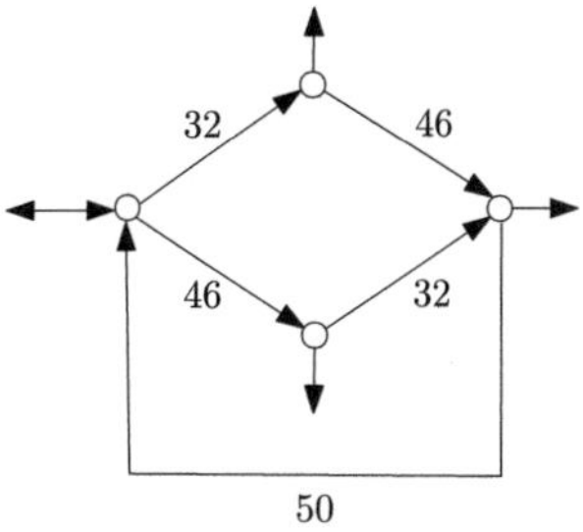

WS1SPEC

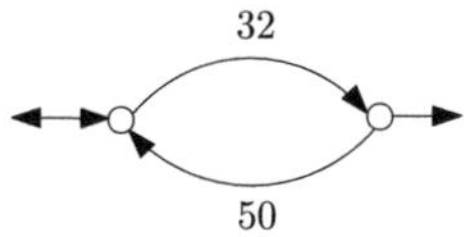

WS13SPEC

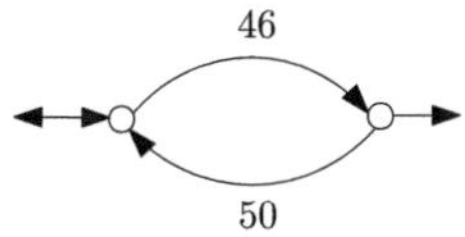

WS14SPEC

We can thereby present the control flow of **IPSWSSPEC** as shown in Fig. 4.6.

In Fig. 4.6, ovals denote specifications (with the suffix 'SPEC' omitted). The WSj ($j = 13, 14, 2, 3$) are buffers, while IPS is considered a 'server' that is shared by **AGV1** and **AGV2** providing to each an independent service: the supply of part type 1 by event 10 to **AGV1**, or type 2 by event 22 to **AGV2**. Notice that the temporary removal of the **ZiSPEC**, and the decomposition of **WS1**, have resulted in a control flow of 'parallel' type that is logically transparent.

In the usual way we begin the modular control design by computing reduced supervisors (suffix **SIM**) for each of the 5 specifications pictured in Fig. 4.6:

AGV35 = **sync(AGV3, AGV5)** (16,32)
WS13SUP = **supcon(AGV35, WS13SPEC)** (24,40)
WS13SUP = **condat(AGV35, WS13SUP)**
WS13SIM = **supreduce(AGV35, WS13SUP, WS13SUP)** (2,12)
AGV45 = **sync(AGV4, AGV5)** (24,48)
WS14SUP = **supcon(AGV45, WS14SPEC)** (34,58)
WS14SUP = **condat(AGV45, WS14SUP)**
WS14SIM = **supreduce(AGV45, WS14SUP, WS14SUP)** (2,15)
AGV13 = **sync(AGV1, AGV3)** (16,32)
WS2SUP = **supcon(AGV13, WS2SPEC)** (24,40)
WS2SUP = **condat(AGV13, WS2SUP)**
WS2SIM = **supreduce(AGV13, WS2SUP, WS2SUP)** (2,12)
AGV24 = **sync(AGV2, AGV4)** (48,96)
WS3SUP = **supcon(AGV24, WS3SPEC)** (62,110)
WS3SUP = **condat(AGV24, WS3SUP)**
WS3SIM = **supreduce(AGV24, WS3SUP, WS3SUP)** (2,21)
AGV12 = **sync(AGV1, AGV2)** (32,64)
IPSSUP = **supcon(AGV12, IPSSPEC)** (24,42)
IPSSUP = **condat(AGV12, IPSSUP)** (24,42)
IPSSIM = **supreduce(AGV12, IPSSUP, IPSSUP)** (2,16)

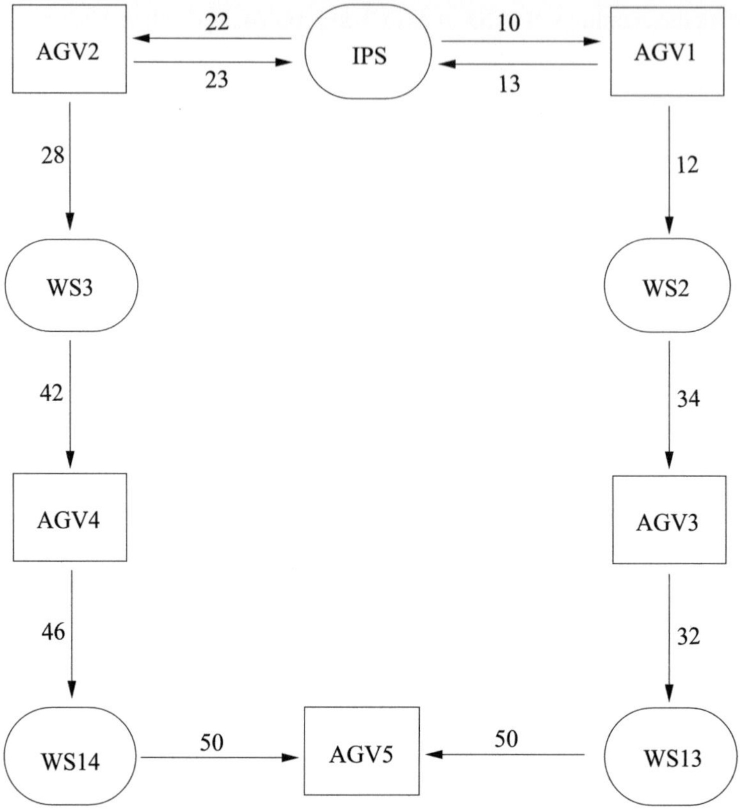

Fig. 4.6 Control flow of IPSWSSPEC

Combining these results we compute

MODSIM0 = **sync**(**WS13SIM, WS14SIM, WS2SIM, WS3SIM, IPSSIM**) (32,416)

but find, as expected,

nonconflict(**AGV, MODSIM0**) = *false*

It remains to construct a coordinator, and Fig. 4.6 suggests how this should be done. Note that the left-hand and right-hand branches

L = **AGV2** - **WS3** - **AGV4** - **WS14**
R = **AGV1** - **WS2** - **AGV3** - **WS13**

can each hold at most 4 workpieces, written $|\mathbf{L}| \leq 4$, $|\mathbf{R}| \leq 4$, say. If $|\mathbf{L}|$ reaches 4 while $|\mathbf{R}| = 0$ then event 50 is blocked. Having just executed event 22, **AGV2** is held at state 4 by disablement of event 23, as otherwise it could uncontrollably execute event 28 and overflow **WS3**. By **IPSSPEC**, event 10 is therefore blocked, so $|\mathbf{R}|$ is held at 0 indefinitely, and the system deadlocks.

By this argument and its symmetric counterpart a coordinator should enforce the condition $||\mathbf{L}| - |\mathbf{R}|| \leq 3$, as expressed by the control-flow specification **CFSPEC** shown below.

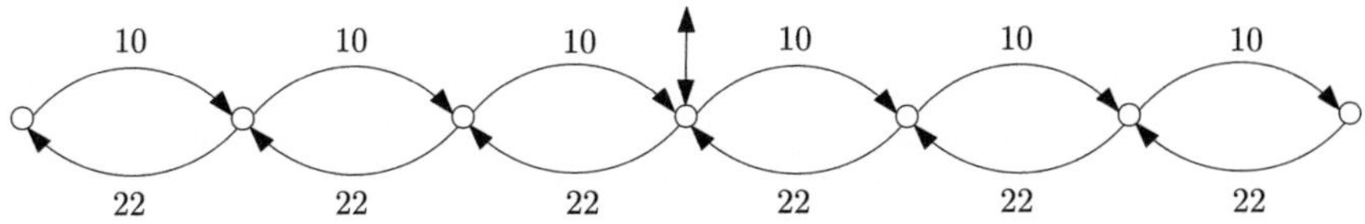

CFSPEC

From this we obtain

CFSUP = **supcon(AGV12, CFSPEC)** (204,396)
CFSUP = **condat(AGV12, CFSUP)**
CFSIM = **supreduce(AGV12, CFSUP, CFSUP)** (7,78)

Now we are ready to compute

MODSIM1 = **sync(MODSIM0, CFSIM)** (224, 2816)

and find, as we hoped,

nonconflict(AGV, MODSIM1) = *true*

Furthermore the controlled behavior

TEST = **meet(AGV, MODSIM1)** (7800, 26248)

is exactly as required, namely

isomorph(TEST, IPSWSSUP) = *true*

We have now justified the full modular supervisor with 10 components

Z1R, Z2R, Z3R, Z4R, WS13SIM, WS14SIM, WS2SIM, WS3SIM, IPSSIM, CFSIM

All but **CFSIM** have state size 2, while |**CFSIM**| = 7.

While straightforward and easily understood, the foregoing approach to modular supervision has involved some large-scale computation at the level of monolithic design. In fact this could be avoided by suitable analysis and the use of abstraction, for which the reader may consult the original work of L. Feng cited in Sect. 4.12 below. The reader is also referred to Sect. 6.8 for an alternative approach using monolithic coordinator abstraction, leading to an equally efficient coordinator.

4.8 Decentralized Supervision by Natural Projection

An approach to decentralized supervision that is often effective is to base the design of each modular component supervisor on a natural projection of the monolithic plant on suitable local components. This will be feasible especially when the plant $\mathbf{G}$ is the synchronous product of components $\mathbf{G_i}$ over pairwise disjoint alphabets Σ_i, and when each specification involves only a small subset of the $\mathbf{G_i}$. Projection onto a subset amounts to forming the synchronous product of $\mathbf{G_i}$ over this subset alone, hence runs no risk of the exponential state explosion attendant on natural projection in general (cf. Exercise 3.3.18). A convincing example is the decentralized AGV supervisor of Sect. 4.7.

In this section we explore 'supervision by projection' in somewhat greater generality. Consider first a DES $\mathbf{G}$ over the alphabet Σ. For simplicity assume that $\mathbf{G}$ is trim, so $L(\mathbf{G}) = \overline{L_m(\mathbf{G})}$. Let $\Sigma_o \subseteq \Sigma$ be a subalphabet of 'observable events' and let $P : \Sigma^* \to \Sigma_o^*$ be the corresponding natural projection. The 'observer's local model' is then the projection $\mathbf{G_o}$ of $\mathbf{G}$ on Σ_o, namely

$$L_m(\mathbf{G_o}) = PL_m(\mathbf{G}),\ \ L(\mathbf{G_o}) = PL(\mathbf{G})$$

With $\Sigma = \Sigma_c \dot{\cup} \Sigma_u$ as usual, we define

$$\Sigma_{oc} = \{\sigma \in \Sigma_c | P\sigma = \sigma\} = \Sigma_o \cap \Sigma_c$$
$$\Sigma_{ou} = \{\sigma \in \Sigma_u | P\sigma = \sigma\} = \Sigma_o \cap \Sigma_u$$

thus assigning control structure to $\mathbf{G_o}$.

Let $E_o \subseteq L_m(\mathbf{G_o})$ be a specification language for the controlled behavior of $\mathbf{G_o}$; in other words, our specification is declared 'locally' (in $\mathbf{G_o}$) rather than 'globally' (in $\mathbf{G}$), and we consider that a 'local supervisor' can be assigned to $\mathbf{G_o}$. To this end we let $\mathcal{C}_o$ be the family of controllable sublanguages of $L(\mathbf{G_o})$ and write

$$K_o := \sup \mathcal{C}_o(E_o)$$

for the supremal controllable sublanguage of E_o in $\mathbf{G_o}$. Of course we have

$$K_o \subseteq E_o \subseteq L_m(\mathbf{G_o}),\ \ \bar{K}_o \subseteq \overline{L_m(\mathbf{G_o})} = L(\mathbf{G_o})$$

The action of a supervisor implementing K_o in $\mathbf{G_o}$ is to disable only events in Σ_{oc}. The corresponding global supervisory action can be described as follows. Let

$$H := P^{-1}(\bar{K}_o) \cap L(\mathbf{G})$$

Then H is closed, $H \subseteq L(\mathbf{G})$, and (using Exercise 1.4.1(iv))

$$PH = \bar{K}_o \cap PL(\mathbf{G}) = \bar{K}_o \cap L(\mathbf{G_o}) = \bar{K}_o$$

We claim that H is globally controllable, namely as a sublanguage of $L(\mathbf{G})$. For this let $s \in H, \sigma \in \Sigma_u, s\sigma \in L(\mathbf{G})$. If $P\sigma = \epsilon$ then

$$P(s\sigma) = Ps \in \bar{K}_o$$

hence $s\sigma \in P^{-1}(\bar{K}_o) \cap L(\mathbf{G}) = H$; while if $P\sigma = \sigma \in \Sigma_{ou}$, then

$$(Ps)\sigma \in \bar{K}_o\Sigma_{ou}$$

and

$$(Ps)\sigma = P(s\sigma) \in PL(\mathbf{G}) = L(\mathbf{G_o})$$

so

$$(Ps)\sigma \in \bar{K}_o\Sigma_{ou} \cap L(\mathbf{G_o}) \subseteq \bar{K}_o$$

by controllability of $\bar{K}_o$. Again we have $s\sigma \in H$, and the claim is proved.

It remains to identify H as the 'induced' implementation of $\bar{K}_o$ in $\mathbf{G}$. Formally, let $s \in H, \sigma \in \Sigma_c$. Then $s\sigma \in H$ implies $P(s\sigma) \in \bar{K}_o$. If $P\sigma = \sigma \in \Sigma_{oc}$ then $(Ps)\sigma \in \bar{K}_o$; i.e. σ is enabled following Ps in $\bar{K}_o$. Conversely if $s_o\sigma \in \bar{K}_o$ with $\sigma \in \Sigma_{oc}$, and $s\sigma \in L(\mathbf{G})$ with $Ps = s_o$, then $P(s\sigma) = s_o\sigma$, and

$$s\sigma \in P^{-1}(\bar{K}_o) \cap L(\mathbf{G}) = H$$

namely $s \in H$ (as H is closed) and σ is enabled following s. The conclusion is that a local supervisor, say $\mathbf{S}_o$, for $\mathbf{G_o}$ with control over events in Σ_{oc} could be implemented (if need be) by a global supervisor $\mathbf{S}$ for $\mathbf{G}$ with controllable event set $\Sigma_{oc} \subseteq \Sigma_c$. However, the information available to $\mathbf{S}$ following a string $s \in L(\mathbf{G})$ is only $Ps \in L(\mathbf{G_o})$. Because of this informational abstraction it cannot be expected that $\mathbf{S}$ is globally 'optimal', as we shall see.

For the same reason, it cannot be expected that $\mathbf{S}$ is nonblocking for $\mathbf{G}$ (i.e. nonconflicting with $\mathbf{G}$), even though (by construction) $\mathbf{S}_o$ is nonblocking for $\mathbf{G_o}$. For global nonblocking we need to impose on P a new condition of observational consistency. Say P is an $L_m(\mathbf{G})$-*observer* if

$$(\forall s \in L(\mathbf{G}), w \in \Sigma_o^*)(Ps)w \in L_m(\mathbf{G_o})$$
$$\Longrightarrow (\exists v \in \Sigma^*)sv \in L_m(\mathbf{G}) \;\&\; P(sv) = (Ps)w$$

Namely, whenever $s \in L(\mathbf{G})$, and $Ps \in L(\mathbf{G_o})$ can be extended to a marked string of $\mathbf{G_o}$, then s can be extended to a marked string of $\mathbf{G}$ with the same projection.

Let the natural projection $P : \Sigma^* \to \Sigma_o^*$ be an $L_m(\mathbf{G})$-observer, and define

$$E := P^{-1}(E_o) \cap L_m(\mathbf{G})$$

$$K := P^{-1}(K_o) \cap L_m(\mathbf{G})$$

Proposition 1

Under the stated conditions, $\bar{K} = H$, K is controllable, and

$$K \subseteq \sup \mathcal{C}(E)$$

Proof

We first show that $P^{-1}(K_o)$ and $L_m(\mathbf{G})$ are nonconflicting. Let

$$s \in \overline{P^{-1}(K_o)} \cap \overline{L_m(\mathbf{G})}$$

so that $s \in L(\mathbf{G})$ and $sx \in P^{-1}(K_o)$ for some $x \in \Sigma^*$. So

$$P(s)P(x) = P(sx) \in K_o \subseteq L_m(\mathbf{G_o})$$

and by the observer property (with $w = Px$) there is $v \in \Sigma^*$ with $sv \in L_m(\mathbf{G})$ and $P(sv) = P(sx)$. Therefore

$$sv \in P^{-1}(K_o) \cap L_m(\mathbf{G})$$

namely

$$s \in \overline{P^{-1}(K_o) \cap L_m(\mathbf{G})}$$

as required. We now have

$$\begin{aligned} \bar{K} &= \overline{P^{-1}(K_o)} \cap \overline{L_m(\mathbf{G})} \\ &= P^{-1}(\bar{K}_o) \cap L(\mathbf{G}) \\ &= H \end{aligned}$$

and therefore K is controllable (because H is). Clearly $K \subseteq E$, and therefore $K \subseteq \sup \mathcal{C}(E)$, as claimed. □

Note that E can be thought of as the global counterpart of the local specification E_o. A marking nonblocking supervisor $\mathbf{S}$ for $\mathbf{G}$ that globally synthesizes K provides the global counterpart of $\mathbf{S}_o$.

As suggested at the beginning of this section, it will often be true in the application that $\mathbf{G}$ is a synchronous product of independent components over disjoint alphabets, say

$$\mathbf{G} = \mathbf{sync}(\mathbf{G}', \mathbf{G}'')$$

i.e.

$$L_m(\mathbf{G}) = L_m(\mathbf{G}') \parallel L_m(\mathbf{G}''), \; L(\mathbf{G}) = L(\mathbf{G}') \parallel L(\mathbf{G}'')$$

Here we denote the corresponding alphabets by Σ', Σ'' with $\Sigma' \cap \Sigma'' = \varnothing$, so the above synchronous products of languages are simply shuffles. Let $\Sigma := \Sigma' \dot{\cup} \Sigma''$,

$\Sigma_o := \Sigma'$ and, as before, $P : \Sigma^* \to \Sigma_o^*$. It is not hard to see that the $L_m(\mathbf{G})$-observer property for P is satisfied automatically. Furthermore

$$\begin{aligned} H &= (\bar{K}_o \parallel \Sigma''^*) \cap \big(L(\mathbf{G}') \parallel L(\mathbf{G}'')\big) \\ &= \bar{K}_o \parallel L(\mathbf{G}'') \\ E &= (E_o \parallel \Sigma''^*) \cap \big(L_m(\mathbf{G}') \parallel L_m(\mathbf{G}'')\big) \\ &= E_o \parallel L_m(\mathbf{G}'') \\ K &= K_o \parallel L_m(\mathbf{G}'') \end{aligned}$$

In summary, supervisor design for $\mathbf{G}'$ is unaffected by the presence of $\mathbf{G}''$ as long as Σ' and Σ'' are disjoint. If $\mathbf{G}''$ as well as $\mathbf{G}'$ is trim, so that $\overline{L_m(\mathbf{G}'')} = L(\mathbf{G}'')$, then $L_m(\mathbf{G}'')$ is automatically controllable, with the pleasant consequence (left to the reader to verify) that $K = \sup \mathcal{C}(E)$, instead of merely ($\subseteq$) as in the general discussion.

To complete this section we briefly describe an extension to the case where several projections of $\mathbf{G}$ are controlled concurrently. Thus assume $\mathbf{G}$ is given, defined over Σ, and several subalphabets $\Sigma_i \subseteq \Sigma$ $(i \in I)$, not necessarily pairwise disjoint, are selected for observation by independent local observers. Define projections $\mathbf{G_i}$ $(i \in I)$ such that

$$L_m(\mathbf{G_i}) = P_i L_m(\mathbf{G}),\ L(\mathbf{G_i}) = P_i L(\mathbf{G}),\ i \in I$$

where $P_i : \Sigma^* \to \Sigma_i^*$ $(i \in I)$ are the natural projections. Assuming as usual $\overline{L_m(\mathbf{G})} = L(\mathbf{G})$, there follows $\overline{L_m(\mathbf{G_i})} = L(\mathbf{G_i})$ $(i \in I)$. Local control structure is assigned to the $\mathbf{G_i}$ as before:

$$\Sigma_{ic} := \Sigma_i \cap \Sigma_c,\ \Sigma_{iu} = \Sigma_i \cap \Sigma_u,\ i \in I$$

and the controllable sublanguage families $\mathcal{C}_i(\cdot)$ are defined for each $\mathbf{G_i}$ in the expected way.

We assume that local specification languages $E_i \subseteq L_m(\mathbf{G_i})$ are given, for which we compute $K_i = \sup \mathcal{C}_i(E_i)$ $(i \in I)$. As before, the ith local supervisor (synthesizing K_i) induces global closed controllable behavior

$$H_i = P_i^{-1}(\bar{K}_i) \cap L(\mathbf{G}),\ i \in I$$

The concurrent action of all local supervisors is then the global closed controllable behavior

$$H = \bigcap \{H_i \mid i \in I\}$$

Clearly $P_i H \subseteq H_i$ $(i \in I)$, but in general equality fails, as the H_i impose synchronization constraints on one another.

To obtain the desired extension of Proposition 1 we shall need a correspondingly strengthened version of the observer property, which applies uniformly across the index set I. For ease of writing let $I = \{1, ..., k\}$. Accordingly, we say that the family

$\{P_i|i \in I\}$ is an *I-fold* $L_m(\mathbf{G})$*-observer* if

$$(\forall s \in L(\mathbf{G}))\ (\forall t_1, ..., t_k)\ ((\forall i \in I)t_i \in L_m(\mathbf{G_i})\ \&\ P_i s \leq t_i)$$
$$\Longrightarrow (\exists v \in \Sigma^*)sv \in L_m(\mathbf{G})\ \&\ (\forall j \in I)P_j(sv) = t_j$$

Namely, whenever $s \in L(\mathbf{G})$, and each $P_i s \in L(\mathbf{G_i})$ can be extended to a marked string of $L(\mathbf{G_i})$, respectively, then s can be extended to a marked string of $\mathbf{G}$ with the same I-fold projection.

Lemma 2

Let $\{P_i|i \in I\}$ be an I-fold $L_m(\mathbf{G})$-observer. Then the family $\{P_i^{-1}(K_i)|i \in I\}$ is *I-fold nonconflicting*, in the sense that

$$\overline{\bigcap\{P_i^{-1}(K_i)|i \in I\}} \cap L(\mathbf{G}) = \bigcap\{P_i^{-1}(\bar{K}_i)|i \in I\} \cap L(\mathbf{G})$$

Proof

It is enough to prove ($\supseteq$). With $s \in P_i^{-1}(\bar{K}_i) \cap L(\mathbf{G})$ $(i \in I)$ there exist $t_i \in \Sigma_i^*$ such that

$$P_i s \leq t_i \in K_i \subseteq L_m(\mathbf{G_i}),\ i \in I$$

By the I-fold observer property there is $v \in \Sigma^*$ with $sv \in L_m(\mathbf{G})$ and $P_j(sv) = t_j$ $(j \in I)$. Therefore

$$sv \in \bigcap\{P_j^{-1}(K_j)|j \in I\}$$

namely

$$s \in \overline{\bigcap\{P_j^{-1}(K_j)|j \in I\}}$$

as required. □

Now define

$$E := \bigcap\{P_i^{-1}(E_i)|i \in I\} \cap L_m(\mathbf{G})$$
$$K := \bigcap\{P_i^{-1}(K_i)|i \in I\} \cap L_m(\mathbf{G})$$

Then we have the following extension of Proposition 1.

Proposition 3

Let $\{P_i|i \in I\}$ be an I-fold $L_m(\mathbf{G})$-observer. Then $\bar{K} = H$, K is controllable, and $K \subseteq \sup \mathcal{C}(E)$.

Proof

Using the I-fold observer property we easily confirm that the two languages

$$\bigcap\{P_i^{-1}(K_i)|i \in I\},\ L_m(\mathbf{G})$$

are nonconflicting. From this it follows that

$$\bar{K} = \overline{\bigcap\{P_i^{-1}(K_i)|i \in I\}} \cap \overline{L_m(\mathbf{G})}$$

By Lemma 2 and the assumption that **G** is trim,

$$\begin{aligned}\bar{K} &= \bigcap\{P_i^{-1}(\bar{K}_i)|i \in I\} \cap L(\mathbf{G}) \\ &= H\end{aligned}$$

as claimed. Thus K is controllable. Clearly $K \subseteq E$ and so $K \subseteq \sup \mathcal{C}(E)$. □

Our discussion has shown that synthesizing each K_i locally in $\mathbf{G_i}$ is equivalent to globally synthesizing K. The I-fold observer property for the P_i will guarantee that the synthesis is globally nonblocking for **G**.

We conclude with a remark on optimality. As suggested before, **G** will often be given as a synchronous product of independent trim components, say $\mathbf{G'_j}$ $(j \in J)$, over disjoint alphabets Σ'_j. In the typical case considered above, each Σ_i will be precisely a disjoint union of form

$$\Sigma_i = \bigcup\{\Sigma'_j|j \in J_i\}$$

for some subset $J_i \subseteq J$. Thus $\mathbf{G_i}$ can be taken as precisely the synchronous product of the $\mathbf{G'_j}$ over $j \in J_i$. Then for each $i \in I$ we can write

$$\mathbf{G} = \mathbf{sync}(\mathbf{G_i}, \mathbf{G''_i})$$

where $\mathbf{G''_i}$ is defined over

$$\Sigma - \Sigma_i = \bigcup\{\Sigma'_j|j \in J - J_i\}$$

and the above synchronous product is simply a shuffle.

It follows that, for instance,

$$\begin{aligned}\tilde{E}_i &:= P_i^{-1}(E_i) \cap L_m(\mathbf{G}) = E_i \parallel \{L_m(\mathbf{G'_j})|j \in J - J_i\},\ i \in I \\ \tilde{K}_i &:= P_i^{-1}(K_i) \cap L_m(\mathbf{G}) = K_i \parallel \{L_m(\mathbf{G'_j})|j \in J - J_i\},\ i \in I\end{aligned}$$

From this it is easy to see that each local control will be globally optimal, in the sense that

$$\tilde{K}_i = \sup \mathcal{C}(\tilde{E}_i),\ i \in I$$

Now

$$E = \bigcap\{\tilde{E}_i|i \in I\},\ K = \bigcap\{\tilde{K}_i|i \in I\}$$

and therefore

$$\begin{aligned}K &= \bigcap \tilde{K}_i = \bigcap \sup \mathcal{C}(\tilde{E}_i) \\ &\supseteq \sup \mathcal{C}\left(\bigcap \tilde{E}_i\right) \\ &= \sup \mathcal{C}(E)\end{aligned}$$

As we already know by Proposition 3 that K is controllable and $K \subseteq E$, it follows that $K = \sup \mathcal{C}(E)$; namely K is globally optimal as well.

In summary: when **G** is the shuffle of trim components, and when the specifications are 'localizable' to component groups (indexed $i \in I$) subject to the I-fold observer property, then concurrent optimal nonblocking local control for each group provides global control that is also optimal and nonblocking. However, it should be recognized that the I-fold observer property is only a sufficient condition, and may well be too strong for many practical applications. In practice other methods of ensuring nonblocking may be preferred which are likely to be case-dependent (cf. Sect. 4.7).

Exercise 4: Taking the index sets $J = \{1, 2, 3\}$, $J_1 = \{1, 2\}$, $J_2 = \{1, 3\}$, $I = \{1, 2\}$, construct examples of $(\mathbf{G}_1', \mathbf{G}_2', \mathbf{G}_3')$ over pairwise disjoint alphabets $(\Sigma_1', \Sigma_2', \Sigma_3')$ for which the I-fold observer property is (resp. is not) satisfied.

Exercise 5: Investigate the applicability of the results of this section to the AGV system of Sect. 4.7.

4.9 Reasoning About Nonblocking

In many applications the verification that the closed-loop languages implemented by individual modular controllers are nonconflicting can be achieved by exploiting plant structure and its relation to the task decomposition on which the modularity is based. For example, in Small Factory the overall supervisory task was decomposed into subtasks corresponding to 'normal operation' and 'breakdown and repair', of which the latter in a natural sense precedes the former: if either or both machines are broken down, then repair them before continuing with production. To verify that modular supervision is nonblocking, it suffices to show, roughly speaking, that at any state of the system **MODSUP/FACT** a breakdown and repair subtask (possibly null) can be completed first, followed by the completion of a normal operation subtask, in such a way that the system is brought to a marker state. The success of this maneuver depends on the fact that the subtasks of the modular decomposition are ordered in a natural sequence.

We present a simple formalization of this idea on which the reader may model his own versions in the context of more elaborate examples. Adopting the notation of Sect. 4.2, let

$$\mathbf{S_i} = (X_i, \Sigma, \xi_i, x_{0i}, X_{mi}) \quad i = 1, 2$$

For simplicity we assume that X_{m2} is a singleton $\{x_{m2}\}$. Now define

$$\Sigma_1 = \{\sigma \in \Sigma | (\forall x \in X_1)\xi_1(x, \sigma)!\}$$

In particular Σ_1 will include the events that are self-looped at each state of $\mathbf{S_1}$, these being the events to which the operation of $\mathbf{S_1}$ is indifferent, namely the events that are irrelevant to the execution of $\mathbf{S_1}$'s subtask; and these events will typically include those that are relevant to $\mathbf{S_2}$. Next define

$$\Sigma_2 = \{\sigma \in \Sigma | \xi_2(x_{m2}, \sigma) = x_{m2}\}$$

Thus Σ_2 is the subset of events that are self-looped at x_{m2} in $\mathbf{S_2}$, hence to which $\mathbf{S_2}$ is indifferent upon completion of its subtask. We impose two structural conditions on $\mathbf{S_1} \wedge \mathbf{S_2}$:

(i) $(\forall s \in L(\mathbf{S_2}/\mathbf{G}))(\exists t \in \Sigma_1^*) st \in L_m(\mathbf{S_2}) \cap L(\mathbf{G})$
(ii) $(\forall s \in L(\mathbf{S_1}/\mathbf{G}) \cap L_m(\mathbf{S_2}))(\exists t \in \Sigma_2^*) st \in L_m(\mathbf{S_1}/\mathbf{G})$

Condition (i) says that any string of $\mathbf{G}$ that is accepted (but not necessarily marked) by $\mathbf{S_2}$ can be completed to a marked string of $\mathbf{S_2}$ by means of a string that is accepted by $\mathbf{G}$ and $\mathbf{S_1}$. Condition (ii) states that any string that is accepted by $\mathbf{G}$ and $\mathbf{S_1}$ and marked by $\mathbf{S_2}$ can be completed to a marked string of both $\mathbf{S_1}$ and $\mathbf{G}$ by means of a string to which $\mathbf{S_2}$ is indifferent (with $\mathbf{S_2}$ resident in x_{m2}).

Theorem 1

Let $\mathbf{S_1}$ and $\mathbf{S_2}$ be proper supervisors for $\mathbf{G}$. Subject to conditions (i) and (ii) above, the supervisor $\mathbf{S_1} \wedge \mathbf{S_2}$ is nonblocking for $\mathbf{G}$.

Proof

Let $s \in L((\mathbf{S_1} \wedge \mathbf{S_2})/\mathbf{G})$. It must be checked that there exists $t \in \Sigma^*$ such that $st \in L_m((\mathbf{S_1} \wedge \mathbf{S_2})/\mathbf{G})$. Since $s \in L(\mathbf{S_2}/\mathbf{G})$, by condition (i) there is $u \in \Sigma_1^*$ such that $su \in L_m(\mathbf{S_2}) \cap L(\mathbf{G})$. By definition of Σ_1 and the fact that $s \in L(\mathbf{S_1}/\mathbf{G})$ it follows that $su \in L(\mathbf{S_1}/\mathbf{G})$; therefore $su \in L(\mathbf{S_1}/\mathbf{G}) \cap L_m(\mathbf{S_2})$. By condition (ii) there is $v \in \Sigma_2^*$ such that $suv \in L_m(\mathbf{S_1}/\mathbf{G})$; and by definition of Σ_2 we also have $suv \in L_m(\mathbf{S_2})$. This shows that

$$suv \in L_m(\mathbf{S_1}) \cap L_m(\mathbf{G}) \cap L_m(\mathbf{S_2}) = L_m((\mathbf{S_1} \wedge \mathbf{S_2})/\mathbf{G})$$

and the result follows on setting $t = uv$. □

As a straightforward illustration, we apply Theorem 1 to Small Factory. Set $\mathbf{G}$ = **FACT**, $\mathbf{S_1}$ = **BUFSUP** and $\mathbf{S_2}$ = **BRSUP**. Then we have

$$\Sigma_1 = \{\lambda_1, \mu_1, \beta_2, \lambda_2, \mu_2\}, \quad \Sigma_2 = \{\mu_1, \alpha_1, \beta_1, \lambda_1, \alpha_2, \beta_2\}$$

Let $s \in L(\mathbf{BRSUP}/\mathbf{FACT})$. Call the states of **BRSUP** 'idle' and 'active'. If **BRSUP** is active (**MACH2** is broken down), then let $u_1 := \mu_2$ (repair **MACH2**), otherwise

$u_1 := \epsilon$ (do nothing). Call the states of **MACHi** 'idle', 'working' and 'down'. If after s was generated **MACH1** was down then let $u_2 := \mu_1$ (repair **MACH1**), otherwise $u_2 := \epsilon$ (do nothing); and set $u := u_1 u_2$. Then u is accepted by **BUFSUP**, and after su **BRSUP** is resident at its marker state 'idle'. Now use the fact that with each of **MACH1** and **MACH2** either idle or working there is a string v accepted by **BUFSUP** that returns both machines to idle and **BUFSUP** to its marker state (where the buffer is empty), while always keeping **BRSUP** idle. The string uv then suffices to show that **BUFSUP** $\wedge$ **BRSUP** is nonblocking.

Exercise 2: Apply Theorem 1 (or a suitable variation thereof) to show that the modular supervisor for Big Factory (Sect. 4.5) is nonblocking. Repeat the exercise for Transfer Line (Sect. 4.6).

Exercise 3: Consider a manufacturing cell consisting of a robot (**ROB**), input conveyor (**INCON**), input buffer (**INBUF**), machining station (**MS**), output buffer (**OUTBUF**), and output conveyor (**OUTCON**). The operations of **MS** are to download and initialize the machining program, accept a workpiece from **INBUF**, machine it to the specified dimensions, and place it in **OUTBUF**. The preconditions for the process to start are that **MS** should be idle and a workpiece should be available at **INBUF**. **ROB** transfers a workpiece from **INCON** to **INBUF**, provided a workpiece is available, **ROB** is free, and **INBUF** is empty. Similarly, **ROB** transfers a completed workpiece from **OUTBUF** to **OUTCON**. **INBUF** (resp. **OUTBUF**) can be in one of the states: empty (full), being loaded (being unloaded) by the robot, or full (empty). A workpiece follows the path: **INCON, INBUF, MS, OUTBUF, OUTCON**.

Develop a DES model for the workcell with plausible assignments of controllable and uncontrollable events. Investigate both centralized and modular supervision, subject (at least) to the specifications that the buffers never overflow or underflow, and that the supervised system is nonblocking.

Exercise 4: Four jobs **A1**, **A2**, **A3**, **A4** are to be done with two tools **T1**, **T2**. Each job is to be done exactly once. **A1** consists of an initial operation using **T1**, then a final operation using both **T1** and **T2**. **A2** consists of an initial operation using **T2**, then a final operation using both **T2** and **T1**. **A3** uses only **T1**; **A4** uses only **T2**. The four jobs can be done in any order; interleaving of several jobs at a time is permitted.

The jobs are identified with corresponding 'agents'; thus '**Ai** does job i'. Model **T1** on two states, with transitions [0,i11,1] (i = 1,2,3) to mean '**Ai** acquires **T1**', and [1,i10,0] to mean '**Ai** releases **T1**'. Similarly **T2** is modeled on two states with transitions [0,i21,1],[1,i20,0] (i = 1,2,4). After a job is finished, the tool or tools are released, in any order, and finally a 'job completed' signal is output (event i00, i = 1,2,3,4). Thus **A1**, **A2** are each modeled on 6 states, **A3** and **A4** on three. The requirement that the ith job be done exactly once is modeled by the two-state automaton **Di** in which only state 1 is marked, and the appropriate events i11, i21 disabled there to prevent a repetition. **PLANT** is the synchronous product of **A1**, **A2**, **A3**, **A4**; **SPEC** is the synchronous product of **T1**, **T2**, **D1**, **D2**, **D3**, **D4**, **ALL**, where **ALL** is obtained from **allevents**.

Find the global supervisor by **supcon**, and then reduce it using **supreduce**. Compute various projections of interest: for instance, focus on tool usage but (by means of **relabel**) blur the identity of agents, or focus on agents but blur the identity of tools.

To construct a modular supervisor, first note that **PLANT** and **T** (=**sync(T1,T2, ALL)**) conflict, since for instance if **A1** takes **T1** (event 111) and immediately afterwards **A2** takes **T2** (event 221) then deadlock occurs; and similarly for the event sequence 221.111. These sequences can be ruled out *a priori* by suitable specifications (conflict resolvers).

Exercise 5: Consider agents **A1**, **A2**, **A3**, each defined on two states with the initial state (only) marked.

A1 has transitions	$[0, \gamma, 0], [0, \alpha, 1], [1, \alpha, 0]$
A2	$[0, \gamma, 0], [0, \beta, 1], [1, \beta, 0]$
A3	$[0, \gamma, 1], [1, \gamma, 0]$

A1 and **A2** can be thought of as operating two switches which control **A3**. If both switches are RESET (state 0) then **A3** can make the transition $[1,\gamma,0]$ and return home, but if either switch is SET then **A3** is blocked at state 1. Clearly **A1** and **A2** can cycle in such a way that, once **A3** has entered its state 1, it remains blocked forever. Despite this, the overall system **A** = **sync(A1, A2, A3)** is nonblocking in the DES sense. Suggest a possible cure.

Exercise 6: **Dining Philosophers**

In this famous problem (due to E.W. Dijkstra) five philosophers (**P1**,..., **P5**), who spend their lives alternately eating and thinking, are seated at a round table at the center of which is placed a bowl of spaghetti. The table is set with five forks (**F1**, ..., **F5**), one between each pair of adjacent philosophers. So tangled is the spaghetti that a philosopher requires both forks, to his immediate right and left, in order to eat; a fork may not be replaced on the table until its user has temporarily finished eating; and reverts to thinking. No *a priori* constraint is placed on the times at which a philosopher eats or thinks.

Design modular supervisors which guarantee that (1) a fork is used by at most one philosopher at a time, and (2) every philosopher who wishes to eat can eventually do so—i.e. no one is starved out by the eating/thinking habits of others. **Hint:** Model each **Pi** on states [0] *Thinking*, [1] *Ready*, and [2] *Eating*, with the transition from [1] to [2] controllable; and each **Fi** on two states [0] *Free* and [1] *In_use*. You may assume that a philosopher can pick up and replace both his forks simultaneously. A fair way to prevent starvation could be to require that no philosopher may commence eating if either of his two neighbors has been ready longer. For this, equip each philosopher with a queue which he and his two neighbors enter when they are ready to eat. Note that the queue for **Pi** should not distinguish between $\mathbf{P(i-1)}$ and $\mathbf{P(i+1)}$, but only enforce priority between **Pi** and one or both of them; it may be modeled on 9 states. Prove that, under your control scheme, anyone who is ready to eat is *guaranteed* eventually to be able to do so: this is a stronger condition than 'nonblocking', as it prohibits 'livelock' behavior such as **P2** and **P5** cycling in such a way as to lock out **P1**.

A *TCT* modular solution along these lines produced a combined online controller size of (1557,5370) with corresponding controlled behavior of size (341,1005).

Next assume that no philosopher dawdles longer than 1 min if *Ready* and enabled to eat, or spends more than 9 mins *Eating*. Show that a philosopher need wait at most 40 mins at *Ready* before *Eating*. **Hint:** As a worst case for **P5**, suppose that initially **P1**, **P2**,...,**P5** enter *Ready* without delay and in that order.

Exercise 7:**Manufacturing cell with interacting loops**

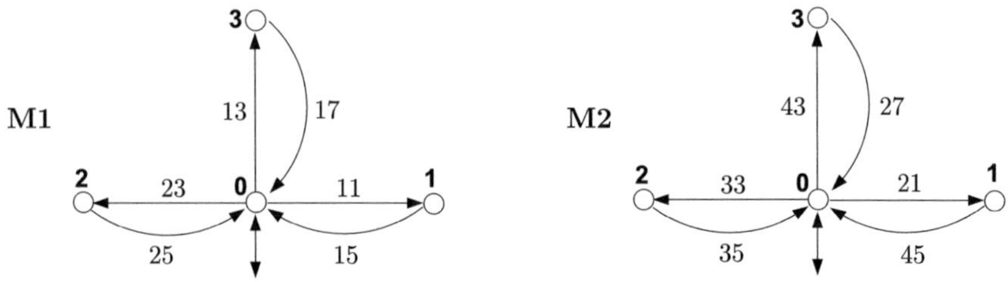

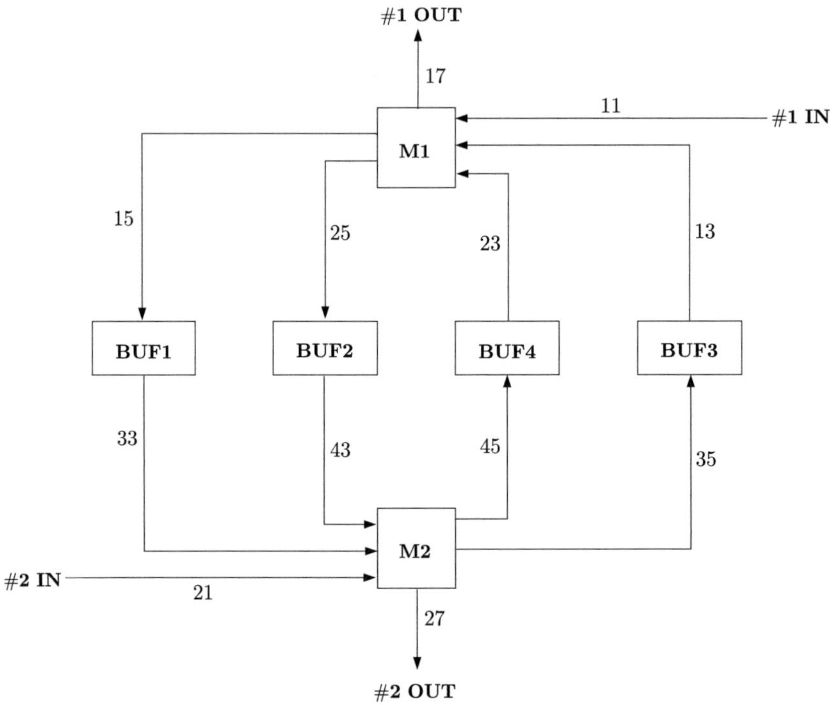

Consider the manufacturing cell displayed, consisting of 2 machines **M1**, **M2**; and 4 buffers **BUF1**,...,**BUF4**. The cell runs 2 production sequences, for workpieces of types 1, 2 respectively. The sequences share use of **M1**, **M2**, but employ separate buffers:

$$\#1: \mathbf{IN} \xrightarrow{11} \mathbf{M1} \xrightarrow{15} \mathbf{BUF1} \xrightarrow{33} \mathbf{M2} \xrightarrow{35} \mathbf{BUF3} \xrightarrow{13} \mathbf{M1} \xrightarrow{17} \mathbf{OUT}$$

$$\#2: \mathbf{IN} \xrightarrow{21} \mathbf{M2} \xrightarrow{45} \mathbf{BUF4} \xrightarrow{23} \mathbf{M1} \xrightarrow{25} \mathbf{BUF2} \xrightarrow{43} \mathbf{M2} \xrightarrow{27} \mathbf{OUT}$$

Numbers on the arrows are event labels; all events are controllable. Each buffer has capacity 1. The specification is no buffer overflow or underflow, together with system nonblocking (with marked initial state for machines and buffers). **M1** and **M2** have similar structure, with events as displayed.

Sequence	Event	Interpretation
1	11	**M1** takes a type #1 workpiece at **IN**
	15	**M1** deposits result in **BUF1**
	33	**M2** takes #1 workpiece from **BUF1**
	35	**M2** deposits result in **BUF3**
	13	**M1** takes #1 workpiece from **BUF3**
	17	**M1** sends finished #1 workpiece **OUT**
2	21	**M2** takes a type #2 workpiece at **IN**
	45	**M2** deposits result in **BUF4**
	23	**M1** takes #2 workpiece from **BUF4**
	25	**M1** deposits result in **BUF2**
	43	**M2** takes #2 workpiece from **BUF2**
	27	**M2** sends finished #2 workpiece **OUT**

In the usual way compute the monolithic supervisor **SUPER** (225,472), which may be reduced to **SIMSUP** (38,258; slb=32). For decentralized control, start with **BUF1**,...,**BUF4** as representations of their individual specifications; combine these with **M1**, **M2** to form **CELL**. As the latter fails to be nonblocking, it turns out that several coordinators are needed. Construct these as follows: **COORD1** (resp. **COORD2**) enforces that at most 3 #1 (resp. #2) workpieces can be in progress at any time; **COORD3** enforces that at most 3 workpieces of either kind can be admitted to the system before a transfer to the second stage of processing (i.e. to **M2** via event 33 for #1 or to **M1** via event 23 for #2); **COORD4** enforces that at most 3 workpieces of either kind can be transferred to the second stage of processing (events 23,33) before a transfer to the output stage (events 13,43). Each coordinator can be modeled on 4 states. Verify that the decentralized architecture comprising the 4 2-state buffer controllers plus the 4 4-state coordinators achieves optimal nonblocking supervision. Analyze the system behavior to explain the rationale underlying this approach.

Repeat this exercise after relabeling the following controllable events to their more 'natural' uncontrollable counterparts: relabel (15,17,25,27,35,45) to, say, (14,16,24,26,34,44) respectively. Explain the (perhaps surprising) difference between the two decentralized architectures.

4.10 Synchronization and Event Hiding

Individual DES can be combined into modules by synchronization followed by projection to achieve event hiding and thus encapsulation. However, care must be taken not to attempt to synchronize an uncontrollable specification with a generator, with respect to an uncontrollable event. The correct procedure would be to compute the supremal controllable sublanguage, and then hide the uncontrollable event. Also, care must be taken not to produce blocking or deadlock.

Example 1:**Small Factory**

Define **MACH1**, **MACH2**, **BUF2** [buffer with 2 slots] as in Example 3.3.19. To plug **MACH1** into **BUF2** requires synchronizing on event 10. Since 10 is uncontrollable, one must compute the supremal controllable sublanguage. For this, take as the specification the (uncontrollable) synchronous product of **MACH1** and **BUF2**: call this **SPEC1**; and as the plant, **MACH1** self-looped with the buffer event 21: call this **SMACH1**.

SPEC1 = sync(MACH1,BUF2) (9,17)
SMACH1 = selfloop(MACH1,[21]) (3,7)
SUPM1B = supcon(SMACH1,SPEC1) (7,12)

MACH1, **BUF2** are now *controllably* synchronized on the shared event 10. Hiding this event, we get

HSUPM1B = project(SUPM1B,null[10])

Thus **HSUPM1B** can be considered as a module with events 11,12,13,21. Let us suppose that the breakdown/repair logic is of no interest, and hide events 12,13. This gives the module

MACH3 = project(HSUPM1B,null[12,13]) (3,5)

Now **MACH3** can be synchronized with **MACH2** on event 21, and events 21,22,23 hidden. This yields the final module **MACH5**, over events 11 ('**MACH1** goes to work') and 20 ('**MACH2** outputs a product').

MACH4 = sync(MACH3,MACH2) (9,20)
MACH5 = project(MACH4,null[21,22,23]) (4,7)

This procedure may be compared with the more standard one of 'monolithic' design:

MACH6 = shuffle(MACH1,MACH2) (9,24)
SPEC2 = selfloop(BUF2,[11,12,13,20,22,23]) (3,22)
MACH7 = supcon(MACH6,SPEC2) (21,49)
MACH8 = project(MACH7,null[10,12,13,21,22,23]) (4,7)

One now can verify that **MACH8**, the standard input–output model for Small Factory under control, is isomorphic with **MACH5**.

Example 2: **Transfer Line**

Systems with feedback loops should be encapsulated by working from inside a loop to the outside. For this system, **M1**, **M2**, **B1**, **B2**, **TU** are created as in Sect. 4.6. From the block diagram, it makes sense to synchronize **M2** with **B2**, then this result with **TU**.

SP1 = **sync(M2,B2)** (4,5)
SM2 = **selfloop(M2**,[5]) (2,4)
SUPM2B2 = **supcon(SM2,SP1)** (3,3)
M3 = **project(SUPM2B2,null**[4]) (2,2)

M3TU = **sync(M3,TU)** (4,7)
M4 = **project(M3TU,null**[5]) (3,6)

At this stage the system can be regarded as the shuffle of plant components **M1** and **M4**, subject to the constraint imposed by **B1** alone as specification. To obtain the corresponding optimal supervisor (**M6**, below) we compute

M5 = **sync(M1**, **M4)** (6,18)
M5ALL = **allevents(M5)** (1,5)
B1SPEC = **sync(B1,M5ALL)** (4,17)
M6 = **supcon(M5,B1SPEC)** (15,36)

Projecting out the internal events 2,3,8 of **M6**, we obtain

M7 = **project(M6,null**[2,3,8]) (4,6)

M7 displays the correct operation of Transfer Line with respect to input event 1 and output event 6, equivalent to a buffer of capacity 3, as shown below.

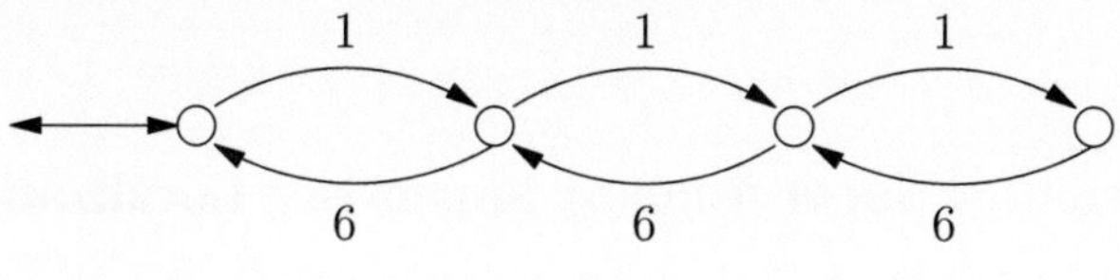

Finally we compute reduced versions of the two modular supervisors **SUPM2B2** and **M6** obtained above.

SUPM2B2 = **condat(SM2,SUPM2B2)**
SIMM2B2 = **supreduce(SM2,SUPM2B2,SUPM2B2)** (2,3;slb=2)
M6 = **condat(M5,M6)**
M6SIM = **condat(M5,M6SIM)**
M6SIM = **supreduce(M5,M6,M6)** (7,25;slb=7)

These can be verified correct in the usual way; for instance

TEST = **meet(M5,M6SIM)** (15,36)
isomorph(M6,TEST) = *true*

We can check the correctness of our decentralized supervisor as follows. Its two modular components yield the equivalent controller

MODSUP = **sync(SIMM2B2,M6SIM)** (14,59)

which is verified to incorporate all events of **TL**:

MODSUPAL = **allevents(MODSUP)** (1,7)
isomorph(MODSUPAL,ALL) = *true*

Furthermore, **MODSUP** is a proper supervisor for **TL**:

TMODSUP = **trim(MODSUP)** (14,59)
isomorph(TMODSUP,MODSUP) = *true*
nonconflict(TL,MODSUP) = *true*
MODSUP = **condat(TL,MODSUP)**

The resulting controlled behavior is

TEST = **meet(TL,MODSUP)** (28,65)

Comparing **TEST** with **CSUP** (Sect. 4.6), we see that **TEST** is optimal, in accordance with

isomorph(CSUP,TEST) = *true*

Finally, it is of interest to compare the reduced supervisor for monolithic control, namely

CSIM = **supreduce(TL,CSUP,CSUP)** (8,31;slb=8)

having (minimal) size 8, with the modular reduced supervisors **SIMM2B2** and **M6SIM**, of (minimal) sizes 2 and 7 respectively.

Exercise 3: Formalize synchronization and event hiding with the aid of the graph-theoretic structure of the plant block diagram.

4.11 Distributed Supervision by Supervisor Localization

As discussed in previous sections of this chapter, decentralized supervision can be viewed as the allocation of control action to distinct specifications. For instance, in Small Factory (Sect. 4.4) we associated a specialized supervisor to each of the specifications concerning buffer control, and priority of repair, respectively.

Each supervisor is decentralized in the sense that it exercises control action with respect to its corresponding specification alone. We have seen that the resulting family of decentralized supervisors must typically (as in Transfer Line, Sect. 4.6) be augmented by a coordinator which oversees their collective action to prevent possible conflict and overall system blocking. A coordinator can be regarded as just one more decentralized supervisor, corresponding to the specification of global nonblocking alone.[5] The goal is to ensure that the resulting family of supervisors is control-equivalent (has the same control action) as the (perhaps hypothetical) optimal and

[5]This specification can be represented in *TCT* by the DES **ALLPLANT** = **allevents(PLANT)**.

nonblocking centralized 'monolithic' supervisor, which for large systems may be infeasible to compute.

In this section we carry the control decomposition process one step further, by an extension and refinement of decentralized supervision that is more properly termed 'distributed supervision'. We start from a plant model that, just as before, is typically the synchronous product of individual component DES that we now term 'agents', for instance in Small Factory the model **FACT = sync(MACH1,MACH2)**. For simplicity we make the assumption (usually the case in decentralized supervision) that the agents $\mathbf{A_i}$ (e.g. **MACH1, MACH2**) are defined over pairwise disjoint alphabets Σ_i; namely in that sense the agents are 'independent'.

Suppose **SUPER** is one of our decentralized supervisors. We propose to decompose **SUPER** into local supervisors, say $\mathbf{LOC_i}$, one for each agent $\mathbf{A_i}$. $\mathbf{LOC_i}$ is 'local' in the sense that it will exercise control (disabling) action only over the controllable events of $\mathbf{A_i}$, although it may very well depend for this action on events 'imported' from various other agents $\mathbf{A_j}$ with $j \neq i$. Thus $\mathbf{LOC_i}$ will be a DES defined over an alphabet formed as the union of Σ_i with some subalphabet of $\cup\{\Sigma_j | j \neq i\}$. The process of computing the components $\mathbf{LOC_i}$ will be called 'supervisor localization'. The desired result (not always achievable) is that each $\mathbf{LOC_i}$ is rather small in size (compared to **SUPER**), and involves relatively few of the Σ_j with $j \neq i$. In any case, the criterion of validity is just that the joint action of the $\mathbf{LOC_i}$ is control-equivalent to **SUPER**; namely if $\mathbf{PLANT} = \mathbf{sync}(\mathbf{A_1}, \ldots, \mathbf{A_N})$ then

$$\mathbf{sync}(\mathbf{A_1}, \mathbf{LOC_1}, \ldots, \mathbf{A_N}, \mathbf{LOC_N}) = \mathbf{SUPER}$$

For a suggestive physical metaphor consider a group of agents—in this case motorists—maneuvering through a congested intersection without benefit of any 'global' supervision by the usual external controls (traffic lights or an officer). Each motorist will exercise direct control only over his own vehicle, in pursuit of his individual objective but obviously in response to visual and auditory cues ('events') from neighboring vehicles, and guided also by commonly accepted protocols. We imagine that a monolithic external controller **SUPER** exists in principle, but is now replaced by its localization into each individual motorist (vehicle and driver).

In general the possible advantages of such a localized control architecture are that the local controllers $\mathbf{LOC_i}$ are relatively simple, hence easily maintainable, and their control action is transparent, namely readily understood.

Returning to Small Factory, say with buffer capacity 1 slot, we obtain the monolithic supervisor **FACTSUPER** (12,24) and its reduced version (using **supreduce**) **FACTSIMSUP** (3,3). We apply *TCT* **localize** to the input arguments (**FACT, MACH1,MACH2,FACTSUPER**) to obtain local supervisors **MACH1LOC** (3,-) and **MACH2LOC** (2,-), as displayed in Fig. 4.7. **MACH1LOC** imports events (21,22,23) from **MACH2**, while **MACH2LOC** imports event 10 from **MACH1**. As a check we may compute the overall behavior under localized control as

$$\mathbf{sync}(\mathbf{MACH1}, \mathbf{MACH1LOC}, \mathbf{MACH2}, \mathbf{MACH2LOC}) = \mathbf{TEST}$$

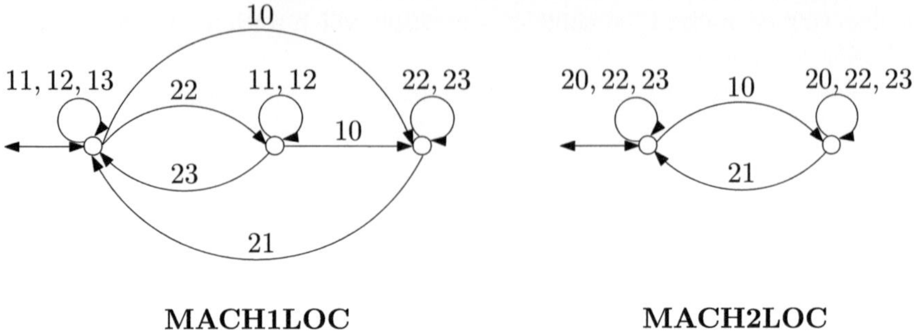

Fig. 4.7 Local supervisors **MACH1LOC** and **MACH2LOC**

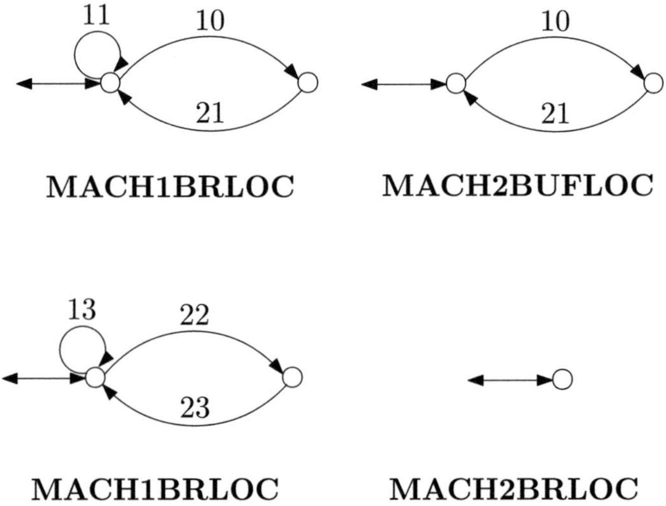

Fig. 4.8 Local supervisors **MACH1BUFLOC**, **MACH1BRLOC**, **MACH2BUFLOC**, **MACH2BRLOC**

say, and confirm that

$$\textbf{isomorph}(\textbf{TEST}, \textbf{FACTSUPER}) = \textit{true}$$

Smaller local supervisors can be obtained by applying localization separately to each of the two decentralized supervisors, say **BUFSUP** (12,25) and **BRSUP** (9,23). As displayed in Fig. 4.8, **MACH1** is equipped with local supervisors **MACH1BUFLOC** (2,-), importing event 21 from **MACH2**, and **MACH1 - BRLOC** (2,-) importing events 22,23 from **MACH2**, while **MACH2** is assigned local supervisors **MACH2BUFLOC** (2,-) importing event 10 from **MACH1** and a 'virtual' supervisor **MACH2BRLOC** (1,0) with no imports and exercising no control action. Again one checks correctness by (re)computing

TEST = sync(MACH1,MACH1BUFLOC,MACH1BRLOC, MACH2,MACH2BUFLOC,MACH2BRLOC)

and confirming

$$\textbf{isomorph}(\textbf{TEST}, \textbf{FACTSUPER}) = true$$

Exercise 1: Verify these results for Small Factory. Interpret and discuss local supervisors' action, paying special attention to the role of imported events.

Exercise 2: Carry out a similar study of localization for Transfer Line (Sect. 4.6, especially Exercise 4.6.2), this time computing 9 local supervisors, namely 3 for each of the plant components **M1,M2,TU** corresponding to the 3 decentralized supervisors obtained from the specifications governing the buffers **B1,B2** and the coordinator **LOOP**.

4.12 Notes

Modular (specifically, decentralized) supervisory control theory, in the sense of this chapter, originated with the doctoral theses of Ramadge (1983), Lin (1987) and Rudie (1992), and related papers (Ramadge and Wonham 1987; Wonham and Ramadge 1988; Lin and Wonham 1988; Rudie and Wonham 1992). The Transfer Line of Sect. 4.6 is adapted from Al-Jaar and Desrochers (1988) and Desrochers and Al-Jaar (1995).

The instructive AGV control problem of Sect. 4.7, including (essentially) Fig. 4.1, is due to Holloway and Krogh (1990), whose approach was based on Petri nets, and was not confirmed to achieve nonblocking. In our version, adding the mutual exclusion Specification 5 and requiring global nonblocking, the problem is appreciably more difficult.

Details of the work by L. Feng cited in the latter part of Sect. 4.7 can be found in Sect. 4.6 of Feng (2007). Summary versions were reported in Feng and Wonham (2006b, 2008), Feng et al. (2009). The key concepts are control flow net, inessential or 'harmless' specifications (here the **ZiSPEC**), and abstraction using natural observers with an output-control-consistency property.

The decentralized projection formulation of Sect. 4.8 is a variation on Lin and Wonham (1988); the observer property is due to K.C. Wong (1994), Wong and Wonham (1996a, b), while its 'I-fold' extension originates here. The computation of local controllers for individual specifications was highlighted by de Queiroz and Cury (2000). For other investigations of decentralized supervisory control see, for instance, Willner and Heymann (1991) and Akesson et al. (2002).

The robotics model of Exercise 4.9.3 was suggested by K.P. Valavanis, while the celebrated problem of the Dining Philosophers (Exercise 4.9.6) originated with Dijkstra (1971) and has been widely reproduced in the literature on concurrency and computer operating systems.

Exercise 4.9.7 is adapted from Lin and Wonham (1990). Our coordinators were obtained using the 'control-flow' decomposition due to Feng and Wonham (2006b; 2008), Feng (2007).

Recent approaches to decentralized supervision, including (Feng and Wonham 2006a; Feng and Wonham 2006b), can be found among the contributions to WODES'06 and WODES'08.

The concept of distributed control by supervisor localization is due to Cai (2008) and derived publications (Cai and Wonham 2009a, b, 2010a, b, 2016). Localization makes essential use of the construct 'control cover' used for supervisor reduction, as introduced in Su and Wonham (2004).

References

Akesson K, Flordal H, Fabian M (2002) Exploiting modularity for synthesis and verification of supervisors. In: Proceedings of the fifteenth IFAC world congress on automatic control, Barcelona, Spain, pp 175–180

Al-Jaar RY, Desrochers AA (1988) A modular approach for the performance analysis of automated manufacturing systems using generalized stochastic Petri nets. Rpt. RAL #116, Robotics and Automation Laboratory, Rensselaer Polytechnic Institute, Troy, New York

Cai K (2008) Supervisor localization: a top-down approach to distributed control of discrete-event systems. MASc thesis, Department of Electrical and Computer Engineering, University of Toronto

Cai K, Wonham WM (2009a) Supervisor localization: a top-down approach to distributed control of discrete-event systems. In: Beji L, Otmane S, Abichou A (eds) Proceedings of the second Mediterranean conference on intelligent systems and automation (CISA '09), Zarzis, Tunisia, March 23–25, pp 302–308

Cai K, Wonham WM (2009b) Supervisor localization for large-scale discrete-event systems. In: Joint forty-eighth IEEE conference on decision and control and twenty-eighth Chinese control conference, Shanghai, China, Dec 16–18, pp 3099–3105

Cai K, Wonham WM (2010a) Supervisor localization: a top-down approach to distributed control of discrete-event systems. IEEE Trans Autom Control 55(3):605–618

Cai K, Wonham WM (2010b) Supervisor localization for large discrete-event systems: case study, production cell. Int J Adv Manufact Technol 50(9–12):1189–1202

Cai K, Wonham WM (2016) Supervisor localization: a top-down approach to distributed control of discrete-event systems. Lecture notes in control and information sciences (LNCIS), vol 459. Springer International Publishing, Switzerland

de Queiroz MH, Cury JER (2000) Modular control of composed systems. In: Proceedings of the American control conference, Chicago, pp 4051–4055

Desrochers AA, Al-Jaar RY (1995) Applications of Petri nets in manufacturing systems. IEEE Press, New York

Dijkstra EW (1971) Hierarchical ordering of sequential processes. Acta Informatica 1:115–138

Feng L (2007) Computationally efficient supervisor design for discrete-event systems. PhD thesis, Department of Electrical and Computer Engineering, University of Toronto

Feng L, Wonham WM (2006a) Computationally efficient supervisor design: abstraction and modularity. In: Proceedings of the eighth international workshop on discrete-event systems (WODES '06), University of Michigan, Ann Arbor, pp 3–8

Feng L, Wonham WM (2006b) Computationally efficient supervisor design: control-flow decomposition. In: Proceedings of the eighth international workshop on discrete-event systems (WODES '06), University of Michigan, Ann Arbor, pp 9–14

Feng L, Wonham WM (2008) Supervisory control architecture for discrete-event systems. IEEE Trans Autom Control 53(6):1449–1461

Feng L, Cai K, Wonham WM (2009) A structural approach to the nonblocking supervisory control of discrete-event systems. Int J Adv Manufact Technol 41(11):1152–1167

Holloway LE, Krogh BH (1990) Synthesis of feedback logic control for a class of controlled Petri nets. IEEE Trans Autom Control 35(5):514–523

Lin F (1987) On controllability and observability of discrete event systems. PhD thesis, Department of Electrical Engineering, University of Toronto

Lin F, Wonham WM (1988) Decentralized supervisory control of discrete-event systems. Inf Sci 44(2):199–224

Lin F, Wonham WM (1990) Decentralized control and coordination of discrete-event systems with partial observation. IEEE Trans Autom Control 35(12):1330–1337

Ramadge PJ (1983) Control and supervision of discrete event processes. PhD thesis, Department of Electrical Engineering, University of Toronto

Ramadge PJ, Wonham WM (1987) Modular feedback logic for discrete event systems. SIAM J Control Optim 25(5):1202–1218

Rudie K (1992) Decentralized control of discrete-event systems. PhD thesis, Department of Electrical Engineering, University of Toronto

Rudie K, Wonham WM (1992) Think globally, act locally: decentralized supervisory control. IEEE Trans Autom Control 37(11):1692–1708. (Reprinted in Sadjadi FA (eds) Selected papers on sensor and data fusion, 1996, ISBN 0-8194-2265-7)

Su R, Wonham WM (2004) Supervisor reduction for discrete-event systems. Discret Event Dyn Syst 14(1):31–53

Willner Y, Heymann M (1991) On supervisory control of concurrent discrete-event systems. Int J Control 54(5):1143–1169

Wong K-C (1994) Discrete-event control architecture: an algebraic approach. PhD thesis, Department of Electrical and Computer Engineering, University of Toronto

Wong KC, Wonham WM (1996a) Hierarchical control of discrete-event systems. Discret Event Dyn Syst 6(3):241–273

Wong KC, Wonham WM (1996b) Hierarchical control of timed discrete-event systems. Discret Event Dyn Syst 6(3):274–306

Wonham WM, Ramadge PJ (1988) Modular supervisory control of discrete event systems. Math Control Signals Syst 1(1):13–30

Chapter 5
Hierarchical Supervision of Discrete-Event Systems

Hierarchy is a layered architecture in which control authority is structured according to scope, as measured by temporal horizon and logical depth of command. The fundamental problem with hierarchy is to guarantee that each layer faithfully executes the commands received from the next higher, as confirmed by the corresponding feedback loop. Such hierarchical consistency is addressed in the framework of controlled DES already introduced; it is achieved by suitable refinement of the information transmitted from a given layer to the next higher.

5.1 Hierarchical Control Structure

Hierarchical structure is a familiar feature of the control of dynamic systems that perform a range of complex tasks. It may be described generally as a division of control action and the concomitant information processing by layering, according to scope. Commonly, the scope of a control action is defined by the extent of its temporal horizon, or by the depth of its logical dependence in a task decomposition. Generally speaking, the broader the temporal horizon of a control and its associated subtask, or the deeper its logical dependence on other controls and subtasks, the higher it is said to reside in the hierarchy. Frequently the two features of broad temporal horizon and deep logical dependency are found together.

In this chapter we formalize hierarchical structure in the control of DES, by means of a mild extension of the framework already introduced. While different approaches to hierarchical control might be adopted even within this restricted framework, the theory to be presented captures the basic feature of scope already mentioned, and casts light on an issue that we call *hierarchical consistency*.

W. M. Wonham and K. Cai, *Supervisory Control of Discrete-Event Systems*, Communications and Control Engineering,
https://doi.org/10.1007/978-3-319-77452-7_5

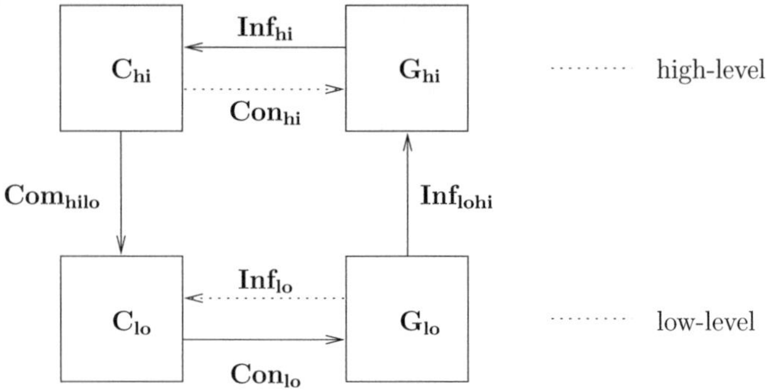

Fig. 5.1 Two-level control hierarchy

In outline our setup will be the following. Consider a two-level hierarchy consisting of a low-level plant $\mathbf{G_{lo}}$ and controller $\mathbf{C_{lo}}$, along with a high-level plant $\mathbf{G_{hi}}$ and controller $\mathbf{C_{hi}}$. These are coupled as shown in Fig. 5.1.

Our viewpoint is that $\mathbf{G_{lo}}$ is the actual plant to be controlled in the real world by $\mathbf{C_{lo}}$, the operator; while $\mathbf{G_{hi}}$ is an abstract, simplified model of $\mathbf{G_{lo}}$ that is employed for decision-making in an ideal world by $\mathbf{C_{hi}}$, the manager. The model $\mathbf{G_{hi}}$ is refreshed or updated every so often via the information channel (or mapping) labeled $\mathbf{Inf_{lohi}}$ ('information-low-to-high') to $\mathbf{G_{hi}}$ from $\mathbf{G_{lo}}$. Alternatively one can interpret $\mathbf{Inf_{lohi}}$ as carrying information sent up by the operator $\mathbf{C_{lo}}$ to the manager $\mathbf{C_{hi}}$: in our model the formal result will be the same. Another information channel, $\mathbf{Inf_{lo}}$ ('low-level information'), provides conventional feedback from $\mathbf{G_{lo}}$ to its controller $\mathbf{C_{lo}}$, which in turn applies conventional control to $\mathbf{G_{lo}}$ via the control channel labeled $\mathbf{Con_{lo}}$ ('low-level control'). Returning to the high level, we consider that $\mathbf{G_{hi}}$ is endowed with control structure, according to which it makes sense for $\mathbf{C_{hi}}$ to attempt to exercise control over the behavior of $\mathbf{G_{hi}}$ via the control channel $\mathbf{Con_{hi}}$ ('high-level control'), on the basis of feedback received from $\mathbf{G_{hi}}$ via the information channel $\mathbf{Inf_{hi}}$ ('high-level information'). In actuality, the control exercised by $\mathbf{C_{hi}}$ in this way is only virtual, in that the behavior of $\mathbf{G_{hi}}$ is determined entirely by the behavior of $\mathbf{G_{lo}}$, through the updating process mediated by $\mathbf{Inf_{lohi}}$. The structure is, however, completed by the command channel $\mathbf{Com_{hilo}}$ linking $\mathbf{C_{hi}}$ to $\mathbf{C_{lo}}$. The function of $\mathbf{Com_{hilo}}$ is to convey the manager's high-level control signals as commands to the operator $\mathbf{C_{lo}}$, which must translate (compile) these commands into corresponding low-level control signals which will actuate $\mathbf{G_{lo}}$ via $\mathbf{Con_{lo}}$. State changes in $\mathbf{G_{lo}}$ will eventually be conveyed in summary form to $\mathbf{G_{hi}}$ via $\mathbf{Inf_{lohi}}$. $\mathbf{G_{hi}}$ is updated accordingly, and then provides appropriate feedback to $\mathbf{C_{hi}}$ via $\mathbf{Inf_{hi}}$. In this way the hierarchical loop is closed. The forward path sequence $\mathbf{Com_{hilo}}$; $\mathbf{Con_{lo}}$ is conventionally designated 'command and control', while the feedback path sequence $\mathbf{Inf_{lohi}}$; $\mathbf{Inf_{hi}}$ could be referred to as 'report and advise'.

As a metaphor, one might think of the command center of a complex system (e.g. manufacturing system, electric power distribution system) as the site of the high-level plant model $\mathbf{G_{hi}}$, where a high-level decision maker (manager) $\mathbf{C_{hi}}$ is in command.

The external (real) world and those (operators) coping with it are embodied in $\mathbf{G_{lo}}$, $\mathbf{C_{lo}}$. The questions to be addressed concern the relationship between the behavior required, or expected, by the manager $\mathbf{C_{hi}}$ of his high-level model $\mathbf{G_{hi}}$, and the actual behavior implemented by the operator $\mathbf{C_{lo}}$ in $\mathbf{G_{lo}}$ in the manner described, when $\mathbf{G_{lo}}$ and $\mathbf{Inf_{lohi}}$ are given at the start. It will turn out that a relationship of *hierarchical consistency* imposes rather stringent requirements on $\mathbf{Inf_{lohi}}$ and that, in general, it is necessary to refine the information conveyed by this channel before consistent hierarchical control structure can be achieved. This result accords with the intuition that for effective high-level control the information sent up by the operator to the manager must be timely, and sufficiently detailed for various critical low-level situations to be distinguished.

Exercise 1: Pyramidal hierarchy

In its simplest form, the 'classical' pyramidal hierarchy can be considered as a population of N 'agents' organized into levels indexed 0, 1, 2, ..., n, starting from the top. Assume that the kth level has population r^k, where the 'scope ratio' r is constant from one level to the next; thus the total population has size $N = 1+r+r^2+\cdots+r^n$. In applications, r could be identified, as here, with the number of subordinate agents reporting directly to a given agent, or alternatively as the ratio of time horizons from one level to the next lower, or the ratio of the average occurrence frequencies of significant events from one level to the next higher. In the analogous structure of nested-loop control systems, r could be thought of as the ratio of bandwidths in passing from a given loop to the next outermost, inner loops being fast and outer ones slow. We conjecture that in real life, the scope ratio r tends to be on the order of 5, ranging from 2 to perhaps 10. Test this claim against examples of your own choosing, say the Mongolian armed forces, or the Pepsi-Cola Co. For instance, you could work out a value of r from an estimate of the total population N and the number of levels $n+1$; or else n from r and N. In a more refined model, r might increase or decrease with rise in level; interpret, and check the numbers again.

5.2 Two-Level Controlled Discrete-Event System

For $\mathbf{G_{lo}}$ we take the usual 5-tuple

$$\mathbf{G_{lo}} = (Q, \Sigma, \delta, q_0, Q_m)$$

Here Σ is the set of *event labels*, partitioned into *controllable* elements ($\Sigma_c \subseteq \Sigma$) and *uncontrollable* elements ($\Sigma_u \subseteq \Sigma$); Q is the *state set*; $\delta : Q \times \Sigma \to Q$ is the *transition function* (in general a partial function, defined at each $q \in Q$ for only a subset of events $\sigma \in \Sigma$: in that case we write $\delta(q, \sigma)!$); q_0 is the *initial state*; and $Q_m \subseteq Q$ is the subset of *marker states*. The *uncontrolled behavior* of $\mathbf{G_{lo}}$ is the language

$$L_{lo} := L(\mathbf{G_{lo}}) \subseteq \Sigma^*$$

consisting of the (finite) strings $s \in \Sigma^*$ for which the (extended) transition map $\delta : Q \times \Sigma^* \to Q$ is defined.

In this section, as well as Sects 5.3–5.5, we only consider the case $Q_m = Q$, namely all the relevant languages are prefix-closed. This assumption is made for simplicity in focusing on the basic issue of hierarchical consistency. The theory will be generalized to include marking and treat nonblocking in Sect. 5.7.

We recall that if $\mathbf{G}$ is a controlled DES over an alphabet $\Sigma = \Sigma_c \dot{\cup} \Sigma_u$, and K is a closed sublanguage of Σ^*, then K is *controllable* (with respect to $\mathbf{G}$) if $K\Sigma_u \cap L(\mathbf{G}) \subseteq K$. To every closed language $E \subseteq \Sigma^*$ there corresponds the (closed) *supremal controllable sublanguage* $\sup \mathcal{C}(E \cap L(\mathbf{G}))$. In this chapter it will be convenient to use the notation

$$\sup \mathcal{C}(M) =: M^{\uparrow}$$

Let T be a nonempty set of labels of 'significant events'. T may be thought of as the events perceived by the manager which will enter into the description of the high-level plant model $\mathbf{G_{hi}}$, of which the derivation will follow in due course. First, to model the information channel (or mapping) $\mathbf{Inf_{lohi}}$ we postulate a map

$$\theta : L_{lo} \to T^*$$

with the properties

$$\theta(\epsilon) = \epsilon,$$

$$\theta(s\sigma) = \begin{cases} \text{either } \theta(s) \\ \text{or } \theta(s)\tau, \text{ some } \tau \in T \end{cases}$$

for $s \in L_{lo}$, $\sigma \in \Sigma$ (here and below, ϵ denotes the empty string regardless of alphabet). Such a map θ will be referred to as *causal*. A causal map is, in particular, *prefix-preserving*: if $s \le s'$ then $\theta(s) \le \theta(s')$. Intuitively, θ can be used to signal the occurrence of events that depend in some fashion on the past history of the behavior of $\mathbf{G_{lo}}$: for instance θ might produce a fresh instance of symbol τ' whenever $\mathbf{G_{lo}}$ has just generated a positive multiple of 10 of some distinguished symbol σ', but 'remain silent' otherwise.

Exercise 1: Prove that $\theta : L_{lo} \to T^*$ is causal iff it commutes with prefix closure, namely for *all* sublanguages $K \subseteq L_{lo}$, not necessarily closed,

$$\overline{\theta(K)} = \theta(\bar{K})$$

Exercise 2: Show that a prefix-preserving map $\theta : L_{lo} \to T^*$ need not be causal.

It is convenient to combine θ with $\mathbf{G_{lo}}$ in a unified description. This may be done in standard fashion by replacing the pair $(\mathbf{G_{lo}}, \theta)$ by a Moore generator having output alphabet

$$T_o = T \cup \{\tau_o\}$$

where τ_o is a new symbol ($\notin T$) interpreted as the 'silent output symbol'. To this end write temporarily

$$\tilde{\mathbf{G}}_{\mathbf{lo}} = (\tilde{Q}, \Sigma, T_o, \tilde{\delta}, \omega, \tilde{q}_0, \tilde{Q}_m)$$

Here the items written with a tilde play the same role as in $\mathbf{G_{lo}}$, while $\omega : \tilde{Q} \to T_o$ is the state output map. $\tilde{\mathbf{G}}_{\mathbf{lo}}$ is constructed so that

$$\tilde{\delta}(\tilde{q}_0, s)! \quad \text{iff} \quad \delta(q_0, s)!, \qquad s \in \Sigma^*$$

Thus $\tilde{\mathbf{G}}_{\mathbf{lo}}$ generates exactly the language L_{lo}. For ω define

$$\omega(\tilde{q}_0) = \tau_o$$

while if $\tilde{\delta}(\tilde{q}_0, s\sigma)!$ then

$$\begin{aligned} \omega(\tilde{\delta}(\tilde{q}_0, s\sigma)) &= \tau_o \text{ if } \theta(s\sigma) = \theta(s) \\ \omega(\tilde{\delta}(\tilde{q}_0, s\sigma)) &= \tau \ \text{ if } \theta(s\sigma) = \theta(s)\tau \end{aligned}$$

Thus ω outputs the silent symbol τ_o if θ outputs 'nothing new', and outputs the 'fresh' symbol $\tau \in T$ otherwise.

An abstract construction of $\tilde{\mathbf{G}}_{\mathbf{lo}}$ is straightforward, using the canonical identification of states with the cells of a suitable right congruence on strings. For $s, s' \in L_{lo}$ define

$$s \equiv s' (\text{mod } L_{lo}) \quad \text{iff} \quad (\forall u \in \Sigma^*) su \in L_{lo} \Leftrightarrow s'u \in L_{lo}$$

Next define $\hat{\omega} : L_{lo} \to T_o$ according to

$$\hat{\omega}(\epsilon) = \tau_o$$

and if $s\sigma \in L_{lo}$, then

$$\hat{\omega}(s\sigma) = \begin{cases} \tau_o \text{ if } \theta(s\sigma) = \theta(s) \\ \tau \ \text{ if } \theta(s\sigma) = \theta(s)\tau \end{cases}$$

Now let, for $s, s' \in L_{lo}$,

$$s \equiv s' (\text{mod } \theta) \quad \text{iff} \quad \hat{\omega}(s) = \hat{\omega}(s') \quad \&$$

$$(\forall u \in \Sigma^*)(\forall t \in T^*) su \in L_{lo} \;\&\; s'u \in L_{lo} \Rightarrow (\theta(su) = \theta(s)t \Leftrightarrow \theta(s'u) = \theta(s')t)$$

It is readily shown that equivalence (mod θ) is a right congruence on L_{lo}. As equivalence (mod L_{lo}) is a right congruence too, so is their common refinement (i.e. their meet in the lattice of right congruences); and the cells of this refinement furnish the states of $\tilde{\mathbf{G}}_{\mathbf{lo}}$.

From this point on we shall assume that the starting point of our hierarchical control problem is the unified description $\tilde{\mathbf{G}}_{\mathbf{lo}}$. So we drop the tilde and write

$$\mathbf{G}_{\mathbf{lo}} = (Q, \Sigma, T_o, \delta, \omega, q_0, Q_m) \tag{5.1}$$

with $Q_m = Q$.

At this stage we temporarily define $\mathbf{G}_{\mathbf{hi}}$. For this we note that, in the absence of any control action, $\mathbf{G}_{\mathbf{lo}}$ generates the uncontrolled language L_{lo}. For now, $\mathbf{G}_{\mathbf{hi}}$ will be taken as the canonical recognizer (in the generator sense) for the image of L_{lo} under θ:

$$L(\mathbf{G}_{\mathbf{hi}}) = \theta(L_{lo}) \subseteq T^*$$

and we write $L(\mathbf{G}_{\mathbf{hi}}) =: L_{hi}$. As yet, however, the event label alphabet T of $\mathbf{G}_{\mathbf{hi}}$ need not admit any natural partition into controllable and uncontrollable subalphabets; that is, $\mathbf{G}_{\mathbf{hi}}$ need not possess any natural control structure. This defect will be remedied in the next section.

The following simple example will be used throughout. Following the integer labeling conventions of *TCT* we define the state set and alphabets of $\mathbf{G}_{\mathbf{lo}}$ in (5.1) according to

$$Q = \{0, 1, 2, 3, 4\}, q_0 = 0$$
$$\Sigma = \{0, 1, 2, 3, 4\}$$
$$T = \{\alpha, \beta\}, \tau_o = o$$

In Σ the odd-numbered elements are controllable, the even-numbered elements uncontrollable. The state transition and output structure of $\mathbf{G}_{\mathbf{lo}}$ is displayed in Fig. 5.2 along with a canonical recognizer for L_{hi}. Observe that, whether or not $\tau \in \{\alpha, \beta\}$ can be disabled as 'next output' by a supervisory controller that can disable the controllable elements of Σ, depends on the current state $q \in Q$ of $\mathbf{G}_{\mathbf{lo}}$: for instance $\tau = \alpha$ can be disabled at $q = 2$ but not at $q = 0, 1, 3$ or 4. Thus $\mathbf{G}_{\mathbf{hi}}$ does not yet possess natural control structure.

Exercise 3: Show that equivalence (mod θ) is a right congruence on L_{lo}.

5.3 High-Level Control Structure

In this section we indicate how to refine the descriptions of $\mathbf{G}_{\mathbf{lo}}$ and $\mathbf{G}_{\mathbf{hi}}$ in order to equip $\mathbf{G}_{\mathbf{hi}}$ with control structure, so that a high-level controller $\mathbf{C}_{\mathbf{hi}}$ that observes only the state of $\mathbf{G}_{\mathbf{hi}}$ can make meaningful control decisions. By way of control structure we adopt the (usual) supervisory structure having the same type as in $\mathbf{G}_{\mathbf{lo}}$. We shall refine the state structure of $\mathbf{G}_{\mathbf{lo}}$, extend the high-level event alphabet T, and partition the extension into controllable and uncontrollable subsets.

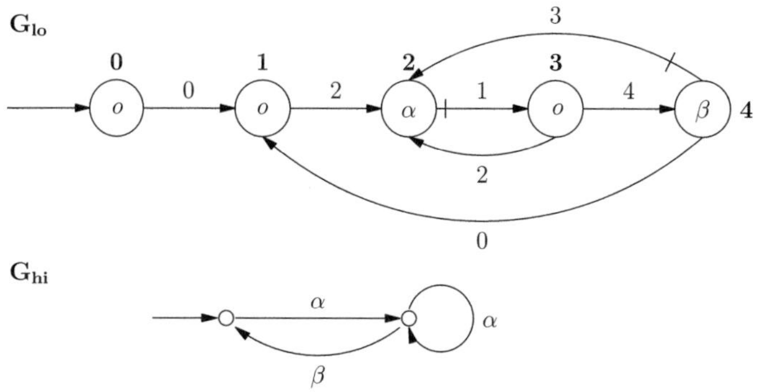

Fig. 5.2 Low- and high-level DES

Conceptually these operations are carried out as follows. Referring to the example in Fig. 5.2, consider a reachability tree for L_{lo} with initial state $q_0 = 0$ as the root. The first few levels of the tree are displayed in Fig. 5.3. Each node of the tree is labeled with the corresponding value $\tau' \in T_o = \{o, \alpha, \beta\}$ of the output map ω, and is then called a τ'-*node*.

In general it will be convenient to write $\hat{\omega} : L_{lo} \to T_o$ for the output map on strings defined by $\hat{\omega}(s) = \omega(\delta(q_0, s))$ whenever $s \in L_{lo}$, i.e. $\delta(q_0, s)!$. With a slight abuse of notation we also write $\hat{\omega} : \mathcal{N} \to T_o$ where $\mathcal{N}$ is the node set of the reachability tree of L_{lo}.

In the tree, τ'-nodes with $\tau' = \tau_o = o$ are *silent*; τ'-nodes with $\tau' \in T = \{\alpha, \beta\}$ are *vocal*. A *silent path* in the tree is a path joining two vocal nodes, or the root to a vocal node, all of whose intermediate nodes (if any) are silent. Schematically a silent path has the form

$$n \xrightarrow{\sigma} s \xrightarrow{\sigma'} s' \longrightarrow \cdots \longrightarrow s'' \xrightarrow{\sigma''} n'$$

where the starting node n is either vocal or the root node, and where the intermediate silent nodes $s, s', ..., s''$ may be absent. Thus to every vocal node n' there corresponds a unique silent path of which it is the terminal node. A silent path is *red* if at least one of its transition labels $\sigma \in \Sigma$ is controllable; otherwise it is *green*. Now color each vocal node red or green according to the color of its corresponding silent path. Create an extended output alphabet T_{ext} as follows. Starting with $T_{ext} = \{\tau_o\}$, for each $\tau \in T$ adjoin a new symbol $\tau_c \in T_{ext}$ if some τ-node in the tree is red; similarly adjoin $\tau_u \in T_{ext}$ if some τ-node in the tree is green. Now define $\hat{\omega}_{ext} : \mathcal{N} \to T_{ext}$ according to

$$\hat{\omega}_{ext}(n) = \tau_o \text{ if } n \text{ is silent}$$
$$\hat{\omega}_{ext}(n) = \tau_c \text{ if } \hat{\omega}(n) = \tau \in T \text{and color}(n) = \text{red}$$
$$\hat{\omega}_{ext}(n) = \tau_u \text{ if } \hat{\omega}(n) = \tau \in T \text{and color}(n) = \text{green}$$

Fig. 5.3 Reachability tree

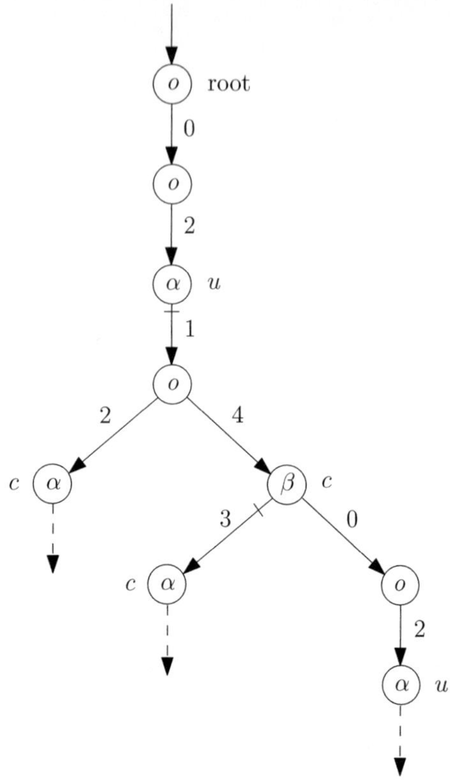

Define the extended tree to be the original tree with the node labeling determined by $\hat{\omega}_{ext}$. In Fig. 5.3, vocal nodes are labeled c or u accordingly. It is clear that $\hat{\omega}_{ext}$ in turn determines an extension

$$\theta_{ext} : L_{lo} \to T^*_{ext}$$

Evidently θ is recovered from θ_{ext} as follows: Define $P : T^*_{ext} \to T^*$ according to

$$P(\tau_c) = P(\tau_u) = \tau, \quad \tau \in T$$

$$P(\epsilon) = \epsilon, \quad P(\tau_0) = \epsilon, \quad P(tt') = P(t)P(t'), \quad t, t' \in T^*_{ext}$$

The line just written expresses the property that P is *catenative*. So P just maps the new output symbols in any string back to where they came from. Then

$$\theta = P \circ \theta_{ext}$$

Finally, define

$$\mathbf{G_{lo,ext}} = (Q_{ext}, \Sigma, T_{ext}, \delta_{ext}, \omega_{ext}, q_0, Q_{ext})$$

from the current transition structure $(Q, \Sigma, \delta, q_0, Q)$ and the map θ_{ext} in just the way $\tilde{\mathbf{G}}_{\mathbf{lo}}$ was defined (Sect. 5.2) in terms of $\mathbf{G_{lo}}$ and θ.

By the construction it is seen that $|T_{ext}| \leq 2|T| + 1$: the number of non-silent output symbols has at most doubled, as each old output symbol has now split into controllable and uncontrollable siblings (in some cases one sibling may be absent). It will be shown that the number of states has at most doubled as well. For this we extend the domain of the color map to include strings of L_{lo}. Returning to the reachability tree of L_{lo}, color the silent nodes by the same rule as used previously for the vocal nodes (and color the root node green). For $s \in L_{lo}$ define node$(s) \in \mathcal{N}$ to be the node reached by s, and define color(s) = color(node(s)). For $s, s' \in L_{lo}$ define $s \equiv s'$ to mean (cf. Sect. 5.2)

$$s \equiv s'(\text{mod } L_{lo}) \quad \text{and} \quad s \equiv s'(\text{mod } \theta)$$

We claim that if $s \equiv s'$ and color(s) = color(s'), then for all $u \in \Sigma^*$ such that $su \in L_{lo}$ it is the case that $su \equiv s'u$ and color(su) = color$(s'u)$. In fact, the first statement follows by the observation (Sect. 5.2) that $\equiv$ is a right congruence on L_{lo}. Thus we know that $s'u \in L_{lo}$, and that for $v \leq u$, if $\theta(sv) = \theta(s)t$ for some $t \in T^*$ then $\theta(s'v) = \theta(s')t$. In other words, 'the output behaviors (under θ) of su (resp. $s'u$) coincide between s (resp. s') and su (resp. $s'u$)'. Since su and $s'u$ share the suffix u, it follows immediately by the definition of color on strings that color(su) = color$(s'u)$, and the second statement follows, as claimed. The upshot of this argument is that the common refinement of the right congruence $\equiv$, and the equivalence defined by equality of color, is again a right congruence on L_{lo}. It is, of course, the right congruence that provides the state structure of $\mathbf{G_{lo,ext}}$. Thus in passing from $\mathbf{G_{lo}}$ to $\mathbf{G_{lo,ext}}$ each state of $\mathbf{G_{lo}}$ is split at most once, into colored siblings. It follows that $|Q_{ext}| \leq 2|Q|$.

In the sections to follow it will be assumed that the foregoing construction has been carried out, namely our new starting point will be the Moore transition structure $\mathbf{G_{lo,ext}}$ as described. The property of $\mathbf{G_{lo,ext}}$ that each output $\tau \in T_{ext}$ is unambiguously controllable or uncontrollable in the sense indicated, will be summarized by saying that $\mathbf{G_{lo,ext}}$ is *output-control-consistent*. While we have not yet presented an algorithm (as distinct from a conceptual procedure) to pass from $\mathbf{G_{lo}}$ to $\mathbf{G_{lo,ext}}$, such an algorithm exists at least in case $|Q| < \infty$ (see Appendix 5.1). The result for our running example is displayed in Fig. 5.4, along with the extended high-level language

$$\theta_{ext}(L(\mathbf{G_{lo,ext}})) \subseteq T^*_{ext}$$

In *TCT*, a DES **GLO** with state outputs is referred to as 'vocalized' and is set up using either the 'vocalize' option in **create** or, more conveniently, *TCT* **vocalize**. State outputs τ can be numbered 10,…,99. The corresponding structure **GHI** is given by **higen(GLO)**. To extend **GLO** to be output-control-consistent, compute

$$\mathbf{OCGLO = outconsis(GLO)}$$

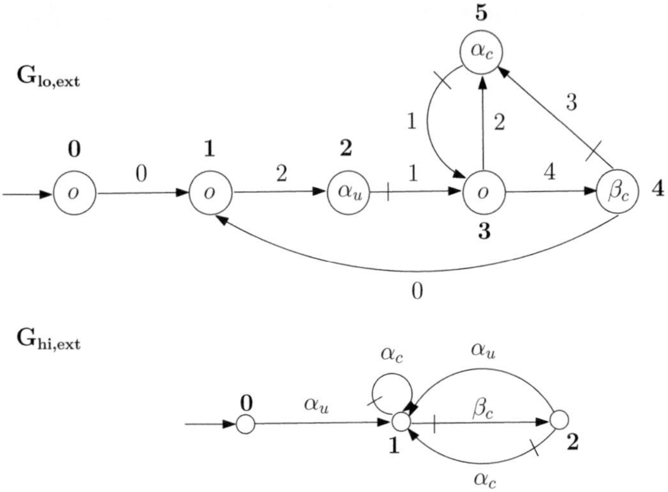

Fig. 5.4 Output-control-consistency

with high-level result

$$\mathbf{OCGHI} = \mathbf{higen}(\mathbf{OCGLO})$$

In this process *TCT* will create event siblings $\tau_u = 100$, $\tau_c = 101$ from $\tau = 10$, and so forth.

For completeness we provide the formal, albeit cumbersome, definition of output-control-consistency of a Moore transition structure

$$\mathbf{G} = (Q, \Sigma, T_o, \delta, \omega, q_0, Q_m) \tag{5.2}$$

where the input alphabet $\Sigma = \Sigma_c \dot{\cup} \Sigma_u$ and the output alphabet $T_o = \{\tau_o\} \cup T$ with $T = T_c \dot{\cup} T_u$. As before write $\hat{\omega}(s)$ for $\omega(\delta(q_0, s))$. Then $\mathbf{G}$ is *output-control-consistent* if, for every string $s \in L(\mathbf{G})$ of the form

$$s = \sigma_1 \sigma_2 \cdots \sigma_k \quad \text{or respectively} \quad s = s' \sigma_1 \sigma_2 \cdots \sigma_k$$

(where $s' \in \Sigma^+$, $\sigma_i \in \Sigma$) with

$$\hat{\omega}(\sigma_1 \sigma_2 \cdots \sigma_i) = \tau_o \quad (1 \le i \le k-1), \quad \hat{\omega}(s) = \tau \neq \tau_o$$

or respectively

$$\hat{\omega}(s') \neq \tau_o, \quad \hat{\omega}(s' \sigma_1 \sigma_2 \cdots \sigma_i) = \tau_o \quad (1 \le i \le k-1), \quad \hat{\omega}(s) = \tau \neq \tau_o$$

it is the case that

$$\text{if } \tau \in T_c \text{ then for some } i\,(1 \le i \le k),\ \sigma_i \in \Sigma_c$$
$$\text{if } \tau \in T_u \text{ then for all } i\,(1 \le i \le k),\ \sigma_i \in \Sigma_u$$

To conclude this section we return to the output-control-consistent structure $\mathbf{G_{lo,ext}}$ and corresponding structure $\mathbf{G_{hi,ext}}$, as above, where from now on the subscript 'ext' will be dropped. While the usefulness of output-control-consistency will be demonstrated in the next section, the following exercise brings out some of its limitations.

Exercise 1: With $\mathbf{G_{lo}}$ output-control-consistent and K_{lo}, K_{hi} prefix-closed, construct examples where

(i) $K_{lo} \subseteq L_{lo}$ is controllable with respect to $\mathbf{G_{lo}}$, but $K_{hi} := \theta(K_{lo})$ is not controllable with respect to $\mathbf{G_{hi}}$.
(ii) $K_{hi} \subseteq L_{hi}$ is controllable with respect to $\mathbf{G_{hi}}$, but $K_{lo} := \theta^{-1}(K_{hi})$ is not controllable with respect to $\mathbf{G_{lo}}$. Furthermore there is no (closed) controllable sublanguage $K'_{lo} \subseteq L_{lo}$ such that $\theta(K'_{lo}) = K_{hi}$.

Exercise 2: With $\mathbf{G_{lo}}$ output-control-consistent, assume $K_{hi} \subseteq L_{hi}$, and that $K_{lo} := \theta^{-1}(K_{hi})$ is controllable with respect to $\mathbf{G_{lo}}$. Show that K_{hi} is controllable with respect to $\mathbf{G_{hi}}$.

5.4 Command and Control

We continue on the assumption that $\mathbf{G_{lo}}$ is output-control-consistent. We shall relate supervisory control defined by the high-level controller (supervisor) $\mathbf{C_{hi}}$ to the appropriate low-level control exercised by $\mathbf{C_{lo}}$, thus defining the command and control path consisting of the command channel $\mathbf{Com_{hilo}}$ followed by the control channel $\mathbf{Con_{lo}}$ shown in Fig. 5.1.

High-level supervisory control is determined by a selection of high-level controllable events to be disabled, on the basis of high-level past history. That is, $\mathbf{C_{hi}}$ is defined by a map

$$\gamma_{hi} : L_{hi} \times T \to \{0, 1\}$$

such that $\gamma_{hi}(t, \tau) = 1$ for all $t \in L_{hi}$ and $\tau \in T_u$. As usual, if $\gamma_{hi}(t, \tau) = 0$ the event (labeled) τ is said to be *disabled*; otherwise τ is *enabled*; of course, only controllable events ($\tau \in T_c$) can be disabled. The result of applying this control directly on the generating action of $\mathbf{G_{hi}}$ would be to synthesize the closed-loop language

$$L(\gamma_{hi}, \mathbf{G_{hi}}) \subseteq T^*$$

say. In the standard theory, implementation of $\mathbf{C_{hi}}$ would amount to the construction of a suitable automaton (supervisor) over T as input alphabet, and the factorization of γ_{hi} through its state space (X, say) to create an equivalent state feedback control $\psi : X \times T \to \{0, 1\}$. In the hierarchical control loop, however, direct implementation of $\mathbf{C_{hi}}$ is replaced by command and control. The action of $\mathbf{C_{hi}}$ on $\mathbf{G_{hi}}$ must be mediated via $\mathbf{Com_{hilo}}$ and $\mathbf{Con_{lo}}$ as already described. To this end, assuming γ_{hi} is given, define the high-level *disabled-event map*

$$\Delta_{hi} : L_{hi} \to Pwr(T_c)$$

($Pwr(\cdot)$ denotes power set) according to

$$\Delta_{hi}(t) = \{\tau \in T_c | \gamma_{hi}(t, \tau) = 0\}$$

Correspondingly we may define the low-level disabled-event map

$$\Delta_{lo} : L_{lo} \times L_{hi} \to Pwr(\Sigma_c)$$

according to

$$\Delta_{lo}(s, t) = \{\sigma \in \Sigma_c | (\exists s' \in \Sigma_u^*) s\sigma s' \in L_{lo} \;\&\; \hat{\omega}(s\sigma s') \in \Delta_{hi}(t) \\ \& \; (\forall s'') s'' < s' \Rightarrow \hat{\omega}(s\sigma s'') = \tau_o\}$$

Observe that the explicit t-dependence of Δ_{lo} factors through the subset evaluation $\Delta_{hi}(t)$; in other words, Δ_{lo} can be evaluated by examination of the structure of $\mathbf{G_{lo}}$ alone, once the subset $\Delta_{hi}(t)$ of high-level events to be disabled has been announced by $\mathbf{C_{hi}}$. The definition says that $\Delta_{lo}(s, t)$ is just the set of low-level controllable events that must be disabled immediately following the generation of s (in $\mathbf{G_{lo}}$) and of t (in $\mathbf{G_{hi}}$) in order to guarantee the nonoccurrence of any $\tau \in \Delta_{hi}(t)$ as the next event in $\mathbf{G_{hi}}$. Of course such a guarantee is actually provided only if, for the given pair (s, t), the set of uncontrollable strings leading to the next occurrence of τ is empty:

$$\{s' \in \Sigma_u^+ | ss' \in L_{lo} \;\&\; \hat{\omega}(ss') = \tau \;\&\; (\forall s'' \in \Sigma^+) s'' < s' \Rightarrow \hat{\omega}(ss'') = \tau_o\} = \varnothing$$

As will be seen, the result of our construction in Sect. 5.3 of an output-control-consistent structure $\mathbf{G_{lo}}$ is that the required guarantee is provided when necessary.

When the hierarchical loop is closed through $\mathbf{Inf_{lohi}}$, a string $s \in L_{lo}$ is mapped to $t = \theta(s) \in L_{hi}$. Then the control implemented by $\mathbf{C_{lo}}$ will be given by

$$\gamma_{lo}(s, \sigma) = \begin{cases} 0 & \text{if } \sigma \in \Delta_{lo}(s, \theta(s)) \\ 1 & \text{otherwise} \end{cases} \tag{5.3}$$

Now suppose that a nonempty closed 'legal' (or specification) language $E_{hi} \subseteq L_{hi}$ is specified to the high-level controller $\mathbf{C_{hi}}$. We assume that E_{hi} is controllable with respect to the high-level model structure; that is,

$$E_{hi} T_u \cap L_{hi} \subseteq E_{hi}$$

In accordance with standard theory, E_{hi} would be synthesized as the controlled behavior of $\mathbf{G_{hi}}$ by use of a suitable control law γ_{hi}. In the standard theory the determination of γ_{hi} is usually not unique; however, γ_{hi} must always satisfy

$$\gamma_{hi}(t, \tau) = 0 \quad \text{if} \quad t \in E_{hi} \quad \& \quad t\tau \in L_{hi} - E_{hi}$$

Define E_{lo} to be the (maximal) behavior in $\mathbf{G_{lo}}$ that would be transmitted by $\mathbf{Inf_{lohi}}$ as behavior E_{hi} in the high-level model $\mathbf{G_{hi}}$:

$$E_{lo} := \theta^{-1}(E_{hi}) \subseteq L_{lo} \tag{5.4}$$

Since $L_{hi} = \theta(L_{lo})$ we have $\theta(E_{lo}) = E_{hi}$. Clearly E_{lo} is closed, but in general it will not be true that E_{lo} is controllable with respect to $\mathbf{G_{lo}}$. The main result of this section states that by use of the control (5.3) the closed-loop language $L(\gamma_{lo}, \mathbf{G_{lo}})$ synthesized in $\mathbf{G_{lo}}$ is made as large as possible subject to the constraint (5.4).

Theorem 1

Under the foregoing assumptions

$$L(\gamma_{lo}, \mathbf{G_{lo}}) = E_{lo}^{\uparrow}$$

Proof

It suffices to show the following:

(i) $L(\gamma_{lo}, \mathbf{G_{lo}})$ is controllable with respect to $\mathbf{G_{lo}}$.
(ii) $L(\gamma_{lo}, \mathbf{G_{lo}}) \subseteq E_{lo}$.
(iii) For any $\mathbf{G_{lo}}$-controllable sublanguage $K \subseteq E_{lo}$ we have $K \subseteq L(\gamma_{lo}, \mathbf{G_{lo}})$.

In the proof we write $L(\mathbf{G_{lo}}) =: L_{lo}$, $L(\mathbf{G_{hi}}) =: L_{hi}$ and $L(\gamma_{lo}, \mathbf{G_{lo}}) =: K_{lo}$. By definition of γ_{lo}, it follows that K_{lo} is nonempty and closed.

(i) By the definition of γ_{lo} we also have

$$\gamma_{lo}(s, \sigma) = \begin{cases} 0 \text{ if } \sigma \in \Delta_{lo}(s, \theta(s)) \subseteq \Sigma_c \\ 1 \text{ otherwise} \end{cases}$$

Since the closed-loop behavior in $\mathbf{G_{lo}}$ is obtained by disabling a subset (possibly null) of controllable events following the generation of any string of L_{lo}, it follows that for all $s \in K_{lo}$, $\sigma \in \Sigma_u$ we have $\gamma_{lo}(s, \sigma) = 1$, and therefore

$$K_{lo}\Sigma_u \cap L_{lo} \subseteq K_{lo}$$

namely K_{lo} is controllable, as claimed.

(ii) Since $E_{lo} = \theta^{-1}(E_{hi})$ it suffices to show that

$$\theta(K_{lo}) \subseteq E_{hi}$$

and we proceed by induction on length of strings. Because both K_{lo} and E_{hi} are nonempty and closed we have

$$\epsilon \in \theta(K_{lo}) \cap E_{hi}$$

Assume that $t \in T^*$, $\tau \in T$, and $t\tau \in \theta(K_{lo})$. Clearly $t \in \theta(K_{lo})$ and $t\tau \in L_{hi}$. Invoking the inductive assumption yields $t \in E_{hi}$. Now if $\tau \in T_u$ then

$$t\tau \in E_{hi} T_u \cap L_{hi}$$

and by controllability of E_{hi}, $t\tau \in E_{hi}$. On the other hand if $\tau \in T_c$ then by the fact that $\mathbf{G_{lo}}$ is output-control-consistent there exist

$$s \in \Sigma^*, \quad \sigma \in \Sigma_c, \quad s' \in \Sigma_u^*$$

such that

$$s\sigma s' \in K_{lo}, \quad \theta(s) = t, \quad \theta(s\sigma s') = t\tau$$

By definition of γ_{lo}, $\sigma \notin \Delta_{lo}(s, t)$; therefore

$$\hat{\omega}(s\sigma s') = \tau \notin \Delta_{hi}(t)$$

so again $t\tau \in E_{hi}$.

(iii) Let $K \subseteq E_{lo}$ be nonempty, and controllable with respect to $\mathbf{G_{lo}}$. Since E_{lo} and K_{lo} are both closed, it can be assumed without loss of generality that K is closed. By induction on length of strings it will be shown that $K \subseteq K_{lo}$. First, $\epsilon \in K \cap K_{lo}$. Now let $s\sigma \in K$. Since K is closed, $s \in K$. Invoking the inductive assumption, $s \in K_{lo}$. Since $K \subseteq E_{lo} \subseteq L_{lo}$ we have $s\sigma \in L_{lo}$. Now if $\sigma \in \Sigma_u$ then $\gamma_{lo}(s, \sigma) = 1$ and therefore $s\sigma \in K_{lo}$. Suppose on the other hand that $\sigma \in \Sigma_c$. To show that $s\sigma \in K_{lo}$ it must be shown that $\gamma_{lo}(s, \sigma) = 1$, or equivalently

$$\sigma \notin \Delta_{lo}(s, \theta(s))$$

Assuming the contrary and setting $t := \theta(s)$ we have by definition of $\Delta_{lo}(s, t)$:

$$(\exists s' \in \Sigma_u^*) s\sigma s' \in L_{lo} \quad \& \quad \hat{\omega}(s\sigma s') \in \Delta_{hi}(t) \quad \& \quad (\forall s'')s'' < s' \Rightarrow \hat{\omega}(s\sigma s'') = \tau_o$$

Since $s\sigma \in K$ and K is controllable it results that $s\sigma s' \in K$. Let $\hat{\omega}(s\sigma s') = \tau$. Then $\theta(s\sigma s') = t\tau$. But $\tau \in \Delta_{hi}(t)$ implies $\gamma_{hi}(t, \tau) = 0$, so $t\tau \in L_{hi} - E_{hi}$. That is $t\tau \notin E_{hi}$, namely *not* $\theta(K) \subseteq E_{hi}$. But this contradicts the fact that $K \subseteq E_{lo} = \theta^{-1}(E_{hi})$. Therefore $\gamma_{lo}(s, \sigma) = 1$ after all, so that $s\sigma \in K_{lo}$ as required. □

Obviously the transmitted high-level behavior will satisfy the required legal constraint:

$$\theta(L(\gamma_{lo}, \mathbf{G_{lo}})) \subseteq E_{hi} \tag{5.5}$$

but in general the inclusion will be proper. That is, while the 'expectation' of the high-level controller $\mathbf{C_{hi}}$ on using the control γ_{hi} might ideally be the synthesis in $\mathbf{G_{hi}}$ of the controllable behavior E_{hi}, only a subset of this behavior can in general actually be realized. The reason is simply that a call by $\mathbf{C_{hi}}$ for the disablement of some high-level event $\tau \in T_c$ may require $\mathbf{C_{lo}}$ (i.e. the control γ_{lo}) to disable paths in $\mathbf{G_{lo}}$ that lead directly to outputs other than τ. However, this result is the best that can be achieved under the current assumptions about $\mathbf{G_{lo}}$.

The condition stated in Theorem 1 will be called *low-level hierarchical consistency*. Intuitively it guarantees that the updated behavior of $\mathbf{G_{hi}}$ will always satisfy the high-level legal constraint, and that the 'real' low-level behavior in $\mathbf{G_{lo}}$ will be as large as possible subject to this constraint. Nonetheless, the high-level behavior expected in $\mathbf{G_{hi}}$ by the manager may be larger than what the operator of $\mathbf{G_{lo}}$ can optimally report.

To conclude this section consider again the running example with $\mathbf{G_{lo}}$ (i.e. $\mathbf{G_{lo,ext}}$) and $\mathbf{G_{hi}}$ (i.e. $\mathbf{G_{hi,ext}}$) as displayed in Fig. 5.4.

First suppose that the transition graph of E_{hi} coincides with that of $\mathbf{G_{hi}}$ except that the (controllable) transition $[2, \alpha_c, 1]$ has been deleted. It is clear that E_{hi} is a controllable sublanguage of L_{hi}. The corresponding control law γ_{lo} requires merely the disablement of event 3 at state 4 in $\mathbf{G_{lo}}$ (in this simple example, state-based control with no additional memory is sufficient). It is evident that

$$\theta(L(\gamma_{lo}, \mathbf{G_{lo}})) = E_{hi}$$

By contrast, suppose instead that E_{hi} is derived from $\mathbf{G_{hi}}$ by deletion of the selfloop $[1, \alpha_c, 1]$. Then γ_{lo} must disable event 1 at state 2 in $\mathbf{G_{lo}}$, with the unwanted side effect that state 4 in $\mathbf{G_{lo}}$, with output β_c, can never be reached. The manager is chagrined to find that the behavior reported by the operator is much less than he expected:

$$\begin{aligned} \theta(L(\gamma_{lo}, \mathbf{G_{lo}})) &= \{\epsilon, \alpha_u\} \\ &\subsetneqq E_{hi} \end{aligned}$$

Exercise 2: Assume that $E_{hi} \subseteq L(\mathbf{G_{hi}})$ is nonempty and closed, but not necessarily controllable, and set $E_{lo} = \theta^{-1}(E_{hi})$. While it is always true that

$$\theta(E_{lo}^{\uparrow}) \subseteq \theta(E_{lo}) = E_{hi}$$

it may be true that

$$\theta(E_{lo}^{\uparrow}) \supsetneqq E_{hi}^{\uparrow}$$

Provide an example to illustrate this situation. In intuitive, 'real world' terms explain why, in general, this result might not be unexpected.

5.5 Hierarchical Consistency

Let $E_{hi} \subseteq L_{hi}$ be closed and controllable and let $\mathbf{G_{lo}}$ be output-control-consistent. It was noted in the previous section that the inclusion

$$\theta((\theta^{-1}(E_{hi}))^{\uparrow}) \subseteq E_{hi} \tag{5.6}$$

may turn out to be strict. Intuitively, the behavior E_{hi} 'expected' by the manager in $\mathbf{G_{hi}}$ may be larger than what the operator can actually realize: the manager is 'overoptimistic' with respect to the efficacy of the command-control process. If equality does hold in (5.6) for every closed and controllable language $E_{hi} \subseteq L_{hi}$, the pair $(\mathbf{G_{lo}}, \mathbf{G_{hi}})$ will be said to possess *hierarchical consistency*. In that case, by Theorem 5.4.1, the command and control process defined in Sect. 5.4 for E_{hi} will actually synthesize E_{hi} in $\mathbf{G_{hi}}$. In the terminology of hierarchical control, every high-level 'task' (represented by a choice of E_{hi}) will be successfully 'decomposed' and executed in the hierarchical control loop.

Achieving equality in (5.6) for arbitrary controllable specification languages E_{hi} in general requires a further refinement of the transition structure of $\mathbf{G_{lo}}$, in other words, the possibly costly step of enhancing the information sent up by $\mathbf{C_{lo}}$ to $\mathbf{C_{hi}}$ (or by $\mathbf{G_{lo}}$ to $\mathbf{G_{hi}}$, depending on one's interpretation of the setup); of course such enhancement might or might not be feasible in an application. Referring to the reachability tree for $L(\mathbf{G_{lo,ext}})$ as described in Sect. 5.3, say that red vocal nodes n_1, n_2, with $\hat{\omega}(n_1) \neq \hat{\omega}(n_2)$, are *partners* if their silent paths start either at the root node or at the same vocal node, say $n = \text{node}(s)$; share an initial segment labeled $s'\sigma$ with $\sigma \in \Sigma_c$; and this shared segment is followed in turn by segments labeled by strings $s''s_1, s''s_2$, respectively, where $s'' \in \Sigma_u^*$ and at least one of the strings s_1, s_2 belongs to Σ_u^* (see Fig. 5.5, where we assume $s_1 \in \Sigma^*\Sigma_c\Sigma_u^*$ and $s_2 \in \Sigma_u^*$). We call node$(ss'\sigma)$ the *antecedent* of the partners n_1, n_2. In this structure the controllable events labeled $\tau_{1c} = \hat{\omega}(n_1)$, $\tau_{2c} = \hat{\omega}(n_2)$ in $\mathbf{G_{hi}}$ cannot be disabled independently by a command to $\mathbf{C_{lo}}$. Thus if E_{hi} requires disabling of τ_{2c} (at some state of its transition structure) then it may be true that $\mathbf{C_{lo}}$ is forced to disable τ_{1c} as well, via direct disablement of σ. So a cure in principle is to break up the occurrence of partners: declare that the hitherto silent antecedent node$(ss'\sigma)$ is now a red vocal node with controllable output any new symbol τ_c'', extend T_c by τ_c'' accordingly, and reevaluate color(n_i), $\hat{\omega}(n_i)$, $(i = 1, 2)$ as appropriate (in Fig. 5.5, n_2 would be recolored green and $\hat{\omega}(n_2)$ redefined to be τ_{2u}).

A formal version of this procedure is provided in Appendix 5.2. The result is embodied in the following definition. Let $\mathbf{G}$ be a Moore transition structure as in (5.2). Then $\mathbf{G}$ is *strictly output-control-consistent* (SOCC) if it is output-control-consistent and if in the reachability tree of $L(\mathbf{G})$ no two red vocal nodes are partners. As in Sect. 5.3, this definition could be formally rephrased in terms of the strings of $L(\mathbf{G})$ if desired. In Appendix 5.2 it is shown that the SOCC property can be obtained by no more than a fourfold increase in state size over that of the original DES $\mathbf{G_{lo}}$ that we started with; in practice, a factor of about 1.5 seems to be more typical.

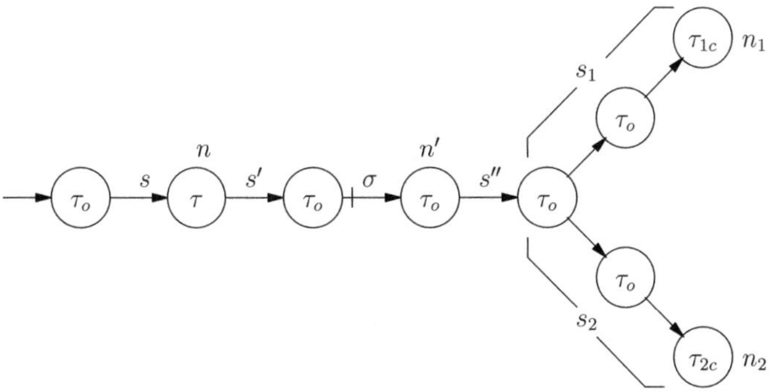

Fig. 5.5 Partners n_1, n_2 with antecedent n'. Note that $\tau_{1c} \neq \tau_{2c}$

Consider again our running example. By inspection of $\mathbf{G_{lo}}$ (i.e. $\mathbf{G_{lo,ext}}$ in Fig. 5.4) we note the partner configuration

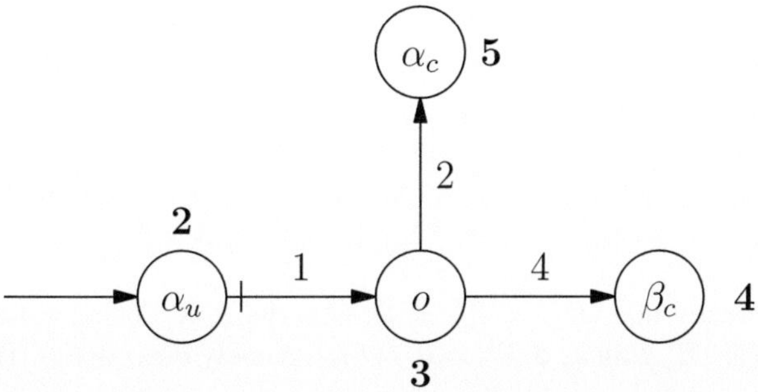

Its cure is the vocalization of the antecedent 3 of states 4 and 5, say with a new controllable event γ_c. The final results are displayed in Fig. 5.6. Notice that in $\mathbf{G_{lo,new}}$ the status of β has changed from β_c to β_u and that $\mathbf{G_{hi,new}}$ is larger than $\mathbf{G_{hi,ext}}$ by one state and transition.

Returning to our hierarchical control structure we finally have the desired result.

Theorem 1

Assume that $\mathbf{G_{lo}}$ is SOCC, and let $E_{hi} \subseteq L(\mathbf{G_{hi}})$ be nonempty, closed and controllable. Then

$$\theta((\theta^{-1}(E_{hi}))^{\uparrow}) = E_{hi}$$

Proof

In the proof write $L(\mathbf{G_{lo}}) =: L_{lo}$, $L(\mathbf{G_{hi}}) =: L_{hi}$, $L(\gamma_{lo}, \mathbf{G_{lo}}) =: K_{lo}$. With $E_{lo} := \theta^{-1}(E_{hi})$, Theorem 5.4.1 can be applied to yield

$$K_{lo} = E_{lo}^{\uparrow} = (\theta^{-1}(E_{hi}))^{\uparrow}$$

which implies

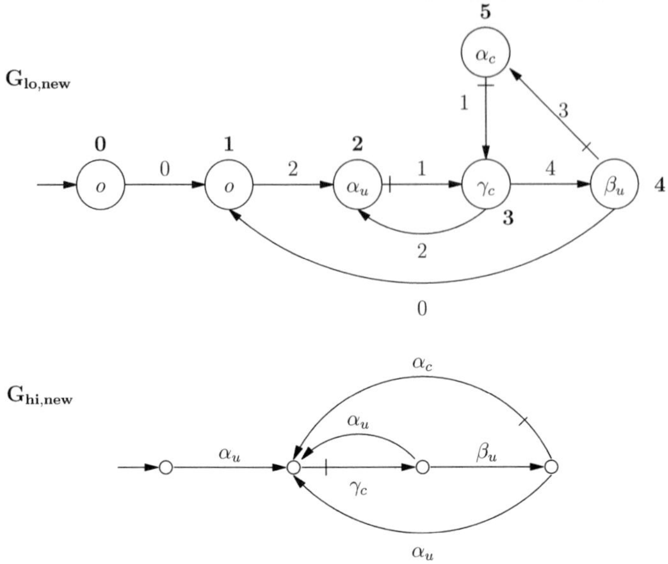

Fig. 5.6 Strict output-control-consistency

$$\theta(K_{lo}) = \theta((\theta^{-1}(E_{hi}))^{\uparrow}) \subseteq \theta(E_{lo}) = E_{hi}$$

We note directly that $(K_{lo} =)E_{lo}^{\uparrow} \neq \varnothing$. Otherwise, K_{lo} being closed, there is a string $s_o \in L_{lo} \cap \Sigma_u^*$ with $s_o \notin E_{lo} = \theta^{-1}(E_{hi})$, namely $\theta(s_o) \notin E_{hi}$. Thus

$$\theta(s_o) \in L_{hi} \cap (T_u^* - E_{hi})$$

which implies that E_{hi} is uncontrollable, contrary to hypothesis.

Now suppose that the inclusion $\theta(K_{lo}) \subseteq E_{hi}$ is strict, and let $t \in E_{hi} - \theta(K_{lo})$. Since $\epsilon \in \theta(K_{lo})$ there exists a maximal prefix $t' < t$ with $t' \in \theta(K_{lo})$. Let $s \in E_{lo}$ with $\theta(s) = t$. Since $\epsilon \in K_{lo}$ we can select a prefix $s'' \leq s$ of maximal (possibly zero) length such that $s'' \in K_{lo}$ and node(s'') is vocal (or is the root node). Then $t'' := \theta(s'')$ satisfies $t'' \leq t'$, where the prefix ordering may be strict. Let $w \in \Sigma^*$ with $s''w \leq s$, node$(s''w)$ vocal, and $\theta(s''w) = \theta(s'')\tau$ for some $\tau \in T$; that is, the path from node(s'') to node$(s''w)$ is silent. Now $w \in \Sigma^*\Sigma_c\Sigma_u^*$, as otherwise $w \in \Sigma_u^*$, which implies by the controllability of K_{lo} that $s''w \in K_{lo}$, contrary to the maximality of s''. Choose $w' \leq w$ to be of maximal length such that $s''w' \in K_{lo}$, namely $s''w'\sigma \notin K_{lo}$, with $w'\sigma \leq w$ and $\sigma \in \Sigma_c$ (so that σ is disabled by γ_{lo}).

We claim that there must exist a string $v \in \Sigma_u^*$ such that (1) node$(s''w'\sigma v)$ is vocal, and (2) the path from node(s'') to node$(s''w'\sigma v)$ is silent, with (3) $\theta(s''w'\sigma v) \notin E_{hi}$. For if there is no string $v \in \Sigma_u^*$ with properties (1)-(3) then, because

$$\theta(s''w) = \theta(s'')\tau \in E_{hi} = E_{hi}^{\uparrow}$$

we have that $s''w \in K_{lo}$ after all, contradicting the maximality of s'' and the fact that K_{lo} is supremal.

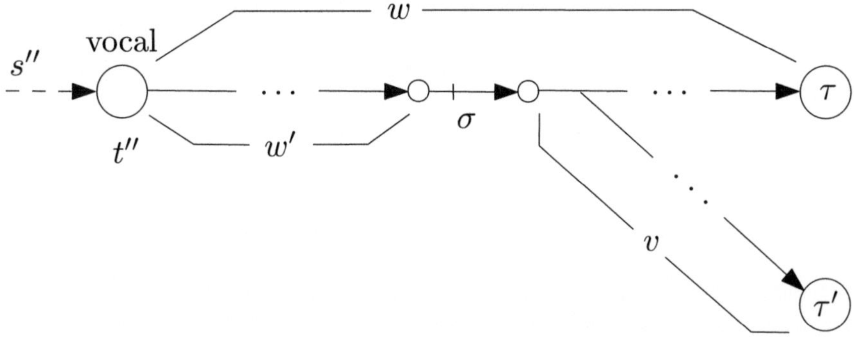

Since

$$t''\tau = \theta(s'')\tau = \theta(s''w) \leq \theta(s) = t \in E_{hi}$$

and $\theta(s''w'\sigma v) \notin E_{hi}$, it results finally that $\theta(s''w'\sigma v) = t''\tau'$, say, with $\tau' \neq \tau$, and therefore node$(s''w)$ and node$(s''w'\sigma v)$ are partners, in contradiction to the main hypothesis of the theorem. □

Thus when $\mathbf{G_{lo}}$ is SOCC, hierarchical consistency is achieved for the pair $(\mathbf{G_{lo}}, \mathbf{G_{hi}})$.

In *TCT*, hierarchical consistency can be achieved by computing either

HCGLO = hiconsis(OCGLO)

or directly as

HCGLO = hiconsis(GLO)

bypassing the intermediate stage of output-control-consistency. The resulting high-level DES is

HCGHI = higen(HCGLO)

More information on **hiconsis** is provided in Appendices 5.1–5.3.

Exercise 2: Use *TCT* to verify the running example of Sects. 5.2–5.5.

Finally it should be noted that our restriction to a hierarchy of two levels was inessential. Once hierarchical consistency has been achieved for the bottom level and first level up, say $(\mathbf{G_0}, \mathbf{G_1})$, the constructions may be repeated on assigning state outputs in $\mathbf{G_1}$ and bringing in a next higher level, $\mathbf{G_2}$. Clearly hierarchical consistency for $(\mathbf{G_1}, \mathbf{G_2})$ can be achieved without disturbing the consistency of $(\mathbf{G_0}, \mathbf{G_1})$. The theory thus possesses the highly desirable attribute of 'vertical modularity'.

To conclude this section we give two results (due to K.C. Wong) which place the property of hierarchical consistency in clear perspective. Let $\mathbf{G_{lo}}$ be OCC. Recall the notation $\mathcal{C}(E)$ for the family of controllable sublanguages of E; thus $\mathcal{C}(L_{lo})$ (resp. $\mathcal{C}(L_{hi})$) is the family of all controllable sublanguages of L_{lo} (resp. L_{hi}).

Let us bring in the

$$\textbf{Main Condition}: \ \theta\mathcal{C}(L_{lo}) = \mathcal{C}(L_{hi})$$

Main Condition (MC) says, not only that θ preserves controllability, but also that every high-level controllable language is the θ-image of some (possibly more than one) low-level controllable language. In other words, equating executable 'tasks' with controllable languages, every task that could be specified in the manager's (aggregated) model $\mathbf{G_{hi}}$ is executable in the operator's (detailed) model $\mathbf{G_{lo}}$; high-level policies can always be carried out operationally. Of course a justification of this interpretation would require that an online hierarchical control mechanism be spelled out; but this was done in Sect. 5.4. Now let $E_{hi} \subseteq L_{hi}$ be a high-level legal specification, not necessarily controllable. Suppose that E_{hi} is 'proposed' to the operator by specification of its preimage $\theta^{-1}(E_{hi})$. The operator may then synthesize $(\theta^{-1}(E_{hi}))^{\uparrow} \subseteq L_{lo}$, with the result that $\theta((\theta^{-1}(E_{hi}))^{\uparrow})$ is implemented in $\mathbf{G_{hi}}$. One would like this implemented sublanguage of L_{hi} to be precisely the language $E_{hi}^{\uparrow}$ that a manager working at the level of $\mathbf{G_{hi}}$ would synthesize directly (if direct control were feasible): this is the essence of hierarchical consistency. The result to follow states that hierarchical consistency in this strong sense is equivalent to MC.

Theorem 3

$$\text{MC} \Leftrightarrow \left((\forall E_{hi}) E_{hi} \subseteq L_{hi} \Rightarrow \theta((\theta^{-1}(E_{hi}))^{\uparrow}) = E_{hi}^{\uparrow}\right) \qquad \square$$

The usefulness of this result resides in the fact that the 'complicated' condition of hierarchical consistency (involving the $(\cdot)^{\uparrow}$ operation) is replaced by the formally simpler MC, which involves only the controllability property.

Along the same lines, on weakening MC slightly the following related result can be proved, as a simpler version of the condition of hierarchical consistency defined earlier in this section.

Theorem 4

$$\theta\mathcal{C}(L_{lo}) \supseteq \mathcal{C}(L_{hi}) \Leftrightarrow \left((\forall E_{hi}) E_{hi} \in \mathcal{C}(L_{hi}) \Rightarrow \theta((\theta^{-1}(E_{hi}))^{\uparrow}) = E_{hi}\right) \qquad \square$$

It is of interest to note that these results depend on nothing more than the fact that the operations $\theta(\cdot)$ and $(\cdot)^{\uparrow} = \sup\mathcal{C}(\cdot)$ are monotone on sublanguages.

Exercise 5: Prove Theorems 3 and 4.

Exercise 6: Construct examples to illustrate Theorems 3 and 4.

5.6 Hierarchical Supervision of Transfer Line

The theory will be illustrated by developing a high-level hierarchical supervisor for Transfer Line (cf. Sect. 4.6). We recall that Transfer Line consists of two machines **M1**, **M2** plus a test unit **TU**, linked by buffers **B1**, **B2** in the sequence: **M1, B1, M2,**

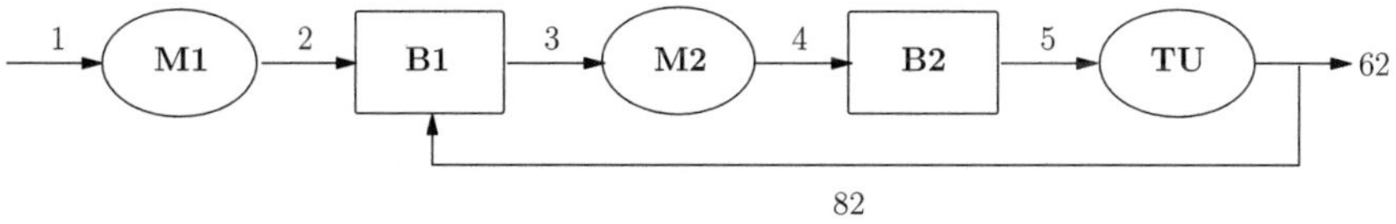

Fig. 5.7 Block diagram of Transfer Line

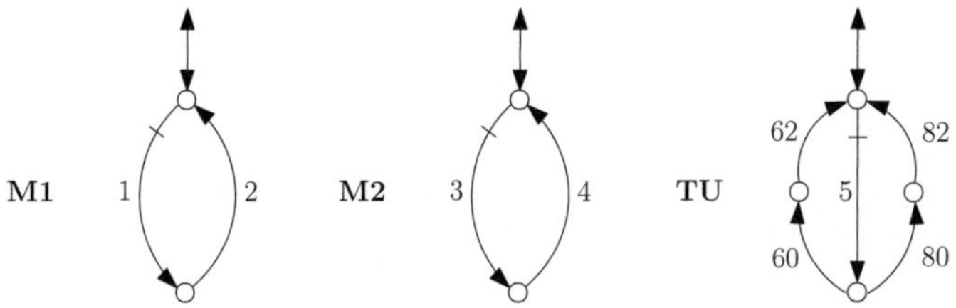

Fig. 5.8 M1, M2, TU

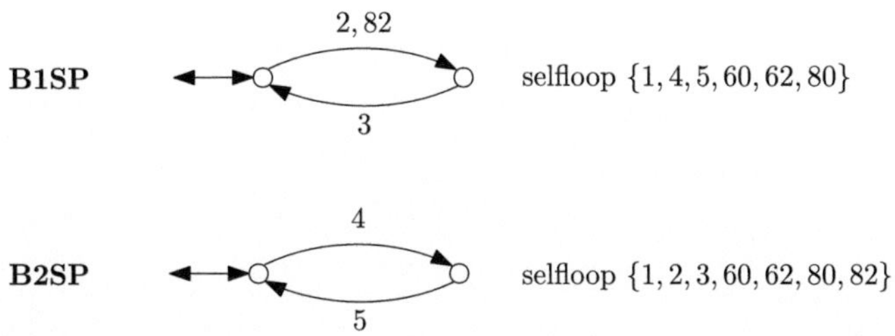

Fig. 5.9 B1SP, B2SP

B2, TU (Fig. 5.7). State transition diagrams of **M1**, **M2**, and **TU** are displayed in Fig. 5.8.

TU either 'passes' or 'fails' each processed workpiece, signaling its decision with events 60, 80 respectively. In case of 'pass test', the workpiece is sent to the system output (event 62); in case of 'fail test', it is returned to **B1** (event 82) for reprocessing by **M2**. There is no limit on the number of failure/reprocess cycles a given workpiece may undergo.

For ease of display we consider only the simplest case, where **B1** and **B2** each has capacity 1. Initially an optimal low-level supervisor is designed by any of the methods of Chaps. 3 or 4, to ensure that neither of the buffers is subject to overflow or underflow. In detail, let

$$\mathbf{PL} = \mathbf{sync}(\mathbf{M1}, \mathbf{M2}, \mathbf{TU})$$

and let **B1SP**, **B2SP** be the buffer specification generators (Fig. 5.9).

Then we set **BSP** = **meet(B1SP,B2SP)**, and

$$\mathbf{PLSUP} = \mathbf{supcon}(\mathbf{PL},\mathbf{BSP})$$

Fig. 5.10 PLSUP

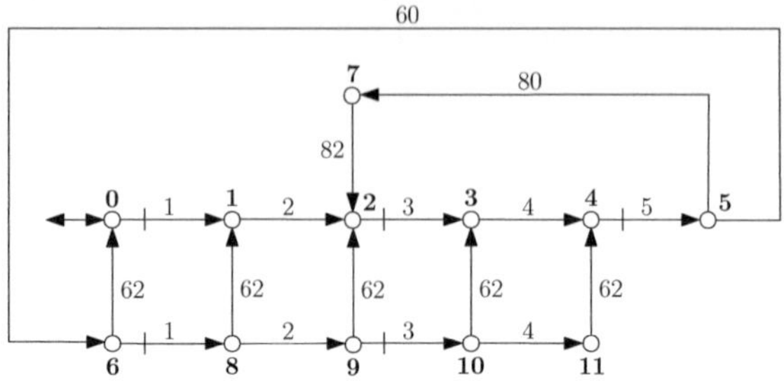

Fig. 5.11 GLO

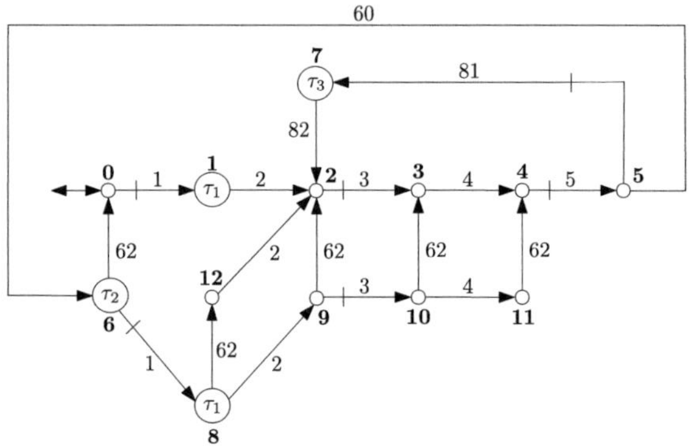

as displayed in Fig. 5.10. With **PLSUP** as the starting point for the development of hierarchical structure, we must first assign the 'significant' events to be signaled to the 'manager'. Let us assume that the manager is interested only in the events corresponding to 'taking a fresh workpiece' (low-level event 1, signaled as high-level event τ_1, say), and to 'pass test' (low-level event 60, signaled as τ_2) or 'fail test' (low-level event 80, signaled as τ_3). If too many failures occur the manager intends to take remedial action, which will start by disabling the failure/reprocess cycle. To this end the uncontrollable event 80 is now replaced in the low-level structure by a new controllable event 81. Furthermore, the meaning of the signaled events τ_1, τ_2, τ_3 must be unambiguous, so a transition entering state 1 like [8,62,1] must not be confused with the 'significant' transition [0,1,1]; namely a new state (say, 12) must be introduced, transition [8,62,1] replaced by [8,62,12], and a new transition [12,2,2] inserted. The final Moore structure, **GLO**, is displayed in Fig. 5.11. Here the vocal [state,output] pairs are [1,τ_1], [8,τ_1], [6,τ_2] and [7,τ_3]. In *TCT*, the foregoing adjustments can be made using **edit** and **vocalize**.

We are now ready to carry out the procedures of the theory. By inspection of Fig. 5.11, it is clear that each of τ_1, τ_2, τ_3 is unambiguously controllable, that is, **GLO** is already output-control-consistent. The corresponding high-level model **GHI** is displayed in Fig. 5.12. In *TCT*, **GHI** = **higen(GLO)**.

However, for the manager to disable τ_2 will require the operator to disable low-level event 5, which in turn disables the high-level event τ_3 as an undesired side effect;

Fig. 5.12 GHI

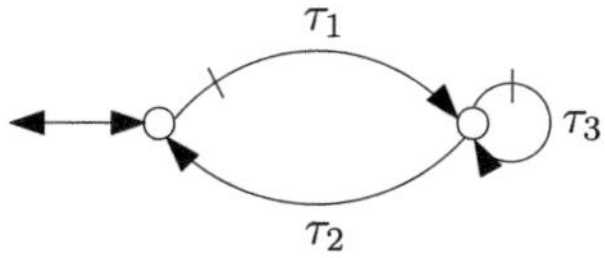

Fig. 5.13 HCGLO

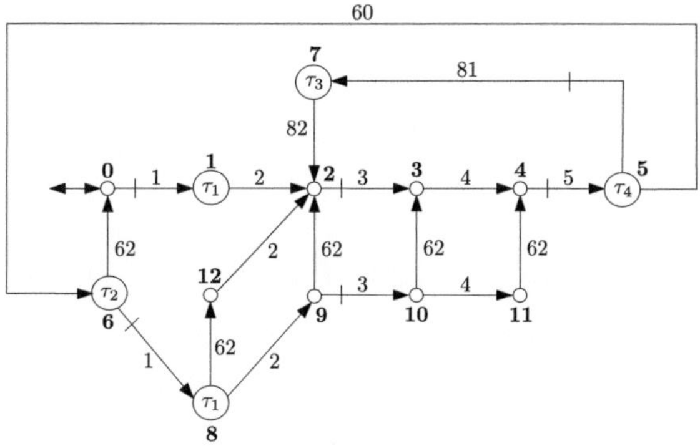

Fig. 5.14 HCGHI

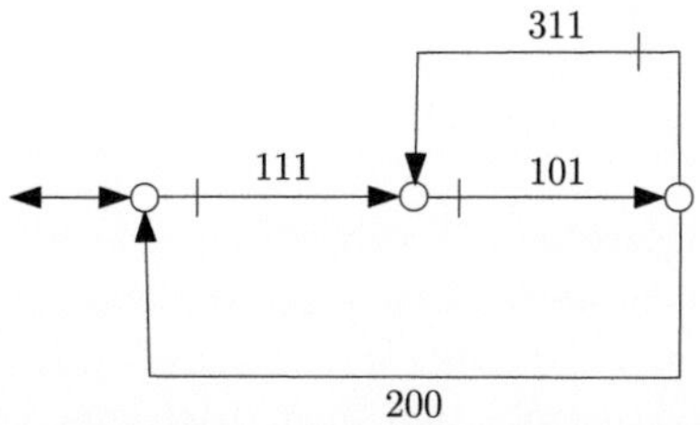

thus **GLO** is not strictly output-control-consistent (SOCC). To improve matters it is enough to vocalize the low-level state 5 with a new high-level output τ_4, signaling the new 'significant' event that 'TU takes a workpiece'. This step incidentally converts the status of τ_2 from controllable to uncontrollable. With this the construction of a SOCC model, say **HCGLO**, from **GLO** is complete (Fig. 5.13). The corresponding high-level model **HCGHI** is displayed in Fig. 5.14, where $\tau_1, \tau_2, \tau_3, \tau_4$ have been coded respectively, as 111,200,311,101. In *TCT*, **HCGLO** = **hiconsis(GLO)** and **HCGHI** = **higen (HCGLO)**.

The simple model **HCGHI** can be supervised by the manager to achieve his objective of 'quality control'. A possible high-level specification might be: 'if two consecutive test failures (311) occur, allow **TU** to operate just once more, then shut down the system'; this is modeled by **SPECHI** as displayed (Fig. 5.15). The resulting supervisor

$$\mathbf{HCGHISUP} = \mathbf{supcon(HCGHI, SPECHI)}$$

is shown in Fig. 5.16. On termination of **HCGHISUP** at state 7, and execution by **TU** of event 62, it can be easily verified that **HCGLO** will have halted at its marker state 0.

Fig. 5.15 SPECHI

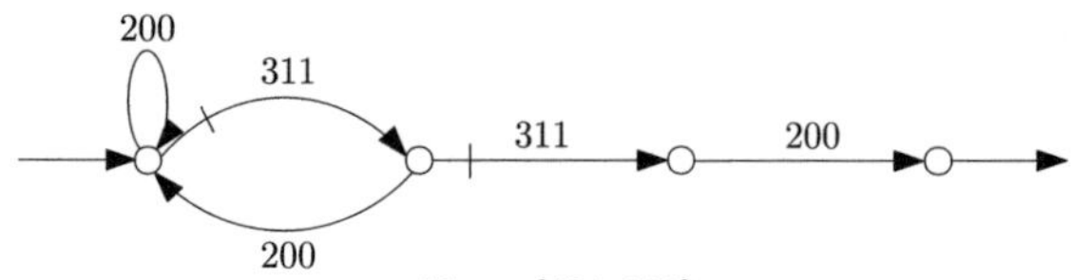

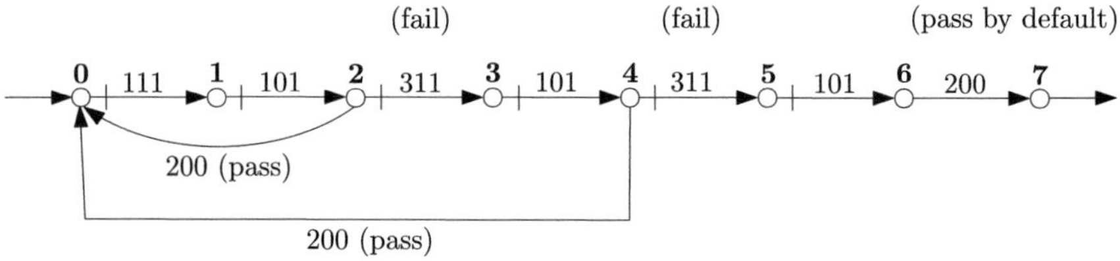

Fig. 5.16 HCGHISUP

Exercise 1: To appraise the utility of hierarchical control for Transfer Line, replace **SPECHI** by an equivalent DES **SPECLO** for **GLO** (with vocalizations removed), compute the corresponding low-level supervisor, and compare its state size with that of **HCGHISUP**.

Exercise 2: For a manageable but nontrivial example with a plausible physical interpretation, carry out the design of a SOCC hierarchical control structure, illustrating the successive refinements involved in first achieving (non-strict) output-control-consistency and then strict consistency. For a particular high-level controllable specification language, define precisely the required command and control γ_{lo}.

5.7 Hierarchical Supervision with Nonblocking

In this section we extend our theory of hierarchical supervision to include marking and nonblocking. Unsurprisingly, nonblocking is not ensured by hierarchical consistency over closed sublanguages, in the sense of Sect. 5.5. For instance, in the example of Fig. 5.17, with α, β, γ uncontrollable, evidently

$$\mathfrak{C}_{lo}(L_{lo}) = \{\varnothing, L_{lo}\}, \quad \mathfrak{C}_{hi}(L_{hi}) = \{\varnothing, L_{hi}\}$$

and $\mathbf{G_{lo}}$ is hierarchically consistent; however, $\mathbf{G_{lo}}$ blocks on executing α.

Throughout this section it will be convenient to write L for $L_{lo} := L(\mathbf{G_{lo}})$, $L_m \subseteq L$ for the marked behavior of $\mathbf{G_{lo}}$, $M := \theta(L)$ for $L_{hi} := L(\mathbf{G_{hi}})$ and $M_m := \theta(L_m)$ for the marked behavior of $\mathbf{G_{hi}}$.

We begin by generalizing Theorem 5.5.3. Our standing assumption is that $\mathbf{G_{lo}}$ is output-control-consistent (Sect. 5.3). The scenario will be that, with $E_{hi} \subseteq M_m$ a given specification language for $\mathbf{G_{hi}}$, the marked behavior 'virtually' synthesized in $\mathbf{G_{hi}}$ is, as usual (cf. Sect. 3.5)

Fig. 5.17 $\mathbf{G_{lo}}$, $\mathbf{G_{hi}}$

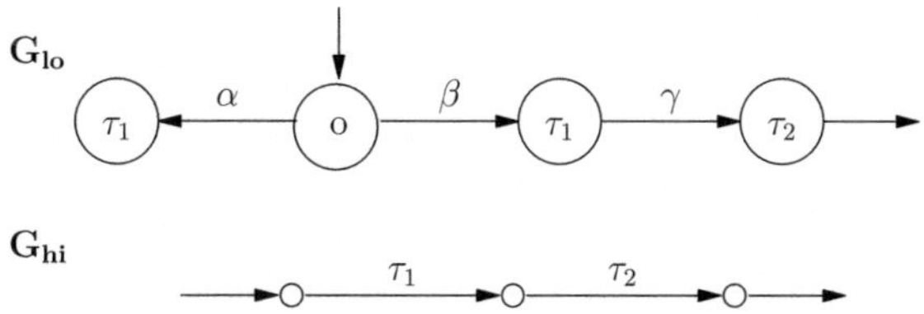

$$K_{hi} := \sup \mathcal{C}_{hi}(E_{hi})$$

The specification 'announced' to the low-level controller is $\theta^{-1}(E_{hi})$, so the marked behavior synthesized in $\mathbf{G_{lo}}$ is, again as usual,

$$K_{lo} := \sup \mathcal{C}_{lo}(L_m \cap \theta^{-1}(E_{hi}))$$

The desired consistency property is then

$$\theta(K_{lo}) = K_{hi}$$

Theorem 1

In the foregoing notation

$$(\forall E_{hi}) E_{hi} \subseteq M_m \Rightarrow \theta(K_{lo}) = K_{hi} \qquad \text{(HCm)}$$

iff

$$\theta \mathcal{C}_{lo}(L_m) = \mathcal{C}_{hi}(M_m) \qquad \text{(MCm)}$$

Proof

(If) Let $E_{hi} \subseteq M_m$ and assume (MCm). Then

$$K_{hi} := \sup \mathcal{C}_{hi}(E_{hi}) \in \mathcal{C}_{hi}(M_m)$$

so by (MCm), $K_{hi} = \theta(K'_{lo})$ for some $K'_{lo} \in \mathcal{C}_{lo}(L_m)$. Therefore

$$\begin{aligned} K'_{lo} &\subseteq \sup \mathcal{C}_{lo}(L_m \cap \theta^{-1}(K_{hi})) \\ &\subseteq \sup \mathcal{C}_{lo}(L_m \cap \theta^{-1}(E_{hi})) = K_{lo} \end{aligned}$$

Thus

$$K_{hi} = \theta(K'_{lo}) \subseteq \theta(K_{lo})$$

But $K_{lo} \subseteq L_m \cap \theta^{-1}(E_{hi})$ implies

$$\theta(K_{lo}) \subseteq M_m \cap E_{hi} = E_{hi}$$

So by (MCm) $\theta(K_{lo}) \subseteq \sup \mathcal{C}_{hi}(E_{hi}) = K_{hi}$. Thus $\theta(K_{lo}) = K_{hi}$ as claimed.

(Only if) Let $K_{hi} \in \mathfrak{C}_{hi}(M_m)$ and set $E_{hi} = K_{hi}$ in (HCm). Then

$$K_{hi} = \theta(K_{lo}) \in \theta(\mathfrak{C}_{lo}(L_m))$$

namely $\mathfrak{C}_{hi}(M_m) \subseteq \theta\mathfrak{C}_{lo}(L_m)$. For the reverse inclusion let $K'_{lo} \in \mathfrak{C}_{lo}(L_m)$ and set $E'_{hi} := \theta(K'_{lo})$. Then $E'_{hi} \subseteq \theta(L_m) = M_m$, so $K'_{hi} := \sup \mathfrak{C}_{hi}(E'_{hi}) \in \mathfrak{C}_{hi}(M_m)$. Now $K'_{lo} \subseteq \theta^{-1}(E'_{hi})$ implies

$$K'_{lo} \subseteq \sup \mathfrak{C}_{lo}\left(L_m \cap \theta^{-1}(E'_{hi})\right)$$

Also (HCm) provides

$$\theta\left(\sup \mathfrak{C}_{lo}(L_m \cap \theta^{-1}(E'_{hi})\right) = \sup \mathfrak{C}_{hi}(E'_{hi})$$

Therefore

$$\theta(K'_{lo}) \subseteq \sup \mathfrak{C}_{hi}(E'_{hi}) \subseteq E'_{hi} = \theta(K'_{lo})$$

giving

$$\theta(K'_{lo}) = \sup \mathfrak{C}_{hi}(E'_{hi}) = K'_{hi} \in \mathfrak{C}_{hi}(M_m)$$

namely $\theta\mathfrak{C}_{lo}(L_m) \subseteq \mathfrak{C}_{hi}(M_m)$ as required. □

While Theorem 1 provides an interesting perspective on hierarchical supervision with nonblocking, the 'Main Condition with marking' (MCm) is not immediately effective. To satisfy (MCm) we shall endow our causal reporter map θ with a certain 'global observer property'; we shall also require a type of 'local controllability' in $\mathbf{G_{lo}}$. These properties seem natural from the viewpoint of a designer with some capability of structuring $\mathbf{G_{lo}}$ in advance.

In the following we use the terminology for reachability tree as in Sect. 5.3. Let

$$L_{voc} := \{s \in L | s = \epsilon \text{ or node}(s) \text{ is vocal}\}$$

Thus L_{voc} is the subset of strings of L that correspond to the root node or a vocal node of the reachability tree of L. Clearly $\theta(L_{voc}) = \theta(L) = M$. To avoid fussy details we shall assume, reasonably, that vocalization is 'complete', namely any string of L can be extended (in L) to a string of L_{voc}, i.e. $\bar{L}_{\text{voc}} = L$. Now we say that $\theta : L \to M$ is an L_{voc}-*observer* if

$$(\forall s \in L_{voc}, t \in M)\theta(s) \leq t \Rightarrow (\exists s' \in \Sigma^*)ss' \in L_{voc} \;\&\; \theta(ss') = t$$

In other words, whenever $\theta(s)$ can be extended to a string $t \in M$, the underlying L_{voc}-string s can be extended to an L_{voc}-string ss' with the same image under θ: 'the manager's expectation can always be executed in $\mathbf{G_{lo}}$', at least when starting from a string in L_{voc}. In the Example of Fig. 5.17, θ is not an L_{voc}-observer, while in that of Figs. 5.11, 5.12 it is.

Write θ_{voc} for the restriction $\theta|L_{voc}$. Thus

$$\theta_{voc} : L_{voc} \to M, \quad \theta_{voc}^{-1} : Pwr(M) \to Pwr(L_{voc})$$

The L_{voc}-observer property can be characterized as commutativity of θ_{voc}^{-1} with prefix-closure. Recall that the operation of prefix-closure maps L_{voc} onto L.

Proposition 2

θ is an L_{voc}-observer iff, for all $E \subseteq M$,

$$\theta_{voc}^{-1}(\bar{E}) = \overline{\theta_{voc}^{-1}(E)} \cap L_{voc}$$

Proof

(If) Let $s \in L_{voc}, t \in M, \theta(s) \leq t$. Then $\theta(s) \in \overline{\{t\}} \subseteq M$, so

$$s \in \theta_{voc}^{-1}(\overline{\{t\}}) = \overline{\theta_{voc}^{-1}(\{t\})} \cap L_{voc}$$

Thus for some $s' \in \Sigma^*$, $ss' \in \theta_{voc}^{-1}(\{t\})$, namely $\theta_{voc}(ss') = t$, as required.

(Only if) The direction $\overline{\theta_{voc}^{-1}(E)} \cap L_{voc} \subseteq \theta_{voc}^{-1}(\bar{E})$ is automatic, for if $ss' \in \theta_{voc}^{-1}(E)$ for $s \in L_{voc}$ and some s' with $ss' \in L_{voc}$, then $\theta_{voc}(s) \leq \theta_{voc}(ss') \in E$, so $\theta_{voc}(s) \in \bar{E}$ and $s \in \theta_{voc}^{-1}(\bar{E})$. For the reverse inclusion, taking $s \in \theta_{voc}^{-1}(\bar{E})$ we have $t := \theta_{voc}(s) \in \bar{E}$, so for some $t', tt' \in E$ and $\theta_{voc}(s) \leq tt'$. Since θ is an L_{voc}-observer, there is some s' with $ss' \in L_{voc}$ and $\theta_{voc}(ss') = tt'$, so $s \leq ss' \in \theta_{voc}^{-1}(E)$, namely $s \in \overline{\theta_{voc}^{-1}(E)} \cap L_{voc}$. □

We now bring in a 'local' description of controllability in $\mathbf{G_{lo}}$. Generalizing the control action of Sect. 5.4, one should now think of control decisions in $\mathbf{G_{lo}}$ (made by the 'operator' of Sect. 5.1) being delegated to 'agents', say to Agent(s) for each $s \in L_{voc}$. The scope of Agent(s) will be the 'local' language $L_{voc}(s)$ linking node(s) to the adjacent downstream vocal nodes (if any) in the reachability tree of L. Formally, for $s \in L_{voc}$ let

$$L_{voc}(s) := \{s' \in \Sigma^* | ss' \in L_{voc} \ \& \ (\exists \tau \in T)\theta(ss') = \theta(s)\tau\}$$

Notice that the empty string ϵ cannot belong to $L_{voc}(s)$, and that possibly $L_{voc}(s)$ is the empty language, $\varnothing$.

Let $L_{lo}(s) := \overline{L_{voc}(s)}$. Thus $L_{lo}(s)$ is either $\varnothing$ or, if not, contains ϵ. Along with $L_{lo}(s)$ define the local reporter map $\theta_s : L_{lo}(s) \to \{\epsilon\} \cup T$ according to

$$\theta_s(s') := \begin{cases} \epsilon, & \text{if } s' = \epsilon, \text{ or } s' \neq \epsilon \text{ and } \hat{\omega}(ss') = \tau_o \\ \tau, & \text{if } \hat{\omega}(ss') = \tau \end{cases}$$

Let $L_{hi}(t)$ be the (at most) one-step closed sublanguage that is postfix to $t := \theta(s)$ in $\mathbf{G_{hi}}$. Thus $L_{hi}(t)$ is the 'next-step' sublanguage seen following t by the manager.

Let $\mathfrak{C}_{lo}(s)$ denote the family of all controllable sublanguages of $L_{lo}(s)$, namely

$$\mathfrak{C}_{lo}(s) := \{K \subseteq L_{lo}(s) | \overline{K}\Sigma_u \cap L_{lo}(s) \subseteq \overline{K}\}$$

Apart from $\varnothing$, these are exactly the sublanguages of $L_{lo}(s)$ that Agent(s) is authorized (and able) to synthesize by supervisory control.

Similarly let $\mathfrak{C}_{hi}(t)$ denote the controllable sublanguages of $L_{hi}(t)$. The local controllability property of interest is that Agent(s) can synthesize (i.e. select from $\mathfrak{C}_{lo}(s)$) a controllable sublanguage for each controllable sublanguage that the manager can see one step ahead in $\mathbf{G_{hi}}$; namely

$$\mathfrak{C}_{hi}(t)\ [:= \mathfrak{C}_{hi}(\theta(s))] = \theta_s\left(\mathfrak{C}_{lo}(s)\right) \tag{5.7}$$

When (5.7) holds we shall say that $\mathbf{G_{lo}}$ is *locally output controllable at* s; and if (5.7) holds for all $s \in L_{voc}$, that $\mathbf{G_{lo}}$ is *globally output controllable*. This says that Agent(s) can always synthesize exactly any controllable local high-level behavior ordered by the manager, no more and no less. Returning to the example of Fig. 5.17, we see that $\mathbf{G_{lo}}$ is locally output controllable at strings ϵ, β, and $\beta\gamma$ but not at string α.

Notice that if $\varnothing \neq \theta_s\,(L_{voc}(s)) \subseteq T_c$, i.e. there exist downstream vocal events and they are all controllable, then the manager can terminate the process, by commanding the controllable sublanguage $\{\epsilon\}$. For this it is sufficient that Agent(s) disable (say) all his immediately downstream controllable events, and this will clearly synthesize a controllable sublanguage of $L_{lo}(s)$, hence an element of $\mathfrak{C}_{lo}(s)$, as required.

To illustrate the notation in detail consider the following two examples from Fig. 5.4.

Example 3

$s = 0.2 \in L_{voc},\ t = \alpha_u$
$L_{voc}(s) = \{1.2,\ 1.4\}, L_{lo}(s) = \{\epsilon,\ 1,\ 1.2,\ 1.4\}$
$L_{hi}(t) = \{\epsilon,\ \alpha_c,\ \beta_c\}$ (by inspection of $\mathbf{G_{hi,ext}}$ in Fig. 5.4)
$\mathfrak{C}_{lo}(s) = \{\varnothing,\ \{\epsilon\},\ \{\epsilon,\ 1,\ 1.2,\ 1.4\},\ \{1,\ 1.2,\ 1.4\},\ \{1.2,\ 1.4\}\}$
$\mathfrak{C}_{hi}(t) = \{\varnothing,\ \{\epsilon\},\ \{\alpha_c\},\ \{\beta_c\},\ \{\alpha_c,\ \beta_c\},\ \{\epsilon,\ \alpha_c\},\ \{\epsilon,\ \beta_c\},\ \{\epsilon,\ \alpha_c,\ \beta_c\}\}$
$\theta_s\left(\mathfrak{C}_{lo}(s)\right) = \{\varnothing,\ \{\epsilon\},\ \{\alpha_c,\ \beta_c\},\ \{\epsilon,\ \alpha_c,\ \beta_c\}\} \subsetneqq \mathfrak{C}_{hi}(t)$
Conclusion: $\mathbf{G_{lo,ext}}$ is not locally controllable at s.

Example 4

$s = 0.2.1.4 \in L_{voc},\ t = \alpha_u\beta_c$
$L_{voc}(s) = \{0.2,\ 3\},\ L_{lo}(s) = \{\epsilon,\ 0,\ 0.2,\ 3\}$
$L_{hi}(t) = \{\epsilon,\ \alpha_u,\ \alpha_c\}$ (by inspection of Fig. 5.4)
$\mathfrak{C}_{lo}(s) = \{\varnothing,\ \{0.2\},\ \{0.2,\ 3\},\ \{\epsilon,\ 0,\ 0.2\},\ \{\epsilon,\ 0,\ 0.2,\ 3\}\}$
$\mathfrak{C}_{hi}(t) = \{\varnothing,\ \{\alpha_u\},\ \{\alpha_u,\ \alpha_c\},\ \{\epsilon,\ \alpha_u\},\ \{\epsilon,\ \alpha_u,\ \alpha_c\}\}$
$\theta_s\left(\mathfrak{C}_{lo}(s)\right) = \{\varnothing,\ \{\alpha_u\},\ \{\alpha_u,\ \alpha_c\},\ \{\epsilon,\ \alpha_u\},\ \{\epsilon,\ \alpha_u,\ \alpha_c\}\} = \mathfrak{C}_{hi}(t)$
Conclusion: $\mathbf{G_{lo,ext}}$ is locally controllable at s.

Notice that the above $\mathfrak{C}$ listings are needlessly cumbersome, since the empty language $\varnothing$ is always listed; if a sublanguage K is listed, it is understood that its closure $\bar{K}$ should be listed also; and if K_1, K_2 are listed then so also should be $K_1 \cup K_2$. Thus in Example 3 one could write, symbolically,

$$\mathfrak{C}_{lo}(s) \approx \{\{\epsilon\},\ \{1.2,\ 1.4\}\},\ \mathfrak{C}_{hi}(t) \approx \{\{\epsilon\},\ \{\alpha_c\},\ \{\beta_c\}\}$$

and in Example 4,

$$\mathfrak{C}_{lo}(s) \approx \{\{0.2\},\ \{0.2,\ 3\}\},\ \mathfrak{C}_{hi}(t) \approx \{\{\alpha_u\},\ \{\alpha_u,\ \alpha_c\}\}$$

We can now state the main result of this section, establishing the condition (MCm) of Theorem 1. As a technical detail, it is convenient to specialize slightly the marking condition $\theta(L_m) = M_m$ as indicated in (iii) below. In this version only strings in L_{voc} will be marked in L.

Theorem 5

Assume that

(i) $\theta : L \to M$ is an L_{voc}-observer;

(ii) $\mathbf{G_{lo}}$ is globally output controllable; and

(iii) $L_m = \theta^{-1}(M_m) \cap L_{voc}$.

Then

$$\mathfrak{C}_{hi}(M_m) = \theta\mathfrak{C}_{lo}(L_m)$$

Proof

For inclusion ($\supseteq$) let $K_{lo} \in \mathfrak{C}_{lo}(L_m)$; we show $K_{hi} := \theta(K_{lo}) \in \mathfrak{C}_{hi}(M_m)$. First, $K_{lo} \subseteq L_m$ and $\theta(L_m) = M_m$ imply $K_{hi} \subseteq M_m$. For controllability, let $t \in \bar{K}_{hi}$, $\tau \in T_u$, $t\tau \in M$. Since $\overline{\theta(K_{lo})} = \theta(\bar{K}_{lo})$ (Exercise 5.2.1) there is $s \in \bar{K}_{lo}$ with $\theta(s) = t$, and we can certainly assume $s \in \bar{K}_{lo} \cap L_{voc}$. As $\theta(s) \le t\tau$, by (i) there exists $s' \in \Sigma^*$ with $ss' \in L_{voc}$ and $\theta(ss') = t\tau$. So $s' \in L_{voc}(s)$ and as $\tau \in T_u$, $s' \in \Sigma_u^*$ (by output-control-consistency, Sect. 5.3); therefore $ss' \in \bar{K}_{lo}$ (by controllability of K_{lo}). Finally

$$t\tau = \theta(ss') \in \theta(\bar{K}_{lo}) = \overline{\theta(K_{lo})} = \bar{K}_{hi}$$

as required.

For the reverse inclusion ($\subseteq$) let $\varnothing \neq K_{hi} \in \mathfrak{C}_{hi}(M_m)$. We construct $K_{lo} \in \mathfrak{C}_{lo}(L_m)$ inductively, such that $\theta(\bar{K}_{lo}) = \bar{K}_{hi}$. By (ii) we know that $\mathbf{G_{lo}}$ is globally output controllable, namely for all $s \in L_{voc}$, $t = \theta_{voc}(s)$,

$$\mathfrak{C}_{hi}(t) = \theta_s\mathfrak{C}_{lo}(s) \tag{5.8}$$

Starting with $\epsilon \in \bar{K}_{hi}$, $\epsilon = \theta_{voc}(\epsilon)$, we have

$$\mathfrak{C}_{hi}(\epsilon) = \theta_\epsilon\mathfrak{C}_{lo}(\epsilon)$$

Since $\bar{K}_{hi}$ is controllable,

$$\mathrm{Elig}(\bar{K}_{hi};\epsilon) := \{\tau \in T | \tau \in \bar{K}_{hi}\} \in \mathfrak{C}_{hi}(\epsilon)$$

and so

$$\mathrm{Elig}(\bar{K}_{hi};\epsilon) = \theta_{\epsilon}(H_{lo}(\epsilon))$$

for some locally controllable sublanguage $H_{lo}(\epsilon) \in \mathfrak{C}_{lo}(\epsilon)$. Thus for every $\tau_1 \in \mathrm{Elig}(\bar{K}_{hi};\epsilon)$ there is at least one string $s_1 \in H_{lo}(\epsilon) \subseteq L_{voc}(\epsilon)$ with $\theta_{voc}(s_1) = \tau_1$. We continue in the evident way: for such τ_1 we have

$$\mathrm{Elig}(\bar{K}_{hi};\tau_1) := \{\tau \in T | \tau_1\tau \in \bar{K}_{hi}\} \in \mathfrak{C}_{hi}(\tau_1)$$

so again by (5.8)

$$\mathrm{Elig}(\bar{K}_{hi};\tau_1) = \theta_{s_1}(H_{lo}(s_1))$$

for some $H_{lo}(s_1) \in \mathfrak{C}_{lo}(s_1)$. In general, for $t \in \bar{K}_{hi}$ and $s \in L_{voc}$ with $\theta_{voc}(s) = t$, we shall have

$$\mathrm{Elig}(\bar{K}_{hi};t) := \{\tau \in T | t\tau \in \bar{K}_{hi}\} = \theta_s(H_{lo}(s))$$

for some locally controllable sublanguage $H_{lo}(s) \in \mathfrak{C}_{lo}(s)$.

Denote by H_{lo} the prefix-closure of all strings of form

$$s = s_1 \cdots s_k, \quad k \in \mathbb{N}$$

such that

$$x_j := s_1 \cdots s_j \in L_{voc}, \quad 1 \le j \le k$$

$$\theta_{voc}(s) \in \bar{K}_{hi}$$

$$Elig(\bar{K}_{hi};\theta_{voc}(x_j)) = \theta_{s_j}(H_{lo}(s_j)), \quad H_{lo}(x_j) \in \mathfrak{C}_{lo}(x_j), \quad 1 \le j \le k$$

$$s_j \in H_{lo}(x_{j-1}), \quad 1 \le j \le k, \; x_0 := \epsilon$$

Clearly

$$\theta(H_{lo}) = \theta_{voc}(H_{lo} \cap L_{voc}) = \bar{K}_{hi}$$

We claim that $H_{lo} \in \mathfrak{C}_{lo}(L)$. Let $s \in H_{lo}, \sigma \in \Sigma_u, s\sigma \in L$, and let s' be the maximal prefix of s such that $s' \in L_{voc}$. If $s' < s$ then $s \in L - L_{voc}$ and $s = s'v$ for some $v \in \bar{H}_{lo}(s')$. Since $H_{lo}(s')$ is locally controllable, $v\sigma \in \bar{H}_{lo}(s')$ so $s\sigma = s'v\sigma \in H_{lo}$. If $s' = s$ and $s\sigma \in L$ then again $\sigma \in \bar{H}_{lo}(s)$ bycontrollability of $H_{lo}(s)$, so $s\sigma \in H_{lo}$. This proves the claim.

Define

$$K_{lo} := H_{lo} \cap L_{voc} \cap \theta^{-1}(K_{hi})$$

Then $K_{lo} \subseteq L_{voc} \cap \theta^{-1}(M_m) = L_m$. To establish $K_{lo} \in \mathfrak{C}_{lo}(L_m)$ it suffices to verify $\bar{K}_{lo} = H_{lo}$, or simply $H_{lo} \subseteq \bar{K}_{lo}$. Let $s \in H_{lo}, t := \theta(s)$. We claim there exists w with $sw \in K_{lo}$, i.e. $sw \in H_{lo} \cap L_{voc} \cap \theta^{-1}(K_{hi})$. If already $s \in K_{lo}$, set $w = \epsilon$. If not, let $u \leq s$ be the maximal prefix of s with $u \in L_{voc}$ and write $s = uv$. By construction of H_{lo}, we know that for some $w_1 \in \Sigma^*$ and locally controllable sublanguage $H_{lo}(u) \in \mathfrak{C}_{lo}(u)$,

$$vw_1 \in H_{lo}(u), \ \text{Elig}(\bar{K}_{hi}; t) = \theta_u(H_{lo}(u))$$

By definition of $\mathfrak{C}_{lo}(u)$, we know $H_{lo}(u) \subseteq L_{voc}(u)$ and thus $\theta_u(vw_1) = \tau_1$ (say) with $t\tau_1 \in \bar{K}_{hi}$. This means $sw_1 = uvw_1 \in H_{lo} \cap L_{voc} \cap \theta^{-1}(\bar{K}_{hi})$. If already $t\tau_1 \in K_{hi}$ then set $w = w_1$; if not, suppose $t\tau_1\tau_2 \cdots \tau_k \in K_{hi}$. Clearly

$$\tau_2 \in \text{Elig}(\bar{K}_{hi}; t\tau_1), \ t\tau_1 = \theta_{voc}(sw_1)$$

Repetition of the previous argument produces $w_2 \in \Sigma^*$ with

$$sw_1w_2 \in H_{lo} \cap L_{voc} \cap \theta^{-1}(\bar{K}_{hi})$$

and we are done after at most k steps.

It only remains to verify that $\theta(K_{lo}) = K_{hi}$. Let $t \in K_{hi} (\subseteq M_m)$. Since $\theta(H_{lo}) = \bar{K}_{hi}$ there is $s \in H_{lo} \cap L_{voc}$ such that $\theta_{voc}(s) = t$, namely

$$s \in H_{lo} \cap L_{voc} \cap \theta^{-1}(K_{hi}) = K_{lo}$$

On the other hand $\theta(K_{lo}) = \theta_{voc}(K_{lo}) \subseteq K_{hi}$. □

Combining Theorems 1 and 5 we have immediately

Corollary 6

Let $E_{hi} \subseteq M_m$, $K_{hi} := \sup \mathfrak{C}_{hi}(E_{hi})$, and

$$K_{lo} := \sup \mathfrak{C}_{lo}(L_m \cap \theta^{-1}(E_{hi}))$$

Assume conditions (i)-(iii) of Theorem 5. Then $(\mathbf{G_{lo}}, \theta)$ is hierarchically consistent, in the sense that $\theta(K_{lo}) = K_{hi}$. □

Exercise 7: By examining the initial state and vocal states in **HCGLO**, Fig. 5.13, show that Transfer Line already satisfies assumptions (i), (ii) of Theorem 5 and can be (mildly) remodeled to satisfy (iii) as well.

The foregoing results can be regarded as a fundamental basis for hierarchical control with nonblocking. Of course, algorithmic design, and verification of the appropriate conditions, remain as challenging issues. To conclude this section we provide a more specialized result, building directly on the theory for closed languages in Sects. 5.4, 5.5, including the property of strict output-control-consistency (SOCC) and the control action of Sect. 5.4. For this we place a much stronger observer condition on θ, and modify our description of nonblocking slightly to facilitate use of Theorem 5.5.1. The price of greater concreteness, in Theorem 9 below, is somewhat involved argumentation in the proof.

With $M_m = \theta(L_m)$ as before we say that $\theta : L \to M$ is an L_m-*observer* if

$$(\forall t \in M_m, s \in L)\theta(s) \leq t \Rightarrow (\exists u \in \Sigma^*)su \in L_m \ \&\ \theta(su) = t$$

In other words, whenever $\theta(s)$ can be extended in T^* to a string $t \in M_m$, the underlying string $s \in L$ can be extended to a string $su \in L_m$ with the same image under θ: 'the manager's expectation is never blocked in $\mathbf{G_{lo}}$'. This property fails for the example of Fig. 5.17, as one sees by taking $t = \tau_1\tau_2$ and $s = \alpha$.

Exercise 8: Show that $\theta : L \to M$ is an L_m-observer iff, for all $E \subseteq M_m$, $\theta^{-1}(\bar{E}) = \theta^{-1}(E) \cap L_m$.

The following more general definition will also be useful. Let $H_{lo} \subseteq L$, $H_{hi} \subseteq M$. We say that H_{lo} is *hierarchically nonblocking (HNB)* with respect to H_{hi} if

$$(\forall t \in \bar{H}_{hi})(\forall s \in \bar{H}_{lo})(\forall t' \in T^*)\theta(s) = t \ \&\ tt' \in H_{hi}$$
$$\Rightarrow (\exists s' \in \Sigma^*)ss' \in H_{lo} \ \&\ \theta(ss') = tt'$$

In words, whenever $\theta(s)$ can be extended to a string in H_{hi}, s can be extended to a string in H_{lo} with the same θ-image. This is essentially the L_m-observer property for θ, but 'parametrized' by H_{lo}, H_{hi} in place of L_m, M_m. We can now state the main result.

Theorem 9

Let $M_m = \theta(L_m)$ and let $\theta : L \to M$ be an L_m-observer. Assume $\mathbf{G_{lo}}$ isSOCC (as in Sect. 5.5). Let $\varnothing \neq K_{hi} \in \mathfrak{C}_{hi}(M_m)$, and define

$$H_{lo} := \sup \mathfrak{C}_{lo}(\theta^{-1}(\bar{K}_{hi}))$$

Then $\theta(H_{lo}) = \bar{K}_{hi}$ and H_{lo} is HNB with respect to K_{hi}.

Proof

Since $\bar{K}_{hi}$ is nonempty, closed and controllable, and $\mathbf{G_{lo}}$ is SOCC, we know by Theorem 5.5.1 that $\theta(H_{lo}) = \bar{K}_{hi}$.

Note that H_{lo} is closed. Suppose H_{lo} is not HNB with respect to K_{hi}. Then for some $t \in \bar{K}_{hi}$, $s \in H_{lo}$ and $w \in T^*$ we have

$$\theta(s) = t, \ tw \in K_{hi}$$

but for all $x \in \Sigma^*$ with $sx \in L$,

$$\theta(sx) = tw \Rightarrow sx \notin H_{lo} \tag{5.9}$$

Table 5.1 String factorizations in proof of Theorem 9

$string \in L$	$\theta(string) \in M$
s	t
$su = su'\sigma u''$	tw
$su'\sigma$	tw'
$su'\sigma u''$	$tw = tw'\tau_1 \cdots \tau_k$
$su'\sigma u'''$	$tw'\tau_1$
$su'\sigma v = su'\sigma v'v''$	$tw'y = tw'\tau_1' \cdots \tau_l'$
$su'\sigma v'$	$tw'\tau_1'$

Note that $w \neq \epsilon$; otherwise, by (5.9) with $x = \epsilon$, $s \notin H_{lo}$, contrary to hypothesis. However, as θ is an L_m-observer, and

$$\theta(s) = t < tw \in K_{hi} \subseteq M_m$$

there is $u \in \Sigma^*$, $u \neq \epsilon$, with $su \in L_m$ and $\theta(su) = tw$ (refer to Table 5.1). By (5.9), $su \notin H_{lo}$. Let u' $(\epsilon \leq u' < u)$ be the maximal prefix of u such that $su' \in H_{lo}$. By controllability of H_{lo}, there is $\sigma \in \Sigma_c$ with $u'\sigma \leq u$ (and $su'\sigma \notin H_{lo}$). Let $u = u'\sigma u''$ (where possibly $u'' = \epsilon$). We have

$$\theta(su'\sigma) \leq \theta(su) = tw \in K_{hi}$$

so $\theta(su'\sigma) \in \bar{K}_{hi}$. Also, since $su' \in H_{lo}$ but $su'\sigma \notin H_{lo}$, and because H_{lo} is supremal controllable with respect to the specification $\theta^{-1}(\bar{K}_{hi})$, there is $v \in \Sigma_u^*$ such that $su'\sigma v \notin \theta^{-1}(\bar{K}_{hi})$, i.e. $\theta(su'\sigma v) \notin \bar{K}_{hi}$. Thus

$$t \leq \theta(su'\sigma) \leq tw$$

so $\theta(su'\sigma) = tw'$ (say) $\in \bar{K}_{hi}$, while

$$\begin{aligned} \theta(su'\sigma v) &= \theta(su'\sigma)y \qquad \text{(say)} \\ &= tw'y \\ &\notin \bar{K}_{hi} \end{aligned}$$

We consider the two possible cases for node$(su'\sigma)$, in the reachability tree of L (as in Sect. 5.5).

1. Node$(su'\sigma)$ is vocal. Since $v \in \Sigma_u^*$, we have that $y \in T_u^*$. Thus $tw' \in \bar{K}_{hi}$ while $tw'y \notin \bar{K}_{hi}$, contradicting the controllability of $\bar{K}_{hi}$.
2. Node$(su'\sigma)$ is silent. We claim there is some vocal node in the path from node$(su'\sigma)$ to node$(su'\sigma u'')$. Otherwise

$$\theta(su') = \theta(su'\sigma) = \theta(su'\sigma u'') = \theta(su) = tw$$

But as $su' \in H_{lo}$ we have a contradiction to (5.9), and conclude that $\theta(su'\sigma u'') > \theta(su'\sigma)$. Let

$$\theta(su'\sigma u'') = \theta(su'\sigma)\tau_1 \cdots \tau_k$$

and let $u''' \leq u''$ with

$$\text{node}(su'\sigma u''') \text{ vocal}, \ \theta(su'\sigma u''') = \theta(su'\sigma)\tau_1$$

Recall that $tw' \in \bar{K}_{hi}$ while $tw'y \notin \bar{K}_{hi}$, so $\epsilon \neq y = \tau_1'\tau_2' \cdots \tau_l'$ (say). Let $v' \leq v$ be such that node$(su'\sigma v')$ is vocal with

$$\theta(su'\sigma v') = \theta(su'\sigma)\tau_1' = tw'\tau_1'$$

We claim that $\tau_1' \neq \tau_1$. Otherwise

$$tw'\tau_1' = tw'\tau_1 = \theta(su'\sigma)\tau_1 = \theta(su'\sigma u''') \leq \theta(su'\sigma u'') = \theta(su) = tw \in K_{hi}$$

i.e. $tw'\tau_1' \in \bar{K}_{hi}$. But then $v = v'v''$ (say), and $v'' \in \Sigma_u^*$, imply $\tau_j' \in T_u$ $(2 \leq j \leq l)$, hence $tw'y \in \bar{K}_{hi}$ (by controllability), a contradiction. So $\tau_1' \neq \tau_1$ as claimed, and therefore

$$\text{node}(su'\sigma u'''), \ \text{node}(su'\sigma v')$$

are partners (in the sense of Sect. 5.5). This contradicts our hypothesis that $\mathbf{G_{lo}}$ is SOCC. □

Corollary 10

With L_{voc} as defined earlier in this section, let

$$L_m := \theta^{-1}(M_m) \cap L_{voc}$$

and (with H_{lo} as in Theorem 9) define

$$K_{lo} := H_{lo} \cap L_m$$

Then $H_{lo} = \bar{K}_{lo}$. □

The L_m-observer property of θ is stronger than necessary for the result in Theorem 9.

In Fig. 5.18, $\theta(\alpha) = \theta(\epsilon) = \epsilon$, $\theta(\epsilon\gamma) = \theta(\gamma) = \tau_2$ but $\theta(\alpha s') \neq \tau_2$ for any s', so θ is not an L_m-observer. However, for any choice of $K_{hi} \in \mathfrak{C}_{hi}(M_m)$, $H_{lo} := \sup \mathfrak{C}_{lo}(\theta^{-1}(\bar{K}_{hi}))$ is HNB.

Exercise 11: Prove Corollary 10 and interpret the result in terms of 'low-level nonblocking'.

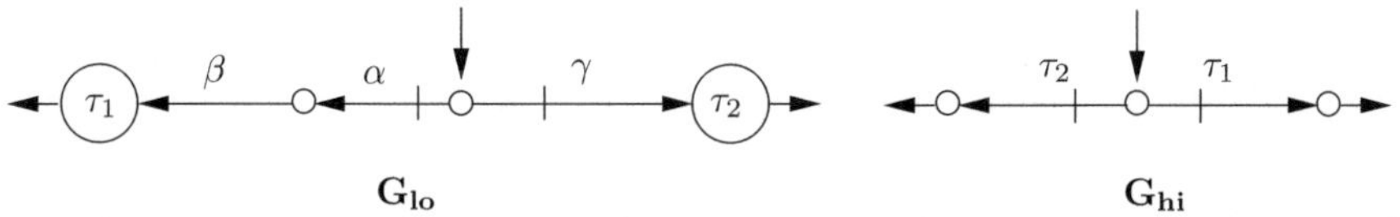

Fig. 5.18 $\mathbf{G_{lo}}$, $\mathbf{G_{hi}}$

Exercise 12: Verify that the assumptions of Theorem 9 are valid for the Transfer Line **HCGLO**, Fig. 5.13.

Exercise 13: Investigate how Theorem 9 might be improved by (possibly) weakening the L_m-observer property but strengthening the SOCC assumption. For instance, the L_m-observer property becomes more 'reasonable' if the local control structure is everywhere as sketched below:

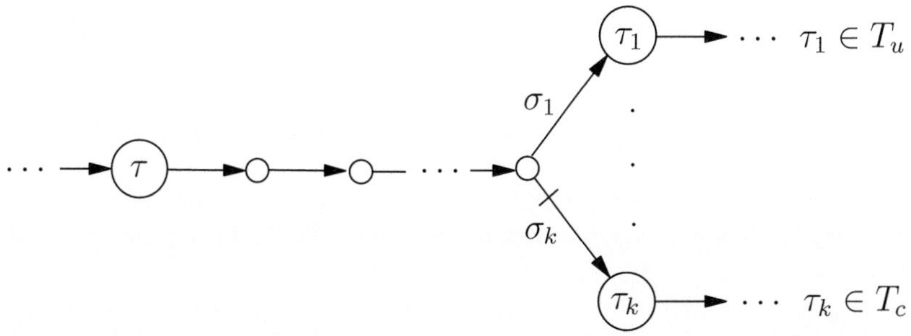

Formalize, relate to Theorem 9, and discuss necessity of the L_m-observer property in this situation.

5.8 Notes

The material of this chapter originates with the theses of Zhong (1987, 1992) and related publications Zhong and Wonham (1989, 1990a, b). Theorems 5.5.3 and 5.5.4 are due to Wong (1990), who also addressed the hierarchical nonblocking problem (Wong 1994; Wong and Wonham 1996a, b). The specific results of Sect. 5.7 are new, but adapted from Zhong (1992), Wong and Wonham (1996a), cf. also Pu (2000). Dual approaches to hierarchical supervisory control, based on state aggregation, have been reported by Schwartz (1992) and Hubbard and Caines (2002); or on state space decomposition, by Wang (1995) and Leduc (2002).

Hierarchy is a long-standing topic in control theory and has been discussed by many authors, notably Mesarovic et al. (1970). For a perceptive and classic essay on the benefits of hierarchical organization the reader is referred to Simon (1962).

For a literary antecedent to Exercise 5.1.1, see L. Tolstoy, *War and Peace*, Epilogue, Part II, Section VI.

Appendix 5.1: Computational Algorithm for Output-Control-Consistency

We provide an algorithm for implementing the procedure of Sect. 5.3 for achievingoutput-control-consistency. It is assumed that $\mathbf{G_{lo}}$ is represented as a finite-state Moore transition structure

$$\mathbf{G_{lo}} = (Q, \Sigma, T_o, \delta, \omega, q_0, Q_m)$$

as defined in Sect. 5.2. The state-transition graph (including state outputs) of $\mathbf{G_{lo}}$ will be denoted simply by $\mathbf{G}$. Recall that $T_o = T \cup \{\tau_o\}$, τ_o being the 'silent' output symbol. Adapting the terminology of Sect. 5.3 to $\mathbf{G}$, we say that a state $q \in Q$ is *silent* if $\omega(q) = \tau_o$ or is *vocal* if $\omega(q) \in T$. The initial state q_0 is silent. A path in $\mathbf{G}$, displayed as

$$q' \xrightarrow{\sigma_1} q_1 \xrightarrow{\sigma_2} q_2 \longrightarrow \cdots \xrightarrow{\sigma_{k-1}} q_{k-1} \xrightarrow{\sigma_k} q$$

is *silent* if (i) q' is either a vocal state or the initial state, and (ii) either $k = 1$, or $k > 1$ and the states $q_1, ..., q_{k-1}$ are silent. A silent path is *red* if at least one of its transition labels $\sigma_1, ..., \sigma_k$ is controllable; otherwise it is *green*. For each $q \in Q$ let $P(q)$ be the set of all silent paths that end at q; because of possible loops in $\mathbf{G}$, $P(q)$ may be infinite. Then $\mathbf{G}$ is *output-control-consistent* (OCC) iff, for each vocal state $q \in Q$, either every $p \in P(q)$ is red or every $p \in P(q)$ is green.

In general $\mathbf{G}$ will fail to be OCC. Our objective is to replace $\mathbf{G}$ by a new version $\mathbf{G_{new}}$ that is OCC, has the same marked and closed behavior as $\mathbf{G}$, and incorporates the modified (split) output structure described for the reachability tree in Sect. 5.3. To this end we define, for any graph $\mathbf{G}$ of the type described, the predecessor and successor functions:

$$\text{in\_set} : Q \rightarrow Pwr(Q \times \Sigma)$$

$$\text{in\_set}(q) := \{(q', \sigma) \in Q \times \Sigma | \delta(q', \sigma) = q\}$$

$$\text{out\_set} : Q \rightarrow Pwr(\Sigma \times Q)$$

$$\text{out\_set}(q) := \{(\sigma, q'') \in \Sigma \times Q | \delta(q, \sigma) = q''\}$$

It can and will be arranged that in_set(q) = $\varnothing$ iff $q = q_0$. For any pair (q', σ) define

$$\text{out\_color}(q', \sigma) \in \{\text{red,green,amber}\}$$

according to the rule

$$\text{out\_color}(q', \sigma) := \begin{cases} \text{red} & \text{if red}(q')\ \&\ \text{silent}(q')\&\text{uncont}(\sigma) \\ & \quad \text{or cont}(\sigma) \\ \text{green} & \text{if green}(q')\ \&\ \text{silent}(q')\&\text{uncont}(\sigma) \\ & \quad \text{or vocal}(q')\&\text{uncont}(\sigma) \\ \text{amber} & \text{if amber}(q')\ \&\ \text{uncont}(\sigma) \end{cases}$$

Here cont(σ) means $\sigma \in \Sigma_c$, uncont(σ) means $\sigma \in \Sigma_u$. Then define, for $q \neq q_0$,

$$\text{new\_color}(q) := \begin{cases} \text{red} & \text{if out\_color}(q', \sigma) = \text{red} \\ & \quad \text{for all } (q', \sigma) \in \text{in\_set}(q) \\ \text{green} & \text{if out\_color}(q', \sigma) = \text{green} \\ & \quad \text{for all } (q', \sigma) \in \text{in\_set}(q) \\ \text{amber} & \text{otherwise} \end{cases}$$

Without essential loss of generality it will be assumed that every state of **G** is coreachable with respect to the vocal states: that is, from every state there is some vocal state that can be reached by a path in **G**. It is then straightforward to show from the foregoing assumptions and definitions that **G** is OCC just in case

(i) for all $q \in Q$, either color(q) = red or color(q) = green, and
(ii) the following stability condition is satisfied:

$$(\forall q \in Q)\text{new\_color}(q) = \text{color}(q) \qquad (stable)$$

The idea of the OCC algorithm is iteratively to modify the graph **G** by a process of recoloring, and elimination of amber states by state-splitting, until (*stable*) is achieved. The formal procedure is summarized as Procedure OCC in the pseudo-Pascal Unit POCC listed in Appendix 5.4. By the procedure Initialize, OCC first assigns to all q color(q) := green. By RecolorStates, each state other than the initial state q_0 (which stays green) is recolored in arbitrary order according to the assignment

$$\text{color(state)} := \text{new\_color(state)} \qquad (*)$$

If thereby any state changes color, (*stable*) is falsified.

In case (*stable*) is false, each amber state (if any), say q_a, is processed by FixAmberStates. Initially the subprocedure SplitAmberStates splits q_a into siblings q_r, q_g with the assignments

$$\text{color}(q_r) := \text{red}, \quad \text{color}(q_g) := \text{green}$$

If $q_a \in Q_m$ then the new states q_r, q_g are declared to be marker states as well, while the subprocedure MakeNewOutputs assigns new outputs

$$\omega(q_r) := [\tau, 1], \quad \omega(q_g) := [\tau, 0] \qquad \text{if} \quad \omega(q_a) = \tau \in T$$

$$\omega(q_r) = \omega(q_g) = \tau_o \qquad \text{if} \quad \omega(q_a) = \tau_o$$

Next the local transition structure at q_a is modified. The subprocedure MakeNewTrans executes the following:

(i) for all $(q', \sigma) \in$ in_set(q_a)
case $q' \neq q_a$

back-connects (q_r, q_g) to create transitions
$[q', \sigma, q_r]$ if out_color(q', σ) = red
$[q', \sigma, q_g]$ otherwise

case $q' = q_a$ & q_a silent

creates transitions

$[q_r, \sigma, q_r]$, $[q_g, \sigma, q_r]$ if cont(σ)
$[q_r, \sigma, q_r]$, $[q_g, \sigma, q_g]$ if uncont(σ)

case $q' = q_a$ & q_a vocal

creates transitions

$[q_r, \sigma, q_r]$, $[q_g, \sigma, q_r]$ if cont(σ)
$[q_r, \sigma, q_g]$, $[q_g, \sigma, q_g]$ if uncont(σ)

(The selfloop cases $q' = q_a$ are treated by first copying q_a as $q_a' \neq q_a$, splitting both q_a' and q_a, back-connecting as in first case, then merging q_r', q_r and q_g', q_g.)

(ii) for all $(\sigma, q'') \in$ out_set(q_a), forward-connects q_r, q_g to create transitions

$$[q_r, \sigma, q''], \quad [q_g, \sigma, q'']$$

(iii) removes q_a with its associated transitions from the database.

A list of split states is maintained to ensure that a state is split at most once. If on a subsequent iteration a state that is a member of a split pair is recolored amber, then instead of SplitAmberStates the subprocedure SwitchOldTrans is invoked and executes the following:

(i) gets the siblings q_r, q_g corresponding to q_a (it will be shown below that necessarily $q_a = q_g$)
(ii) for all $(q', \sigma) \in$ in_set(q_a) such that out_color(q', σ) = red, creates $[q', \sigma, q_r]$ and deletes $[q', \sigma, q_a]$.

The foregoing process is iterated by the **repeat** loop until (*stable*) (hence condition (∗)) becomes true.

It must be shown that OCC terminates. For this we first note

Property 1

Once a state (or a state sibling) has been colored red, it remains red in subsequent iterations.

Proof

Suppose some red state reverts to green or amber on a subsequent iteration. Consider the first instance (state q and iteration #N) of such a change. At iteration #$(N-1)$, for each element

$$(q', \sigma) \in \text{in\_set}(q)$$

it was the case that either $\sigma \in \Sigma_c$ or (if $\sigma \in \Sigma_u$) q' was silent and red. Since the controllable status of σ and the vocal status of q' are both invariant, q' must have changed from red to green or amber, contradicting the assumption that q is the first state to have so changed. □

Since a state is split at most once, Property 1 implies

Property 2

Eventually the state set size, and the subsets of red states and of green states on reentry to the **repeat** loop, remain invariant.

To prove termination it now suffices to show that a state can change from green to amber at most finitely often. By Property 2 and the action of SwitchOldTrans, eventually all transitions

$$[q', \sigma, q] \quad \text{with out\_color}(q', \sigma) = \text{red}$$

will have been switched to states q with color(q) = red. Under this condition, on reentry to the **repeat** loop it will be true for any transition $[q', \sigma, q]$ that

$$\text{color}(q) = \text{green} \Rightarrow \text{out\_color}(q', \sigma) = \text{green}$$

It follows that for all states q, $(*)$ is satisfied and therefore (*stable*) is true, as claimed.

Because a state is split at most once, when OCC terminates the state set will have at most doubled in size; it is also clear that the closed and marked behaviors of the generator described by the graph remain unchanged.

Computational effort can be estimated as follows. Suppose **G** has n states and m transitions. Both RecolorStates and FixAmberStates will require $O(nm)$ steps (comparisons and assignments), while the number of iterations is at worst $O(n+m)$, giving $O(nm(n+m))$ steps for the overall computation.

Appendix 5.2: Conceptual Procedure for Strict-Output-Control-Consistency

To achieve strict-output-control-consistency, a conceptual procedure based on the reachability tree of $\mathbf{G}_{\mathbf{lo}}$ (say, *tree*) can be organized as follows.

Starting at the root node, order the nodes of *tree* level-by-level (i.e. breadth first) as 0,1,2,...; and write $n < n'$ if n precedes n' in the ordering. Let $\zeta \notin T_o$ be a new output symbol, and let

$$T_{o,new} = T_o \dot{\cup} \{\zeta\}$$

be the extension of T_o. We write path(nsn'), and refer to the *path* nsn', if there is a path in *tree* starting at n and leading via $s \in \Sigma^*$ to n'. Say the node n is *vocalizable* if

(i) n is silent;
(ii) $(\exists n_o < n)(\exists \sigma \in \Sigma_c)\text{path}(n_o \sigma n)$;
(iii) $(\exists n_1 > n)(\exists s_1 \in \Sigma_u^*)\text{path}(ns_1n_1)$, ns_1n_1 is silent, and n_1 is vocal; and
(iv) $(\exists n_2 > n)(\exists s_2 \in \Sigma^*)\text{path}(ns_2n_2)$, ns_2n_2 is silent, n_2 is vocal, and $\hat{\omega}(n_2) \neq \hat{\omega}(n_1)$.

Conditions (i)–(iv) express the fact that n_1, n_2 are partners with respect to n. Examining each node in order, modify *tree* as follows. If node n is vocalizable, then *vocalize* n by redefining $\hat{\omega}$ at n as $\hat{\omega}_{new}(n) = \zeta$. [If $\hat{\omega}(n_1) = \tau_c \in T_c$ then ultimately we shall redefine $\hat{\omega}$ at n_1 as $\hat{\omega}_{new}(n_1) = \tau_u$, but at this stage no change of state output will be introduced.] Otherwise, if n is not vocalizable, go on to the successor node of n.

Since vocalization of n has no effect on nodes $n' < n$, the procedure is well defined, transforming *tree* to *newtree*, say. Furthermore if n is vocalizable it remains so as the procedure moves to nodes $n'' > n$ (vocalization at a given level is never rendered superfluous by vocalization later on), because the uncontrollable silent path nsn_1 is never modified. Define str : $\mathcal{N} \rightarrow \Sigma^*$ by node(str(n)) = n. Write s = str(n), s' = str(n') and define $n \equiv n'(\text{mod } tree)$ to mean (cf. Sect. 5.3)

$$s \equiv s'(\text{mod } L_{lo}) \text{ and } s \equiv s'(\text{mod } \theta)$$

Note that $n \equiv n'$ iff the subtrees of *tree* rooted at n and n' respectively, are identical. Corresponding to *newtree* we shall have the map

$$\theta_{new} : \Sigma^* \rightarrow T_{o,new}^*$$

The equivalence $\equiv$ (mod θ_{new}) is then defined in similar fashion (Sect. 5.3), as is equivalence $\equiv$ (mod *newtree*).

Now suppose that n, n' are vocalizable. If $n < n'$ and there is $s \in \Sigma^*\Sigma_c$ such that path(nsn'), it is clear that vocalization of n can have no effect on the subtree rooted at n'; and the same is true *a fortiori* if there is no path from n to n'. Consequently, if nodes n, n' are vocalizable, and $n \equiv n'(\text{mod } tree)$, then also $n \equiv n'$(mod *newtree*). Next suppose that neither n nor n' is vocalizable and that $n \equiv n'$(mod *tree*).

Since the subtrees of *tree* rooted respectively at n and at n' are identical, and neither n nor n' gets vocalized, the vocalization procedure applied to these subtrees must yield the same result; that is, the subtrees of *newtree* rooted at n and n' must be identical, and so again $n \equiv n'(\text{mod } newtree)$.

The foregoing discussion can be summarized by the assertions that the cells of $\equiv$ (mod *newtree*) are formed by splitting the cells of $\equiv$ (mod *tree*) according to the partition of nodes in *tree* into vocalizable and nonvocalizable; and that only cells of the silent nodes of *tree* are so affected. Thus, in the regular case, if the canonical Moore automaton $\mathbf{G_{lo}}$ (corresponding to *tree*) has N_s silent states and N_v vocal states, the canonical Moore automaton $\mathbf{G_{lo,new}}$ (corresponding to *newtree*) will have no more than N_s silent states and $N_v + N_s$ vocal states.

The vocalization procedure in no way depended on a prior assumption that $\mathbf{G_{lo}}$ wasoutput-control-consistent: in (iv) above the condition $\hat{\omega}(n_1) \neq \hat{\omega}(n_2)$ is true after the OCC output assignment procedure (Sect. 5.4) iff it was true beforehand. Thus vocalization could be carried out initially, before the OCC procedure itself, with no difference to the result. Suppose this is done. It is clear that *newtree* can then be rendered OCC by the OCC procedure, and that this process will not introduce any vocalizable nodes (by the remark just made about (iv)). The final result is therefore SOCC, as required, and the final state count in terms of the parameters above is bounded by $2(N_v + 2N_s)$, or less than four times the state count of the Moore structure provided at the start.

Appendix 5.3: Computational Algorithm for Hierarchical Consistency

While the property of strict-output-control-consistency is sufficient for hierarchical consistency (Theorem 5.5.1) and, as described in Appendix 5.2, is conceptually straightforward to achieve, it falls short of being necessary. In this section a somewhat weaker (albeit still not necessary) condition that ensures the desired result will be introduced, together with an effective algorithm to achieve hierarchical consistency in the regular case. This algorithm is slightly more efficient than the procedure of Appendix 5.2 in that possibly fewer states need to be newly vocalized. Regrettably, the approach is rather complicated to describe. A high-level pseudo-Pascal version is listed as Program PSHC in Appendix 5.4. PSHC **uses** Procedure OCC from Unit POCC, as well as Procedure HCC described below.

Consider a (finite) state-transition graph $\mathbf{G}$ for $\mathbf{G_{lo}}$, where we assume that $\mathbf{G_{lo}}$ is already output-control-consistent. Thus, in the terminology for state-transition graphs introduced in Appendix 5.1, each vocal state of $\mathbf{G}$ is unambiguously either red (controllable output) or green (uncontrollable output). The following definitions will be needed. A *silent path suffix (sps)* is a path

$$q \xrightarrow{\sigma_1} q_1 \xrightarrow{\sigma_2} q_2 \longrightarrow \cdots \xrightarrow{\sigma_{k-1}} q_{k-1} \xrightarrow{\sigma_k} q'$$

with $q_1, ..., q_{k-1}$ silent and q' vocal (but q unrestricted). An *agent* is a pair

$$(q, \sigma) \in Q \times \Sigma_c$$

such that there is an *sps* as displayed, with $\sigma_1 = \sigma \in \Sigma_c$ and $\sigma_2, ..., \sigma_k \in \Sigma_u$ (thus q' is red). Say q' is *critical* for (q, σ), and denote the subset of critical states by $C(q, \sigma)$. According to the definition of γ_{lo}, we have $\gamma_{lo}(\_, \sigma) = 0$ (i.e. an event $\sigma \in \Sigma_c$ is disabled under command and control) just at states $q \in Q$ such that (q, σ) is an agent and $C(q, \sigma)$ contains a state q' such that $\omega(q') = \tau'$ with $\gamma_{hi}(\_, \tau') = 0$. Next, the *blocking set* $B(q, \sigma)$ of an agent (q, σ) is the subset of (red, vocal) states $q' \in Q$ such that an *sps* exists as displayed above, with $\sigma_1 = \sigma$ but $\sigma_2, ..., \sigma_k$ unrestricted; thus $B(q, \sigma) \supseteq C(q, \sigma)$, and the inclusion may be strict. An agent (q, σ) is *unary* if $|B(q, \sigma)| = 1$, i.e. contains just one state, which is thus critical. Let $p, q \in Q$ with p red and vocal but q arbitrary. An *sps* from q to p is *dedicated* if each of its transitions $[a, \lambda, b]$ is such that either (a, λ) is not an agent, or (a, λ) is a unary agent with $B(a, \lambda) = \{p\}$. Define the set

$$D(q) := \{p \in Q | \text{red}(p) \,\&\, \text{vocal}(p) \\ \&\text{ there exists a dedicated } sps \text{ joining } q \text{ to } p\}$$

An agent (q, σ) will be called *admissible* if

$$B(q, \sigma) \subseteq D(q)$$

that is, each state p (potentially) blocked by (q, σ) is reachable from q along a dedicated *sps*. Otherwise (q, σ) is *inadmissible*. Note that if $q' = \delta(q, \sigma)$ (with q' vocal, $\sigma \in \Sigma_c$) then (q, σ) is always admissible; so if (q, σ) is inadmissible then $\delta(q, \sigma)$ is silent.

The idea of the algorithm (Procedure HCC) is to identify the inadmissible agents (q, σ), then vocalize the states $\delta(q, \sigma)$. The connection with the conceptual procedure of Appendix 5.2 is provided by

Proposition 1

If $\mathbf{G_{lo}}$ is SOCC then every agent of $\mathbf{G_{lo}}$ is admissible. If $\mathbf{G_{lo}}$ is OCC and every agent in some finite transition graph (FTG) of $\mathbf{G_{lo}}$ is unary, then $\mathbf{G_{lo}}$ is SOCC.

Proof

Suppose an agent (q, σ) is inadmissible. Then there is $p \in B(q, \sigma)$ with $p \notin D(q)$, so there is no dedicated *sps* from q to p. By definition of $B(q, \sigma)$ there is an *sps*

$$q \xrightarrow{\sigma} q_1 \xrightarrow{\sigma_2} q_2 \longrightarrow \cdots \xrightarrow{\sigma_{k-1}} q_{k-1} \xrightarrow{\sigma_k} p$$

Since this *sps* is not dedicated, it includes a transition $[a, \lambda, b]$ such that (a, λ) is an agent and it is not the case that (a, λ) is a unary agent with $B(a, \lambda) = \{p\}$. Thus either (a, λ) is a unary agent with $B(a, \lambda) \neq \{p\}$, or (a, λ) is a non-unary agent. The first case is impossible since the *sps* ends at p. In the second case, since (a, λ) is an agent there is an *sps*

$$a \xrightarrow{\lambda} a_1 \xrightarrow{\lambda_2} a_2 \longrightarrow \cdots \xrightarrow{\lambda_{h-1}} a_{h-1} \xrightarrow{\lambda_h} a'$$

with a' red and vocal and $\lambda_2, ..., \lambda_h \in \Sigma_u$; and since (a, λ) is non-unary there is an *sps*

$$a \xrightarrow{\lambda} a_1 \xrightarrow{\mu_2} b_2 \longrightarrow \cdots \xrightarrow{\mu_{j-1}} b_{j-1} \xrightarrow{\mu_j} a''$$

with a'' red and vocal, $a'' \neq a'$, and the μ's unrestricted. Let n be a node in the reachability tree of $\mathbf{G_{lo}}$ corresponding to a, and let the strings $\lambda\lambda_2 \cdots \lambda_h$ (resp. $\lambda\mu_2 \cdots \mu_j$) lead from n to nodes n_1 (resp. n_2). Clearly n_1, n_2 are partners, namely $\mathbf{G_{lo}}$ is not SOCC.

For the second assertion note that the reachability tree of $\mathbf{G_{lo}}$ can be constructed inductively from an FTG for $\mathbf{G_{lo}}$ by tracing all silent paths in the FTG from the root node to a vocal node or from one vocal node to another. If every agent $a = (q, \sigma)$ in the FTG is unary then all *sps* in the FTG which start with a terminate at the same (red, vocal) state, hence the corresponding silent paths in the tree terminate at equivalent nodes. But this fact rules out the existence of partners. □

We note that $\mathbf{G_{lo}}$ need not be SOCC even though every agent of its FTG is admissible. For instance, consider the structures **G** and **T** shown in Fig. 5.19; **T** is the reachability tree corresponding to **G**. In **G** the agent (0,1) is admissible because the *sps* $0 \xrightarrow{3} 2$ and $0 \xrightarrow{5} 3$ are dedicated, and obviously (0,3), (0,5) are admissible. Hence every agent is admissible, but in **T** nodes 4 and 5 are partners.

The counterpart of Theorem 5.5.1 that we need is the following.

Proposition 2

Assume that $\mathbf{G_{lo}}$ is OCC and that every agent in the (finite) state-transition graph of $\mathbf{G_{lo}}$ is admissible. Let $E_{hi} \subseteq L(\mathbf{G_{hi}})$ be nonempty, closed and controllable. Then

$$\theta((\theta^{-1}(E_{hi}))^{\uparrow}) = E_{hi}$$

Proof

The proof will just be summarized, with details left to the reader. Write $L(\mathbf{G_{lo}}) =: L_{lo}$, $L(\mathbf{G_{hi}}) =: L_{hi}$ and $L(\gamma_{lo}, \mathbf{G_{lo}}) =: K_{lo}$. As in the proof of Theorem 5.5.1 it

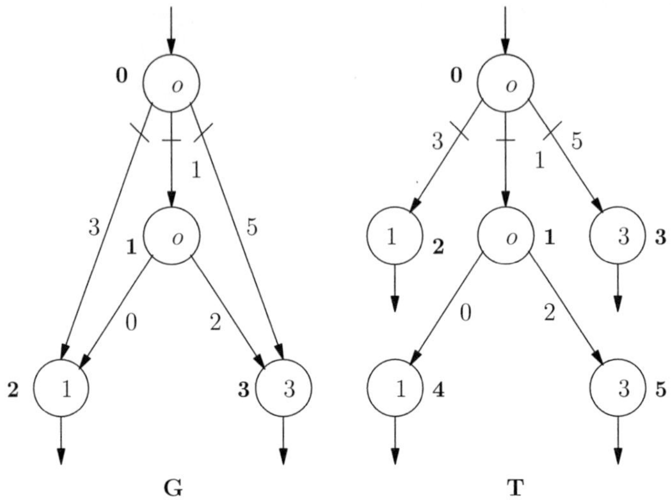

Fig. 5.19 DES **G** and its reachability tree **T**

suffices to show that

$$\theta(K_{lo}) = E_{hi}$$

To the contrary, assume that $E_{hi} - \theta(K_{lo}) \neq \emptyset$. We begin with the same construction as in the proof of Theorem 5.5.1, and select $t, t', t'', s, s'', w, w', \sigma$ and τ as before, so that $t \in E_{hi} - \theta(K_{lo})$, $t' < t$ is maximal for $t' \in \theta(K_{lo})$, $s \in E_{lo}$ with $\theta(s) = t$, $\theta(s'') = t'' \leq t'$, $\theta(s''w) = t''\tau$, $s''w' \in K_{lo}$, $s''w'\sigma \notin K_{lo}$, $w'\sigma \leq w$, $s''w \leq s$, and $\sigma \in \Sigma_c$. This time we work in the finite state-transition graph of $\mathbf{G_{lo}}$ and write $q'' := \delta(q_0, s'')$, $q' := \delta(q_0, s''w')$ and $q := \delta(q_0, s''w)$. Like node(s'') and node$(s''w)$ in the former proof, here q'' and q are vocal. Note that (q', σ) is an agent, with $q \in B(q', \sigma)$. By hypothesis, there is a dedicated *sps* joining q' to each state p in the set $B(q', \sigma)$, hence there is an *sps* from q'' to q via q', along which no element $\sigma \in \Sigma_c$ is disabled by γ_{lo}. That is, there is $v \in \Sigma^*$ of the form $v = w'v'$ such that

$$\delta(q_0, s''v) = \delta(q_0, s''w) = q, \quad s''v \in K_{lo}, \quad \text{and } \theta(s''v) = t''\tau$$

In case $t'' = t'$, we have that $t'\tau \in \theta(K_{lo})$, contradicting the maximality of t', and we are done. Otherwise, suppose $t'' < t'$. Since $s''w \leq s$, we have $s''wx = s$ for some $x \in \Sigma^*$. Now replace s by $s_{new} := s''vx$, s'' by $s''v$, t'' by $t''_{new} := t''\tau$, q'' by $q''_{new} := q$, and repeat the argument starting from state q''_{new}. At each stage we extend $t'' < t'$ by one element τ at a time, to eventually achieve that $t''_{new} = t'$. After one more, final stage we obtain $t'\tau \leq t$ with $t'\tau \in \theta(K_{lo})$. Since this contradicts the maximality of t', we must have $\theta(K_{lo}) = E_{hi}$ after all. □

Observe that an agent (q, σ) that is inadmissible can be rendered admissible simply by vocalizing the state $q' := \delta(q, \sigma)$ (cf. Appendix 5.2). It is clear that the totality of such vocalizations cannot introduce further agents, i.e. convert a pair $(\hat{q}, \hat{\sigma})$ (with $\hat{\sigma} \in \Sigma_c$ and $\delta(\hat{q}, \hat{\sigma})!$) to an agent if it was not an agent before. In fact, if a new agent $(\hat{q}, \hat{\sigma})$ were thereby created, a new *sps* would appear, of the form

$$\hat{q} \xrightarrow{\hat{\sigma}} q_1 \xrightarrow{\sigma_2} q_2 \longrightarrow \cdots \xrightarrow{\sigma_{k-1}} q_{k-1} \xrightarrow{\sigma_k} q'$$

where $\sigma_2, ..., \sigma_k \in \Sigma_u$. Because (q, σ) was already an agent, there is an *sps*

$$q' \xrightarrow{\sigma_1'} q_1 \xrightarrow{\sigma_2'} q_2 \longrightarrow \cdots \xrightarrow{\sigma_{k-1}'} q_{k-1} \xrightarrow{\sigma_k'} p$$

with $\sigma_1', ..., \sigma_k' \in \Sigma_u$. Clearly the catenation of these two *sps* was, prior to the vocalization of q', an *sps* joining $\hat{q}$ to p, namely $(\hat{q}, \hat{\sigma})$ was previously an agent after all.

In HCC each agent is inspected in turn, with any agent that is inadmissible immediately vocalized; as shown above, in this process no new agents are created. However, it may result from the process that an admissible agent is converted to one that is inadmissible. But as any agent, once vocalized, remains admissible, by repetition of the process a number of times at most equal to the number of agents (i.e. no more than the number of transitions) all agents will eventually be rendered admissible. HCC therefore loops until this condition is satisfied.

In general, vocalization will destroy the property that $\mathbf{G_{lo}}$ is output-control-consistent. If the algorithm of Appendix 5.1 (Procedure OCC) is executed once more, OCC will be restored. It remains to show that in this final step, involving state-splitting and recoloring, no inadmissible agents are created. Indeed, a new agent is created only if a former agent (q, σ) is split, into (q_r, σ), (q_g, σ) say. Consider a dedicated *sps* formerly joining q to q' (say). If q' is not split, the *sps* will get replaced by a dedicated *sps* to q' from each of q_r, q_g, so the two new agents are again admissible; while if q' is split into q_r', q_g' this conclusion holds with q_r' in place of q'. We infer that the inclusion $B(q, \sigma) \subseteq D(q)$ is true for each agent in the new transition structure provided it was true for each agent in the old.

Computational effort can be estimated as follows. Suppose $\mathbf{G}$ has n states and m transitions. For a pair (q, σ), the subsets $D(q)$ and $B(q, \sigma)$ can each be identified by examining all (state, transition) pairs, namely in $O(nm)$ steps. Checking the inclusion $B(q, \sigma) \subseteq D(q)$ requires at most $O(n^2)$ steps. As there are at most m agents, Procedure HCC therefore requires $O(n^3m + n^2m^2)$ steps. Combining this result with that of Appendix 5.1 for Procedure OCC, we obtain a complexity bound of

$$O(n^3m + n^2m^2 + nm^2) = O(n^3m + n^2m^2)$$

for the overall Program PSHC for achieving high-level hierarchical consistency.

Appendix 5.4: Listing for Pseudo-Pascal Unit POCC and Program PSHC

```
UNIT POCC;

  {pseudo-Pascal Unit with procedure OCC for achieving
  output-control-consistency; operates on database defining
  transition structure of G_lo, to create G_lo_new}

  INTERFACE

  CONST
    {specify integers MaxStateSize, MaxAlphabetSize}

  TYPE
    States = 0..MaxStateSize;
    Events = 0..MaxAlphabetSize;
    Colors = (red, green, amber);
    In_Events = array[States, Events] of Integer;
    Out_Events = array[Events, States] of Integer;

  PROCEDURE OCC;

  FUNCTION GetSize: States;

  IMPLEMENTATION

  VAR
    stable: Boolean;

  FUNCTION GetSize;
    begin
      {reads size of state set from database}
    end;

  FUNCTION GetColor(state: States): Colors;
    begin
      {reads state color from database}
    end;

  PROCEDURE SetColor(state: States; color: Colors);
    begin
      {assigns state color to database}
    end;

  PROCEDURE SetSibling(state, sibling: States);
    begin
      {assigns sibling to state in database}
    end;

  FUNCTION GetSplit(state: States): Boolean;
    begin
      {returns true if state listed as split in database}
    end;
```

```
PROCEDURE SetSplit(state: States; split: Boolean);
  begin
    {lists state as split in database}
  end;

PROCEDURE InSet(state: States; var in_set: In_Events);
  begin
    {collects in_set(state) from database}
  end;

PROCEDURE OutSet(state: States; var out_set: Out_Events);
  begin
    {collects out_set(state) from database}
  end;

PROCEDURE Initialize;
  var size, state: States;
  begin
    size:= GetSize;
    for state:= 0 to size-1 do
      begin
        SetColor(state, green);
        SetSplit(state, false)
      end
  end;

PROCEDURE RecolorStates;

  FUNCTION NewColor(in_set: In_Events): Colors;
    begin
      {returns color determined by in_set}
    end;

  VAR old_color, new_color: Colors;
      in_set: In_Events;
      size, state: States;

  begin {RecolorStates}
    size:= GetSize;
    for state:= 0 to size-1 do
      begin
        old_color:= GetColor(state);
        InSet(state, in_set);
        new_color:= NewColor(in_set);
        if new_color <> old_color then
          begin
            stable:= false;
            SetColor(state, new_color)
          end
      end
  end;

PROCEDURE FixAmberStates;

  var size, state: States;
      count: Integer;
```

```
    PROCEDURE SplitAmberStates(state: States; var count: Integer);

    PROCEDURE MakeSibling(var count: Integer);
      var sibling: States;
      begin
        sibling:= size+count;
        count:= count+1;
        SetSibling(state, sibling);
        SetColor(state, green);
        SetSplit(state, true);
        SetColor(sibling, red);
        SetSplit(sibling, true)
      end;

    PROCEDURE MakeNewOutputs(state: States);
      begin
        {writes to database:
         tags a vocal state and its sibling with appropriate
         controllable or uncontrollable version of state output}
      end;

    PROCEDURE MakeNewTrans(state: States;in_set: In_Events;
                           out_set: Out_Events);
      begin
        {writes to database:
         redistributes incoming transitions
         between state and its sibling}
      end;

    var in_set: In_Events;
        out_set: Out_Events;

    begin {SplitAmberStates}
      MakeSibling(count);
      MakeNewOutputs(state);
      InSet(state, in_set);
      OutSet(state, out_set);
      MakeNewTrans(state, in_set, out_set)
    end;

  PROCEDURE SwitchOldTrans(state: States);
    begin
      {writes to database:
       redistributes outgoing transitions between
       state and its sibling}
    end;

    begin {FixAmberStates}
      size:= GetSize;
      count:= 0;
      for state:= 0 to size-1 do
        if GetColor(state) = amber then
          begin
            if not GetSplit(state) then SplitAmberStates(state, count)
            else SwitchOldTrans(state)
```

```
        end
    end;

  PROCEDURE OCC;

    BEGIN {PROCEDURE OCC}
      Initialize;
      repeat
        stable:= true;
        RecolorStates; {returns stable = false unless
                        OCC achieved}
        if stable = false then FixAmberStates
      until stable
    END;

  END.

PROGRAM PSHC;

  {pseudo-Pascal program for achieving
  strict-output-control-consistency; operates on database defining
  transition structure of G_lo, to create G_lo_new}

  USES POCC;

  PROCEDURE HCC;

  TYPE
    Array_States = array[States] of Integer;

  FUNCTION Agent(state: States; event: Events): Boolean;
    begin
      {returns true if (state,event) is an agent}
    end;

  PROCEDURE GetDedicatedSet(state: States; var dedicated_set: Array_States);
    begin
      {collects dedicated set from database}
    end;

  PROCEDURE GetBlockingSet(state: States; event: Events;
                                    var blocking_set: Array_States);
    begin
      {collects blocking set from database}
    end;

  FUNCTION Subset(blocking_set, dedicated_set: Array_States): Boolean;
    begin
      {returns true if blocking_set is a subset of dedicated_set}
    end;

  PROCEDURE Vocalize(state: States; event: Events);
    begin
      {finds transition (state,event,new_state) in database;
      vocalizes new_state}
    end;
```

```
var flag: Boolean;
    size, state: States;
    event: Events;
    dedicated_set, blocking_set: Array_States;

begin {PROCEDURE HCC}
    size:= GetSize;
    repeat
    flag:= true;
    for state:= 0 to size-1 do
      begin
        for event:= 0 to MaxAlphabetSize do
          ifAgent(state,event) then
            begin
              GetDedicatedSet(state,dedicated_set);
              GetBlockingSet(state,event,blocking_set);
              ifnot Subset(blocking_set,dedicated_set) then
                begin
                  flag:= false; {agent is inadmissible}
                  Vocalize(state,event)
                end
            end
      end
    until flag = true {all agents are admissible}
end;

BEGIN {PROGRAM PSHC}
    OCC;
    HCC;
    OCC
END.
```

References

Hubbard P, Caines PE (2002) Dynamical consistency in hierarchical supervisory control. IEEE Trans Autom Control 47(1):37–52

Leduc RJ (2002) Hierarchical interface-based supervisory control. PhD thesis, Department of Electrical and Computer Engineering, University of Toronto

Mesarovic MD, Macko D, Takahara Y (1970) Theory of hierarchical, multilevel, systems. Academic Press

Pu KQ (2000) Modeling and control of discrete-event systems with hierarchical abstraction. MASc thesis, Department of Electrical and Computer Engineering, University of Toronto

Schwartz B (1992) State aggregation of controlled discrete-event systems. MASc thesis, Department of Electrical and Computer Engineering, University of Toronto

Simon H (1962) The architectecture of complexity. In: Proceedings of the American philosophical society, vol 106, December 1962 pp 467–482. Reprinted In: Simon HA, The sciences of the artificial (Chap. 7, pp 193–229), 2nd ed. The MIT Press, Cambridge (1981)

Wang B (1995) Top-down design for RW supervisory control theory. MASc thesis, Department of Electrical and Computer Engineering, University of Toronto

Wong KC (1990) An algebraic description of hierarchical control in discrete-event systems. MASc thesis, Department of Electrical and Computer Engineering, University of Toronto

Wong K-C (1994) Discrete-event control architecture: an algebraic approach. PhD thesis, Department of Electrical and Computer Engineering, University of Toronto
Wong KC, Wonham WM (1996a) Hierarchical control of discrete-event systems. Discret Event Dyn Syst 6(3):241–273
Wong KC, Wonham WM (1996b) Hierarchical control of timed discrete-event systems. Discret Event Dyn Syst 6(3):274–306
Zhong H (1987) Control of discrete-event systems: decentralized and hierarchical control. MASc thesis, Department of Electrical and Computer Engineering, University of Toronto
Zhong H (1992) Hierarchical control of discrete-event systems. PhD thesis, Department of Electrical Engineering, University of Toronto
Zhong H, Wonham WM (1989) Hierarchical control of discrete-event systems: computation and examples. In: Proceedings of the twenty-seventh annual allerton conference on communication, control and computing, University of Illinois, pp 511–519
Zhong H, Wonham WM (1990a) Hierarchical coordination. In: Proceedings of the fifth IEEE international symposium on intelligent control, Philadelphia, 5–7 Sept 1990, pp 8–14
Zhong H, Wonham WM (1990b) On consistency of hierarchical supervision in discrete-event systems. IEEE Trans Autom Control 35(10):1125–1134. [Figures 1–5, 10–12, 14–19 are reprinted, with permission, from IEEE.]

Chapter 6
Supervisory Control with Partial Observations

A natural generalization of the control paradigm explored so far allows for feedback in which information may be lost in the channel linking plant to controller. It will be assumed that the channel behaves like a natural (zero-memory) projection, transmitting observable and erasing unobservable events symbol by symbol. A concept of language observability is introduced to express that control decisions based on such partial feedback information are always correct. It is shown that control synthesis of a given language is possible just when the language is controllable and observable; constructive procedures are provided to design and implement the resultant feasible controller. It is then effectively computable how control performance depends on the information available to the supervisor, as parametrized by the subset of observable events.

6.1 Natural Projections and Normal Languages

In this chapter we consider the problem of supervisory control under the assumption that only a subset of the event labels generated by the plant can actually be observed by the supervisor. This subset in general need have no particular relation to the subset of controllable events. Our model will lead to a natural definition of observable language, in terms of which the existence of a supervisor for the standard type of control problem considered in previous chapters can be usefully discussed. It will turn out that, while observability is a somewhat difficult property to work with, two stronger properties called, respectively, normality and relative observability, provide effective alternatives and often lead to satisfactory solutions of the problem of supervision.

W. M. Wonham and K. Cai, *Supervisory Control of Discrete-Event Systems*, Communications and Control Engineering,
https://doi.org/10.1007/978-3-319-77452-7_6

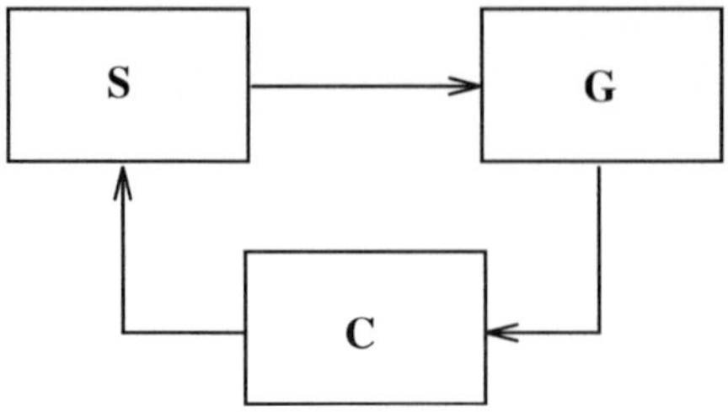

The general setup is shown in the figure. The DES generator **G** modeling the system to be controlled is of the form considered in previous chapters. The new feature here is the communication channel **C** linking **G** to the supervisor **S**. We consider only the simplest case, where the events visible to **S** form a subset Σ_o of the alphabet Σ associated with **G**. Apart from this feature, **S** operates in the usual way, enabling or disabling events in the controllable subset Σ_c of Σ. No particular relation is postulated to hold between Σ_c and Σ_o: in particular, **S** can potentially disable controllable events that are not observable; namely $\Sigma_c - \Sigma_o$ need not be empty. To model the channel **C** we employ the natural projection

$$P : \Sigma^* \to \Sigma_o^*$$

defined inductively according to

$$P(\epsilon) = \epsilon$$
$$P(\sigma) = \begin{cases} \sigma \text{ if } \sigma \in \Sigma_o \\ \epsilon \text{ otherwise} \end{cases}$$
$$P(s\sigma) = P(s)P(\sigma) \quad \text{for } s \in \Sigma^*, \sigma \in \Sigma$$

Thus the action of P on a string s is just to erase from s the events that do not belong to Σ_o, leaving the order of Σ_o-events in s unchanged. If $s_1, s_2 \in \Sigma^*$ then $P(s_1 s_2) = P(s_1)P(s_2)$; namely P is *catenative*. For $s, s' \in \Sigma^*$ define

$$s \equiv s' (\text{mod ker } P) \quad \text{if} \quad Ps = Ps'$$

Because P is catenative, ker P is a right congruence on Σ^* (in general with infinitely many cells).

In *TCT*, P is implemented by **project**. Let **E** be a DES over Σ, and **null**(P) (resp. **image**(P)) a list of the events erased by P, i.e. $\sigma \in \Sigma - \Sigma_o$ (resp. preserved by P, i.e. $\sigma \in \Sigma_o$). Then

$$\mathbf{PE} := \mathbf{project}(\mathbf{E}, \mathbf{null}(P)/\mathbf{image}(P))$$

is a (minimal-state) DES with

$$L_m(\mathbf{PE}) = PL_m(\mathbf{E}), \quad L(\mathbf{PE}) = PL(\mathbf{E})$$

Because **project** uses the subset construction (Exercise 2.5.9) to ensure that **PE** is deterministic, this procedure requires, in the worst case, computational effort (both time and computer memory) that is exponential in the state size of **E**. Fortunately, such exponential blowup seems to be rarely encountered in practice.

Denote by

$$P^{-1} : Pwr(\Sigma_o^*) \rightarrow Pwr(\Sigma^*)$$

the usual inverse image function of P (cf. Sect. 1.4). In *TCT* P^{-1} is implemented by **selfloop**. Thus let **GO** be a DES over Σ_o, and let

$$\mathbf{PINVGO} := \mathbf{selfloop}(\mathbf{GO}, \mathbf{null}(P))$$

Then

$$L_m(\mathbf{PINVGO}) = P^{-1}L_m(\mathbf{GO})$$
$$L(\mathbf{PINVGO}) = P^{-1}L(\mathbf{GO})$$

Exercise 1: Let $K \subseteq \Sigma^*$, $\Sigma_o \subseteq \Sigma$, and $P : \Sigma^* \rightarrow \Sigma_o^*$ the natural projection. Write $[s]$ for the cell of s (mod ker P) i.e.

$$[s] := \{s' \in \Sigma^* | Ps' = Ps\}$$

Denote by $P_K(s)$ the Nerode K-cell of $s \in \Sigma^*$, and define the function

$$f : \Sigma^* \rightarrow Pwr\left(\Sigma^*/\text{Ner}(K)\right)$$

according to

$$f(s) = \{P_K(s') | s' \in [s]\}, \; s \in \Sigma^*$$

As in Exercise 1.4.21, write

$$\wp(\text{Ner}(K)) = \ker f$$

for the exponentiation of $\text{Ner}(K)$ with respect to ker P. Show that $\wp(\text{Ner}(K))$ is a right congruence on Σ^*. Next let $\text{Ner}(PK)$ be the Nerode equivalence relation of PK on Σ_o^*, and denote by $\text{Ner}(PK) \cdot P$ the equivalence relation on Σ^* given by

$$s_1 \equiv s_2 \text{ iff } Ps_1 \equiv Ps_2 (\text{mod Ner}(PK))$$

Show that $\text{Ner}(PK) \cdot P$ is a right congruence on Σ^*, with

$$\wp(\text{Ner}(K)) \leq \text{Ner}(PK) \cdot P = \text{Ner}(P^{-1}(PK))$$

From this result deduce that

$$\|PK\| \leq 2^{\|K\|}$$

where $\|\cdot\|$ denotes Nerode index.

Exercise 2: In Exercise 1 let κ be a right congruence on Σ^* with $\text{Ner}(K) \geq \kappa$. Show that $\wp(\kappa)$ is also a right congruence, and that

$$\|PK\| \leq |\wp(\kappa)| \leq 2^{|\kappa|}$$

Here $|\alpha|$ denotes the number of cells of the equivalence relation α.

Exercise 3:Marking and reduction

Let ν be a right congruence on Σ^* and $N \subseteq \Sigma^*$ with $\text{Ner}(N) \geq \nu$. Define $n : \Sigma^* \to \{0, 1\}$ according to $n(s) = 1$ iff $s \in N$. If $P_\nu : \Sigma^* \to \Sigma^*/\nu$ is the canonical projection, show that $\ker P_\nu \leq \ker n$, so there exists $\bar{n} : \Sigma^*/\nu \to \{0, 1\}$ such that $n = \bar{n} \circ P_\nu$. We say ν is *marked by* $\bar{n}$, and that the pair $(\nu, \bar{n})$ *marks* N. Since $\Sigma^* - \bar{N}$ is (if nonempty) a cell of $\text{Ner}(N)$, it is a union of cells of ν. We say ν is *reduced with respect to* N if $\Sigma^* - \bar{N}$ is a single cell of ν, which we call the *dump cell* of ν. If $(\nu, \bar{n})$ marks N and ν is reduced for N, we say $(\nu, \bar{n})$ *represents* N.

With reference to Exercises 1 and 1.4.21, observe that a cell of $\wp(\nu)$, or element of $\Sigma^*/\wp(\nu)$, is naturally identified with a subset of elements of Σ^*/ν, according to

$$[s]_{\wp(\nu)} = f(s), \ s \in \Sigma^*$$

where

$$f(s) = \{P_\nu(s') | s' \in [s]\}$$

Define $\bar{\mathfrak{n}} : \Sigma^*/\wp(\nu) \to \{0, 1\}$ according to

$$\bar{\mathfrak{n}}([s]_{\wp(\nu)}) = 1 \ \text{ iff } \ \bar{n}([s']_\nu) = 1$$

for at least one string $s' \in [s]$. Show that $(\wp(\nu), \bar{\mathfrak{n}})$ marks $P^{-1}(PN)$.

In case ν is reduced for N, so that $\Sigma^* - \bar{N}$ is a cell of ν, obtain a reduced representation of $P^{-1}(PN)$ by aggregating f as follows. For $s \in \Sigma^*$ define

$$\bar{f}(s) = \begin{cases} \{P_\nu(s') | s' \in [s] \cap \bar{N}\} & \text{if } [s] \cap (\Sigma^* - \bar{N}) = \varnothing \\ \{P_\nu(s') | s' \in [s] \cap \bar{N}\} \dot{\cup} \{\Sigma^* - \bar{N}\} & \text{if } [s] \cap (\Sigma^* - \bar{N}) \neq \varnothing \end{cases}$$

In the first case all look-alike strings s' belong to $\bar{N}$; in the second at least one s' does not. Write $\bar{\wp}(\nu) = \ker \bar{f}$, and show that $\bar{\wp}(\nu)$ is a right congruence on Σ^* which is reduced for $P^{-1}(P\bar{N})$. In $\bar{\wp}$, $\Sigma^* - \bar{N}$ again plays the role of dump cell, for $s \in \Sigma^* - P^{-1}(P\bar{N})$ iff $[s] \subseteq \Sigma^* - \bar{N}$, namely $\bar{f}(s) = \{\Sigma^* - \bar{N}\}$. Thus $(\bar{\wp}(\nu), \bar{\mathfrak{n}})$ represents $P^{-1}(PN)$.

Exercise 4: Let $K \subseteq \Sigma^*$, $\bar{\kappa} := \{\bar{K}, \Sigma^* - \bar{K}\}$, and let ν be a right congruence on Σ^*. If $\text{Ner}(K) \geq \bar{\kappa} \wedge \nu$, show that $\bar{\kappa} \wedge \nu$ is also a right congruence on Σ^*. Suppose $N \subseteq \Sigma^*$, Ner $N \geq \nu$, and $(\nu, \bar{n})$ marks N. Define $\bar{k} : \Sigma^*/\bar{\kappa} \wedge \nu \to \{0, 1\}$ as follows: If $c \subseteq \Sigma^*$ is a cell of $\bar{\kappa} \wedge \nu$, then $\bar{k}(c) = 1$ iff $c \subseteq \bar{K} \cap N$. Assuming $(\bar{\kappa} \wedge \nu, \bar{k})$ marks K, show that $K = \bar{K} \cap N$.

Definition 5 With $K, \bar{\kappa}, \nu, N$, and $\bar{k}$ as in Exercise 4, we say that K *inherits (its marking) from* ν. In case $\nu = \text{Ner}(N)$, we may say that K is *marked by* N.

Associated with any projection P is a useful property of languages defined as follows. Let

$$N \subseteq M \subseteq \Sigma^*$$

Define N to be (M, P)-*normal* if

$$N = M \cap P^{-1}(PN)$$

Note that in this equality the inclusion $\subseteq$ is automatic, while the reverse inclusion is not. Thus N is (M, P)-normal iff it can be recovered from its projection along with a procedure for deciding membership in M. Equivalently, from $s \in N$, $s' \in M$ and $P(s') = P(s)$ one may infer that $s' \in N$. An (M, P)-normal language N is the largest sublanguage $\hat{N}$ of M with the property that $P\hat{N} = PN$. It is easily seen that both M and the empty language $\emptyset$ are (M, P)-normal. In fact, if $K \subseteq \Sigma_o^*$ is arbitrary, then the language

$$N := M \cap P^{-1}K$$

is always (M, P)-normal. Notice also that if N is (M, P)-normal, so is $M - N$.

Exercise 6: Let $[s]$ denote the cell of ker P containing $s \in \Sigma^*$. If $N \subseteq \Sigma^*$ show that

$$P^{-1}(PN) = \bigcup \{[s] | s \in N\}$$

Then show that N is (M, P)-normal iff

$$N = \bigcup \{[s] \cap M | s \in N\}$$

Illustrate with a sketch of Σ^* partitioned by ker P.

With P fixed in the discussion, and the language $E \subseteq \Sigma^*$ arbitrary, bring in the family of languages

$$\mathcal{N}(E; M) = \{N \subseteq E | N \text{ is } (M, P)\text{-normal}\},$$

the class of (M, P)-normal sublanguages of E. Then $\mathcal{N}(E; M)$ is nonempty ($\emptyset$ belongs) and enjoys the following algebraic closure property.

Proposition 7

The class of languages $\mathcal{N}(E; M)$ is a complete sublattice (with respect to sublanguage inclusion) of the lattice of sublanguages of E. In particular, the intersection and union of (M, P)-normal sublanguages are (M, P)-normal. □

From Proposition 7 it follows by the usual argument that the language

$$\sup \mathcal{N}(E; M)$$

exists in $\mathcal{N}(E; M)$: that is, any language E contains a unique supremal (M, P)-normal sublanguage, the 'optimal' (M, P)-normal approximation to E from below.

Exercise 8: With $[s]$ as in Exercise 6, show that

$$\sup \mathcal{N}(E; M) = \bigcup \{[s] \cap M | [s] \cap M \subseteq E\}$$

Illustrate with a sketch.

In supervisory control the two most important classes of (M, P)-normal languages result from setting $M = L_m(\mathbf{G})$ or $M = L(\mathbf{G})$, respectively. A simple relation between them is the following.

Proposition 9

Assume that $\mathbf{G}$ is trim (in particular $\overline{L_m(\mathbf{G})} = L(\mathbf{G})$), $N \subseteq L_m(\mathbf{G})$ is $(L_m(\mathbf{G}), P)$-normal, and the languages $L_m(\mathbf{G})$, $P^{-1}(PN)$ are nonconflicting. Then $\bar{N}$ is $(L(\mathbf{G}), P)$-normal. □

Again in connection with supervisory control the following result will find application later.

Proposition 10

Let $E \subseteq L_m(\mathbf{G})$. The class of languages

$$\bar{\mathcal{N}}(E; L(\mathbf{G})) = \{N \subseteq E | \bar{N} \text{ is } (L(\mathbf{G}), P)\text{-normal}\}$$

is nonempty and closed under arbitrary unions. □

As before, from Proposition 10 we infer that the language

$$\sup \bar{\mathcal{N}}(E; L(\mathbf{G}))$$

exists in $\bar{\mathcal{N}}(E; L(\mathbf{G}))$.

Examples show that in general the supremal $(L(\mathbf{G}), P)$-normal sublanguage of a closed language need not be closed. However, we have the following more detailed result, which states that if $C \subseteq L(\mathbf{G})$ is closed, N is the supremal $(L(\mathbf{G}), P)$-normal sublanguage of C, and B is the supremal closed sublanguage of N, then in fact B is the supremal sublanguage of C whose closure is $(L(\mathbf{G}), P)$-normal.

Proposition 11 (Feng Lin)

Let $C \subseteq L(\mathbf{G})$ be closed. Then

$$\sup \bar{\mathcal{N}}(C; L(\mathbf{G})) = \sup \mathcal{F}(\sup \mathcal{N}(C; L(\mathbf{G})))$$

Proof

In the proof write 'normal' for '$(L(\mathbf{G}), P)$-normal', $N := \sup \mathcal{N}(C; L(\mathbf{G}))$, $B := \sup \mathcal{F}(N)$, $\hat{B} := L(\mathbf{G}) \cap P^{-1}(PB)$. Clearly $\hat{B}$ is closed and normal. Since $B \subseteq N$ and N is normal,

$$\hat{B} \subseteq L(\mathbf{G}) \cap P^{-1}(PN) = N$$

Therefore $\hat{B} \subseteq B$. But automatically $\hat{B} \supseteq B$ (since $B \subseteq L(\mathbf{G})$), so $\hat{B} = B$; i.e. B is normal. Let $D := \sup \bar{\mathcal{N}}(C; L(\mathbf{G}))$. Now $B \subseteq C$ and $\bar{B} = B$ is normal, so $B \subseteq D$. Also $\bar{D} \subseteq \bar{C} = C$ and $\bar{D}$ normal imply $\bar{D} \subseteq N$, and then $\bar{D}$ closed implies that $\bar{D} \subseteq B$, so $D \subseteq B$. That is, $D = B$, as claimed. □

For $\sup \mathcal{N}(E; M)$ with $E \subseteq M$ but E, M otherwise arbitrary, we have the following explicit description.

Proposition 12 (Lin-Brandt formula)[1]

Let $E \subseteq M$. Then

$$\sup \mathcal{N}(E; M) = E - P^{-1}P(M - E)$$

Proof

In the proof write 'normal' for '(M, P)-normal', $S := P^{-1}P(M - E)$ and $N := E - S$. To see that N is normal, suppose

$$u \in M, \quad Pu \in PN$$

For some $v \in N$, $Pu = Pv$. We claim $u \notin S$: otherwise, there exists $t \in M - E$ with $Pu = Pt$, so $Pv = Pt$, i.e. $v \in S$, a contradiction. We also claim that $u \in E$: otherwise $u \in M - E$ and therefore $u \in S$, a contradiction. Thus $u \in E - S = N$, namely

$$M \cap P^{-1}PN \subseteq N$$

and hence N is normal. Now let $K \subseteq E$ be normal. We claim that $K \subseteq N$: otherwise there is $s \in K$ (so $s \in E$) with $s \in S$, and so there is $t \in M - E$ with $Ps = Pt$, i.e. $Pt \in PK$. But $t \in M$ and $t \in P^{-1}PK$, so by normality $t \in K$, a contradiction to $K \subseteq E$. □

An implementation of the Lin-Brandt formula is available in *TCT*:

$$\mathbf{N} = \mathbf{supnorm}(\mathbf{M}, \mathbf{E}, \mathbf{null/image}(P))$$

---

[1] So-called here for brevity: see Brandt et al. (1990).

Here **E**, **M** are representative DES for E and M, with E arbitrary. Then **N** represents $\sup \mathcal{N}(E \cap M; M)$; thus the user need not arrange in advance that $E \subseteq M$. Like **project**, **supnorm** is computationally expensive.

Exercise 13: Illustrate the Lin-Brandt formula with a sketch showing Σ^* partitioned by $\ker P$, along with sublanguages $E \subseteq M \subseteq \Sigma^*$. In light of Exercises 6 and 8, the formula should now be 'obvious'; clearly it is valid for sets and functions in general.

As the last topic of this section we introduce the following related property that is sometimes useful. Let $R \subseteq \Sigma^*$. Say that R is $(L(\mathbf{G}), P)$-*paranormal* if

$$\bar{R}(\Sigma - \Sigma_o) \cap L(\mathbf{G}) \subseteq \bar{R}$$

Thus R is $(L(\mathbf{G}), P)$-paranormal if the occurrence of an unobservable event never results in exit from the closure of R. By analogy with controllability it is clear, for instance, that the class of $(L(\mathbf{G}), P)$-paranormal sublanguages of an arbitrary sublanguage of Σ^* is nonempty, closed under union (but not necessarily intersection), and contains a (unique) supremal element.

Proposition 14

If the closure of R is $(L(\mathbf{G}), P)$-normal, then R is $(L(\mathbf{G}), P)$-paranormal. □

The converse of Proposition 14 is false: an $(L(\mathbf{G}), P)$-paranormal sublanguage of $L(\mathbf{G})$, closed or not, need not be $(L(\mathbf{G}), P)$-normal. However, the result can be useful in showing that a given closed language R is not $(L(\mathbf{G}), P)$-normal, by showing that an unobservable event may cause escape from R.

To illustrate these ideas we consider three examples.

Example 15

Let $\Sigma = \{\alpha, \beta\}$, $\Sigma_o = \{\alpha\}$, $L(\mathbf{G}) = \{\epsilon, \alpha, \beta\}$, $C = \{\epsilon, \alpha\}$. Then C is closed and $PC = C$. However

$$L(\mathbf{G}) \cap P^{-1}(PC) = \{\epsilon, \alpha, \beta\} = L(\mathbf{G}) \supsetneqq C$$

and C is not $(L(\mathbf{G}), P)$-normal. C is not $(L(\mathbf{G}), P)$-paranormal either:

$$\bar{C}(\Sigma - \Sigma_o) \cap L(\mathbf{G}) = C\beta \cap L(\mathbf{G}) = \{\beta, \alpha\beta\} \cap \{\epsilon, \alpha, \beta\} = \{\beta\} \subsetneqq C$$

namely the unobservable event β catenated with the string $\epsilon \in C$ results in escape from C. On the other hand, for the sublanguage $A := \{\alpha\}$,

$$L(\mathbf{G}) \cap P^{-1}P\{\alpha\} = L(\mathbf{G}) \cap \beta^*\alpha\beta^* = \{\alpha\} = A$$

so that A is $(L(\mathbf{G}), P)$-normal, and therefore $A = \sup \mathcal{N}(C; L(\mathbf{G}))$. It can be checked that $\sup \mathcal{N}(C; L(\mathbf{G}))$ is correctly calculated by the Lin-Brandt formula (Proposition 12). Note also that $\bar{A} = C$, showing that the supremal $(L(\mathbf{G}), P)$-normal sublanguage of a closed language need not be closed; here, in fact, $\sup \bar{\mathcal{N}}(C; L(\mathbf{G})) = \varnothing$, in agreement with Proposition 11.

Now let $B := \{\alpha, \beta\}$. Whereas B is $(L(\mathbf{G}), P)$-paranormal, we have

$$L(\mathbf{G}) \cap P^{-1}(PB) = L(\mathbf{G}) \cap P^{-1}\{\alpha, \epsilon\} = L(\mathbf{G}) \supsetneqq B$$

so B is not $(L(\mathbf{G}), P)$-normal.

Example 16

As another example let $\Sigma = \{\alpha, \beta, \gamma\}$, $\Sigma_o = \{\gamma\}$,

$$L(\mathbf{G}) = \{\epsilon, \alpha, \alpha\gamma, \alpha\gamma\gamma, \beta, \beta\gamma, \beta\gamma\gamma\} = \overline{(\alpha + \beta)\gamma^2}$$
$$C = \{\epsilon, \alpha, \alpha\gamma, \beta, \beta\gamma, \beta\gamma\gamma\} = \overline{(\alpha + \beta\gamma)\gamma}$$

Then

$$L(\mathbf{G}) \cap P^{-1}(PC) = L(\mathbf{G}) \supsetneqq C$$

so C is not $(L(\mathbf{G}), P)$-normal; in fact Lin-Brandt yields

$$\sup \mathcal{N}(C; L(\mathbf{G})) = \{\epsilon, \alpha, \beta, \alpha\gamma, \beta\gamma\} = \overline{(\alpha + \beta)\gamma}$$

so in this case $\sup \bar{\mathcal{N}}(C; L(\mathbf{G}))$ and $\sup \mathcal{N}(C; L(\mathbf{G}))$ coincide. On the other hand C is $(L(\mathbf{G}), P)$-paranormal, since the occurrence of unobservable events α or β does preserve membership in C.

Example 17

Let $\Sigma = \{\alpha, \beta\}$, $\Sigma_o = \{\alpha\}$, $L(\mathbf{G}) = \{\epsilon, \alpha, \beta, \beta\alpha\}$, $A = \{\beta\alpha\}$. Then A is $(L(\mathbf{G}), P)$-paranormal:

$$\bar{A}(\Sigma - \Sigma_o) \cap L(\mathbf{G}) = \{\epsilon, \beta, \beta\alpha\}\beta \cap L(\mathbf{G}) = \{\beta\} \subseteq \bar{A}$$

However, the closure $\bar{A}$ is not $(L(\mathbf{G}), P)$-normal, because

$$L(\mathbf{G}) \cap P^{-1}(P\bar{A}) = \{\epsilon, \alpha, \beta, \beta\alpha\} \supsetneqq \bar{A}$$

Exercise 18: With $\Sigma_o \subseteq \Sigma$, let $P : \Sigma^* \to \Sigma_o^*$ be the natural projection, and let $A \subseteq \Sigma^*$, $B \subseteq \Sigma_o^*$. Show that

$$\overline{PA} = P\bar{A}, \quad \overline{P^{-1}B} = P^{-1}\bar{B}$$

Exercise 19: Supply proofs of Propositions 7, 9, 10, and 14.

Exercise 20: Show by example that $\bar{\mathcal{N}}(E; L(\mathbf{G}))$ is not in general closed under intersection.

Exercise 21: Does there exist a language A that is normal but not paranormal? If so, what can be said about $\bar{A}$?

Exercise 22: Show that, with $E \subseteq M$,

$$P \sup \mathcal{N}(E; M) = PE - P(M - E)$$

Exercise 23: With $L \subseteq \Sigma^*$ write

$$\mathcal{F}(L) := \{H \subseteq L | H = \bar{H}\}$$

for the family of closed sublanguages of L, and let M be its supremal element:

$$\begin{aligned} M &:= \sup \mathcal{F}(L) \\ &= \{s \in L | \overline{\{s\}} \subseteq L\} \end{aligned}$$

Prove the inductive rules

$$M \neq \varnothing \Leftrightarrow \epsilon \in L$$

$$(\forall s \in \Sigma^*)(\forall \sigma \in \Sigma) s\sigma \in M \Leftrightarrow s \in M \;\&\; s\sigma \in L$$

Referring to Sect. 2.8, show that $M \underline{\vdash} L$, but in general $\text{Ner}(M) \not\geq \text{Ner}(L)$. More generally, suppose λ is a right congruence on Σ^* with $\text{Ner}(L) \geq \lambda$. Show that $\bar{\mu} \wedge \lambda$ is a right congruence and $\text{Ner}(M) \geq \bar{\mu} \wedge \lambda$, where

$$\bar{\mu} = \{M, \Sigma^* - M\} \in \mathcal{E}(\Sigma^*)$$

Show also that

$$M = L - (\Sigma^* - L)\Sigma^*$$

In particular, L is closed iff

$$L \cap (\Sigma^* - L)\Sigma^* = \varnothing$$

For $i = 1, 2$ let $L_i \subseteq \Sigma_i^*$. We say that L_1, L_2 are *synchronously nonconflicting over* $(\Sigma_1 \cup \Sigma_2)^*$ if

$$\overline{L_1 \parallel L_2} = \bar{L}_1 \parallel \bar{L}_2$$

Let $P_i : (\Sigma_1 \cup \Sigma_2)^* \to \Sigma_i^*$ $(i = 1, 2)$ be the natural projections. By Exercise 18, $\overline{P_i^{-1} L_i} = P_i^{-1} \bar{L}_i$ $(i = 1, 2)$, so L_1, L_2 are synchronously nonconflicting over $(\Sigma_1 \cup \Sigma_2)^*$ iff $P_1^{-1} L_1, P_2^{-1} L_2 \subseteq (\Sigma_1 \cup \Sigma_2)^*$ are nonconflicting in the usual sense (Sect. 3.6).

Exercise 24: (Lei Feng) For $i = 1, 2$ let $L_i \subseteq \Sigma_i^*$. Also let $\Sigma_1 \cap \Sigma_2 \subseteq \Sigma_o \subseteq \Sigma_1 \cup \Sigma_2$, and $P_o : (\Sigma_1 \cup \Sigma_2)^* \to \Sigma_o^*$ be the natural projection. Call P_o an L_i-*observer* if

$$(\forall t \in P_o L_i, s \in \bar{L}_i) P_o s \leq t \;\Rightarrow\; (\exists u \in \Sigma_i^*) su \in L_i \;\&\; P_o(su) = t$$

In other words, whenever $P_o s$ can be extended to a string $t \in P_o L_i$, the underlying string $s \in \bar{L}_i$ can be extended to a string $su \in L_i$ with the same projection. Assume that P_o is an L_i-observer ($i = 1, 2$) and that

$$P_o L_1 \subseteq (\Sigma_o \cap \Sigma_1)^*, \ P_o L_2 \subseteq (\Sigma_o \cap \Sigma_2)^*$$

are synchronously nonconflicting over Σ_o^*. Show that L_1, L_2 are synchronously nonconflicting over $(\Sigma_1 \cup \Sigma_2)^*$. Specialize to each of the cases $\Sigma_1 = \Sigma_2$, $\Sigma_o = \Sigma_1 \cap \Sigma_2$, and $\Sigma_1 \cap \Sigma_2 = \varnothing$. **Hint**: Make use of Exercise 3.3.8.

Exercise 25: With reference to Exercises 6 and 8, write $N := \sup \mathcal{N}(E; M)$; and $[s]$ for the cell of s (mod ker P), namely $[s] = \{s' \in \Sigma^* | Ps' = Ps\}$. Recall that

$$N = \bigcup \{[s] \cap M \,|\, s \in \Sigma^*, \ [s] \cap M \subseteq E\}$$

Let $\mu = \text{Ner}(M)$, $\eta = \text{Ner}(E)$ be the Nerode equivalence relations for M, E, respectively, and write

$$P_\mu : \Sigma^* \to \Sigma^*/\mu, \quad P_\eta : \Sigma^* \to \Sigma^*/\eta$$

for the corresponding canonical projections. Define

$$f : \Sigma^* \to Pwr(\Sigma^*/\mu \times \Sigma^*/\eta)$$

according to

$$f(s) := (P_\mu \times P_\eta)[s] = \{(P_\mu s', P_\eta s') | s' \in [s]\}$$

Show that $\wp(\mu \wedge \eta) := \ker f$ is a right congruence on Σ^* and that

$$\text{Ner}(N) \geq \mu \wedge \wp(\mu \wedge \eta) \tag{6.1}$$

From this conclude that

$$\|N\| \leq \|M\| \cdot 2^{\|M\| \cdot \|E\|}$$

Finally sharpen (6.1) as follows. Define

$$f(\cdot; \bar{M}) : \Sigma^* \to Pwr(\Sigma^*/\mu \wedge \eta)$$

according to

$$f(s; \bar{M}) := \{(P_\mu(s'), P_\eta(s')) | s' \in [s] \cap \bar{M}\}$$

and let $\wp(\mu \wedge \eta; \bar{M}) := \ker f(\cdot, \bar{M})$. Show that

$$\text{Ner}(N) \geq \mu \wedge \wp(\mu \wedge \eta; \bar{M})$$

Hint: First note that P is a congruence of catenation:

$$[sw] = [s][w] := \{s'w' | s' \in [s],\ w' \in [w]\}$$

Thus

$$f(sw) = \{(P_\mu(s'w'), P_\eta(s'w')) | s' \in [s], w' \in [w]\}$$

If $f(s_2) = f(s_1)$ then for every pair

$$(P_\mu(s_2'), P_\eta(s_2')) \in f(s_2)$$

there is an identical pair

$$(P_\mu(s_1'), P_\eta(s_1')) \in f(s_1)$$

Since μ, η are right congruences

$$P_\mu(s_2'w') = P_\mu(s_1'w'), \quad P_\eta(s_2'w') = P_\eta(s_1'w')$$

so

$$(P_\mu(s_2'w'), P_\eta(s_2'w')) \in f(s_1 w)$$

i.e. $f(s_2 w) \subseteq f(s_1 w)$. Similarly $f(s_1 w) \subseteq f(s_2 w)$, and thus $\ker f$ is a right congruence.

Write $\pi := \mu \wedge \ker f$. Then π is a right congruence, and it suffices to show $\pi \leq \{N, \Sigma^* - N\}$. Let $s_1 \equiv s_2 (\mathrm{mod}\ \pi)$ and $s_1 \in N$, so that $[s_1] \cap M \subseteq E$. Since $N \subseteq M$ and $s_2 \equiv s_1 (\mathrm{mod}\ \mu)$, we have $s_2 \in M$, and it remains to show that $[s_2] \cap M \subseteq E$. Let $s_2' \in [s_2] \cap M$. Using the fact that $f(s_2) \subseteq f(s_1)$, note that there is $s_1' \in [s_1]$ with

$$P_\mu s_2' = P_\mu s_1', \quad P_\eta s_2' = P_\eta s_1'$$

So $s_2' \in M$ implies $s_1' \in M$, hence $s_1' \in E$, and in turn $s_2' \in E$, as required.

Exercise 26: Let $N \subseteq \Sigma^*$, and suppose ν is a right congruence on Σ^*, with $P_\nu : \Sigma^* \to \Sigma^*/\nu$. Let $t \in \Sigma^*$, and define

$$K = \{s \in N | (\forall u \leq s) u \not\equiv t (\mathrm{mod}\ \nu)\}$$

Thus K is the sublanguage of N whose strings never visit the ν-cell of t. Show that $K = \bar{K} \cap N$. Define $T \subseteq \Sigma^*/\nu$ according to

$$T := \{P_\nu(s\sigma) | s \in \bar{K}, \sigma \in \Sigma, s\sigma \notin \bar{K}\}$$

and verify

$$(\forall s \in \Sigma^*)(\forall \sigma \in \Sigma) s\sigma \in \bar{K} \Leftrightarrow s \in \bar{K}\ \&\ P_\nu(s\sigma) \notin T$$

Show next that

$$\text{Ner}(\bar{K}) \geq \bar{\kappa} \wedge \nu$$

where $\bar{\kappa} = \{\bar{K}, \Sigma^* - \bar{K}\}$. Conclude that

$$\text{Ner}(K) \geq \bar{\kappa} \wedge \nu \wedge \text{Ner}(N)$$

Exercise 27: Let $M, E \subseteq \Sigma^*$, $\Sigma_o \subseteq \Sigma$, and let $P : \Sigma^* \to \Sigma_o^*$ be the natural projection. Consider a chain

$$\Sigma^* \supseteq K_0 \supseteq N_0 \supseteq K_1 \supseteq N_1 \supseteq \cdots$$

where $K_0 = E$, $N_0 = \sup \mathcal{N}(K_0; M)$, and for $j \geq 1$,

$$K_j = \bar{K}_j \cap N_{j-1}, \;\; N_j = \sup \mathcal{N}(K_j; M)$$

Let $[s]$ denote the cell of s (mod ker P). Show by induction on j that

$$N_j = \bigcup \{[s] \cap M | s \in \Sigma^*, [s] \cap M \subseteq \bar{K}_j \cap E\}, \;\; j \geq 0$$

Write $\mu := \text{Ner}(M)$, $\eta := \text{Ner}(E)$, and $\bar{\kappa}_j := \{\bar{K}_j, \Sigma^* - \bar{K}_j\}$. Assuming further

$$\text{Ner}(K_j) \geq \bar{\kappa}_j \wedge \mu \wedge \eta \wedge \wp(\mu \wedge \eta), \;\; j \geq 1$$

show that

$$\text{Ner}(N_j) \geq \mu \wedge \wp(\bar{\kappa}_j \wedge \mu \wedge \eta), \;\; j \geq 1$$

Conclude that the chain is finitely convergent, say to a limit N, where

$$\|N\| \leq \|M\| \cdot 2^{\|M\| \cdot (\|E\|+1)}$$

Hint: For the latter estimate justify the replacement of η, in the inequality for $\text{Ner}(N)$, by the reduction of η with respect to $\bar{K}$ (for 'reduction' see Exercise 3).

Exercise 28: Let

$$\begin{aligned} K &:= \sup \bar{\mathcal{N}}(E; L(\mathbf{G})) \\ &:= \sup\{F \subseteq E | \bar{F} \text{ is } (L(\mathbf{G}), P)\text{−normal}\} \end{aligned}$$

Define the operator $T : Pwr(\Sigma^*) \to Pwr(\Sigma^*)$ according to

$$T(J) = J \cap \sup \mathcal{F}(\sup \mathcal{N}(\bar{J}; L(\mathbf{G}))), \;\; J \subseteq \Sigma^*$$

Set $K_0 := E$, $K_j := T(K_{j-1})$, $j \geqslant 1$. Show that, provided E and $L(\mathbf{G})$ are regular, the chain $K_0 \supseteq K_1 \supseteq K_2 \supseteq \cdots$ is finitely convergent to K. Furthermore

$$\|K\| = O(\|L(\mathbf{G})\| \cdot \|E\| \cdot 2^{\|L(\mathbf{G})\| \cdot \|E\|})$$

Hint: Let $\mu = \text{Ner}(L(\mathbf{G}))$, $\eta = \text{Ner}(E)$, and for $j \geq 0$,

$$N_j := \sup \mathcal{N}(\bar{K}_j; L(\mathbf{G}))$$
$$F_j := \sup \mathcal{F}(N_j)$$

So

$$K_{j+1} = K_j \cap F_j$$

Show that the sequences $\{N_j\}$, $\{F_j\}$, $\{K_j\}$ are descending chains; by the results of Exercises 23 and 25,

$$\text{Ner}(F_j) \geq \varphi_j \wedge \mu \wedge \wp(\varphi_{j-1} \wedge \cdots \wedge \varphi_0 \wedge \mu \wedge \eta)$$

where $\varphi_j = \{F_j, \Sigma^* - F_j\}$. Noting that, for all $j \geq 0$,

$$F_j = \bigcup \{[s] \cap L(\mathbf{G}) | s \in \Sigma^*, \ (\forall t \leq s)[t] \cap L(\mathbf{G}) \subseteq \bar{K}_j\}$$

show that

$$s \in F_j, \ s' \in [s] \cap L(\mathbf{G}) \Rightarrow s' \in F_j$$

from which there follows

$$\text{Ner}(F_j) \geq \varphi_j \wedge \mu \wedge \wp(\mu \wedge \eta)$$

where the right-hand side is a right congruence. It follows that

$$\text{Ner}(K_{j+1}) \geq \varphi_j \wedge \mu \wedge \eta \wedge \wp(\mu \wedge \eta)$$

with the right-hand side a right congruence. This yields the asserted bound, together with the finite convergence of $\{K_j\}$ to a limit, say $\hat{K}$. Confirm by induction on j that $\hat{K} = K$.

Exercise 29: Let $K \subseteq M \subseteq \Sigma^*$, $\Sigma = \Sigma_c \dot{\cup} \Sigma_u$, $\Sigma_o \subseteq \Sigma^*$, and let $P : \Sigma^* \to \Sigma_o^*$ be the natural projection. Suppose PK is controllable with respect to PM, namely

$$(\overline{PK})(P\Sigma_u) \cap \overline{PM} \subseteq \overline{PK}$$

Show that if $\bar{K}$ is $(\bar{M}, P)$-normal then K is controllable with respect to M, namely

$$\bar{K}\Sigma_u \cap \bar{M} \subseteq \bar{K}$$

6.2 Observable and Relatively Observable Languages

Let $\mathbf{G} = (\_, \Sigma, \_, \_, \_)$ be a DES over alphabet Σ. Let $K \subseteq \Sigma^*$ be arbitrary. In order to define observability, it is convenient to associate with each string $s \in \Sigma^*$ two distinguished subsets of events, as follows. Define the *active* event set

$$A_K(s) = \begin{cases} \{\sigma \in \Sigma | s\sigma \in \bar{K}\}, & s \in \bar{K} \\ \varnothing & \text{otherwise} \end{cases}$$

and the *inactive* event set

$$IA_K(s) = \begin{cases} \{\sigma \in \Sigma | s\sigma \in L(\mathbf{G}) - \bar{K}\}, & s \in \bar{K} \\ \varnothing & \text{otherwise} \end{cases}$$

Thus $A_K(s)$ consists of just those events whose occurrence following a prefix s of K preserves the prefix property; while events in $IA_K(s)$ could occur in $\mathbf{G}$, but destroy the prefix property. Next we define the binary relation K-*active* on Σ^*, denoted by act_K, according to: $(s, s') \in \text{act}_K$ iff

(i) $A_K(s) \cap IA_K(s') = \varnothing = A_K(s') \cap IA_K(s)$, and
(ii) $s \in \bar{K} \cap L_m(\mathbf{G})$ & $s' \in \bar{K} \cap L_m(\mathbf{G}) \Rightarrow (s \in K \Leftrightarrow s' \in K)$

Equivalently, for all $s, s' \in \Sigma^*$, $(s, s') \in \text{act}_K$ iff

(i′) $(\forall \sigma \in \Sigma) s\sigma \in \bar{K}$ & $s' \in \bar{K}$ & $s'\sigma \in L(\mathbf{G}) \Rightarrow s'\sigma \in \bar{K}$,
(ii′) $s \in K \cap L_m(\mathbf{G})$ & $s' \in \bar{K} \cap L_m(\mathbf{G}) \Rightarrow s' \in K$, and
(iii′) conditions (i′) and (ii′) hold with s and s' interchanged.

Note that a pair $(s, s') \in \text{act}_K$ if either of s or s' does not belong to $\bar{K}$, because if $s \notin \bar{K}$ then $A_K(s) = IA_K(s) = \varnothing$. Otherwise (the nontrivial case) membership of a pair of strings (s, s') in act_K means, roughly, that prefixes s and s' of K have identical one-step continuations with respect to membership in $\bar{K}$; if each is in $L_m(\mathbf{G})$ and one actually belongs to K, then so does the other. It should be noted that act_K is a tolerance relation on Σ^*; namely it is reflexive and symmetric but need not be transitive. Notice finally that if K is closed, or $L_m(\mathbf{G})$-closed, then conditions (ii) and (ii′) are satisfied automatically and may be dropped.

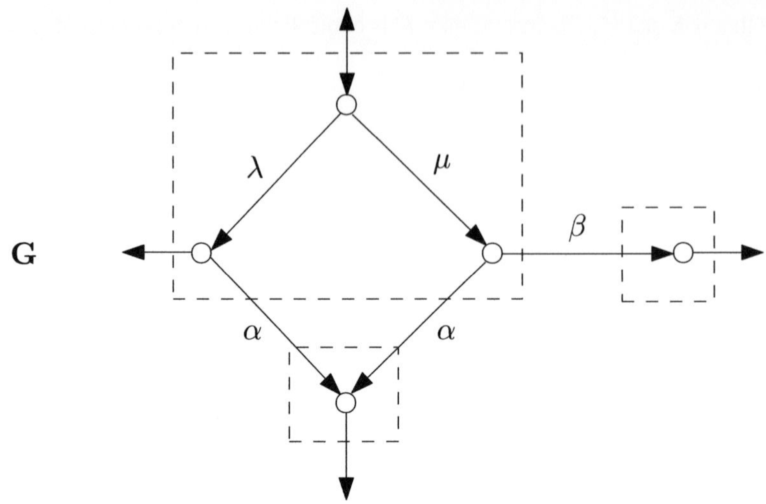

Example 1

Let $\Sigma = \{\alpha, \beta, \lambda, \mu\}$, $\Sigma_o = \{\alpha, \beta\}$,

$$L(\mathbf{G}) = \overline{\lambda\alpha + \mu(\alpha + \beta)}, \quad K = \overline{\lambda\alpha + \mu\beta}$$

Then, for instance

$$A_K(\epsilon) = \{\lambda, \mu\},\ IA_K(\epsilon) = \varnothing$$
$$A_K(\lambda) = \{\alpha\},\ IA_K(\lambda) = \varnothing$$
$$A_K(\mu) = \{\beta\},\ IA_K(\mu) = \{\alpha\}$$

So the string pairs $(\epsilon, \lambda), (\epsilon, \mu) \in \text{act}_K$, but $(\lambda, \mu) \notin \text{act}_K$. Suppose

$$L_m(\mathbf{G}) = \epsilon + \lambda(\epsilon + \alpha) + \mu(\alpha + \beta), \quad J = \epsilon + \lambda\alpha + \mu\beta$$

Then $\bar{J} = K$ but J is not even $L_m(\mathbf{G})$-closed, since

$$\lambda \in \bar{J} \cap L_m(\mathbf{G}), \quad \lambda \notin J$$

Now $\epsilon \in J \cap L_m(\mathbf{G})$ and $\lambda \in \bar{J} \cap L_m(\mathbf{G})$, but $\lambda \notin J$, so in this case $(\epsilon, \lambda) \notin \text{act}_J$.

We can now frame the definition desired. With $\Sigma_o \subseteq \Sigma$ and $P : \Sigma^* \to \Sigma_o^*$ as before, say that a language $K \subseteq \Sigma^*$ is $(\mathbf{G}, P)$-*observable*, or simply *observable*, if

$$\ker P \le \text{act}_K$$

Explicitly, on substituting the definitions of $\ker P$ and act_K, we see that $K \subseteq \Sigma^*$ is $(\mathbf{G}, P)$-observable provided

$$(\text{i}'')(\forall s, s' \in \Sigma^*)(\forall \sigma \in \Sigma) s\sigma \in \bar{K} \;\; \& \;\; s' \in \bar{K} \;\; \& \;\; s'\sigma \in L(\mathbf{G}) \;\; \& \;\; Ps' = Ps \Rightarrow s'\sigma \in \bar{K}$$

$$(\text{ii}'')(\forall s, s' \in \Sigma^*) s \in K \cap L_m(\mathbf{G}) \;\; \& \;\; s' \in \bar{K} \cap L_m(\mathbf{G}) \;\; \& \;\; Ps' = Ps \Rightarrow s' \in K \cap L_m(\mathbf{G})$$

Thus observability declares that the equivalence relation ker P refines act_K, namely that P preserves at least the information required to decide consistently the question of continuing membership of a string $s\sigma$ in $\bar{K}$ after the hypothetical occurrence of an event σ, as well as to decide membership of s in K when membership in $\bar{K} \cap L_m(\mathbf{G})$ happens to be known. If two strings 'look the same' (have the same projections), then a decision rule that applies to one can be used for the other. By contrast, if K is not $(\mathbf{G}, P)$-observable, then an event (observable or not) may have different consequences for look-alike strings. For example, in case $K \subseteq L(\mathbf{G})$ is closed, there would exist $s, \;\; s' \in K$ with $Ps = Ps'$, and $\sigma \in \Sigma$, such that $s\sigma \in K$ but $s'\sigma \in L(\mathbf{G}) - K$. Nevertheless, observability does not preclude the existence of $s \in K$ and $\sigma \in \Sigma - \Sigma_o$ (hence $Ps = P(s\sigma)$) such that $s\sigma \in L(\mathbf{G}) - K$: see the remark following Example 7 below.

In the transition graph for Example 1, the nodes are grouped to display ker P. Since

$$P\epsilon = P\lambda = P\mu$$

neither J nor K is observable.

Exercise 2: Show that the main condition that K be $(\mathbf{G}, P)$-observable (from (i$''$) above) can be written

$$(\forall s')(\forall \sigma) s' \in \bar{K} \; \& \; s'\sigma \in L(\mathbf{G}) \; \& \; [(\exists s) s\sigma \in \bar{K} \; \& \; Ps = Ps'] \Rightarrow s'\sigma \in \bar{K}$$

Roughly, the test for $s'\sigma \in \bar{K}$ is the existence of at least one look-alike s such that $s\sigma \in \bar{K}$. **Hint:** Use the predicate logic tautology

$$(\forall x, y) P(y) \; \& \; Q(x, y) \Rightarrow R(y) \;\; \equiv \;\; (\forall y) P(y) \; \& \; [(\exists x) Q(x, y)] \Rightarrow R(y)$$

Exercise 3: The main condition that K be $(\mathbf{G}, P)$ observable, i.e. either (i), or (i$'$) and (iii$'$), or (i$''$) above, can be displayed as follows. For brevity of notation write $L(\mathbf{G}) = L$ and assume $K = \bar{K}$. For $\sigma \in \Sigma$ let $\text{cat}_\sigma : \Sigma^* \to \Sigma^* : s \mapsto s\sigma$ and define

$$K_\sigma := K \cap \text{cat}_\sigma^{-1}(L)$$

Thus $K_\sigma \subseteq L$ since L is closed. Also let $\overline{\text{cat}_\sigma} : \Sigma_o^* \to \Sigma_o^*$ according to

$$s \mapsto s\sigma \text{ if } \sigma \in \Sigma_o, \quad s \mapsto s \text{ if } \sigma \notin \Sigma_o$$

Let $\rho_K : L \to \{0, 1\}$ with $\rho_K(s) = 1$ iff $s \in K$. For each $\sigma \in \Sigma$ consider the diagram.

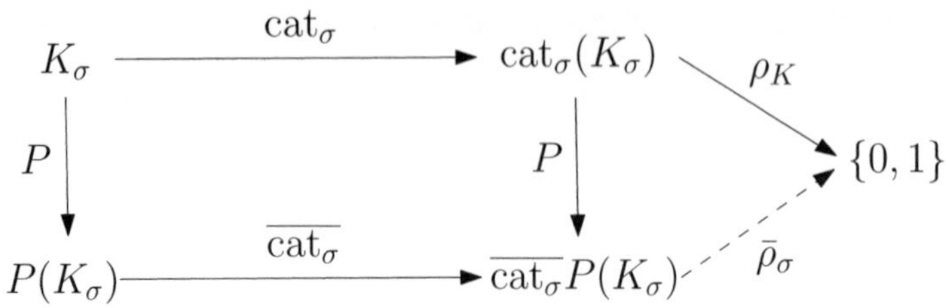

The square always commutes, by definition. Show that $\bar{\rho}_\sigma$ exists for each $\sigma \in \Sigma$ and makes the right side of the diagram commute, iff $\ker P \le \mathrm{act}_K$. In other words observability means that for every $\sigma \in \Sigma$, the 'decision function' $\rho_K|\mathrm{cat}_\sigma(K_\sigma)$ for membership in K factors through $P|\mathrm{cat}_\sigma(K_\sigma)$.

Of some interest in practice is a condition intermediate between observability and normality. We first recall from Sect. 2.2 the definition of Nerode equivalence: if $H \subseteq T^*$ then strings $t, t' \in T^*$ are *Nerode equivalent with respect to* H, written $t \equiv t' (\mathrm{mod}\ H)$ if, for all $v \in T^*$, $tv \in H$ iff $t'v \in H$. This means that t, t' lead to the same state of a canonical recognizer for H. Now let $K \subseteq \Sigma^*$. We say that K is *strongly* $(\mathbf{G}, P)$*-observable* provided the following hold:

$$(\mathrm{i}''')(\forall s, s' \in \Sigma^*)(\forall \sigma \in \Sigma) s\sigma \in \bar{K} \ \ \&\ \ s' \in \bar{K} \ \ \&\ \ s'\sigma \in L(\mathbf{G}) \\ \&\ \ Ps' \equiv Ps (\mathrm{mod}\ PK) \Rightarrow s'\sigma \in \bar{K}$$

$$(\mathrm{ii}''')(\forall s, s' \in \Sigma^*) s \in K \cap L_m(\mathbf{G}) \ \ \&\ \ s' \in \bar{K} \cap L_m(\mathbf{G}) \\ \&\ \ Ps' \equiv Ps (\mathrm{mod}\ PK) \Rightarrow s' \in K$$

Here conditions (i‴) and (ii‴) agree with (i″) and (ii″) above, except for the crucial replacement of the condition $Ps' = Ps$ by the much weaker condition $Ps' \equiv Ps(\mathrm{mod}\ PK)$. The practical consequence is that a correct decision as to whether $s'\sigma \in \bar{K}$ in condition (i‴), or $s' \in K$ in condition (ii‴), can be based on the 'current' state reached by the observed string Ps' in any transition structure for PK, for instance a canonical recognizer: no further information about the past history of Ps' is necessary.

Exercise 4: Define the equivalence relation $(\equiv (\mathrm{mod}\ PK)) \circ P$ on Σ^* according to

$$s \equiv s'(\mathrm{mod}\ (\equiv (\mathrm{mod}\ PK)) \circ P)$$

just in case $Ps \equiv Ps'(\mathrm{mod}\ PK)$. Show that K is strongly $(\mathbf{G}, P)$-observable iff

$$(\equiv (\mathrm{mod}\ PK)) \circ P \le \mathrm{act}_K$$

This order relation succinctly compares strong observability with the concise form of the observability definition, $\ker P \le \mathrm{act}_K$.

It is clear that observability is implied by strong observability. Our next result shows that, under mild restrictions on marking, strong observability is implied in turn by normality.

Proposition 5

Assume that either $K \subseteq L(\mathbf{G})$ and K is closed, or $K \subseteq L_m(\mathbf{G})$ and K is $L_m(\mathbf{G})$-closed. If $\bar{K}$ is $(L(\mathbf{G}), P)$-normal then K is strongly $(\mathbf{G}, P)$-observable. □

The converse statement is false.

Example 6: Strong observability does not imply normality

Let $\Sigma = \{\alpha, \beta\}$, $\Sigma_o = \{\alpha\}$, so $P\alpha = \alpha$ and $P\beta = \epsilon$. Let $\mathbf{G}, \mathbf{K}$ be as displayed.

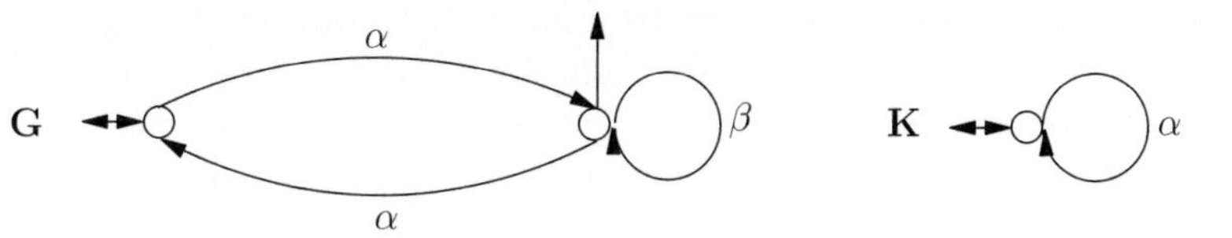

Taking $K = L_m(\mathbf{K})$ we have $PK = K = \alpha^*$. Therefore K is closed, but

$$L(\mathbf{G}) \cap P^{-1}PK = L(\mathbf{G}) \cap \Sigma^* = L(\mathbf{G}) \supsetneqq K$$

so K is not $(L(\mathbf{G}), P)$-normal. To see that K is strongly observable note that $s\sigma \in \bar{K} = K$ implies $\sigma = \alpha$, and then $s' \in K$ implies $s'\alpha \in K$ as required by (i′′′). Of course (ii′′′) holds since K is closed.

Example 7: Observability does not imply strong observability

Let $\Sigma = \{\alpha_1, \alpha_2, \beta_1, \beta_2, \beta_3\}$, $\Sigma_o = \{\alpha_1, \alpha_2\}$, so $P\alpha_i = \alpha_i$ and $P\beta_j = \epsilon$. Let $\mathbf{G}, \mathbf{K}$ be as displayed, with all states marked, so $K = L_m(\mathbf{K})$ is closed.

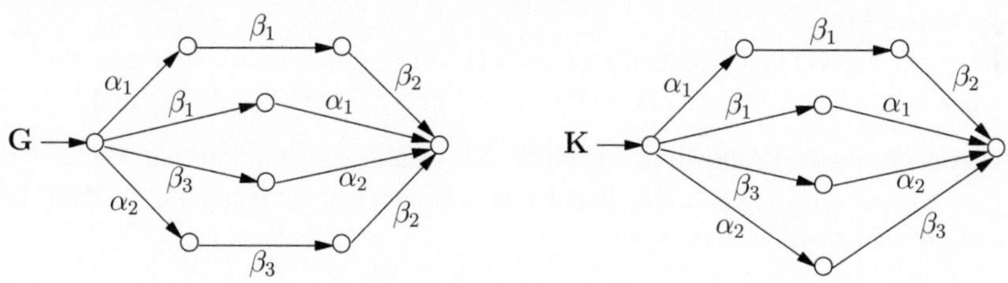

Thus PK is represented by $\mathbf{PK}$ as shown:

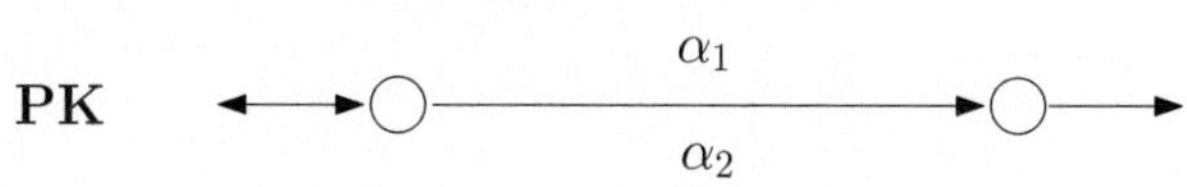

Let

$$s = \alpha_1\beta_1, \ s' = \alpha_2\beta_3, \ \sigma = \beta_2$$

Then

$$s\sigma = \alpha_1\beta_1\beta_2 \in K, \ s' \in K, \ s'\sigma = \alpha_2\beta_3\beta_2 \in L(\mathbf{G})$$
$$Ps = \alpha_1, \ Ps' = \alpha_2, \ Ps' \equiv Ps(\text{mod } PK)$$

As $s'\sigma \in L(\mathbf{G}) - K$, K is not strongly observable. However, by inspecting the groups of strings $s \in L(\mathbf{G})$ with $Ps = \epsilon$, $Ps = \alpha_1$, $Ps = \alpha_2$, respectively, the reader may confirm that K is observable.

From the viewpoint of an observing agent, the essential difference between a $(\mathbf{G}, P)$-observable language and a closed $(L(\mathbf{G}), P)$-normal language is that with a normal language one can always tell, by watching the projection Ps of an evolving string s, if and when the string exits from the language; but with an observable language in general this is not the case. For instance in Example 6, the occurrence of β would represent an unobservable exit from K. As we shall see in Sect. 6.5 (Propositions 6.5.4, 6.5.6) this difference between observability and normality has the following implication for supervisory control. Suppose the (closed, controllable) language to be synthesized is normal. Then no unobservable event will cause a string to exit, in particular no controllable unobservable event. Thus no such event will ever be disabled by the supervisor.

Exercise 8: Formalize (i.e. provide a rigorous version of) the preceding statement, and supply a proof.

Exercise 9: Prove Proposition 5.

Exercise 10: As a variation on Proposition 5, show that K is strongly $(\mathbf{G}, P)$-observable if K is $(L_m(\mathbf{G}), P)$-normal and $\bar{K}$ is $(L(\mathbf{G}), P)$-normal.

The following is a partial converse to Proposition 5.

Proposition 11

Let $K \subseteq L_m(\mathbf{G})$ be $\mathbf{G}$-controllable and $(\mathbf{G}, P)$-observable. Assume that $P\sigma = \sigma$ for all $\sigma \in \Sigma_c$. Then K is $(L_m(\mathbf{G}), P)$-normal and $\bar{K}$ is $(L(\mathbf{G}), P)$-normal. □

Exercise 12: Prove Proposition 11, under the weaker assumption that $P\sigma = \sigma$ for all $\sigma \in \Sigma_c$ such that σ is actually disabled by a supervisor synthesizing K. **Hint**: Use structural induction on strings.

Exercise 13: **Checking observability by 'observation consistency'**

Let

$$\mathbf{G} = (Q, \Sigma, \delta, q_0, Q_m)$$
$$\mathbf{K} = (X, \Sigma, \xi, x_0, X_m)$$

be DES generators over the alphabet Σ, representing languages $L, K \subseteq \Sigma^*$, respectively; i.e. $L = L_m(\mathbf{G})$, $\bar{L} = L(\mathbf{G})$, $K = L_m(\mathbf{K})$, $\bar{K} = L(\mathbf{K})$. For $\Sigma_o \subseteq \Sigma$ let $P : \Sigma^* \to \Sigma_o^*$ be the natural projection. Whether or not K is $(\mathbf{G}, P)$-observable may in principle be checked as follows. For every $s \in \Sigma^*$ compute the set of all state pairs reached by look-alike strings s', namely

$$T(s) := \{(\delta(q_0, s'), \xi(x_0, s')) | Ps' = Ps, \delta(q_0, s')!, \xi(x_0, s')!\}$$
$$\subseteq Q \times X$$

and write $\mathcal{T} := \{T(s) | s \in \Sigma^*, T(s) \neq \varnothing\}$. The family $\mathcal{T}$ of subsets T can be computed in exponential time inasmuch as $|\mathcal{T}| \leq 2^{|Q|\cdot|X|}$. Now say $T \in \mathcal{T}$ is *observation-consistent* (*o.c.*) if

$$(\forall (q, x), (q', x') \in T)\big((\forall \sigma \in \Sigma)[\delta(q', \sigma)! \ \& \ \xi(x, \sigma)! \Rightarrow \xi(x', \sigma)!]$$
$$\& \ [q \in Q_m \ \& \ q' \in Q_m \ \& \ x \in X_m \Rightarrow x' \in X_m]\big)$$

Show that K is $(\mathbf{G}, P)$-observable iff T is o.c. for every $T \in \mathcal{T}$. Provide simple examples to illustrate both outcomes.

In *TCT* observability (resp. strong observability) can be checked by the procedure **(s)observ**.

Exercise 14: **Uncertainty sets**

Let $\mathbf{G} = (Q, \Sigma, \delta, q_0, Q_m)$ be a recognizer for $K := L_m(\mathbf{G})$, so the function $\delta : Q \times \Sigma \to Q$ is total. Let $\Sigma_o \subseteq \Sigma$ and $P : \Sigma^* \to \Sigma_o^*$ be the natural projection. For $s \in \Sigma^*$, observation of Ps alone results in uncertainty as to the state of $\mathbf{G}$ given by the *uncertainty set*

$$U(s) := \{\delta(q_0, s') | Ps' = Ps, s' \in \Sigma^*\} \subseteq Q$$

We can use the uncertainty sets to obtain a recognizer for the projected language PK, as follows. First note that, with $U : \Sigma^* \to Pwr(Q)$, $\ker U \in \mathcal{E}(\Sigma^*)$ is a right congruence on Σ^*; namely if $s_1, s_2 \in \Sigma^*$ and $U(s_1) = U(s_2)$ then for all $w \in \Sigma^*$ we have $U(s_1 w) = U(s_2 w)$. Let $P_U : \Sigma^* \to \Sigma^* / \ker U$ be the corresponding canonical projection, as displayed below.

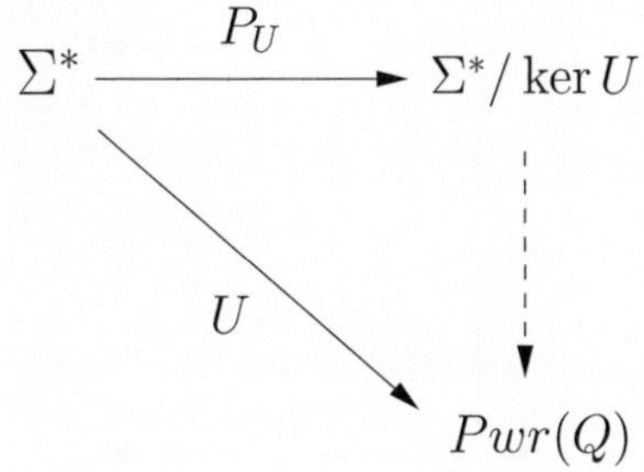

Here the (injective) dashed arrow identifies each cell of $\ker U$ with an uncertainty subset of Q. Now define $\mu \in \mathcal{E}(P\Sigma^*)$ according to $\bar{s}_1 \equiv \bar{s}_2 (\mathrm{mod}\ \mu)$ (where $\bar{s}_i = Ps_i$) iff $U(s_1) = U(s_2)$. Check that μ is well-defined and is a right congruence on $P\Sigma^*$ that is finer than the binary partition $\{PK, P\Sigma^* - PK\}$. Let $P_\mu : P\Sigma^* \to P\Sigma^*/\mu$. With the help of μ and P_μ verify that there exist unique maps $\bar{\delta}$, h, and $\bar{\eta}$ such that the following diagram commutes.

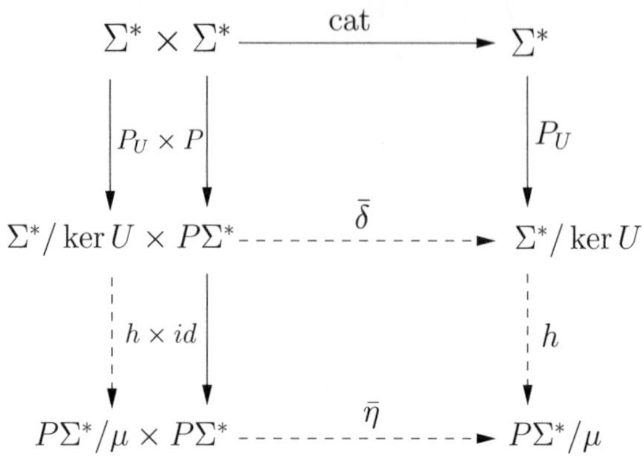

Fig. 6.1 Commutative diagram for Exercise 14

In Fig. 6.1 $\bar{\delta}$ is defined by $P_U \circ \text{cat} = \bar{\delta} \circ (P_U \times P)$, h by $P_\mu \circ P = h \circ P_U$, and $\bar{\eta}$ by $h \circ \bar{\delta} = \bar{\eta} \circ (h \times id)$. As h is bijective (verify!) we obtain for PK the equivalent (isomorphic) recognizers

$$(P_U \Sigma^*,\ \Sigma_o,\ \bar{\delta},\ P_U(\epsilon),\ P_U K)$$

and

$$(P_\mu(P\Sigma^*),\ \Sigma_o,\ \bar{\eta},\ P_\mu(\epsilon),\ P_\mu(PK))$$

In general these are not canonical (i.e. minimal-state), even if **G** is canonical for K. Nevertheless they provide a 'universal' basis for supervisor design, as explained in Example 6.3.4 below.

In this representation of PK show that a state $P_U(s) \in \Sigma^*/\ker U$ is marked iff q is marked for some state $q \in U(s)$.

Generalize the above discussion when **G** is a generator and $\delta : Q \times \Sigma \to Q$ is a partial function. Then work out details for the example displayed below. Assume all states are marked, $\Sigma = \{0, 1, 2\}$, and $\Sigma_o = \{0\}$.

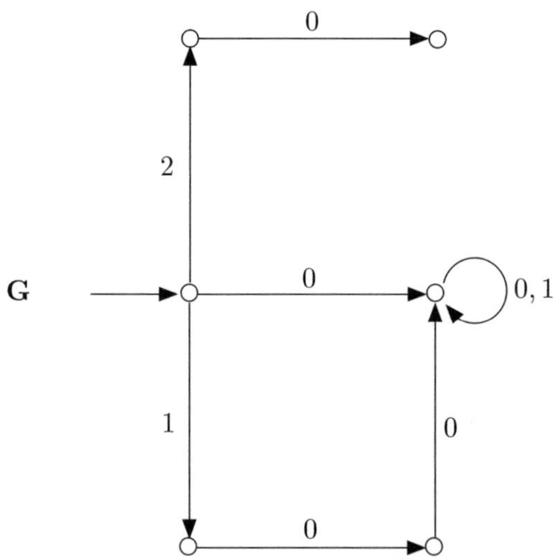

To conclude this section we briefly introduce the concept of *relative observability*, which will prove to be an effective alternative to (standard) observability as hitherto defined. With relative observability a fixed, *ambient* language is given (or selected by the designer), relative to which the standard observability property is tested. Relative observability is stronger than observability, but enjoys the important property that it is preserved under sublanguage union; hence there exists the supremal relatively observable sublanguage of a given (say, specification) language. Relative observability is weaker than normality and, unlike normality in the control context (Sect. 6.5, below), places no constraint on the disablement of unobservable controllable events.

Relative observability is formally defined as follows. Given a (plant) DES $\mathbf{G}$ as usual, with alphabet Σ, fix a sublanguage $C \subseteq L_m(\mathbf{G})$: its closure $\bar{C} \subseteq L(\mathbf{G})$ will be called the *ambient* language. Fix $\Sigma_o \subseteq \Sigma$ as observable subalphabet, and let $P : \Sigma^* \to \Sigma_o^*$ be the natural projection. Consider a sublanguage $K \subseteq C$. We say that K is *relatively observable with respect to* C, $\mathbf{G}$, and P, or simply C*-observable*, if the following hold:

(i) $(\forall s, s' \in \Sigma^*)(\forall \sigma \in \Sigma) s\sigma \in \bar{K}$ & $s' \in \bar{C}$ & $s'\sigma \in L(\mathbf{G})$
& $Ps' = Ps \Rightarrow s'\sigma \in \bar{K}$

(ii) $(\forall s, s' \in \Sigma^*) s \in K$ & $s' \in \bar{C} \cap L_m(\mathbf{G})$ & $Ps' = Ps \Rightarrow s' \in K$

Note that a pair of look-alike strings (s, s') trivially satisfies (i) and (ii) if either s or s' does not belong to $\bar{C}$. For a look-alike pair (s, s') both in $\bar{C}$, relative observability requires that (i) s and s' have identical one-step continuations if allowed in $L(\mathbf{G})$, with respect to membership in $\bar{K}$ and (ii) if each string is in $L_m(\mathbf{G})$ and one actually belongs to K, then so does the other.

In the special case $C = K$ the definition above becomes that of observability of K as given at the beginning of this section. This implies that if K is C-observable then K is also observable. Examples show, however, that the reverse statement need not be true. Thus the set of C-observable sublanguages of C is in general a proper subset of the observable sublanguages of C.

We can now state our main results.

Proposition 15: **Closure of relative observability under union**

Let $K_i \subseteq C, i \in I$ (some index set) be C-observable. Then

$$K := \bigcup \{K_i | i \in I\}$$

is also C-observable. □

Now let $K \subseteq C$ be an arbitrary sublanguage of C, and write $\mathcal{O}(K)$ for the family of C-observable sublanguages of K. Trivially the empty language $\varnothing$ is a member of $\mathcal{O}(K)$.

Proposition 16: **Existence of the supremal C-observable sublanguage**

The family $\mathcal{O}(K)$ is nonempty and contains a unique supremal element, namely

$$\sup \mathcal{O}(K) := \bigcup \{K' | K' \in \mathcal{O}(K)\}$$

□

Exercise 17: Prove Propositions 15 and 16.

In *TCT*, a representation of $\sup \mathcal{O}(K)$ can be computed by the procedure

KSUP := **suprobs**(**G**, **K**, **null**(P)/**image**(P))

where **K** represents the language K. In **suprobs** the ambient language $\bar{C}$ is taken to be $\bar{K}$. In computations it can (and should) be verified by use of *TCT* **(s)observ** that L_m(**KSUP**) is indeed observable.

As to the relation between relative observability and normality, it is not difficult to show the following.

Proposition 18: **Relative observability is weaker than normality**

Let $K \subseteq C$ be $(L_m(\mathbf{G}), P)$-normal, and suppose $\bar{K}$ is $(L(\mathbf{G}), P)$-normal. Then K is C-observable. □

Exercise 19: Prove Proposition 18. Provide an example where K is C-observable but $\bar{K}$ is not $(L(\mathbf{G}), P)$-normal.

6.3 Feasible Supervisory Control

We now introduce the concept of supervisory control, proceeding just as in Sect. 3.4, except for taking into account the constraint that control must be based purely on the result of observing the strings generated by **G** through the channel **C**, namely on information transmitted by $P : \Sigma^* \to \Sigma_o^*$. With

$$\mathbf{G} = (\_, \Sigma, \_, \_, \_), \quad \Sigma = \Sigma_c \dot{\cup} \Sigma_u$$

as usual, define as before the set of all control patterns

$$\Gamma = \{\gamma \in Pwr(\Sigma) | \gamma \supseteq \Sigma_u\}$$

A *feasible supervisory control* for **G** is any map $V : L(\mathbf{G}) \to \Gamma$ such that

$$\ker(P|L(\mathbf{G})) \leq \ker V$$

Here $P|L(\mathbf{G})$ denotes the restriction of P to $L(\mathbf{G})$. As before we write $V/\mathbf{G}$ to suggest '**G** under the supervision of V'. The *closed behavior* $L(V/\mathbf{G})$ and *marked behavior* $L_m(V/\mathbf{G})$ are defined exactly as in Sect. 3.4, as is the property that V is nonblocking for **G**. Our first main result is the expected generalization of Theorem 3.4.3.

Theorem 1

Let $K \subseteq L_m(\mathbf{G})$, $K \neq \varnothing$. There exists a nonblocking feasible supervisory control V for $\mathbf{G}$ such that $L_m(V/\mathbf{G}) = K$ iff

(i) K is controllable with respect to $\mathbf{G}$,
(ii) K is observable with respect to $(\mathbf{G}, P)$, and
(iii) K is $L_m(\mathbf{G})$-closed.

Proof

(If) The proof follows the same lines as that of Theorem 3.4.3, but extended to ensure the feasibility property. First bring in the function

$$Q : \bar{K} \to Pwr(\Sigma)$$

according to

$$Q(s) := \{\sigma \in \Sigma | (\forall s' \in \bar{K}) Ps' = Ps \ \& \ s'\sigma \in L(\mathbf{G}) \Rightarrow s'\sigma \in \bar{K}\}$$

Now define $V : L(\mathbf{G}) \to \Gamma$ as follows. If $s \in \bar{K}$ then

$$V(s) := \Sigma_u \cup (\Sigma_c \cap Q(s))$$

while if $s \in L(\mathbf{G}) - \bar{K}$ and $Ps = Pv$ for some $v \in \bar{K}$, let

$$V(s) := V(v)$$

$V(s)$ is well defined in the latter case, for if also $Ps = Pw$ with $w \in \bar{K}$ then $Pv = Pw$, hence by definition of Q, $Q(v) = Q(w)$, so $V(v) = V(w)$. Finally if $s \in L(\mathbf{G}) - \bar{K}$ and there is no $v \in \bar{K}$ such that $Ps = Pv$ then let

$$V(s) := \Sigma_u$$

Next we show that V is feasible, namely $\ker(P|L(\mathbf{G})) \leq \ker V$. Let $s_1, s_2 \in L(\mathbf{G})$, $Ps_1 = Ps_2$. We consider the three cases (i) $s_1, s_2 \in \bar{K}$, (ii) $s_1 \in \bar{K}$, $s_2 \in L(\mathbf{G}) - \bar{K}$, and (iii) $s_1, s_2 \in L(\mathbf{G}) - \bar{K}$. As to (i) it is easily checked that $Q(s_1) = Q(s_2)$, so $V(s_1) = V(s_2)$, namely $(s_1, s_2) \in \ker V$ as claimed. For (ii), by definition $V(s_2) = V(s_1)$, so $(s_1, s_2) \in \ker V$. In case (iii), if $Ps_1 = Pv$ for some $v \in \bar{K}$, then by definition $V(s_1) = V(v)$, and $Ps_2 = Ps_1$ implies similarly $V(s_2) = V(v)$; while if $Ps_1 = Pv$ for no $v \in \bar{K}$ then

$$V(s_1) = V(s_2) = \Sigma_u$$

so in either subcase $(s_1, s_2) \in \ker V$, as required.

To complete the proof it may be shown by induction on length of strings that

$$L(V/\mathbf{G}) = \bar{K}$$

and then directly that $L_m(V/\mathbf{G}) = K$. As the argument is similar to the proof of Theorem 3.4.3, we just provide the inductive step. Thus suppose $s \in L(V/\mathbf{G})$, $s \in \bar{K}$, and $s\sigma \in L(V/\mathbf{G})$, i.e. $\sigma \in V(s)$ and $s\sigma \in L(\mathbf{G})$. If $\sigma \in \Sigma_u$ then $s\sigma \in \bar{K}$ by controllability; while if $\sigma \in \Sigma_c \cap Q(s)$ then $s\sigma \in \bar{K}$ by definition of Q. Conversely suppose $s\sigma \in \bar{K}$. If $\sigma \in \Sigma_u$ then clearly $\sigma \in V(s)$ so $s\sigma \in L(V/\mathbf{G})$. Suppose $\sigma \in \Sigma_c$. We claim $\sigma \in Q(s)$: for if $s' \in \bar{K}$ with $Ps' = Ps$ then by observability $(s, s') \in \mathrm{act}_K$, and then $s'\sigma \in L(\mathbf{G})$ implies $s'\sigma \in \bar{K}$, the required result. Thus it follows that $\sigma \in V(s)$, and as $s\sigma \in L(\mathbf{G})$ we again conclude that $s\sigma \in L(V/\mathbf{G})$. This shows that $L(V/\mathbf{G}) = \bar{K}$, as claimed.

(Only if) Let V be a nonblocking feasible supervisory control for $\mathbf{G}$ with $L_m(V/\mathbf{G}) = K$. As the proof that K is controllable and $L_m(\mathbf{G})$-closed is unchanged from the proof of Theorem 3.4.3, it suffices to show that K is observable. So let $(s, s') \in \ker P$, $s\sigma \in \bar{K}$, $s' \in \bar{K}$, and $s'\sigma \in L(\mathbf{G})$. Since $s, s' \in L(\mathbf{G})$ and $\ker(P|L(\mathbf{G})) \le \ker V$, there follows $V(s) = V(s')$. Therefore $s\sigma \in \bar{K}$ implies in turn $\sigma \in V(s)$, $\sigma \in V(s')$, and $s'\sigma \in \bar{K}$. This verifies the observability condition (i′) of Sect. 6.2; condition (ii′) is automatic since K is $L_m(\mathbf{G})$-closed, while condition (iii′) is true by symmetry of the argument. □

Corollary 2

Let $K \subseteq L(\mathbf{G})$ be nonempty and closed. There exists a feasible supervisory control V for $\mathbf{G}$ such that $L(V/\mathbf{G}) = K$ iff K is controllable with respect to $\mathbf{G}$ and observable with respect to $(\mathbf{G}, P)$. □

For brevity we refer to a nonblocking feasible supervisory control (for $\mathbf{G}$, P) as an NFSC. As before we may generalize this idea to incorporate marking as well as control in the supervisory action. Thus if $M \subseteq L_m(\mathbf{G})$ we define a *marking nonblocking feasible supervisory control* for the triple $(M, \mathbf{G}, P)$, or MNFSC, as a map $V : L(\mathbf{G}) \to \Gamma$ as defined above, but now with marked behavior given by

$$L_m(V/\mathbf{G}) = L(V/\mathbf{G}) \cap M$$

However, for this definition to satisfy the intended interpretation of 'marking' we must place a further restriction on M. For instance, in a manufacturing system a string $s \in L_m(\mathbf{G})$ might correspond to 'completion of a finished workpiece', while $s \in M$ might mean 'completion of a batch of finished workpieces'. If a batch consists of 10 workpieces, then we would not want the supervisor to confuse a string s corresponding to 6 batches with a string s' corresponding to 61 workpieces. It is natural, then, to require that s, s' be distinguishable, namely look different when viewed through P. In general terms we require

$$(\forall s, s' \in \Sigma^*) s \in M \,\&\, s' \in \bar{M} \cap L_m(\mathbf{G}) \,\&\, s' \notin M \Rightarrow Ps \neq Ps'$$

or more directly

$$(\forall s, s' \in \Sigma^*) s \in M \,\&\, s' \in \bar{M} \cap L_m(\mathbf{G}) \,\&\, Ps = Ps' \Rightarrow s' \in M$$

or succinctly

$$\ker(P|(\bar{M} \cap L_m(\mathbf{G}))) \leq \{M, \bar{M} \cap L_m(\mathbf{G}) - M\}$$

If this condition is satisfied then we shall say that M is $(\mathbf{G}, P)$-*admissible*. As the counterpart to Theorem 3.4.5 we now have

Theorem 3

Let $K \subseteq L_m(\mathbf{G})$, $K \neq \varnothing$, and let K be $(\mathbf{G}, P)$-admissible. There exists an MNFSC V for $(K, \mathbf{G}, P)$ such that

$$L_m(V/\mathbf{G}) = K$$

iff K is controllable with respect to $\mathbf{G}$ and observable with respect to $(\mathbf{G}, P)$.

Proof

The proof of sufficiency may be left to the reader (cf. the proof of Theorem 3.4.5). As to necessity, the proof that K is controllable is unchanged from the proof of Theorem 1. For observability let $s, s' \in L(\mathbf{G})$ with $Ps = Ps'$. The proof of condition (i′) (or (i″)) of Sect. 6.2 is unchanged, while condition (ii′) (or (ii″)) is just the property that K is $(\mathbf{G}, P)$-admissible. □

Example 4: **Implementation of a feasible supervisor based on uncertainty sets**

We show how to implement a feasible supervisor based on the uncertainty sets defined in Exercise 6.2.14. The method succeeds whenever the (controllable) behavior to be synthesized is observable. Initially assume (for the sake of implementation) that all events are observable, and construct a proper supervisor $\mathbf{K} = (X, \Sigma, \xi, x_0, X_m)$ that represents the controlled behavior $K \subseteq L_m(\mathbf{G})$ of the plant DES $\mathbf{G}$ (cf. Sect. 3.6). For instance, $\mathbf{K}$ might be obtained as **supcon(G,E)** for a specification DES $\mathbf{E}$. With the natural projection $P : \Sigma^* \to \Sigma_o^*$ given, assume that K is $(\mathbf{G}, P)$-observable. Let $[s]$ denote the cell of s (mod $\ker P$), and let P_U be the canonical projection on uncertainty subsets of X, as defined in Exercise 6.2.14; the state set for our feasible supervisor can then be taken to be $\Sigma^*/\ker U$, as displayed in Fig. 6.1. Finally let

$$D : \Sigma^* \times \Sigma \to \{0, 1\}$$

be the supervisory control disablement function, defined according to

$$D(s, \sigma) = \begin{cases} 1 & \text{if } s\sigma \in L(\mathbf{G}) \ \& \ s \in \bar{K} \ \& \ s\sigma \notin \bar{K} \\ 0 & \text{if } s\sigma \notin L(\mathbf{G}), \text{or } s \notin \bar{K}, \ \text{or } s\sigma \in \bar{K} \end{cases}$$

Thus $D(s, \sigma) = 1$ iff σ must be disabled following string s by the supervisor **K**. We first claim that, with K observable, D factors through $P_U \times id_\Sigma$, as displayed below.[2]

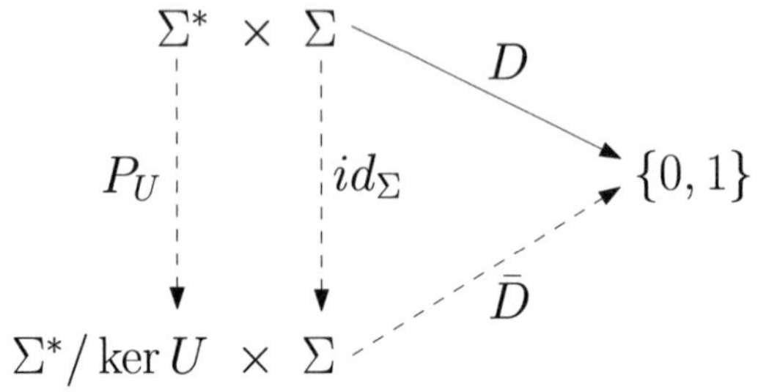

It must be shown that $\ker(P_U \times id_\Sigma) \le \ker D$. For this let $\sigma \in \Sigma$, $s_1, s_2 \in \Sigma^*$, $P_U(s_1) = P_U(s_2)$, and $D(s_1, \sigma) = 1$. It will be verified that $D(s_2, \sigma) = 1$. We have

$$\begin{aligned} P_U(s_1) = P_U(s_2) &\Rightarrow U(s_1) = U(s_2) \\ &\Rightarrow \{P_K(s_1')|s_1' \in [s_1]\} = \{P_K(s_2')|s_2' \in [s_2]\} \end{aligned} \tag{6.2}$$

where $P_K : \Sigma^* \to \Sigma^*/\mathrm{Ner}(K)$ is the canonical state projection for K. By observability

$$\begin{aligned} D(s_1, \sigma) = 1 \Rightarrow & s_1\sigma \in L(\mathbf{G}) \;\&\; s_1 \in \bar{K} \;\&\; s_1\sigma \notin \bar{K} \;\& \\ & (\forall s_1' \in [s_1])[s_1'\sigma \in L(\mathbf{G}) \;\&\; s_1' \in \bar{K} \Rightarrow s_1'\sigma \notin \bar{K}] \end{aligned}$$

By (6.2), with $s_2' = s_2 \in [s_2]$ there exists $s_1' \in [s_1]$ with $P_K(s_1') = P_K(s_2)$. Therefore,

$$\begin{aligned} & s_2\sigma \in L(\mathbf{G}) \;\&\; s_1' \in \bar{K} \;\&\; s_1'\sigma \notin \bar{K} \\ \Rightarrow & s_2\sigma \in L(\mathbf{G}) \;\&\; s_2 \in \bar{K} \;\&\; s_2\sigma \notin \bar{K} \\ \Rightarrow & D(s_2, \sigma) = 1 \quad \text{(by definition of} D(.,.)) \end{aligned}$$

as claimed.

It follows immediately that the induced disablement mapping

$$\bar{D} : (\Sigma^*/\ker U) \times \Sigma \to \{0, 1\}$$

displayed above provides the required feasible control action corresponding to the given supervisor **K**.

Exercise 5: **Verification of observability using uncertainty sets**

Given a DES **G**, and a sublanguage $K \subseteq L_m(\mathbf{G})$, let K be represented by the DES $\mathbf{K} = (X, \Sigma, \xi, x_0, X_m)$, where (for simplicity) we assume that **K** is a recognizer with transition function $\xi : X \times \Sigma \to X$ a total function. Let $\Sigma_o \subseteq \Sigma$ and $P : \Sigma^* \to \Sigma_o^*$

---

[2]Strictly speaking, the asserted factorization needs to hold only on pairs (s, σ) with $s\sigma \in L(\mathbf{G})$, but we refrain from complicating the notation accordingly.

the natural projection. As in Exercise 6.2.14 we define the uncertainty mapping U to state subsets of **K** according to

$$U : \Sigma^* \to Pwr(X) : s \mapsto U(s) := \{\xi(x_0, s')|s' \in \Sigma^*, Ps' = Ps\} \subseteq X$$

Denote $\Sigma^*/\ker U = \bar{X}$, and write $P_U : \Sigma^* \to \bar{X}$ for the canonical projection. Show that there is a well-defined induced transition map $\bar{\xi} : \bar{X} \times \Sigma \to \bar{X}$ such that

$$\bar{\xi}(P_U(s), \sigma) = P_U(s\sigma)$$

for all $s \in \Sigma^*$, $\sigma \in \Sigma$. As before, define

$$D : \Sigma^* \times \Sigma \to \{0, 1\}$$

with respect to K. Check that K is observable (with respect to **G** and P) only if D factors through $P_U \times id_\Sigma$ as displayed above. For a more complete statement we need in addition that the mapping U distinguish between K and its complement; namely, confirm the following

Proposition: Characterization of observability by uncertainty sets
K is observable with respect to (**G**, P) iff

$$\ker(P_U \times id_\Sigma) \le \ker D \;\;\&\;\; \ker P_U \le \{K, \Sigma^* - K\}$$

Example 6: **Mutual exclusion under partial observation**
The previous construction is illustrated by the following problem of mutual exclusion under partial observation. Consider agents **A1**, **A2** as shown in Fig. 6.2. The state name Use refers to a single shared resource, subject to mutual exclusion, so simultaneous occupancy of the state pair (2,2) is prohibited. An additional specification is that resource usage be 'fair' in the sense of 'first-request-first-use', implemented by means of a queue. It is assumed that events 11, 21 (transitions from Request to Use) are unobservable. To attempt a solution, start by constructing **A = sync(A1,A2)**, then **ASPEC** (left to the reader), and finally the supervisor **ASUPER = supcon(A,ASPEC)**, with the result displayed in Fig. 6.3.

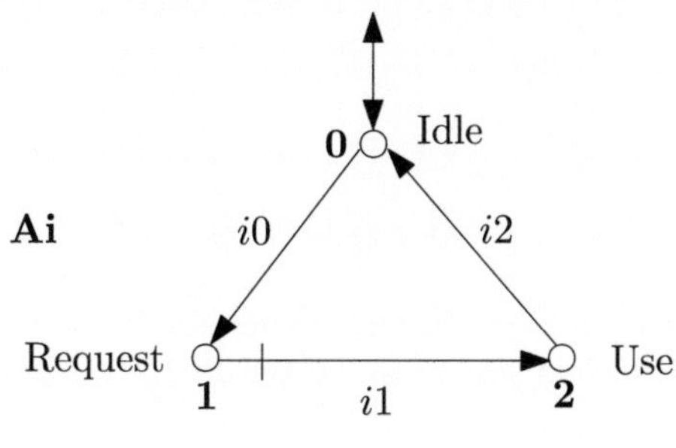

Fig. 6.2 Agents subject to mutual exclusion ($i = 1, 2$)

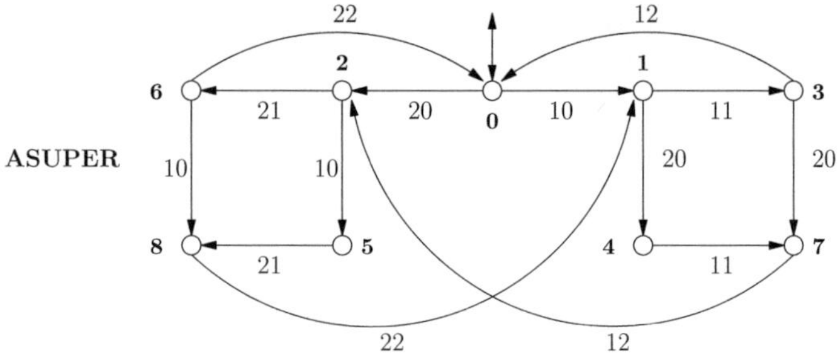

Fig. 6.3 Supervisor with full observation

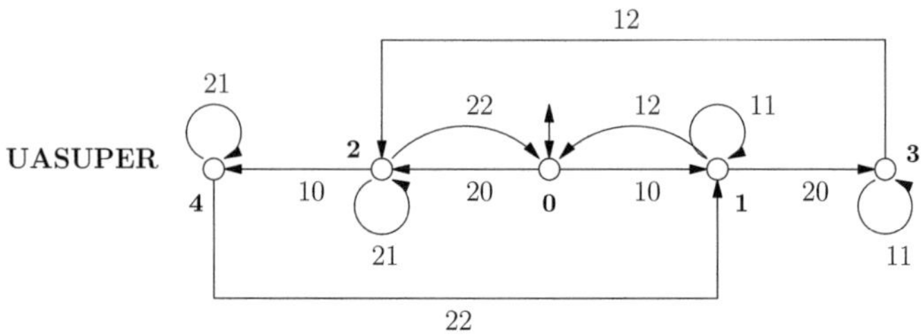

Fig. 6.4 Projection of **ASUPER** under P_U (state set $Y = \{0, 1, 2, 3, 4\}$)

By construction, **ASUPER** is controllable and is easily seen to be (**A**, P)-observable. With events 11, 21 unobservable, inspection of Fig. 6.3 yields the uncertainty sets

$$y_0 := \{0\}, \quad y_1 := \{1, 3\}, \quad y_2 := \{2, 6\}, \quad y_3 := \{4, 7\}, \quad y_4 := \{5, 8\}$$

and corresponding state transition structure in Fig. 6.4, where $U : \Sigma^* \to Pwr$ ($\{0, ..., 8\}$) is identified with the projection P_U of Exercise 6.2.14. Note that the unobservable events 11, 21 appear in the transition graph of **UASUPER** only as selfloops.

Now the table **condat(A, ASUPER)** shows that event 11 is disabled at $x = 5, 8$ while 21 is disabled at $x = 4, 7$. From this we assert

$$D(5, 11) = D(8, 11) = D(4, 21) = D(7, 21) = 1$$

and obtain

$$\bar{D}(y_3, 21) = \bar{D}(y_4, 11) = 1;$$
$$\bar{D}(y_1, 11) = \bar{D}(y_2, 21) = \bar{D}(y_3, 11) = \bar{D}(y_4, 21) = 0$$

The disablement logic expressed by the function $\bar{D}$ is already incorporated in the transition structure of **UASUPER**. It is easily checked that **UASUPER** and **A** are nonconflicting, so the result is nonblocking for **A**, and in fact the controlled behavior of **A** under supervision by **UASUPER** is identical with the behavior by **ASUPER**. Intuitively it is clear that observation of events 11 and 21 ought to be irrelevant to control, subject to the given specification.

Exercise 7: Show that $K = L_m(\mathbf{ASUPER})$ in Example 6 is strongly ($\mathbf{A}$,P)-observable.

Exercise 8: **Supervision based on observation consistency**

In case the language K to be synthesized is controllable and observable (but not necessarily strongly observable) a supervisor can be constructed as follows. We adopt the notation of Exercise 6.2.13. Assume that K (say by construction) is already controllable, and $\varnothing \neq K \subseteq L = L_m(\mathbf{G})$. First show that $T \in \mathcal{T}$ is observation-consistent (o.c.) iff

$$(\forall (q,x), (q',x') \in T)\big((\forall \sigma \in \Sigma_c)[\delta(q',\sigma)! \;\&\; \xi(x,\sigma)! \Rightarrow \xi(x',\sigma)!]$$
$$\& \; [q \in Q_m \;\&\; q' \in Q_m \;\&\; x \in X_m \Rightarrow x' \in X_m]\big)$$

namely the definitional test for o.c. needs to be applied only for $\sigma \in \Sigma_c$.

Assume next that every $T \in \mathcal{T}$ is o.c., so K is ($\mathbf{G}$, P)-observable. For (feasible) supervision we start with a representation of PK derived as follows. For each $T \in \mathcal{T}$ define $X_T \subseteq X$ according to

$$X_T := \{x \in X | (\exists q \in Q)(q,x) \in T\}$$

namely X_T is the projection of $T \subseteq Q \times X$ on X. Clearly $X_T \neq \varnothing$. To represent PK we adopt the state set

$$\mathcal{X} := \{X_T | T \in \mathcal{T}\}$$

and adjoin the obvious definitions

$$X_0 := \{\xi(x_0,s) | Ps = \epsilon, \delta(q_0,s)!, \xi(x_0,s)!\}$$
$$\mathcal{X}_m := \{X_T | X_T \cap X_m \neq \varnothing, T \in \mathcal{T}\}$$
$$X \stackrel{\sigma}{\longmapsto} X' := \{\xi(x,\sigma) | x \in X, \xi(x,\sigma)!\}, \;\; X \in \mathcal{X}, \;\; \sigma \in \Sigma_o$$

In the above definitions, check that X_0 and (when nonempty) X' are indeed elements of $\mathcal{X}$. Writing $\Xi : \mathcal{X} \times \Sigma_o \to \mathcal{X}$ for the transition rule as just defined, let

$$\mathcal{S}_\mathbf{o} = \{\mathcal{X}, \Sigma_o, \Xi, X_0, \mathcal{X}_m\}$$

Supply the details of checking that $\mathcal{S}_\mathbf{o}$ represents PK. In general—and this is significant—$\mathcal{S}_\mathbf{o}$ will *not* be a minimal-state representation of PK (in the Nerode sense).[3]

[3] It can be replaced by one only if K is *strongly* observable.

It remains to extend $\mathcal{S}_0$ to a feasible supervisor (over Σ), say $\mathcal{S}$. For this it is enough to self-loop each state X of $\mathcal{S}_0$ by $\sigma \in \Sigma - \Sigma_o$, exactly when there exists $x \in X$ such that $\xi(x, \sigma)!$. It is clear that $\mathcal{S}$ is feasible, and the reader may complete the detailed verification that $\mathcal{S}$ has the fundamental supervisory properties required, namely

$$L_m(\mathbf{G}) \cap L_m(\mathcal{S}) = K$$
$$L(\mathbf{G}) \cap L(\mathcal{S}) = \bar{K}$$

Example 9: Consider the plant **G** and specification **E**, defined over $\Sigma = \{0, 1, 2\}$, as shown below.

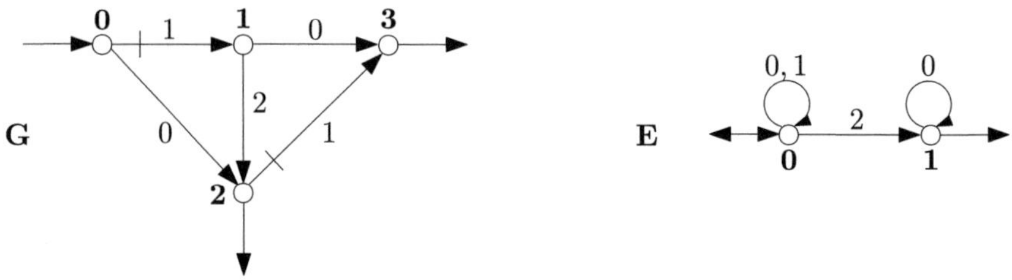

Thus $L(\mathbf{G}) = \{\epsilon, 0, 1, 0.1, 1.0, 1.2, 1.2.1\}$. Computing

$$\mathbf{K} = \mathbf{minstate}(\mathbf{supcon}(\mathbf{G}, \mathbf{E}))$$
$$\mathbf{K}(.\mathbf{DAT}) = \mathbf{condat}(\mathbf{G}, \mathbf{K})$$

we obtain:

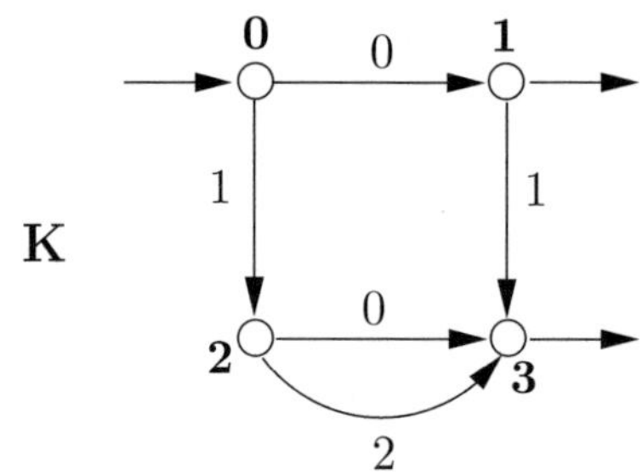

K(.DAT)

State	Disabled event
3	1

Let the observable event subset $\Sigma_o = \{0, 2\}$. We shall verify that **K** is observable, and obtain the corresponding feasible supervisor.

Referring to Exercise 6.2.13, we first record the state sets Q, X of **G**, **K**, respectively:

$$Q = \{0, 1, 2, 3\}, \ X = \{0, 1, 2, 3\}$$

Corresponding to the possible observed strings

$$PL(\mathbf{G}) = \{\epsilon, 0, 2\}$$

we have the partition induced by look-alike strings:

$$\ker(P|L(\mathbf{G})) = \{\{\epsilon, 1\}, \{0, 0.1, 1.0\}, \{1.2, 1.2.1\}\}$$

The resulting state pair subsets of $Q \times X$ determine the family

$$\mathcal{T} = \{T_0, T_1, T_2\}$$

where

$$T_0 = \{(0, 0), (1, 2)\},\ T_1 = \{(2, 1), (3, 3)\},\ T_2 = \{(2, 3)\}$$

We also need the state subsets

$$Q_j := \{q \in Q|\delta(q, j)!\},\ X_j := \{x \in X|\xi(x, j)!\},\ j = 0, 1, 2$$

namely

$$Q_0 = \{0, 1\},\ X_0 = \{0, 2\}$$
$$Q_1 = \{0, 2\},\ X_1 = \{0, 1\}$$
$$Q_2 = \{1\},\ X_2 = \{2\}$$

along with

$$Q_m = \{2, 3\},\ X_m = \{1, 3\}$$

	σ		0		1		2		mark	
$\mathcal{T}$	q	x	Q_0	X_0	Q_1	X_1	Q_2	X_2	Q_m	X_m
T_0	0	0	$\surd$	$\surd$	$\surd$	$\surd$	$\times$	$-$	$\times$	$-$
	1	2	$\surd$	$\surd$	$\times$	$-$	$\surd$	$\surd$	$\times$	$-$
T_1	2	1	$\times$	$-$	$\surd$	$\surd$	$\times$	$-$	$\surd$	$\surd$
	3	3	$\times$	$-$	$\times$	$-$	$\times$	$-$	$\surd$	$\surd$
T_2	2	3	$\times$	$-$	$\surd$	$\times$	$\times$	$-$	$\surd$	$\surd$

Observability can be verified by inspection of the table displayed. A cell in column Q_j (resp. X_j) is checked ($\surd$) if in the corresponding row *both* $q \in Q_j$ *and* $x \in X_j$ $(j = 0, 1, 2)$. If $q \in Q_j$ but $x \notin X_j$, the corresponding X_j-cell is crossed ($\times$); while if $q \notin Q_j$ its cell is crossed ($\times$) and the corresponding X_j-cell is considered 'don't care' ($-$). If $q \in Q_m$ then its Q_m-cell is checked ($\surd$); the corresponding X_m-cell is checked ($\surd$) if $x \in X_m$ or crossed ($\times$) if $x \notin X_m$. Finally if $q \notin Q_m$ its cell is crossed ($\times$) and the X_m-cell is labeled ($-$).

The observability condition depends only on $T \in \mathcal{T}$ with $|T| \geq 2$, here just T_0, T_1. Within each such T, it must hold (for observation consistency) that if *some* (q, x)-cell pair is checked $(\surd, \surd)$ then *every* cell pair (q', x') with q' checked $(\surd)$ must have x' checked $(\surd)$. This condition clearly holds for the table shown, from which we conclude that the language $L_m(\mathbf{K})$ is $(\mathbf{G}, P)$-observable.

The desired feasible supervisor is obtained as in Exercise 8. Our representation $\mathcal{S}_\mathbf{o}$ of $PL_m(\mathbf{K})$ will be based on the projected state subsets

$$X_{T_0} = \{0, 2\},\ X_{T_1} = \{1, 3\},\ X_{T_2} = \{3\}$$

which we denote by Y_0, Y_1, Y_2 say. So the initial state (corresponding to $Ps = \epsilon$) is Y_0, and the marker subset is $\mathcal{Y}_m = \{Y_1, Y_2\}$. The transition function (Ξ in Exercise 8) is then

$$Y_0 \stackrel{0}{\longmapsto} \{1, 3\} = Y_1,\ Y_0 \stackrel{2}{\longmapsto} \{3\} = Y_2$$

and is undefined otherwise. Thus we have

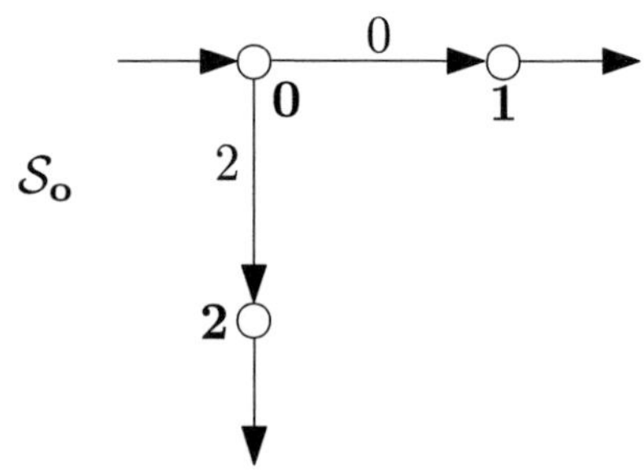

It remains to extend $\mathcal{S}_\mathbf{o}$ (over Σ_o) to a feasible supervisor $\mathcal{S}$ (over Σ). As $\Sigma - \Sigma_o = \{1\}$ and (as above) $X_1 = \{0, 1\}$, we see that $Y_0 (= \{0, 2\})$ and $Y_1 (= \{1, 3\})$ should be self-looped by $\sigma = 1$, but not $Y_2 (= \{3\})$. Our final result for $\mathcal{S}$ is displayed below.

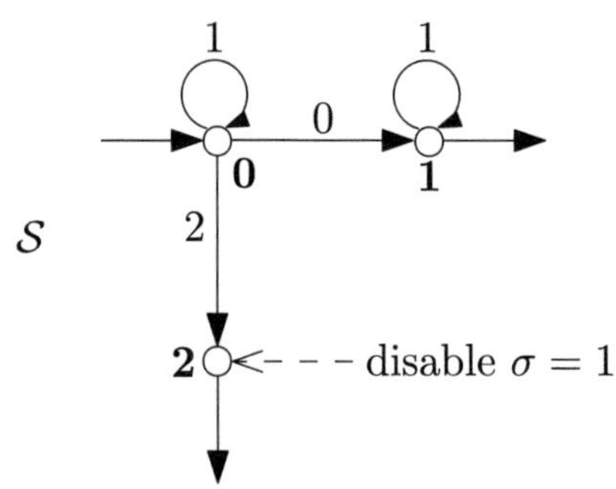

The reader can easily check that $\mathcal{S}$ as supervisor for $\mathbf{G}$ synthesizes $L_m(\mathbf{K})$.

Exercise 10: With $\mathbf{G}, \mathbf{E}, \mathbf{K}$, and P as in Example 9, use the method of uncertainty sets (Example 4) to carry through the design of a feasible supervisor corresponding to

$$\mathbf{SUPER} = \mathbf{supcon}(\mathbf{G}, \mathbf{E})$$

Also investigate the reduced versions, say **SIMK, SIMSUP** of **K, SUPER**, respectively, computed as usual using **supreduce**.

Example 11: **Relative observability in feasible supervision**

Let **G** be given as usual, together with a specification language E, and a natural projection P defining observable events. Taking the ambient language for relative observability to be $\bar{K}$, where K is the supremal controllable sublanguage of $E \cap L_m(\mathbf{G})$, we may consider the two sublanguage families: (1) controllable sublanguages of K and (2) relatively K-observable sublanguages of K. Since both families contain $\emptyset$ and are closed under arbitrary unions, the same is true of their intersection, and therefore there exists the supremal controllable relatively observable sublanguage of K, say H. With **G** finite and E regular, it can be shown that H is regular. With E represented by DES **E**, a representation **H** of H can be computed in *TCT* as

H = supconrobs(G, E, null(P)**/image**(P))

Use **supconrobs** to explore a range of examples of supervisory control under partial observation, such as those considered in this text so far. Of interest is Small Factory (Example 3.3.19), taking event α_2 (in *TCT*, 21) as unobservable. In your examples, appraise the price paid for event unobservability in terms of diminished performance.

Example 12: **Local observability for distributed control**

We consider a rudimentary decentralized control problem which is not solvable unless the agents involved are provided with a mechanism to communicate. By repeated observability checking, the 'channel' by which they communicate is found as part of the problem solution. We develop the problem as a *TCT* computation, providing an annotated MAKEIT.TXT file.

Let DES **G1** and **G2** be independent agents, that is, have disjoint alphabets. It is assumed that **G1** and **G2** can observe and control only events in their respective alphabets.

Suppose that **G1** can execute a sequence of 3 actions, namely the string 10.12.11, while **G2** can execute only a single action 20.

G1 = **create(G1**,[mark all],[tran [0,10,1],[1,12,2],[2,11,3]]) (4,3)
G2 = **create(G2**,[mark all],[tran [0,20,1]]) (2,1)

We may write Σ_1, Σ_2 to refer to the alphabets of **G1, G2** or their later enhanced versions. To start with, $\Sigma_1 = \{10, 11, 12\}$, $\Sigma_2 = \{20\}$. In fact only Σ_1, but not Σ_2, will be changed later.

SPEC declares that **G1** should execute its third action 11 iff **G2** executes 20 before **G1** executes its first action 10.

SPEC = **create(SPEC**,[mark all],[tran [0,10,1],[0,20,2],[1,20,3],[2,10,4], [4 ,11,4]]) (5,5)

This global supervision problem is solved as usual.

PLANT = **sync(G1,G2)** (8,10) Blocked_events = None
ALLPLANT = **allevents(PLANT)** (1,4)
SPEC = **sync(SPEC,ALLPLANT)** (5,10) Blocked_events = None
SUPER = **supcon(PLANT,SPEC)** (9,9)
MSUPER = **minstate(SUPER)** (8,9)
MSUPER = **condat(PLANT,MSUPER)** Controllable
MSUPER disables 11 at states [3] and [6]
SIMSUP = **supreduce(PLANT,MSUPER,MSUPER)** (3,7;slb=2)
SIMSUP = **condat(PLANT,SIMSUP)** Controllable

It is clear that the supervisor **SUPER** or its reduced version **SIMSUP** cannot be implemented locally by **G1** on the basis of its current state, as **G1** has no means of sensing whether or not the priority condition embodied in the specification is satisfied. To make local supervision possible by **G1**, information must be transferred in some manner to **G1** from **G2**. We know that **G1** can locally synthesize only behavior that is controllable and observable within its own structure. Use of this fact will lead us to a mechanism for the required information transfer.

We begin by testing observability of **MSUPER** with respect to each of the alphabets seen (i.e. owned) by **G1** and **G2**, respectively, namely Σ_1, Σ_2.

obs(PLANT,MSUPER,image[10,11,12]) = *false*

The diagnostic file for observability failure displays

$$3:11 \qquad 6:11$$

namely observability fails at states [3] and [6] of **MSUPER** for event 11.

obs(PLANT,MSUPER,image[20]) = *false*

In this case the diagnostic file for observability failure displays

$$6:11$$

or (again) that observability fails at state [6] of **MSUPER** for event 11.

Since event 20 (the sole event of **G2**) does not appear as a failure event in either diagnostic file, we conclude that **MSUPER** is 'locally' observable with respect to **G2** but not **G1**. For the latter, observability fails at **MSUPER** states [3] and [6], where 11 is disabled. A 'witness' pair of **MSUPER** strings confirming observability failure at [3] is

Witness: s = [0]10.12[3]x11, s' = [0]20.10.12[7]v11

Here the notation [3]x11 means '11 is disabled at [3]', while [7]v11 means '11 is enabled at [7]'. Obviously $P_1(s) = P_1(s')$, where P_1 is natural projection with **image**[10,11,12] = Σ_1. Thus **G1** cannot distinguish s, s', and therefore cannot execute conflicting control decisions based on their observation via P_1.

To resolve the situation at state [3], we note that [3] is entered only by event 12, so we first split 12 into 121 and 123. This will allow **G1** to distinguish the string 10.123, following which 11 will be disabled, from the string 20.10.121, after which 11 will be enabled.

We bring in the appropriate enhancements to **PLANT** and **SPEC**.

PLANT1 = **edit(PLANT**,[trans +[1,121,3],+[1,123,3],+[4,121,6],
+[4,123,6],−[1,12,3],−[4,12,6]]) (8,12)
ALLPLANT1 = **allevents(PLANT1**) (1,5)

We check that the corresponding enhancement to **G1** leads to the same result.

EG1 = **edit(G1**,[trans +[1,121,2],+[1,123,2],−[1,12,2]]) (4,4)
EPLANT1 = **sync(EG1,G2**) (8,12) Blocked_events = None
isomorph(PLANT1,EPLANT1;[[3,4],[4,3],[5,6],[6,5]]) = *true*

Similarly we enhance **SPEC**.

SPEC1 = **create(SPEC1**,[mark all],[tran [0,10,1],[0,20,2],[1,20,3],
[2,10,4],[4,11,4]]) (5,5)
SPEC1 = **sync(SPEC1,ALLPLANT1**) (5,15) Blocked_events = None
SUPER1 = **supcon(PLANT1,SPEC1**) (9,12)
MSUPER1 = **minstate(SUPER1**) (8,12)
QCMSUPER1 = **supqc(MSUPER1,image**[10,11,20]) (5,5) Deterministic

The behavior of **MSUPER1** projects[4] exactly to specification, as required.

Next we retest local observability.

obs(PLANT1,MSUPER1,image[10,11,121,123]) = *false*
obs(PLANT1,MSUPER1,image[20]) = *false*

Local observability holds with respect to Σ_2, but fails with respect to Σ_1: [4]x11, [6]x11.

Witness: s = [0]10.(121/123)[4]x11, s' = [0]20.10.(121/123)[7]v11

Cure: Disable 121 at [1] in s, i.e. delete [1,121,4]; disable 123 at [5] in s', i.e. delete [5,123,7].

SUPER2 = **edit(SUPER1**,[trans −[1,121,4],−[5,123,7]]) (9,10)
SUPER2 = **minstate(SUPER2**) (8,10)

Continue with a new check for local observability which, unsurprisingly, again fails with respect to Σ_1.

obs(PLANT1,SUPER2,image[10,11,121,123]) = *false*
obs(PLANT1,SUPER2,image[20]) = *false*

Local observability holds with respect to Σ_2, but fails with respect to Σ_1: [1]x121, [5]x123, [6]x11.

Witness: s = [0]10[1]x121v123, s' = [0]20.10[5]v121x123

Cure: Create channel in which initial occurrence of 20 enables 121 and disables 123, while initial occurrence of 10 disables 121 and enables 123. By this 'semantic' split the channel becomes equivalent to an encoder, or translator, from **G2** to **G1**.

[4]For **supqc** see Sect. 6.7.

CHNL2 = **create**(**CHNL2**,[mark all],[tran [0,10,1],[0,20,2],[1,20,1],
[1,123,1],[2,10,2],[2,121,2]]) (3,6)

If we attempt to treat **CHNL2** as an auxiliary controller, we fail to solve the local control problem, because of course **CHNL2** is not locally observable. We are therefore forced to 'hardwire' **CHNL2** into **PLANT1** as an auxiliary *encoder* as distinct from a *controller*. Just as 'controllability' is the basis of *controller* design, so 'observability' can be seen here as the basis of *encoder* design. 'Controlling' and 'encoding' can be considered as operational dualities reflecting the structural dualities of controllability and observability.

The hardwiring yields

PLANT2 = **sync**(**PLANT1**,**CHNL2**) (11,12) Blocked_events = None
SPEC2 = **edit**(**SPEC1**) (5,15)
SUPER3 = **supcon**(**PLANT2**,**SPEC2**) (9,9)
MSUPER3 = **minstate**(**SUPER3**) (8,9)
isomorph(**SUPER3**,**MSUPER3**) = *false*
QCMSUPER3 = **supqc**(**MSUPER3**,**image**[10,11,20]) (5,5) Deterministic
MSUPER3 projects to plant behavior in accordance with specification, as required.
obs(**PLANT2**,**MSUPER3**,**image**[10,11,121,123]) = *true*
obs(**PLANT2**,**MSUPER3**,**image**[20]) = *false*
MSUPER3 is locally observable with respect to both Σ_1 and Σ_2. The goal has been achieved.

It remains to confirm that the projection of **MSUPER3** on Σ_1 leads to a correct local supervisor.

P1MSUPER3 = **project**(**MSUPER3**,**image**[10,11,121,123]) (4,4)
P2MSUPER3 = **project**(**MSUPER3**,**image**[20]) (2,1)

The controlled behavior of **EG1** under **P1MSUPER3** is given by

SUPEG1 = **meet**(**EG1**,**P1MSUPER3**) (5,4)

We compute the overall system behavior **SYSTEM2** from that of the controlled **EG1**, together with **CHNL2** and **G2**.

SYSTEM2 = **sync**(**SUPEG1**,**G2**) (10,13) Blocked_events = None
SYSTEM2 = **sync**(**SYSTEM2**,**CHNL2**) (9,9) Blocked_events = None
QCSYSTEM2 = **supqc**(**SYSTEM2**,**image**[10,11,20]) (5,5) Deterministic

Overall system behavior is correct, in accordance with **SPEC2**.

TESTEG1 = **supcon**(**EG1**,**SUPEG1**) (5,4)
isomorph(**TESTEG1**,**SUPEG1**) = *true*
SUPEG1 = **condat**(**EG1**,**SUPEG1**) Controllable
SIMEG1 = **supreduce**(**EG1**,**SUPEG1**,**SUPEG1**) (2,4;slb=2)
SIMEG1 = **condat**(**EG1**,**SIMEG1**) Controllable
TEST1 = **meet**(**EG1**,**SIMEG1**) (5,4)
isomorph(**TEST1**,**SUPEG1**) = *true*

Thus the reduced local supervisor for **EG1** is confirmed correct.

This problem works out satisfactorily as an example of top-down encoder design starting from the initial global optimal supervisor. The principles are that (1) event splitting is used to provide an adequate information structure for achieving local observability, and then (2) the requisite encoding to resolve unobservability is implemented by a channel. This entity must (3) be controllable, namely no further control action is required to implement it, and (4) be hardwired into the plant, to represent exactly the intercommunication action between the independent plant components. Thus *channels* play a new role, alongside (local) *controllers*. Essentially, channels are encoders through which distinct agents communicate. What they encode is event symbol strings from one or both alphabets into some event of the other alphabet.

This might suggest an alternative approach to the present example, where the channel simply encodes event 20 in **G2** into a new event in **G1**. It is easily seen, however, that this maneuver will not work, as what is needed is an encoding into yes/no of whether or not event 20 occurs before or after event 10. Thus our approach is correct as it stands, with no further economy apparently possible.

6.4 Infimal Closed Observable Sublanguages

Let **G**, P be as before, and let $A \subseteq L(\mathbf{G})$. Consider the class of languages

$$\bar{\mathcal{O}}(A) = \{K \supseteq A | K \text{ is closed and } (\mathbf{G}, P)\text{-observable}\}$$

We have the following (dual) analog of Theorem 3.5.6.

Theorem 1

$\bar{\mathcal{O}}(A)$ is nonempty and is closed under arbitrary intersections. In particular, $\bar{\mathcal{O}}(A)$ contains a (unique) infimal element [which we denote by inf $\bar{\mathcal{O}}(A)$].

Proof

Clearly $L(\mathbf{G}) \in \bar{\mathcal{O}}(A)$. Let $K_\beta \in \bar{\mathcal{O}}(A)$ for all β in some index set B, and let

$$K = \bigcap\{K_\beta | \beta \in B\}$$

Then K is closed. Suppose $(s, s') \in \ker P$, $s\sigma \in K$, $s' \in K$, and $s'\sigma \in L(\mathbf{G})$. We have for each β that $s\sigma \in K_\beta$ and $s' \in K_\beta$, so $s'\sigma \in K_\beta$. Hence $s'\sigma \in K$, and K is observable. In particular

$$\inf \bar{\mathcal{O}}(A) = \bigcap\{K | K \in \bar{\mathcal{O}}(A)\}$$

□

In general the conclusion of Theorem 1 fails if the observable languages are not closed, nor does it help to require them to be $L_m(\mathbf{G})$-closed.

Example 2

Let $\Sigma = \{\alpha, \beta, \gamma, \delta\}$, $\Sigma_o = \{\beta, \gamma, \delta\}$ and

$$L(\mathbf{G}) = \overline{\delta + \alpha\delta(\beta + \gamma)}, \qquad L_m(\mathbf{G}) = \{\alpha, \delta, \alpha\delta(\beta + \gamma)\}$$

$$K_1 = \alpha + \delta + \alpha\delta\beta, \qquad K_2 = \alpha + \delta + \alpha\delta\gamma$$

The reader may verify that K_1 and K_2 are both $L_m(\mathbf{G})$-closed and observable. Now

$$\bar{K}_1 = \{\epsilon, \alpha, \delta, \alpha\delta, \alpha\delta\beta), \qquad \bar{K}_2 = \{\epsilon, \alpha, \delta, \alpha\delta, \alpha\delta\gamma\}$$

$$K_1 \cap K_2 = \{\alpha, \delta\}, \qquad \overline{K_1 \cap K_2} = \{\epsilon, \alpha, \delta\}$$

$$\bar{K}_1 \cap \bar{K}_2 = \{\epsilon, \alpha, \delta, \alpha\delta\}$$

$$\bar{K}_1 \cap \bar{K}_2 - \overline{K_1 \cap K_2} = \{\alpha\delta\}$$

Taking $s = \epsilon$, $s' = \alpha$ gives $s, s' \in \overline{K_1 \cap K_2}$, while

$$s\delta = \delta \in \overline{K_1 \cap K_2}, \qquad s'\delta = \alpha\delta \in \bar{K}_1 \cap \bar{K}_2 - \overline{K_1 \cap K_2}$$

Thus $K_1 \cap K_2$ is not observable.

Furthermore, in general it is not true that the union of observable languages (closed or not) is observable.

Example 3

Let $\Sigma = \{\alpha, \beta\}$, $\Sigma_o = \{\beta\}$, with

$$L(\mathbf{G}) = \overline{(\epsilon + \alpha)\beta}, \qquad K_1 = \{\alpha\}, \qquad K_2 = \{\beta\}$$

Then K_1 and K_2 are both observable, but for $K = K_1 \cup K_2$ we have

$$\epsilon, \alpha \in \bar{K}, \quad P(\epsilon) = P(\alpha), \quad \epsilon\beta = \beta \in \bar{K}, \quad \alpha\beta \notin \bar{K}$$

and thus K is not observable.

We conclude from these results that the class of observable languages containing a given language (and with no closure requirement), despite its seemingly natural definition from the viewpoint of system theory, is algebraically rather badly behaved. A more satisfactory approach will be described in the section to follow. In the meantime we can, however, solve a problem of optimal supervision that addresses only the closed behavior of the resulting system. The result will be applicable provided nonblocking is not an issue.

Without essential loss of generality, we assume for the remainder of this section that

$$\Sigma_o \cup \Sigma_c = \Sigma$$

namely every event is either observable or controllable. As a consequence, every uncontrollable event is observable. Let A and E be closed sublanguages with

$$A \subseteq E \subseteq L(\mathbf{G})$$

We interpret E as 'legal behavior' and A as 'minimally adequate behavior'. Our objective is:

Obtain a feasible supervisory control V such that

$$A \subseteq L(V/\mathbf{G}) \subseteq E \tag{$*$}$$

First suppose that $A = \varnothing$. The supervisory control V defined by permanently disabling all controllable events is feasible, and it is enough to check that $L(V/\mathbf{G}) \subseteq E$. If $A \neq \varnothing$, bring in the language class $\bar{\mathcal{O}}(A)$ as before, and the class $\bar{\mathcal{C}}(E)$ defined by

$$\bar{\mathcal{C}}(E) = \{K \subseteq E | K \text{ is closed and controllable}\}$$

Recall from Sect. 3.5 that $\bar{\mathcal{C}}(E)$ is closed under arbitrary intersections. We now have the following abstract solvability condition.

Theorem 4 (Feng Lin)

Assume $A \neq \varnothing$. The problem $(*)$ is solvable iff

$$\inf \bar{\mathcal{O}}(A) \subseteq \sup \bar{\mathcal{C}}(E)$$

Proof

(Only if) Let $K = L(V/\mathbf{G})$. Then K is closed. Taking $L_m(\mathbf{G}) = L(\mathbf{G})$ in Corollary 6.3.2 we obtain that K is controllable and observable, so

$$\inf \bar{\mathcal{O}}(A) \subseteq K \subseteq \sup \bar{\mathcal{C}}(E)$$

from which the condition follows.

(If) The family of sublanguages

$$\bar{\mathcal{K}} = \{K' | K' \supseteq \inf \bar{\mathcal{O}}(A) \,\&\, K' \in \bar{\mathcal{C}}(E)\}$$

is nonempty ($\sup \bar{\mathcal{C}}(E)$ belongs). Since $\bar{\mathcal{C}}(E)$ is closed under intersections, the language

$$\hat{K} := \inf \bar{\mathcal{K}} = \bigcap \{K' | K' \in \bar{\mathcal{K}}\}$$

belongs to $\bar{\mathcal{K}}$ and is thus closed and controllable.

Write $\hat{A} = \inf \bar{\mathcal{O}}(A)$. It will be shown that $\hat{K}$ is given explicitly by

$$\hat{K} = \hat{A}\Sigma_u^* \cap L(\mathbf{G})$$

Denote the right side of the proposed equality by $K^\#$. Clearly $K^\#$ is closed and contains $\hat{A}$. Also, $K^\#$ is controllable: for if $s = s_1 s_2$ with $s_1 \in \hat{A}$, $s_2 \in \Sigma_u^*$, $\sigma \in \Sigma_u$, and $s\sigma \in L(\mathbf{G})$, then

$$s\sigma \in \hat{A}\Sigma_u^* \cap L(\mathbf{G}) = K^\#$$

and it is easy to see (by induction on strings) that any closed controllable language containing $\hat{A}$ must contain $K^\#$.

We claim that $\hat{K}$ is even observable. With this established, it only remains to invoke Corollary 6.3.2 for the desired result.

Because $\hat{K}$ is closed, to prove the claim it suffices to show

$$(\forall s, s')(\forall \sigma) s\sigma \in \hat{K} \;\&\; s' \in \hat{K} \;\&\; s'\sigma \in L(\mathbf{G}) \;\&\; Ps = Ps' \Rightarrow s'\sigma \in \hat{K}$$

Taking $\sigma \in \Sigma_u$ in the antecedent yields the result by controllability of $\hat{K}$. Suppose $\sigma \in \Sigma_c$ and assume, for a proof by contradiction, that $s'\sigma \notin \hat{K}$. We must have $s\sigma \in \hat{A}$. Clearly $s'\sigma \notin \hat{A}$, for if $s'\sigma \in \hat{A}$ and $s'\sigma \in L(\mathbf{G})$ then $s'\sigma \in L(\mathbf{G}) \cap \hat{A}\Sigma_u^* = \hat{K}$, contrary to our assumption.

It will be shown that $s' \in \hat{A}$. Otherwise, if $s' \notin \hat{A}$, let $w'\sigma'$ be the longest prefix of s' such that $w' \in \hat{A}$ and $w'\sigma' \notin \hat{A}$: because $\hat{A}$ is nonempty and closed, we have at least $\epsilon \in \hat{A}$, so if $s' \notin \hat{A}$ then $|s'| > 0$, and a prefix of the form described must surely exist. Furthermore $\sigma' \in \Sigma_u$: in fact, the assumption $s' \in \hat{K}$ implies $w'\sigma' \in \hat{K} = \hat{A}\Sigma_u^* \cap L(\mathbf{G})$, and then $w'\sigma' \notin \hat{A}$ requires $\sigma' \in \Sigma_u$ as stated. Now

$$\Sigma = \Sigma_c \cup \Sigma_o$$

implies $\Sigma_u \subseteq \Sigma_o$, so that $\sigma' \in \Sigma_o$. Since $Ps = Ps'$ by hypothesis, there is a prefix $w\sigma'$ of s such that $Pw = Pw'$. Since $s \in \hat{A}$ so is $w\sigma'$. Therefore

$$w\sigma' \in \hat{A}, \quad w' \in \hat{A}, \quad w'\sigma' \in L(\mathbf{G})$$

and by observability of $\hat{A}$ there follows $w'\sigma' \in \hat{A}$. This contradicts the supposition above that $w'\sigma' \notin \hat{A}$. Therefore $s' \in \hat{A}$ after all. Finally we have

$$Ps = Ps', \quad s\sigma \in \hat{A}, \quad s' \in \hat{A}, \quad s'\sigma \in L(\mathbf{G}) - \hat{A}$$

in contradiction to the fact that $\hat{A}$ is observable. The claim is proved, and with it the desired result. □

Example 5

The requirement in Theorem 4 that the relevant languages be closed cannot be dropped. Suppose, for instance, we replace $\bar{\mathcal{C}}(E)$ by $\mathcal{C}(E)$, the family of all controllable sublanguages of E, and replace $\bar{\mathcal{K}}$ by

$$\mathcal{K} = \{K | K \supseteq \inf \bar{\mathcal{O}}(A) \;\&\; K \in \mathcal{C}(E)\}$$

Then inf $\mathcal{K}$ need not exist. As an example, let

$$\Sigma = \Sigma_o = \{\alpha, \beta, \gamma\}, \quad \Sigma_c = \{\beta, \gamma\}, \quad \Sigma_u = \{\alpha\}$$

$$L_m(\mathbf{G}) = \{\epsilon, \alpha\beta, \alpha\gamma\}, \quad A = \{\epsilon\}, \quad E = L_m(\mathbf{G})$$

Since all events are observable, inf $\bar{\mathcal{O}}(A) = A$. Since $\alpha \in A\Sigma_u \cap L(\mathbf{G})$ and $\alpha \notin A$, A is not controllable. Because $L_m(\mathbf{G})$ is controllable, if inf $\mathcal{K}$ exists then $A \subseteq \inf \mathcal{K} \subseteq L_m(\mathbf{G})$. Therefore the possible candidates for inf $\mathcal{K}$ are

$$\{\epsilon, \alpha\beta\}, \quad \{\epsilon, \alpha\gamma\}, \quad \text{or} \quad \{\epsilon, \alpha\beta, \alpha\gamma\}$$

but none of these is infimal.

Example 6

If A and E are not closed, a solution to our problem $(*)$ need not exist, even if A is observable and E is controllable. Let

$$\Sigma = \{\alpha, \beta\}, \quad \Sigma_o = \{\alpha\}, \quad \Sigma_c = \{\beta\}$$

$$L(\mathbf{G}) = \{\epsilon, \alpha, \alpha\beta, \beta, \beta\alpha, \beta\alpha\beta\}, \quad L_m(\mathbf{G}) = L(\mathbf{G}) - \{\epsilon\}$$

We take

$$A = \{\beta\}, \quad E = \{\alpha, \beta, \beta\alpha\beta\}$$

Now $\bar{A} = \{\epsilon, \beta\}$ and $\beta \in L_m(\mathbf{G})$ so $A = \bar{A} \cap L_m(\mathbf{G})$; i.e. A is $L_m(\mathbf{G})$-closed. Also (in the active/inactive set notation of Sect. 6.2)

$$A_A(\epsilon) = \{\beta\}, \, IA_A(\epsilon) = \{\alpha\}$$
$$A_A(\beta) = \varnothing, \, IA_A(\beta) = \{\alpha\}$$

hence A is observable. However, as

$$\beta \in \bar{A}, \quad \alpha \in \Sigma_u, \quad \beta\alpha \in L(\mathbf{G}) - \bar{A}$$

A is not controllable. Next, it can be verified that E is controllable; however, as

$$\alpha, \beta\alpha \in \bar{E}, \, P\alpha = \alpha = P(\beta\alpha)$$
$$A_E(\alpha) = \varnothing, \, IA_E(\alpha) = \{\beta\}$$
$$A_E(\beta\alpha) = \{\beta\}, \, IA_E(\beta\alpha) = \varnothing$$

it follows that E is not observable. Thus neither A nor E is a solution of the problem $(*)$. Finally, if

$$A \subsetneqq K \subsetneqq E$$

then

$$K = K_1 := \{\alpha, \beta\} \quad \text{or} \quad K = K_2 := \{\beta, \beta\alpha\beta\}$$

but neither K_1 nor K_2 is controllable, and we conclude that (∗) is not solvable. On the other hand, if E is replaced by $\bar{E}$ then the problem (∗) is solved by

$$K = \{\epsilon, \alpha, \beta, \beta\alpha\}$$

In general, if E is not closed then (∗) may fail to be solvable simply because E has too few sublanguages.

6.5 Supervisory Control and Normality

As we saw in the previous section, the observability property can be conveniently exploited in supervisory control only when the relevant languages are all closed. Even then, because observability is not preserved under union, in general an optimal (minimally restrictive) supervisory control will not exist. We obtain a better structured problem if we replace observability by the stronger requirement of normality. To this end we set up our problem anew, in such a way that this section is independent of Sects. 6.2, 6.3 and 6.4.

Let the controlled DES $\mathbf{G}$ over $\Sigma = \Sigma_c \dot{\cup} \Sigma_u$ be given, along with the observing agent's projection $P : \Sigma^* \to \Sigma_o^*$. As in Sect. 3.4, define the set of control patterns

$$\Gamma = \{\gamma \in Pwr(\Sigma) | \gamma \supseteq \Sigma_u\}$$

Just as before, we bring in the concept of a supervisory control $V : L(\mathbf{G}) \to \Gamma$. However, V must now 'respect' the observational constraint that control be based purely on the result of observing the strings generated by $\mathbf{G}$ through the channel $\mathbf{C}$, namely on the information transmitted by P. We say V is *feasible* if

$$(\forall s, s' \in L(\mathbf{G}))Ps = Ps' \Rightarrow V(s) = V(s')$$

namely 'look-alike strings result in the same control decision'. Succinctly,

$$\ker(P|L(\mathbf{G})) \le \ker V$$

As to marking, we require as usual that

(i) $L_m(V/\mathbf{G}) \subseteq L_m(\mathbf{G})$

It is natural to require as well that marking 'respect' the observational constraint, namely 'look-alike strings in $L_m(\mathbf{G}) \cap L(V/\mathbf{G})$ are either both marked or both unmarked':

(ii) $(\forall s, s')s \in L_m(V/\mathbf{G})\ \&\ s' \in L_m(\mathbf{G}) \cap L(V/\mathbf{G})\ \&\ Ps' = Ps \Rightarrow s' \in L_m(V/\mathbf{G})$

If both (i) and (ii) hold we shall say that V is *admissible*. Admissibility is related to normality as follows.

Lemma 1

(i) If $L_m(V/\mathbf{G})$ is $(L_m(\mathbf{G}), P)$-normal then V is admissible.
(ii) If $L(V/\mathbf{G})$ is $(L(\mathbf{G}), P)$-normal and V is admissible then $L_m(V/\mathbf{G})$ is $(L_m(\mathbf{G}), P)$-normal. □

Thus if $L(V/\mathbf{G})$ is $(L(\mathbf{G}), P)$-normal then V admissible means that $L_m(V/\mathbf{G})$ is a union of sublanguages of the form $[s] \cap L_m(\mathbf{G})$, with $[s]$ a cell of ker P.

Exercise 2: Prove Lemma 1.

Now let $E \subseteq L_m(\mathbf{G})$ be a specification language. We introduce

Supervisory Control and Observation Problem (SCOP)
Find nonblocking, feasible, admissible V such that

$$\varnothing \neq L_m(V/\mathbf{G}) \subseteq E$$

To investigate SCOP we bring in the following three families of languages.

$$\mathcal{C}(E) := \{K \subseteq E | K \text{ is controllable with respect to } \mathbf{G}\}$$

$$\mathcal{N}(E; L_m(\mathbf{G})) := \{K \subseteq E | K \text{ is } (L_m(\mathbf{G}), P)\text{-normal}\}$$

$$\bar{\mathcal{N}}(E; L(\mathbf{G})) := \{K \subseteq E | \bar{K} \text{ is } (L(\mathbf{G}), P)\text{-normal}\}$$

Each family is nonempty ($\varnothing$ belongs) and is closed under arbitrary unions. Let

$$\mathcal{S}(E) := \mathcal{C}(E) \cap \mathcal{N}(E; L_m(\mathbf{G})) \cap \bar{\mathcal{N}}(E; L(\mathbf{G}))$$

Then $\mathcal{S}(E)$ is nonempty and closed under arbitrary unions, so that sup $\mathcal{S}(E)$ exists in $\mathcal{S}(E)$. Now we can provide a sufficient condition for the solution of SCOP.

Theorem 3
Let $K \neq \varnothing$ and $K \in \mathcal{S}(E)$. Define $V : L(\mathbf{G}) \to \Gamma$ according to:

$$V(s) := \begin{cases} \Sigma_u \cup \{\sigma \in \Sigma_c | P(s\sigma) \in P\bar{K}\} & \text{if } Ps \in P\bar{K} \\ \Sigma_u & \text{if } Ps \in PL(\mathbf{G}) - P\bar{K} \end{cases}$$

and define

$$L_m(V/\mathbf{G}) := L(V/\mathbf{G}) \cap K$$

Then V solves SCOP, with

$$L_m(V/\mathbf{G}) = K$$

Proof

Clearly V is feasible. We first claim that

$$L(V/\mathbf{G}) = \bar{K}$$

Notice that

$$K \neq \varnothing \Rightarrow \bar{K} \neq \varnothing \Rightarrow \epsilon \in \bar{K}$$

To show $L(V/\mathbf{G}) \subseteq \bar{K}$, let $s \in L(V/\mathbf{G})$, $s \in \bar{K}$, and $s\sigma \in L(V/\mathbf{G})$. By definition of $L(V/\mathbf{G})$, we have $s\sigma \in L(\mathbf{G})$ and $\sigma \in V(s)$; $s \in \bar{K}$ implies $Ps \in P\bar{K}$. If $\sigma \in \Sigma_u$ then, since $K \in \mathcal{C}(E)$,

$$s\sigma \in \bar{K}\Sigma_u \cap L(\mathbf{G}) \subseteq \bar{K}$$

If $\sigma \in \Sigma_c$, then $P(s\sigma) \in P\bar{K}$, which implies

$$s\sigma \in L(\mathbf{G}) \cap P^{-1}(P\bar{K}) = \bar{K}$$

since $K \in \bar{\mathcal{N}}(E; L(\mathbf{G}))$.

Next we show $\bar{K} \subseteq L(V/\mathbf{G})$. Let $s \in \bar{K}$, $s \in L(V/\mathbf{G})$, and $s\sigma \in \bar{K}$. Then $Ps \in P\bar{K}$; also $s\sigma \in \overline{L_m(\mathbf{G})} \subseteq L(\mathbf{G})$. If $\sigma \in \Sigma_u$ then $\sigma \in V(s)$. If $\sigma \in \Sigma_c$ then, since $P(s\sigma) \in P\bar{K}$, again $\sigma \in V(s)$. Thus $s \in L(V/\mathbf{G})$, $s\sigma \in L(\mathbf{G})$, and $\sigma \in V(s)$, so $s\sigma \in L(V/\mathbf{G})$, and our claim is proved.

To see that V is nonblocking, note that

$$\begin{aligned} L_m(V/\mathbf{G}) &:= L(V/\mathbf{G}) \cap K \\ &= \bar{K} \cap K \\ &= K \end{aligned}$$

namely $\overline{L_m(V/\mathbf{G})} = L(V/\mathbf{G})$.

Finally, V is admissible, by the fact that

$$L_m(V/\mathbf{G}) = K \in \mathcal{N}(E; L_m(\mathbf{G}))$$

and Lemma 1. □

It is well to note that the replacement of observability by normality will restrict the resulting supervisory control by prohibiting the disablement of any controllable event that happens not to be observable: i.e. only observable events will be candidates for disablement. We may state this fact precisely as follows.

Proposition 4

Let $K \subseteq L(\mathbf{G})$ be controllable. If $\bar{K}$ is $(L(\mathbf{G}), P)$-normal, then

$$(\forall s \in \Sigma^*)(\forall \sigma \in \Sigma)s \in \bar{K} \;\&\; s\sigma \in L(\mathbf{G}) - \bar{K} \Rightarrow \sigma \in \Sigma_o \cap \Sigma_c \qquad \square$$

Exercise 5: Prove Proposition 4.

On a more positive note, observability is tantamount to normality in the pleasant circumstance that all controllable events are observable under P. For convenience we restate Proposition 6.2.11.

Proposition 6

Let $K \subseteq L_m(\mathbf{G})$ be controllable and observable. Assume $P\sigma = \sigma$ for all $\sigma \in \Sigma_c$. Then K is $(L_m(\mathbf{G}), P)$-normal and $\bar{K}$ is $(L(\mathbf{G}), P)$-normal. $\square$

Exercise 7: Prove Proposition 6, under the weaker assumption (suitably formalized) that $P\sigma = \sigma$ for all $\sigma \in \Sigma_c$ except possibly for those σ that are never disabled in the synthesis of K.

Now let $\mathbf{G}_o$ be defined over the alphabet Σ_o, with

$$L_m(\mathbf{G}_o) = PL_m(\mathbf{G}), \quad L(\mathbf{G}_o) = PL(\mathbf{G})$$

Thus $\mathbf{G}_o$ is the 'observer's local model of $\mathbf{G}$'. Let

$$\Sigma_{c,o} := \{\sigma \in \Sigma_c | P\sigma = \sigma\} = \Sigma_c \cap \Sigma_o$$

and

$$\Sigma_{u,o} := \Sigma_u \cap \Sigma_o$$

be, respectively, the controllable and uncontrollable event subsets in $\mathbf{G}_o$. For $E_o \subseteq \Sigma_o^*$, let

$$\mathcal{C}_o(E_o) := \{K_o \subseteq E_o | K_o \text{ is controllable with respect to } \mathbf{G}_o\}$$

Fix a specification language $E \subseteq \Sigma^*$, and bring in languages

$$\begin{aligned} N_o &:= P \sup \mathcal{N}(E; L_m(\mathbf{G})) \\ K_o &:= \sup \mathcal{C}_o(N_o) \\ J &:= P^{-1} K_o \\ K &:= L_m(\mathbf{G}) \cap J \end{aligned}$$

Theorem 8

Assume $\mathbf{G}$ is nonblocking, i.e. $\overline{L_m(\mathbf{G})} = L(\mathbf{G})$. If $L_m(\mathbf{G}), J$ are nonconflicting and $K \neq \varnothing$, then SCOP is solvable with

$$L_m(V/\mathbf{G}) = K$$

Proof

It will be shown that $K \in \mathcal{S}(E)$. For this, let

$$N := \sup \mathcal{N}(E; L_m(\mathbf{G}))$$

Then

$$\begin{aligned} K &\subseteq L_m(\mathbf{G}) \cap P^{-1}(N_o) \\ &= L_m(\mathbf{G}) \cap P^{-1}(PN) \\ &= N \quad \text{(by normality)} \\ &\subseteq E \end{aligned}$$

Also (cf. remark preceding Exercise 6.1.1),

$$K = L_m(\mathbf{G}) \cap P^{-1} K_o$$

implies that K is $(L_m(\mathbf{G}), P)$-normal, i.e.

$$K \in \mathcal{N}(E; L_m(\mathbf{G})) \tag{6.3}$$

Since $L_m(\mathbf{G})$, J are nonconflicting, we have

$$\begin{aligned} \bar{K} &= \overline{L_m(\mathbf{G}) \cap J} \\ &= \overline{L_m(\mathbf{G})} \cap \bar{J} \\ &= L(\mathbf{G}) \cap P^{-1} \bar{K}_o \qquad \text{(by Exercise 6.1.18)} \end{aligned}$$

i.e. $\bar{K}$ is $(L(\mathbf{G}), P)$-normal, namely

$$K \in \bar{\mathcal{N}}(E; L(\mathbf{G})) \tag{6.4}$$

To see that K is controllable, let $s \in \bar{K}$, $\sigma \in \Sigma_u$, and $s\sigma \in L(\mathbf{G})$. Then $Ps \in \bar{K}_o$. If $P\sigma = \sigma$, then

$$(Ps)\sigma = P(s\sigma) \in P(L(\mathbf{G})) = L(\mathbf{G}_o)$$

By $\mathbf{G}_o$-controllability of $\bar{K}_o$, we have $P(s\sigma) \in \bar{K}_o$, i.e.

$$s\sigma \in L(\mathbf{G}) \cap P^{-1}(\bar{K}_o) = \bar{K}$$

If $P\sigma = \epsilon$, then

$$P(s\sigma) = Ps \in \bar{K}_o$$

so again

$$s\sigma \in L(\mathbf{G}) \cap P^{-1}(\bar{K}_o) = \bar{K}$$

as required. Thus

$$K \in \mathcal{C}(E) \tag{6.5}$$

and by (6.3)–(6.5), $K \in \mathcal{S}(E)$. The result now follows by Theorem 3. □

Exercise 9: Under the conditions of Theorem 8, with K as previously defined, show that $K = \sup \mathcal{S}(E)$. **Hint**: By the proof of the theorem, $K \in \mathcal{S}(E)$. If $H \in \mathcal{S}(E)$ use $H \in \mathcal{C}(E) \cap \bar{\mathcal{N}}(E; L(\mathbf{G}))$ to show that $PH \in \mathcal{C}_o(PE)$, and then $H \subseteq K$.

Exercise 10: Write $L_m = L_m(\mathbf{G})$, $L = L(\mathbf{G})$, and $P_L := P|L$; thus $P_L : L \to \Sigma_o^*$. Call P_L an L_m-*observer* if

$$(\forall t \in PL_m, s \in L) Ps \le t \Rightarrow \big((\exists u \in \Sigma^*) su \in L_m \;\&\; P(su) = t\big)$$

In other words, whenever Ps can be extended in Σ_o^* to a string $t \in PL_m$, the underlying string $s \in L$ can be extended to a string $su \in L_m$ with the same projection; the 'local observer's expectation is never blocked in **G**'. Show that P_L is an L_m-observer iff, for every sublanguage $E \subseteq PL_m$,

$$\overline{L_m \cap P^{-1}(E)} = L \cap P^{-1}(\bar{E})$$

In particular, if $K \subseteq L_m$ and P_L is an L_m–observer then

$$\overline{L_m \cap P^{-1}(PK)} = L \cap P^{-1}(P\bar{K})$$

If K is (L_m, P)-normal conclude that $\bar{K}$ is (L, P)-normal, in which case

$$\mathcal{S}(E) = \mathcal{C}(E) \cap \mathcal{N}(E; L_m(\mathbf{G}))$$

With J as defined prior to Theorem 8, show that $L_m(\mathbf{G})$, J are necessarily nonconflicting, and so it is enough to assume in Theorem 8 that **G** is nonblocking and $K \neq \emptyset$.

The foregoing results are brought together in the following (somewhat restricted) step-by-step design method. For a more general algorithm which is always effective, see Exercises 16, 17 below.

TCT **Procedure for SCOP**

0. Given **G**, **E**, and the list **NULL** of P-unobservable events
1. **N := supnorm(G, E, null(NULL))**
2. **NO := project(N, null(NULL))**[5]
3. **GO := project(G, null(NULL))**
4. **KO := supcon(GO,NO)** [proposed 'observer's supervisor']
5. **KODAT := condat(GO,KO)**

[5]Note that steps 1 and 2 could be combined by use of Exercise 6.1.22.

6. **PINVKO** := **selfloop(KO,NULL)**
7. **nonconflict(G,PINVKO)** = *true*?
8. **K** = **meet(G,PINVKO)**
9. **K** nonempty?

If this procedure terminates successfully (with 'yes' at steps 7 and 9), then **PINVKO** provides a solution to SCOP, and **K** is the corresponding controlled behavior. If the procedure fails (or in general), the optimal solution to SCOP in the normality framework can be computed by TCT **supscop** as described in Exercise 16 and can be implemented as in Exercise 17, below.

Example 11: **SCOP for Small Factory**

Take **MACH1**, **MACH2** as in Small Factory (Example 3.3.19), with specification

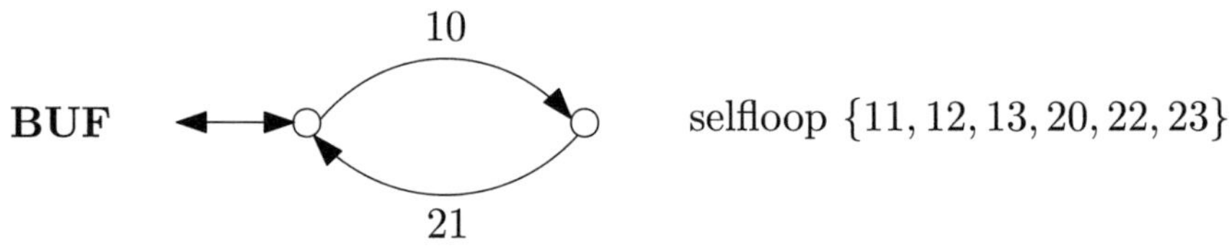

$$\mathbf{FACT} = \mathbf{sync}(\mathbf{MACH1}, \mathbf{MACH2})$$
$$\Sigma_o = \{10, 11, 20, 21\}$$

Thus the unobservable events pertain to breakdown and repair. By the design procedure we obtain the following.

0. **FACT, BUF, NULL** := [12,13,22,23]
1. **N** = **supnorm(FACT, BUF, null(NULL))** (18,42)
2. **NO** = **project(N, null(NULL))** (8,18)
3. **FACTO** = **project(FACT, null(NULL))** (4,12)
4. **KO** = **supcon(FACTO, NO)** (6,11)
5. **KODAT** = **condat(FACTO, KO)**
6. **PINVKO** = **selfloop(KO, NULL)** (6,35)
7. **nonconflict(FACT, PINVKO)** = *true*
8. **K** = **meet(FACT, PINVKO)** (20,37)
9. **K** is nonempty

Thus termination is successful. In the following additional steps we compare our result with the ideal case where all events are observable.

10. **MK** = **minstate(K)** (12,25)
11. **BUFSUP** = **supcon(FACT,BUF)** (12,25)
12. **isomorph(MK,BUFSUP)** = *true*

From this we conclude (rigorously if unsurprisingly) that optimal control of the buffer does not require observation of breakdowns and repairs.

Exercise 12: A possible drawback of the foregoing, restricted *TCT* design procedure for SCOP is that the observer's supervisor, being based on the projected plant model (**GO** in step 4), may fail to account for possible blocking (cf. the discussion in Sect. 2.5). As a result, conflict may result at step 7, whereupon the design procedure fails. For instance, consider the plant model **G** displayed below with alphabet $\Sigma = \{0, 1, 2, 3, 5\}$, in which event 0 is unobservable.

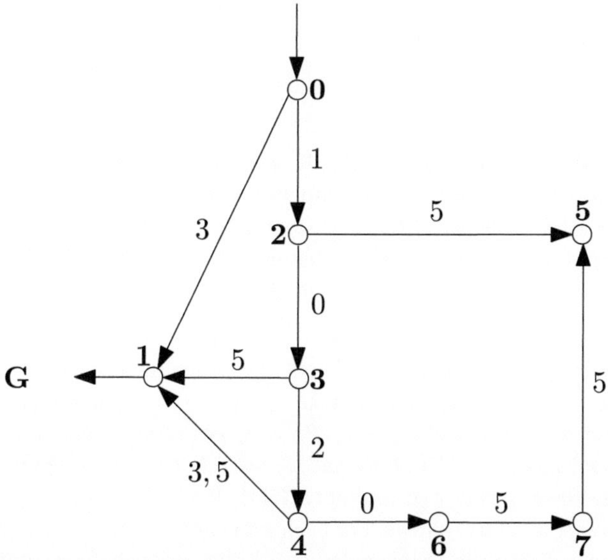

The observer's plant model **GO** = **project**(**G**, **null**[0]) is the following:

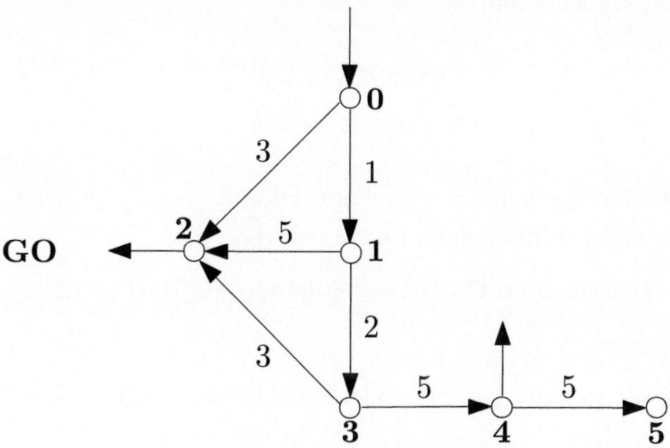

Note that the closed and marked behaviors of **GO** are exactly the projections of the corresponding behaviors of **G** as required. However, **GO** has obliterated the information that the event sequence 1.0.2.0 in **G** leads to a non-coreachable state in **G**. With reference to Exercise 10, verify that here P_L is *not* an L_m-observer.

Attempt the *TCT* design procedure using the specification language Σ^* (i.e. requiring only nonblocking), verify that it fails at step 7, and explain how the proposed supervisor could cause blocking. Start again, enhancing the plant model by selflooping its non-coreachable states (5, 6, 7) with an auxiliary uncontrollable but (hypothetically) observable event (4, say). Repeating the design procedure (with Σ unchanged, so that non-coreachable states are prohibited), verify that it now succeeds, and describe the control action. Finally, generalize this approach for arbitrary specifications, and provide conditions for its validity.

Exercise 13: It is almost obvious *a priori* that the approach of this section will fail for the two-agent problem in Example 6.3.6. Why?

Exercise 14: Given $\mathbf{G}, \mathbf{E}$, and P, take $\mathbf{K} = \mathbf{supcon}(\mathbf{G}, \mathbf{E})$, assumed observable, and proceed as in Example 6.3.4 to compute the feasible supervisor, say **S1**. Also compute **S2** = **PINVKO** as in SCOP. Assuming that both **S1** and **S2** turn out to be nonconflicting with respect to **G**, show that the (marked) controlled behavior using **S1** contains that using **S2**. **Hint**: If V_1, V_2 are the corresponding supervisory controls, show that $L_m(V_1/\mathbf{G}) \supseteq L_m(V_2/\mathbf{G})$.

Exercise 15: For a mild generalization of the setup in this chapter, suppose T is an alphabet disjoint from Σ, but that $P : \Sigma^* \to T^*$ remains catenative, and that for each $\sigma \in \Sigma$, either $P\sigma \in T$ or $P\sigma = \epsilon$. For 'control consistency' assume also that the subsets $T_c := T \cap P\Sigma_c$ and $T_u := T \cap P\Sigma_u$ are disjoint, while $T = T_c \dot{\cup} T_u$. Revise the exposition accordingly. **Hint**: If $\mathbf{G} = (Q, \Sigma, \delta, q_0, Q_m)$, define $\tilde{\mathbf{G}} = (\tilde{Q}, \tilde{\Sigma}, \tilde{\delta}, \tilde{q}_0, \tilde{Q}_m)$ as follows. For each transition $[q, \sigma, q']$ such that $P\sigma \in T$ introduce a new state x, and replace $[q, \sigma, q']$ with two new transitions $[q, \sigma, x]$, $[x, P\sigma, q']$. If X is the set of all such x, let $\tilde{Q} := Q \dot{\cup} X$. Define $\tilde{\Sigma} := \Sigma \dot{\cup} T$, $\tilde{q}_0 := q_0$, $\tilde{Q}_m := Q_m$, and $\tilde{\delta} : \tilde{Q} \times \tilde{\Sigma} \to \tilde{Q}$ (pfn) accordingly. Let $\tilde{\Sigma}_o := T$, and define $\tilde{P} : \tilde{\Sigma} \to \tilde{\Sigma}_o \cup \{\epsilon\}$ according to

$$\tilde{P}\sigma = \epsilon, \ \sigma \in \Sigma$$

$$\tilde{P}\tau = \tau, \ \tau \in T$$

Clearly $\tilde{P}$ determines a (unique) natural projection $\tilde{\Sigma}^* \to \tilde{\Sigma}_o^*$.[6] Finally, consider the (standard) problems of this chapter for the pair $(\tilde{\mathbf{G}}, \tilde{P})$.

Exercise 16: In the notation of SCOP, define the operator $T : Pwr(\Sigma^*) \to Pwr(\Sigma^*)$ according to

$$T(J) := \sup \mathcal{C}(\sup \mathcal{N}(\sup \bar{\mathcal{N}}(J; L(\mathbf{G})); L_m(\mathbf{G}))), \ J \subseteq \Sigma^*$$

Set $K_0 := E$, $K_j = T(K_{j-1})$, $j \geqslant 1$. Show that, provided E and $L_m(\mathbf{G})$ are regular, the chain $K_0 \supseteq K_1 \supseteq K_2 \supseteq \cdots$ is finitely convergent, to

[6]The foregoing steps could be carried out more directly using the *TCT* hierarchical procedures **vocalize** and **higen** described in Chap. 5.

$$K := \sup \mathcal{S}(E)$$

Hint: Use the results of Exercises 6.1.25, 6.1.28, 3.5.11 to obtain estimates of the form 'Ner$(\cdot) \geq$' for $\sup \bar{\mathcal{N}}$, $\sup \mathcal{N}$, $\sup \mathcal{C}$, respectively, and then for $\text{Ner}(K_j)$ $(j \geq 1)$. Proceed as in Exercise 6.1.28 to show that the state sizes $|\text{Ner}(K_j)|$ are bounded and hence the chain $\{K_j\}$ is finitely convergent.

The result of Exercise 16 is implemented in the *TCT* procedure **supscop**. Thus if **E**, **G** and **K** represent, respectively, E, $L_m(\mathbf{G})$ and K, and if **null/image**(P) lists the events erased (or resp. retained) by P, then

$$\mathbf{K} = \mathbf{supscop}(\mathbf{G}, \mathbf{E}, \mathbf{null}(P)/\mathbf{image}(P))$$

Since $\|K\|$ may be exponentially large in $\|E\| \cdot \|L_m(\mathbf{G})\|$, **supscop** is computationally intensive.

Exercise 17: Verify the following 'feedback implementation' of the solution **K** of SCOP computed above. The 'observer's model' of the supervisor, driven by strings Ps at the channel output, can be represented by $\mathbf{PK} := \mathbf{project}(\mathbf{K}, \mathbf{null}(P)/\mathbf{image}(P))$. Also the observer's model of the plant is $\mathbf{PG} = \mathbf{project}(\mathbf{G}, \mathbf{null}(P)/\mathbf{image}(P))$. Thus the observer will disable controllable events on the basis of

$$\mathbf{PK}(.\text{DAT}) = \mathbf{condat}(\mathbf{PG}, \mathbf{PK})$$

Show that this controller is consistent with **K**. **Hint:** Compute

$$\mathbf{QPK} = \mathbf{selfloop}(\mathbf{PK}, \mathbf{null}(P))$$
$$\mathbf{TEST} = \mathbf{meet}(\mathbf{G}, \mathbf{QPK})$$

and verify

$$\mathbf{isomorph}(\mathbf{TEST}, \mathbf{K}) = \textit{true}$$

(possibly after application of **minstate** to the arguments). Justify these steps on the basis of *TCT* semantics.

Exercise 18: **State size reduction under natural observer**

Let L_m, $L := \bar{L}_m \subseteq \Sigma^*$, $\Sigma_o \subseteq \Sigma$, and $P : \Sigma^* \to \Sigma_o^*$ be the natural projection. Suppose P is an L_m-observer, namely

$$(\forall t \in PL_m, s \in L) Ps \leq t \Rightarrow \big((\exists u \in \Sigma^*) su \in L_m \;\&\; P(su) = t\big)$$

Let $\lambda = \text{Ner}(L_m)$, $\pi = \text{Ner}(PL_m)$, and write $\pi \circ P \in \mathcal{E}(\Sigma^*)$ for $\ker(P_\pi \circ P)$. Show that $\lambda \leq \pi \circ P$, and deduce $\|PL_m\| \leq \|L_m\|$.

The following definition will be useful. Given a DES **G** over Σ, a subalphabet $\Sigma_o \subseteq \Sigma$, and the natural projection $P : \Sigma^* \to \Sigma_o^*$, say that P is a *natural observer* for **G** if P is both an $L_m(\mathbf{G})$-observer and an $L(\mathbf{G})$-observer.

Exercise 19: Let **G** be nonblocking, and assume $P : \Sigma^* \to \Sigma_o^*$ is a natural projection. Show that if P is an observer for $L_m(\mathbf{G})$ then it is necessarily an observer for $L(\mathbf{G})$; i.e. P is a natural observer for **G**. Show by example that the conclusion may fail if **G** is blocking.

6.6 Control of a Guideway

The following example illustrates the ideas of this chapter in an intuitively simple setting, as well as the computation using *TCT* of the various DES required. Of course, this 'universal' approach to the example problem is far from being the most efficient; furthermore the piecemeal *TCT* computations could certainly be combined into higher-level procedures if desired.

Stations A and B on a guideway are connected by a single one-way track from A to B. The track consists of 4 sections, with stoplights ($\star$) and detectors (!) installed at various section junctions.

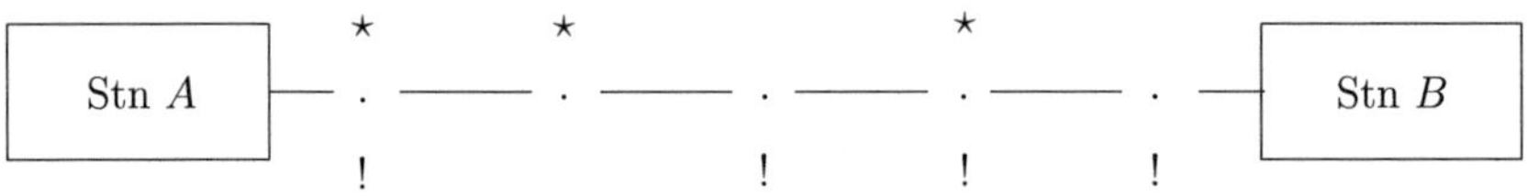

Two vehicles, **V1**, **V2**, use the guideway simultaneously. Either vehicle is in state 0 (at A), state i (while travelling in section i, $i = 1, ..., 4$), or state 5 (at B).

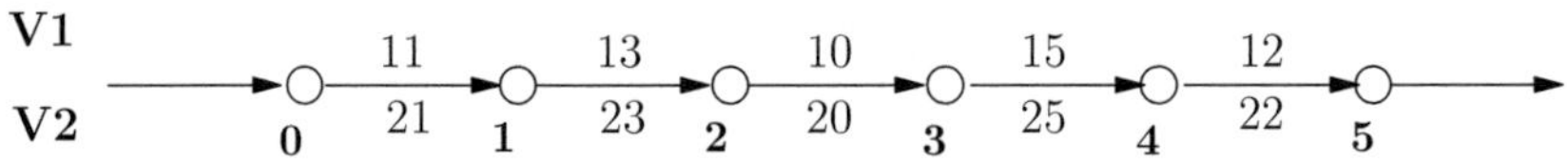

To prevent collision, control of the stoplights must ensure that **V1** and **V2** never travel on the same section of track simultaneously: i.e. the **V**'s are subject to mutual exclusion of the state pairs (i, i), $i = 1, ..., 4$. Controllable events are odd-numbered; the unobservable events are {13, 23}.

By *TCT* the solution can be carried out as follows. Bracketed numbers (m, n) report the state size m and number of transitions n of the corresponding DES.

Following the procedure and notation in Sect. 6.5 (cf. Theorem 6.5.3), steps 0 to 9 compute the plant (generator) **G** = **V**, the legal specification language (generator) **E**, then the proposed feasible supervisor **K**.

0. **create(V1)** (6,5)
   create(V2) (6,5)
   V = sync(V1,V2) (36,60)
   E = mutex(V1,V2,[(1,1),(2,2),(3,3),(4,4)]**)** (30,40)
   NULL = [13,23]
1. **N = supnorm(V,E, null(NULL))** (26,32)

2. **NO = project(N, null(NULL))** (20,24)
3. **VO = project(V, null(NULL))** (25,40)
4. **KO = supcon(VO,NO)** (20,24)
5. **KODAT = condat(VO,KO)**
   [**KODAT** could instead be named **KO**, as in step 4 **KO** is filed as **KO.DES**, but in step 5 the result of **condat** is filed with suffix **.DAT**]
6. **PIKO = selfloop(KO,NULL)** (20,64)
7. **nonconflict(V,PIKO)** = *true*
8. **K = meet(V,PIKO)** (26,32)
9. **K** is nonempty by step 8.

It can be verified that in this example **K** turns out to be isomorphic to **N**.

The supervisory action of **K** can be read from the tabulated transition structure or from the transition graph and is the following (where tsi stands for 'track section i'): if **V2** starts first (event 21), it must enter $ts4$ before **V1** may start (event 11: disabled by light #1). **V1** may then continue into $ts3$ (event 10), but may not enter $ts4$ (event 15: disabled by light #3) until **V2** enters Stn B (event 22). Light #2 is not used. In fact, switching light #2 to red would mean disabling event 13 or 23; but these events are unobservable, while **K** is normal. If all events were observable, supervision could be based on **E**, allowing **V1** to start when **V2** has entered $ts2$. But then **V1** must halt at light #2 until **V2** has entered $ts4$.

The transition graph for **K** when **V2** starts first is displayed in Fig. 6.5 (for **E** adjoin the events shown dashed).

This example illustrates that the replacement of observability by normality as the property to be sought in control synthesis results in general in some 'loss of performance' in the sense of a restriction on control behavior that is not strictly necessary. For brevity write E for $L_m(\mathbf{E})$. We claim first that E is not observable. For let $s = 21$, $s' = 21.23$. The projection P nulls the event set $\{13,23\}$, so $Ps = Ps' = 21$, i.e. $(s, s') \in \ker P$. By inspection of the transition structure of **V1,V2** we see that

$$\begin{aligned}
A_E(s) &= \{\sigma | s\sigma \in \bar{E}\} = \{23\} \\
IA_E(s) &= \{\sigma | s\sigma \in L(\mathbf{V}) - \bar{E}\} = \{11\} \\
A_E(s') &= \{11, 20\} \\
IA_E(s') &= \varnothing
\end{aligned}$$

The fact that $A_E(s') \cap IA_E(s) = \{11\} \neq \varnothing$ proves the claim.

To obtain a controllable and observable sublanguage of E, delete from E the transitions [4,11,7] and [7,20,11], along with their mirror image counterparts [3,21,6], [6,10,10]; call the resulting language COB (Fig. 6.6). It is clear that COB is controllable, since the first transition in each of these pairs is controllable. Now the conditions

$$s, s' \in \overline{\mathrm{COB}}, \quad (s, s') \in \ker P$$

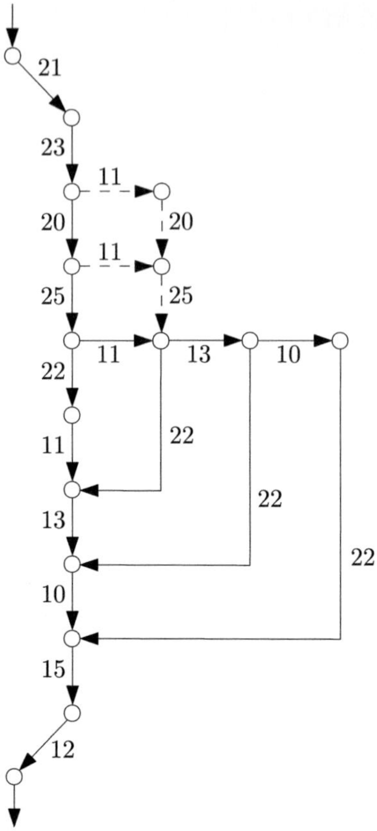

Fig. 6.5 Transition graph for **K** when **V2** starts first (for **E** adjoin events shown dashed)

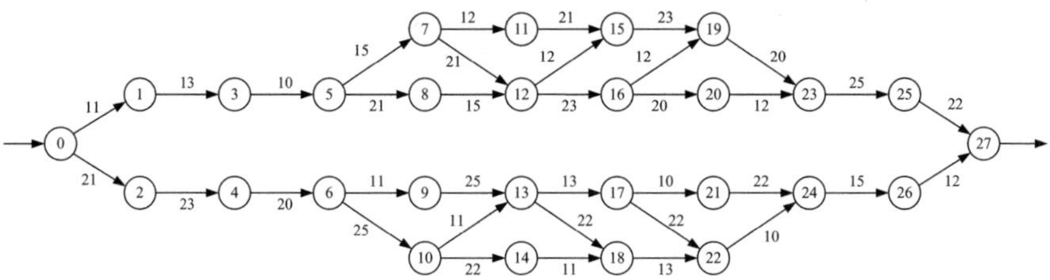

Fig. 6.6 Generator representing COB

plus the assumption that $s \neq s'$ hold (in the displayed graph for E) for

$$s = 21, \quad s' = 21.23$$

or vice versa, and this time we have

$$A_{\text{COB}}(s) = \{23\}, \quad IA_{\text{COB}}(s) = \{11\}$$
$$A_{\text{COB}}(s') = \{20\}, \quad IA_{\text{COB}}(s') = \{11\}$$

for which the null intersection requirement is satisfied; similarly for the remaining pairs $(s, s') \in \ker P$. So COB is $(\mathbf{V}, P)$-observable. Therefore COB can be synthesized by a feasible supervisor; by inspection the supervisory control requires the supervisor to: (i) disable event 11 after 21, keeping 11 disabled until after the next observable event, and (ii) enable 11 but disable 13 after 21.23.20, and so on, as in the synthesis of E. Note that control calls for the disablement of the unobservable event 13, whereas in the synthesis of a closed normal language (cf. K, above) only observable events ever need to be disabled. Further details are provided in Exercise 2. We check directly that $\bar{E}$ is not normal. Let

$$t = 21.23.11 \in \bar{E}$$
$$s = Pt = 21.11 \in L(\mathbf{V})$$

so $s = Ps = Pt \in P\bar{E}$ and $s \in L(\mathbf{V}) \cap P^{-1}(P(\bar{E}))$. But $s \notin \bar{E}$, i.e.

$$\bar{E} \subsetneqq L(\mathbf{V}) \cap P^{-1}(P(\bar{E}))$$

We can also check directly that $\overline{\text{COB}}$ is not normal, by exhibiting a string terminating with an unobservable event that must be disabled. Let

$$s = 21.23.20.11.13$$
$$t = 21.23.20.11$$

Then

$$Ps = 21.20.11$$
$$Pt = 21.20.11$$

Now $t \in \overline{\text{COB}}$ and $Ps = Pt$, so $Ps \in P(\overline{\text{COB}})$, and so

$$s \in P^{-1}P(\overline{\text{COB}}) \cap L(\mathbf{V})$$

but $s \notin \overline{\text{COB}}$. Summarizing, we have

$$K \subsetneqq \text{COB} \subsetneqq E$$

Exercise 1: Complete the detailed verification that COB is $(\mathbf{V}, P)$-observable. In addition show that COB is strongly $(\mathbf{V}, P)$-observable.

Exercise 2: As in Example 6.3.4, apply the method of uncertainty sets to the guideway example of this section, in order to implement the controllable observable lan-

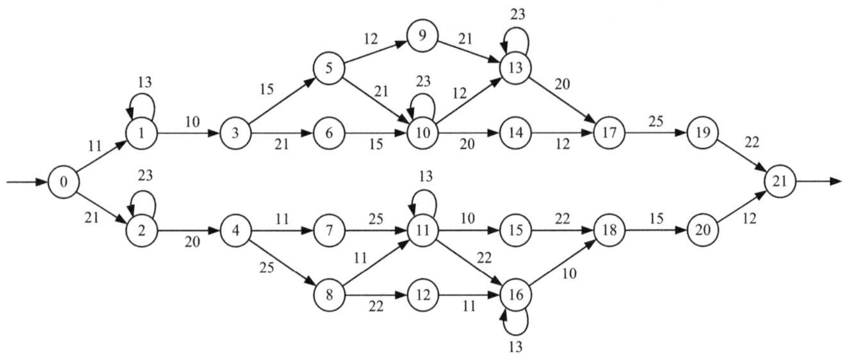

Fig. 6.7 **UCOB**

guage COB by use of the feasible supervisor **UCOB**. **Hint**: With **null**$(P) = [13, 23]$ compute **UCOB** (Fig. 6.7) either directly from the definition or in this case as the projection **PCOB** = **project(COB**, **null**[13,23]), self-looped with event 13 at states 1, 11, 16 and event 23 at 2, 10, 13. To verify correctness let **TEST** = **meet(V**, **UCOB)**, **MTEST** = **minstate(TEST)**, **MCOB** = **minstate(COB)**, and confirm

$$\textbf{isomorph}(\textbf{MTEST}, \textbf{MCOB}) = true$$

Exercise 3: Using *TCT* **supscop**, verify directly the solution **K** computed at step 9, above. Obtain **supscop** solutions for some unobservable event subsets other than $\{13, 23\}$. For what maximal subsets does a (nontrivial) solution of this type exist? Confirm that for any such solution no unobservable controllable event is ever disabled. For each of your solutions carry out the consistency check of Exercise 6.5.17. Finally, compute the reduced supervisors using **supreduce**.

Exercise 4: As in Exercise 3, suppose the only observable events are $\{11, 12, 21, 22\}$. Explain why performance is degraded relative to the original case where the unobservable events were $\{13, 23\}$.

Exercise 5: As above, suppose $\{12, 22\}$ are the only unobservable events. Explain why the result for **supscop** is empty. Is the optimal controlled behavior subject to partial observation (but with no restriction to normality) empty?

Exercise 6: As above, suppose $\{11, 21\}$ are the only unobservable events. Explain why the result for **supscop** is empty. But without the normality restriction there do exist nonempty controlled behaviors subject to partial observation. Show that there are two such maximal behaviors—what are they?

Exercise 7: Referring to Example 6.3.11 on relative observability, explore guideway using *TCT* **supconrobs** with various choices for unobservable events. Compare your results with the corresponding (normality-based) results using **supscop**. As you now know, the latter results prohibit the disablement of unobservable controllable events, whereas the **supconrobs** results impose no such restriction. In some applications, however, this restriction could be considered a desirable safety feature. Illustrate with an example.

6.7 Nondeterminism, Quasi-congruences, and the Observer Property

In this section we extend the definitions of nondeterministic dynamic system and quasi-congruence (Exercise 1.4.15), and nondeterministic generator (Sect. 2.5), to treat partial observation under a natural projection P. This will provide a state-based characterization and computational test for the property that P is an L_m-observer (as defined in Exercise 6.5.10).

Consider initially the (deterministic) DES $\mathbf{G} = (Q, \Sigma, \delta, q_0, Q_m)$. Assume $\mathbf{G}$ is reachable and coreachable. Let $\mu \notin \Sigma$ be a new event label, and bring in a new self-looped transition $[q, \mu, q]$ to 'flag' each $q \in Q_m$ (cf. Remark 3.3.16). We now replace $\mathbf{G}$ with

$$\mathbf{G}' = (Q, \Sigma', \delta', q_0, Q_m)$$

where $\Sigma' := \Sigma \cup \{\mu\}$ and δ' is δ extended to $Q \times \Sigma'$ as just described: thus

$$Q_m = \{q \in Q | \delta'(q, \mu)!\}$$

Clearly $\mathbf{G}'$ is reachable and coreachable. For the time being, to simplify notation we drop primes and employ the usual symbols $\mathbf{G}, \Sigma, \delta$ to stand for their modifications $\mathbf{G}', \Sigma', \delta'$ as just described. We shall return to the original $\mathbf{G}, \Sigma, \delta$ after stating Theorem 11 below.

Next let $\Sigma_o \subseteq \Sigma$ and $P : \Sigma^* \to \Sigma_o^*$ be the natural projection. Define

$$\mathbf{H} = (Q, \Sigma_o, \eta, q_0, Q_m)$$

to be the nondeterministic generator with $\eta : Q \times \Sigma_o^* \to Pwr(Q)$ given by

$$\eta(q, s_o) = \{\delta(q, s) | s \in \Sigma^*,\ \delta(q, s)!,\ Ps = s_o\}$$

Of course, if *not* $\delta(q, s)!$ for every s with $Ps = s_o$ then $\eta(q, s_o) = \varnothing$. Note that, in general,

$$\begin{aligned}\eta(q, \epsilon) &= \{\delta(q, s) | Ps = \epsilon\} \\ &\supsetneqq \{q\}\end{aligned}$$

namely **H** generalizes the definition in Sect. 2.5. In **H**, $\eta(q, \epsilon)$ is the subset of states 'silently reachable' from q in **G**. Evidently **H** is reachable, in the sense that

$$(\forall q \in Q)(\exists s_o \in \Sigma_o^*) q \in \eta(q_0, s_o)$$

For η we have the usual composition property, as follows.

Proposition 1

For all $q \in Q$ and $s_o, t_o \in \Sigma_o^*$,

$$\eta(q, s_o t_o) = \bigcup \{\eta(q', t_o) | q' \in \eta(q, s_o)\} \tag{6.6}$$

Proof

Note first that, P being catenative,

$$\{u \in \Sigma^* | Pu = s_o t_o\} = \{st | s, t \in \Sigma^*, Ps = s_o, Pt = t_o\}$$

With the restriction to those s, t for which $\delta(., .)!$ understood, (6.6) is equivalent to

$$\{\delta(q, st) | Ps = s_o, Pt = t_o\} = \bigcup \left\{ \{\delta(q', t) | Pt = t_o\} \,\middle|\, q' \in \{\delta(q, s) | Ps = s_o\} \right\}$$

For inclusion ($\subseteq$) let $Ps = s_o$, $Pt = t_o$, $q' := \delta(q, s)$. By the composition rule for δ (Sect. 2.5),

$$\delta(q, st) = \delta(\delta(q, s), t) = \delta(q', t)$$

For ($\supseteq$) let $q' \in \{\delta(q, s') | Ps' = s_o\}$, i.e. $q' = \delta(q, s)$ for some s with $Ps = s_o$. Then if $Pt = t_o$,

$$\delta(q', t) = \delta(\delta(q, s), t) = \delta(q, st)$$

□

For economy of notation it will be convenient to write (6.6) in the form of a composition rule

$$\eta(q, s_o t_o) = \eta(\eta(q, s_o), t_o)$$

thus extending the first argument of $\eta(., .)$ to a subset of Q.

Let $\pi \in \mathcal{E}(Q)$ be an equivalence relation on Q and $P_\pi : Q \to Q/\pi =: \bar{Q}$ the corresponding canonical projection. If $R \subseteq Q$ then as usual

$$P_\pi(R) := \{P_\pi(q) | q \in R\} \subseteq \bar{Q}$$

In this case, to emphasize the domain we may write $P_{\pi *}(R) (:= P_\pi(R))$, with

$$P_{\pi *} : Pwr(Q) \to Pwr(\bar{Q})$$

We say that π is a *quasi-congruence* for **H** (with respect to the natural projection P) if, for all $q, q' \in Q$,

$$P_\pi(q) = P_\pi(q') \Rightarrow (\forall \alpha \in \Sigma_o) P_\pi \eta(q, \alpha) = P_\pi \eta(q', \alpha)$$

Thus

$$(\forall \alpha \in \Sigma_o) \ker P_\pi \leq \ker P_{\pi *}(\eta(., \alpha))$$

so there exists a unique 'induced' map

$$\bar{\eta} : \bar{Q} \times \Sigma_o \to Pwr(\bar{Q})$$

such that the diagram below commutes:

$$\begin{array}{ccc}
Q \times \Sigma_o & \xrightarrow{\ \eta\ } & Pwr(Q) \\
P_\pi \downarrow \ \downarrow id & & \downarrow P_{\pi *} \\
\bar{Q} \times \Sigma_o & \xrightarrow{\ \bar{\eta}\ } & Pwr(\bar{Q})
\end{array}$$

Trivially, $\perp$ is a quasi-congruence for **H**; but in general $\top$ is not, by Proposition 2 below.

It is useful to extend the induced map $\bar{\eta}$ to $\bar{Q} \times \Sigma_o^*$, by adjoining

$$\bar{\eta}(\bar{q}, \epsilon) := \bigcup \{P_\pi(\eta(q', \epsilon)) | P_\pi q' = \bar{q}\}, \ \bar{q} \in \bar{Q}$$

Unfortunately the composition rule extends only partially. We have

$$\begin{aligned}
\bar{\eta}(\bar{\eta}(\bar{q}, \epsilon), \alpha) &= \bigcup_{\bar{x}} \{\bar{\eta}(\bar{x}, \alpha) | \bar{x} \in \bar{\eta}(\bar{q}, \epsilon)\} \\
&= \bigcup_{\bar{x}} \left\{ \{P_\pi \eta(x, \alpha) | P_\pi x = \bar{x}\} \;\middle|\; \bar{x} \in \bar{\eta}(\bar{q}, \epsilon) \right\} \\
&= \bigcup_{q'} \bigcup_{x} \left\{ P_\pi \{\eta(x, \alpha) | x \in \eta(q', \epsilon)\} \;\middle|\; P_\pi q' = \bar{q} \right\} \\
&= \bigcup_{q'} \left\{ P_\pi \left\{ \bigcup_{x} \eta(x, \alpha) | x \in \eta(q', \epsilon) \right\} \;\middle|\; P_\pi q' = \bar{q} \right\} \\
&= \bigcup_{q'} P_\pi \{\eta(q', \epsilon\alpha) | P_\pi q' = \bar{q}\} \\
&= \bigcup_{q'} P_\pi \{\eta(q', \alpha) | P_\pi q' = \bar{q}\} \\
&= \bigcup_{q'} \{P_\pi \eta(q', \alpha) | P_\pi q' = \bar{q}\} \\
&= \bar{\eta}(\bar{q}, \alpha)
\end{aligned}$$

In general, however,

$$\bar{\eta}(\bar{\eta}(\bar{q}, \alpha), \epsilon) \supsetneqq \bar{\eta}(\bar{q}, \alpha)$$

Clearly for $s_o \in \Sigma_o^+$ we have by induction on $|s_o|$,

$$\bar{\eta}(\bar{q}, s_o) = P_\pi \eta(q, s_o)$$

for any q with $P_\pi q = \bar{q}$. Thus we have the commutative diagram

$$\begin{array}{ccc} Q \times \Sigma_o^+ & \xrightarrow{\ \eta\ } & Pwr(Q) \\ P_\pi \Big\downarrow \ \Big\downarrow id & & \Big\downarrow P_{\pi*} \\ \bar{Q} \times \Sigma_o^+ & \xrightarrow{\ \bar{\eta}\ } & Pwr(\bar{Q}) \end{array} \tag{6.7}$$

If $t_o \in \Sigma_o^+$, then by induction on $|t_o|$,

$$\bar{\eta}(\bar{q}, s_o t_o) = \bar{\eta}(\bar{\eta}(\bar{q}, s_o), t_o)$$

Note, however, that in general, for $P_\pi q = \bar{q}$,

$$P_{\pi*}\eta(q, \epsilon) \subsetneqq \bar{\eta}(\bar{q}, \epsilon)$$

With π a quasi-congruence for **H**, set $\bar{q}_0 = P_\pi(q_0)$, $\bar{Q}_m = P_\pi Q_m$, and let

$$\bar{\mathbf{H}} = (\bar{Q}, \Sigma_o, \bar{\eta}, \bar{q}_0, \bar{Q}_m)$$

We refer to $\bar{\mathbf{H}}$ as the *reduction of* **H** (mod π).

For $\alpha \in \Sigma_o$, let $E_\alpha = \{q \in Q | \eta(q, \alpha) \neq \varnothing\}$ and $\pi_\alpha = \{E_\alpha, Q - E_\alpha\} \in \mathcal{E}(Q)$. Note that E_α is the subset of states $q \in Q$ either at which **G** can execute α (i.e. $\delta(q, \alpha)!$) or from which such a state is silently reachable:

$$E_\alpha = \{q \in Q | (\exists u \in \Sigma^*) Pu = \epsilon \ \&\ \delta(q, u\alpha)!\}$$
$$Q - E_\alpha = \{q \in Q | (\forall u \in \Sigma^*) Pu = \epsilon \Rightarrow \textit{not}\ \delta(q, u\alpha)!\}$$

Define

$$\pi_o := \bigwedge\{\pi_\alpha | \alpha \in \Sigma_o\}$$

Proposition 2

Let π be a quasi-congruence for **H**. Then $\pi \leq \pi_o$.

Proof

Let $P_\pi(q) = P_\pi(q')$ and $q \in Q - E_\alpha$. Now

$$P_\pi[\eta(q', \alpha)] = P_\pi[\eta(q, \alpha)] = P_\pi(\emptyset) = \emptyset$$

so $\eta(q', \alpha) = \emptyset$, i.e. $q' \in Q - E_\alpha$. Therefore

$$(\forall \alpha \in \Sigma_o) P_\pi(q) = P_\pi(q') \Rightarrow q \equiv q' (\text{mod } \pi_\alpha)$$

namely $\pi \leq \pi_o$ as claimed. □

Proposition 3

Let ψ, ω be quasi-congruences for **H** with respect to P. Then so is $\pi := \psi \vee \omega$.

Proof

Let $x, y \in Q$ with $P_\pi(x) = P_\pi(y)$, and let $\alpha \in \Sigma_o$. It must be shown that

$$P_\pi \eta(x, \alpha) = P_\pi \eta(y, \alpha)$$

For this let $q_0, q_1, \ldots, q_n \in Q$ with $q_0 = x, q_n = y, q_{i-1} \equiv q_i (\text{mod } \psi \text{ or mod } \omega)$ $(i = 1, \ldots, n)$. Since ψ, ω are quasi-congruences, we have for each i

$$P_\varphi \eta(q_{i-1}, \alpha) = P_\varphi \eta(q_i, \alpha) \tag{6.9}$$

where either $\varphi = \psi$ or $\varphi = \omega$ as appropriate. Let $a \in \eta(x, \alpha)$; we claim there is $b \in \eta(y, \alpha)$ with $P_\pi(b) = P_\pi(a)$. By (6.9) there is $c_1 \in \eta(q_1, \alpha)$ with $c_1 \equiv a (\text{mod } \varphi)$, then $c_2 \in \eta(q_2, \alpha)$ with $c_2 \equiv c_1 (\text{mod } \varphi)$, and finally $b \in P(y, \alpha)$ with $b \equiv c_{n-1} (\text{mod } \varphi)$. Thus $b \equiv a (\text{mod } \psi \vee \omega)$, i.e. $b \equiv a (\text{mod } \pi)$. This proves $P_\pi \eta(x, \alpha) \subseteq P_\pi \eta(y, \alpha)$, and the reverse inclusion follows similarly. □

Proposition 3 extends easily to arbitrary finite joins of quasi-congruences; hence

$$\rho := \sup\{\pi \in \mathcal{E}(Q) | \pi \text{ is a quasi-congruence for } \mathbf{H}\}$$

is the supremal (maximally coarse) quasi-congruence for **H**.

In computing ρ we use the following notation. If $\xi \in \mathcal{E}(Q)$ and $\alpha \in \Sigma_o$, then $\xi \circ \eta(\cdot, \alpha) \in \mathcal{E}(Q)$ is defined by

$$q \equiv q' (\text{mod } \xi \circ \eta(\cdot, \alpha)) \text{ iff } P_\xi \eta(q, \alpha) = P_\xi \eta(q', \alpha)$$

Proposition 4

Let $\rho_0 = \pi_o$ and for $n \geq 1$,

$$\rho_n = \rho_{n-1} \wedge \bigwedge \{\rho_{n-1} \circ \eta(\cdot, \alpha) | \alpha \in \Sigma_o\}$$

Then $\rho_\infty := \lim \rho_n (n \to \infty)$ exists in $\mathcal{E}(Q)$. If $|Q| < \infty$, the limit is achieved in finitely many steps, and in that case $\rho_\infty = \rho$.

Proof

Since $\rho_n \le \rho_{n-1}$, and the lattice $\mathcal{E}(Q)$ is complete, the limit ρ_∞ exists and is given by

$$\rho_\infty = \inf\{\rho_n | n \geqslant 0\}$$

In case $|Q| < \infty$, so is $\mathcal{E}(Q)$, and the ρ_n sequence must converge finitely. In that case, clearly

$$\rho_\infty \le \bigwedge\{\rho_\infty \circ \eta(\cdot, \alpha) | \alpha \in \Sigma_o\}$$

Write $P_\infty : Q \to Q/\rho_\infty$. Then for all $q, q' \in Q$, and all $\alpha \in \Sigma_o$, $P_\infty q = P_\infty q'$ implies

$$P_\infty \eta(q, \alpha) = P_\infty \eta(q', \alpha)$$

namely ρ_∞ is a quasi-congruence for **H**. We claim that $\rho_\infty = \rho$. In fact if π is any quasi-congruence for **H** then by Proposition 2 and induction on n there results $\pi \le \rho_n$ $(n \ge 1)$, so $\pi \le \rho_\infty$; namely ρ_∞ is supremal. □

The proposition shows that ρ can be determined effectively provided, as usual, $|Q| < \infty$ and $|\Sigma| < \infty$.

Let π be any quasi-congruence. By the commutative diagram (6.7),

$$(\forall q, q' \in Q)(\forall s_o \in \Sigma_o^+) P_\pi(q) = P_\pi(q') \Rightarrow P_\pi \eta(q, s_o) = P_\pi \eta(q', s_o) \qquad (6.10)$$

The following provides a useful explicit description of supremal quasi-congruence. First write $\mathcal{P}_o$ for the (nonempty) cells of π_o; thus $\mathcal{P}_o$ partitions Q. Now define the binary relation $\sim$ on Q according to $q_1 \sim q_2$ iff

$$(\forall s_o \in \Sigma_o^+)(\forall C_o \in \mathcal{P}_o) \eta(q_1, s_o) \cap C_o \neq \varnothing \Leftrightarrow \eta(q_2, s_o) \cap C_o \neq \varnothing \qquad (6.11)$$

Thus $q_1 \sim q_2$ iff every string in Σ_o^+ leads both q_1 and q_2 (in **H**) into the same block of cells defining enablement/disablement patterns of the events of Σ_o.

Proposition 5

The relation $\sim$ is a quasi-congruence for **H**.

Proof

Clearly $\sim \in \mathcal{E}(Q)$. Let $q_1 \sim q_2$ and $\alpha \in \Sigma_o$. It must be shown that $\eta(q_1, \alpha) \sim \eta(q_2, \alpha)$; namely for every $x_1 \in \eta(q_1, \alpha)$ there is $x_2 \in \eta(q_2, \alpha)$ with $x_1 \sim x_2$ (and vice versa, which will follow by symmetry). Supposing the contrary, there are $x_1 \in \eta(q_1, \alpha)$, $s_o \in \Sigma_o^+$, and $C_o \in \mathcal{P}_o$ such that

$$\eta(x_1, s_o) \cap C_o \neq \varnothing, \quad (\forall x_2 \in \eta(q_2, \alpha)) \eta(x_2, s_o) \cap C_o = \varnothing$$

or vice versa with indices 1, 2 interchanged. As $\eta(x_1, s_o) \subseteq \eta(q_1, \alpha s_o)$ it follows that $\eta(q_1, \alpha s_o) \cap C_o \neq \varnothing$ whereas

$$\eta(q_2, \alpha s_o) \cap C_o = \left[\bigcup\{\eta(x_2, s_o) | x_2 \in \eta(q_2, \alpha)\}\right] \cap C_o = \varnothing$$

in contradiction to $q_1 \sim q_2$. □

From this we obtain

Proposition 6

Let π be any quasi-congruence for **H**. Then $\pi \leq \sim$, namely $\sim = \rho$.

Proof

Let $q_1, q_2 \in Q$ with $P_\pi q_1 = P_\pi q_2$. We must verify (6.11). Let $s_o \in \Sigma_o^+$. By (6.10) above,

$$P_\pi \eta(q_1, s_o) = P_\pi \eta(q_2, s_o)$$

Thus if $q_1' \in \eta(q_1, s_o)$ there is $q_2' \in \eta(q_2, s_o)$ with $q_1' \equiv q_2' (\text{mod } \pi)$ and so $q_1' \equiv q_2' (\text{mod } \pi_o)$. Thus for all $\alpha \in \Sigma_o$,

$$q_1' \equiv q_2' (\text{mod } \{E_\alpha, Q - E_\alpha\})$$

In particular $q_1' \in E_\alpha$ implies $q_2' \in E_\alpha$, so

$$\eta(q_1, s_o) \cap E_\alpha \neq \varnothing \Rightarrow \eta(q_2, s_o) \cap E_\alpha \neq \varnothing$$

Similarly for all $\beta \in \Sigma_o$, $q_1' \in Q - E_\beta$ implies $q_2' \in Q - E_\beta$, namely

$$\eta(q_1, s_o) \cap (Q - E_\beta) \neq \varnothing \Rightarrow \eta(q_2, s_o) \cap (Q - E_\beta) \neq \varnothing$$

and (6.11) (with $\Rightarrow$) follows. The reverse implication is true by symmetry; thus $\pi = \ker P_\pi \leq \sim$. □

We next discuss nondeterminism. While **H** is formally nondeterministic by definition of η, we can nevertheless distinguish as follows between 'actual' nondeterminism and determinism. We shall say that **H** is *structurally deterministic* if, for all $q \in Q$ and $s_o \in \Sigma_o^*$, we have

$$\eta(q, s_o) \neq \varnothing \Rightarrow |\eta(q, s_o)| = 1$$

namely s_o leads from q to at most one state q'. Otherwise, **H** is *structurally nondeterministic*. Of principal relevance is the application to the reduction of **H** (mod ρ).

Let $\rho \in \mathcal{E}(Q)$ be the supremal quasi-congruence for **H**, as above, and $\bar{\mathbf{H}}$ the reduction of **H** (mod ρ). One can think of $\bar{\mathbf{H}}$ as the canonical form of **H** with respect to quasi-congruence. Since ρ is supremal (maximally coarse), and $\rho = \sim$, no coarser reduction is possible for which (6.11) is valid.

Proposition 7

Let $\bar{\mathbf{H}}$ be structurally deterministic. Then for every $s_o \in \Sigma_o^*$ there is a cell C_o of π_o such that $\eta(q_0, s_o) \subseteq C_o$.

Proof

Assume $\bar{\eta}(\bar{q}_0, s_o) \neq \varnothing$, so that $|\bar{\eta}(\bar{q}_0, s_o)| = 1$. If $s_o = \epsilon$ it follows from

$$\bar{\eta}(\bar{q}_0, \epsilon) = \bigcup \{P_\rho(\eta(q_0', \epsilon)) | P_\rho q_0' = \bar{q}_0\}$$

that

$$P_\rho(\eta(q_0, \epsilon)) = \{\bar{q}_0\}$$

In case $s_o \in \Sigma_o^+$ then

$$\bar{\eta}(\bar{q}_0, s_o) = P_\rho \eta(q_0, s_o) = \{\bar{q}_1\}$$

say, so in either case

$$\eta(q_0, s_o) \subseteq C$$

for some cell C of ρ. Since $\rho \leq \pi_o$, we have $C \subseteq C_o$ for some cell C_o of π_o, as claimed. □

Remark 8: The proposition holds for the reduction of **H** with respect to any quasi-congruence π for **H**.

Proposition 9

Assume $\bar{\mathbf{H}}$ is structurally deterministic, and let $s_o, t_o \in \Sigma_o^*$. If $q \in \eta(q_0, s_o)$ and $\eta(q, t_o) \neq \varnothing$, then $\eta(q', t_o) \neq \varnothing$ for every $q' \in \eta(q_0, s_o)$.

Proof

If $t_o = \epsilon$, $\eta(q', t_o) \supseteq \{q'\}$. Suppose $t_o \in \Sigma_o^+$. By Proposition 7, $\eta(q_0, s_o) \subseteq C_o$ for some cell C_o of π_o, namely $q' \equiv q (\text{mod } \pi_o)$. First assume $t_o = \alpha \in \Sigma_o$. Since $\eta(q, \alpha) \neq \varnothing$ we have $q \in E_\alpha$, hence $q' \in E_\alpha$, so $\eta(q', \alpha) \neq \varnothing$. The general result follows easily by induction on $|t_o|$. □

For the converse to Proposition 7, the restriction to ρ is essential.

Proposition 10

If $\bar{\mathbf{H}}$ is structurally nondeterministic, there are a string $s_o \in \Sigma_o^*$ and distinct cells C_1, C_2 of π_o such that

$$\eta(q_0, s_o) \cap C_1 \neq \varnothing \text{ and } \eta(q_0, s_o) \cap C_2 \neq \varnothing$$

Proof

By hypothesis there are $\bar{q} \in \bar{Q}$ and $v_o \in \Sigma_o^*$ such that $|\bar{\eta}(\bar{q}, v_o)| \geq 2$.

First assume $v_o = \epsilon$, $|\bar{\eta}(\bar{q}, \epsilon)| \geq 2$, where $\bar{q} = P_\rho q$. There are $q' \neq q$ with $P_\rho q' = P_\rho q$, and also $x' \in \eta(q', \epsilon)$ with $P_\rho x' = \bar{x}' \neq \bar{q}$, namely $x' \not\equiv q (\text{mod } \rho)$. Since $\rho = \sim$ (by Proposition 6) we have that (6.11) fails for the pair x', q; namely for some $w_o \in \Sigma_o^+$ and some cell $C_o \in \mathcal{P}_o$, either

$$\eta(q, w_o) \cap C_o \neq \varnothing, \ \eta(x', w_o) \cap C_o = \varnothing \tag{6.12}$$

or (6.12) holds with q, x' interchanged. Assuming (6.12) as displayed, suppose first that $\eta(x', w_o) \neq \varnothing$. There is then $C_o' \in \mathcal{P}_o$, $C_o' \neq C_o$, such that $\eta(x', w_o) \cap C_o' \neq \varnothing$. Since

$$\eta(q', w_o) = \eta(\eta(q', \epsilon), w_o) \supseteq \eta(x', w_o)$$

there follows $\eta(q', w_o) \cap C_o' \neq \varnothing$. As $q' \equiv q (\mathrm{mod}\ \rho)$ and $\eta(q, w_o) \cap C_o \neq \varnothing$ we also have $q' \sim q$ and so

$$\eta(q', w_o) \cap C_o \neq \varnothing$$

By reachability of **H** there is $u_o \in \Sigma_o^*$ such that $q' \in \eta(q_0, u_o)$. Setting $s_o = u_o v_o w_o = u_o w_o$ we obtain

$$\begin{aligned} \eta(q_0, s_o) \cap C_o &= \eta(\eta(q_0, u_o), w_o) \cap C_o \\ &\supseteq \eta(q', w_o) \cap C_o \\ &\neq \varnothing \end{aligned}$$

and similarly $\eta(q_0, s_o) \cap C_o' \neq \varnothing$, as required.

Now suppose that in (6.12), $\eta(x', w_o) = \varnothing$, and let $w_o' < w_o$ be the maximal prefix of w_o such that $\eta(x', w_o') \neq \varnothing$. Then for some $\alpha \in \Sigma_o$ we have $w_o'\alpha \leq w_o$, so $\eta(x', w_o'\alpha) = \varnothing$, namely $\eta(x', w_o') \cap E_\alpha = \varnothing$, or $\eta(x', w_o') \subseteq Q - E_\alpha$. Clearly $\eta(q', w_o) \neq \varnothing$ implies $\eta(q', w_o') \cap E_\alpha \neq \varnothing$; hence there are distinct C_o, C_o' in $\mathcal{P}_o$ such that

$$\eta(q', w_o') \cap C_o \neq \varnothing, \ \ \eta(x', w_o') \cap C_o' \neq \varnothing$$

With w_o' in place of w_o, the argument is completed as before.

If (6.12) holds with q, x' interchanged, we have

$$\eta(q, w_o) \cap C_o = \varnothing, \ \ \eta(x', w_o) \cap C_o \neq \varnothing$$

Suppose $\eta(q, w_o) \neq \varnothing$. There is $C_o' \neq C_o$ such that $\eta(q, w_o) \cap C_o' \neq \varnothing$. Since $q' \sim q$ we have $\eta(q', w_o) \cap C_o' \neq \varnothing$ as well, and the argument is completed as above. If $\eta(q, w_o) = \varnothing$ we proceed using suitable $w_o' < w_o$ just as before.

It remains to treat the case $v_o \in \Sigma_o^+$. Again let $\bar{q}_1, \bar{q}_2 \in \bar{\eta}(\bar{q}, v_o)$ with $\bar{q}_1 \neq \bar{q}_2$. If $\bar{q} = P_\rho(q)$ then, as $v_o \in \Sigma_o^+$,

$$P_\rho(\eta(q, v_o)) \supseteq \{\bar{q}_1, \bar{q}_2\} = \{P_\rho q_1, P_\rho q_2\}$$

for some $q_1, q_2 \in \eta(q, v_o)$. We claim that there exist a string $w_o \in \Sigma_o^*$ and an element $\beta \in \Sigma_o$ such that

$$\eta(q_1, w_o) \cap E_\beta \neq \varnothing \ \text{ and } \ \eta(q_2, w_o) \cap E_\beta = \varnothing$$

or vice versa, with indices 1, 2 interchanged. Otherwise, $q_1 \sim q_2$ and, by Proposition 6, $P_\rho q_1 = P_\rho q_2$ or $\bar{q}_1 = \bar{q}_2$, a contradiction.

We may assume $\eta(q_2, w_o) \neq \varnothing$. Otherwise, for some prefix $w'_o < w_o$ and element $\beta' \in \Sigma_o$ with $w'_o\beta' \leq w_o$, we would have $\eta(q_2, w'_o) \neq \varnothing$ together with

$$\eta(q_1, w'_o) \cap E_{\beta'} \neq \varnothing,\ \eta(q_2, w'_o) \subseteq Q - E_{\beta'}$$

and we could replace w_o with w'_o.

Let

$$q'_1 \in \eta(q_1, w_o) \cap E_\beta \text{ and } q'_2 \in \eta(q_2, w_o) \subseteq Q - E_\beta$$

Clearly q'_1, q'_2 belong to distinct cells C_1, C_2 of π_o. By reachability of $\mathbf{H}$, there is $u_o \in \Sigma_o^*$ with $q \in \eta(q_0, u_o)$. Set $s_o = u_o v_o w_o$. Then $q'_i \in \eta(q_0, s_o)$ $(i = 1, 2)$, namely

$$\eta(q_0, s_o) \cap C_1 \neq \varnothing \text{ and } \eta(q_0, s_o) \cap C_2 \neq \varnothing$$

and the proof is complete. □

We can now identify the property (of P) that $\bar{\mathbf{H}}$ is structurally deterministic with the property that P is an $L_m(\mathbf{G})$-observer. Recall that the latter means

$$(\forall s \in \Sigma^*, t_o \in \Sigma_o^*) s \in L(\mathbf{G})\ \&\ (Ps)t_o \in PL_m(\mathbf{G})$$
$$\Rightarrow (\exists t \in \Sigma^*) Pt = t_o\ \&\ st \in L_m(\mathbf{G})$$

As already assumed, $\mathbf{G}$ is trim (reachable and coreachable) and we now require that the flag event $\mu \in \Sigma_o$. Recall that $q \in E_\mu$ iff

$$(\exists v \in \Sigma^*) Pv = \epsilon\ \&\ \delta(q, v\mu)!$$

Thus $q \in E_\mu$ iff there is q' silently reachable from q where $\delta(q', \mu)!$, namely $q' \in Q_m$.

Let $s \in L(\mathbf{G})$, $s_o := Ps$. Suppose $t_o \in \Sigma_o^*$ with $s_o t_o \in PL_m(G)$, so $s_o t_o = Pu$ for some $u \in L_m(G)$. Thus

$$\eta(q_0, s_o t_o) = \{\delta(q_0, u') | Pu' = s_o t_o\}$$

satisfies $\eta(q_0, s_o t_o) \cap Q_m \neq \varnothing$. Since $Q_m \subseteq E_\mu$, there follows

$$\eta(q_0, s_o t_o) \cap E_\mu \neq \varnothing$$

Now suppose $\bar{\mathbf{H}}$ is structurally deterministic. By Proposition 7 we must have

$$\eta(q_0, s_o t_o) = \eta(\eta(q_0, s_o), t_o) \subseteq E_\mu$$

Since $q := \delta(q_0, s) \in \eta(q_0, s_o)$ there results $\eta(q, t_o) \subseteq E_\mu$. By Proposition 9, $\eta(q, t_o) \neq \varnothing$, hence there exists t' with $Pt' = t_o$ such that $q_1 := \delta(q, t') \in E_\mu$. As noted previously, there exists $v \in \Sigma^*$ with $Pv = \epsilon$ and $\delta(q_1, v) \in Q_m$. On setting $t = t'v$ we have $Pt = Pt' = t_o$ as well as

$$\begin{aligned}\delta(q_0, st) &= \delta(\delta(q_0, s), t) = \delta(q, t'v) \\ &= \delta(\delta(q, t'), v) = \delta(q_1, v) \\ &\in Q_m\end{aligned}$$

namely $st \in L_m(\mathbf{G})$, confirming the observer property for P.

Suppose on the other hand that $\bar{\mathbf{H}}$ is structurally nondeterministic. By Proposition 10 there are a string $u_o \in \Sigma_o^*$ and distinct cells C_1, C_2 of π_o such that

$$\eta(q_0, u_o) \cap C_1 \neq \varnothing \text{ and } \eta(q_0, u_o) \cap C_2 \neq \varnothing$$

hence strings $u_i \in \Sigma^*$ $(i = 1, 2)$ with $Pu_i = u_o$ such that

$$q_1 := \delta(q_0, u_1) \in C_1 \text{ and } q_2 := \delta(q_0, u_2) \in C_2$$

Since $C_1 \neq C_2$ there is $\alpha \in \Sigma_o$ with $C_1 \subseteq E_\alpha$, $C_2 \subseteq Q - E_\alpha$, so

$$\eta(q_1, \alpha) \neq \varnothing, \ \eta(q_2, \alpha) = \varnothing$$

Thus for some v_1 with $Pv_1 = \epsilon$, $q_1' := \delta(q_1, v_1)!$ and $\delta(q_1', \alpha)!$, while for all v_2 with $Pv_2 = \epsilon$, $q_2' := \delta(q_2, v_2)!$ implies *not* $\delta(q_2', \alpha)!$. Since $\mathbf{G}$ is coreachable, there exists w_1 with $\delta(q_1', \alpha w_1) \in Q_m$, namely $\delta(q_0, u_1 v_1 \alpha w_1) \in Q_m$, hence $u_1 v_1 \alpha w_1 \in L_m(\mathbf{G})$, so $u_o \alpha(Pw_1) \in PL_m(\mathbf{G})$. On the other hand for every string $w_2 \in \Sigma^*$ with $\delta(q_2, v_2 w_2)!$ the equality $P(v_2 w_2) = \alpha P(w_1)$ must fail, and thus P is not an $L_m(\mathbf{G})$-observer.

We have now proved the main result of this section.

Theorem 11

Let $\mathbf{G} = (Q, \Sigma, \delta, q_0, Q_m)$ be reachable and coreachable, $\Sigma_o \subseteq \Sigma$, and $P : \Sigma^* \to \Sigma_o^*$ the natural projection. Assume Q_m is flagged by an (observable) event $\mu \in \Sigma_o$. Let $\mathbf{H}$ be defined from $\mathbf{G}$ as in the text and $\bar{\mathbf{H}}$ be the canonical form of $\mathbf{H}$ with respect to quasi-congruence. Then P is an $L_m(\mathbf{G})$-observer iff $\bar{\mathbf{H}}$ is structurally deterministic. □

At this point we recall that the DES $\mathbf{G}$ originally given can be recovered from the current 'working' version by replacing Σ with $\Sigma - \{\mu\}$, Σ_o with $\Sigma_o - \{\mu\}$, and dropping the μ-selfloops from δ. As no change was made in Q_m, it is not difficult to see that the validity of Theorem 11 is not affected, namely also provides a criterion for the $L_m(\mathbf{G})$-observer property of P, now with the original $\mathbf{G}$ and $P : \Sigma^* \to \Sigma_o^*$.

Exercise 12: Justify the above remark in detail.

Exercise 13: In Fig. 6.8, unlabeled transitions are 'silent', namely correspond to strings $s \in \Sigma^*$ with $Ps = \epsilon$. With $\Sigma_o = \{\alpha, \mu\}$ we have

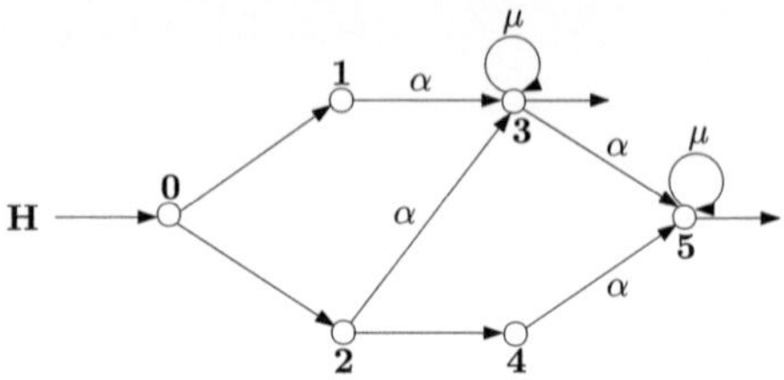

Fig. 6.8 Transition diagram, Exercise 13

$$
\begin{aligned}
E_\alpha &= \{0, 1, 2, 3, 4\},\ Q - E_\alpha = \{5\} \\
E_\mu &= \{3, 5\},\ Q - E_\mu = \{0, 1, 2, 4\} \\
\pi_o &= \{E_\alpha, Q - E_\alpha\} \wedge \{E_\mu, Q - E_\mu\} \\
&= \{\{0, 1, 2, 4\}, \{3\}, \{5\}\} \\
&= \{A, B, C\},\ \text{say}
\end{aligned}
$$

q	$\eta(q,\alpha)$	$\eta(q,\mu)$	$P_{\rho_0}(\eta(q,\alpha))$	$P_{\rho_0}(\eta(q,\mu))$
0	3,5	$\varnothing$	B, C	$\varnothing$
1	3	$\varnothing$	B	$\varnothing$
2	3,5	$\varnothing$	B, C	$\varnothing$
3	5	3	C	B
4	5	$\varnothing$	C	$\varnothing$
5	$\varnothing$	5	$\varnothing$	C

Setting $\rho_0 = \pi_o$, we compute as shown in the table. Then

$$
\begin{aligned}
\rho_1 &= \pi_o \wedge \pi_o \circ \eta(\cdot, \alpha) \wedge \pi_o \circ \eta(\cdot, \mu) \\
&= \{\{0, 1, 2, 4\}, \{3\}, \{5\}\} \\
&\qquad \wedge \{\{0, 2\}, \{1\}, \{3, 4\}, \{5\}\} \\
&\qquad\quad \wedge \{\{0, 1, 2, 4\}, \{3\}, \{5\}\} \\
&= \{\{0, 2\}, \{1\}, \{3\}, \{4\}, \{5\}\} \\
&= \{A, B, C, D, E\},\ \text{say}
\end{aligned}
$$

Repeating the process with ρ_1 we obtain

q	$\eta(q,\alpha)$	$\eta(q,\mu)$	$P_{\rho_1}(\eta(q,\alpha))$	$P_{\rho_1}(\eta(q,\mu))$
0	3,5	$\varnothing$	C, E	$\varnothing$
1	3	$\varnothing$	C	$\varnothing$
2	3,5	$\varnothing$	C, E	$\varnothing$
3	5	3	E	C
4	5	$\varnothing$	E	$\varnothing$
5	$\varnothing$	5	$\varnothing$	E

$$\begin{aligned}\rho_2 &= \rho_1 \wedge \rho_1 \circ \eta(\cdot, \alpha) \wedge \rho_1 \circ \eta(\cdot, \mu)\\ &= \{\{0,2\},\{1\},\{3\},\{4\},\{5\}\}\\ &\quad \wedge \{\{0,2\},\{1\},\{3,4\},\{5\}\}\\ &\qquad \wedge \{\{0,1,2,4\},\{3\},\{5\}\}\\ &= \{\{0,2\},\{1\},\{3\},\{4\},\{5\}\}\\ &= \rho_1\end{aligned}$$

Thus the computation terminates, to yield $\rho = \rho_1$. Using the same cell labels as for ρ_1 we obtain the transition structure of $\bar{\mathbf{H}}$ displayed below.[7] For instance

$$\begin{aligned}\bar{\eta}(A, \epsilon) &= P_\rho(\eta(0, \epsilon)) \cup P_\rho(\eta(2, \epsilon))\\ &= P_\rho(\{0,1,2,4\}) \cup P_\rho(\{2,4\})\\ &= \{A, B, D\}\\ \bar{\eta}(A, \alpha) &= P_\rho(\eta(0, \alpha)) \cup P_\rho(\eta(2, \alpha))\\ &= P_\rho(\{3,5\}) \cup P_\rho(\{3,5\})\\ &= \{C, E\}\end{aligned}$$

$\bar{q}$	$\bar{\eta}(\bar{q}, \epsilon)$	$\bar{\eta}(\bar{q}, \alpha)$	$\bar{\eta}(\bar{q}, \mu)$
A	A, B, D	C, E	$\varnothing$
B	B	C	$\varnothing$
C	C	E	C
D	D	E	$\varnothing$
E	E	$\varnothing$	E

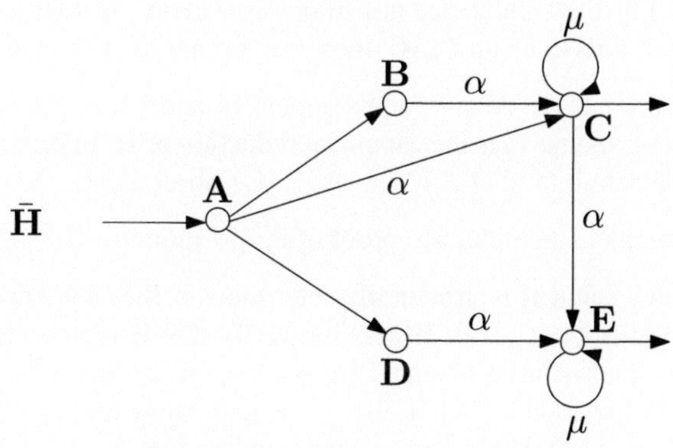

[7] For clarity only direct transitions under ϵ or $\alpha \in \Sigma_o$ are displayed in transition graphs. Thus in the graph for $\bar{\mathbf{H}}$ the compositional transition $\mathbf{A} \to \mathbf{E}$ under α corresponding to $\mathbf{0} \to \mathbf{5}$ under $\epsilon.\epsilon.\alpha$ in $\mathbf{H}$ (and which appears in the table) is omitted, as it can be derived from $\mathbf{A} \xrightarrow{\epsilon} \mathbf{D}$, $\mathbf{D} \xrightarrow{\alpha} \mathbf{E}$ by composition.

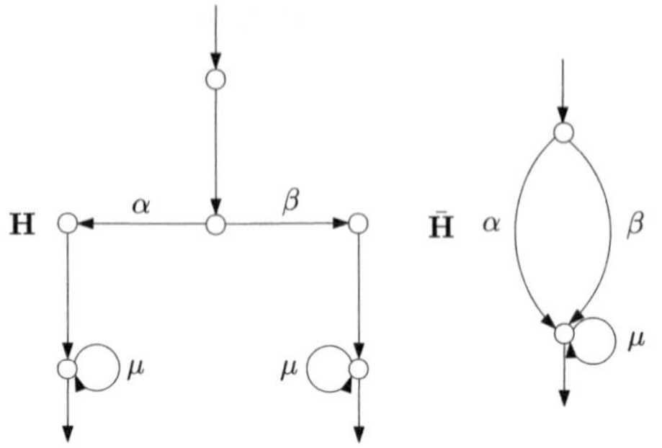

Fig. 6.9 Transition diagram, Example 14

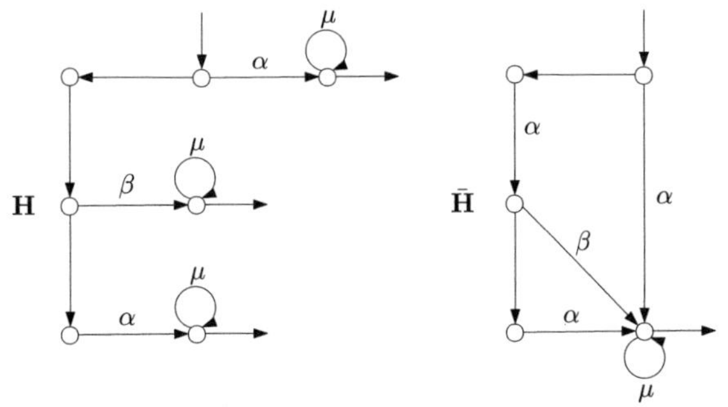

Fig. 6.10 Transition diagram, Example 15

Evidently $\bar{\mathbf{H}}$ is structurally nondeterministic, and P fails to be an $L_m(\mathbf{G})$-observer for any $\mathbf{G}$ with $\mathbf{H}$ as first displayed.

Example 14: In Fig. 6.9, unlabeled transitions of $\mathbf{H}$ are 'silent'. Furthermore, $\bar{\mathbf{H}}$ is structurally deterministic, and P is an $L_m(\mathbf{G})$-observer for any $\mathbf{G}$ with $\mathbf{H}$ as displayed.

Example 15: In Fig. 6.10, unlabeled transitions are 'silent'. $\bar{\mathbf{H}}$ is structurally nondeterministic, and P fails to be an $L_m(\mathbf{G})$-observer for any $\mathbf{G}$ with $\mathbf{H}$ as displayed.

Exercise 16: Work out the details for Examples 14 and 15. In particular calculate π_o and ρ, and confirm that $\bar{\mathbf{H}}$ is the canonical reduction of $\mathbf{H}$. In Example 15, verify directly from the definition that P is not an $L_m(\mathbf{G})$-observer.

Exercise 17: Complete the inductive proof of Proposition 9.

Exercise 18: Show that if $\bar{\mathbf{H}}$ is structurally deterministic, then it is a canonical (state-minimal) generator for $PL_m(\mathbf{G})$. **Hint**: First verify that $\bar{\mathbf{H}}$ represents $PL_m(\mathbf{G})$, so the required generator can be obtained from a suitable congruence, say ν, on the state set $\bar{Q}$ (cf. Proposition 2.5.5). If $P_\nu : \bar{Q} \to \bar{Q}/\nu$ show that $P_\nu \circ P_\rho$ determines a quasi-congruence on Q that is coarser than ρ unless $\nu = \bot$.

Exercise 19: Construct an example of **H** where $\bar{\mathbf{H}}$, the canonical form of **H** with respect to quasi-congruence, is nondeterministic in the sense of having both silent transitions and transitions $[q, \alpha, q'], [q, \alpha, q'']$ with $\alpha \in \Sigma_o$ and $q' \neq q''$.

Exercise 20: **Strong quasi-congruence**

Say that $\varphi \in \mathcal{E}(Q)$ is a *strong quasi-congruence* (sqc) for **H** (with respect to the natural projection $P : \Sigma^* \to \Sigma_o^*$) if, for all $q, q' \in Q$,

$$P_\varphi(q) = P_\varphi(q') \Rightarrow \big(P_\varphi \eta(q, \epsilon) = P_\varphi \eta(q', \epsilon)\big) \;\&\; \big((\forall \alpha \in \Sigma_o) P_\varphi \eta(q, \alpha) = P_\varphi \eta(q', \alpha)\big)$$

Thus every sqc is a quasi-congruence which, in addition, preserves state equivalence under silent drift. Evidently an sqc φ will satisfy the strengthened version of the commutative diagram (6.7) obtained by replacing Σ_o^+ by Σ_o^*:

$$\begin{array}{ccc}
Q \times \Sigma_o^* & \xrightarrow{\;\eta\;} & Pwr(Q) \\
{\scriptstyle P_\varphi}\downarrow \;\; \downarrow{\scriptstyle id} & & \downarrow{\scriptstyle P_{\varphi *}} \\
\hat{Q} \times \Sigma_o^* & \xrightarrow{\;\hat{\eta}\;} & Pwr(\hat{Q})
\end{array}$$

In this diagram $\widehat{Q} = Q/\varphi$ and $\hat{\eta}$ is the map induced by η in the usual way. Show that every sqc φ satisfies $\varphi \leq \pi_o$ and that the supremal sqc, say ψ, always exists and satisfies $\psi \leq \rho$. Compute the corresponding reductions $\widehat{\mathbf{H}}$ for Examples 13–15. Show that if $\bar{\mathbf{H}}$ is structurally deterministic then ρ is an sqc. Finally show that $\widehat{\mathbf{H}}$ is structurally deterministic iff $\bar{\mathbf{H}}$ is.

Canonical (i.e. supremal) quasi-congruence and canonical strong quasi-congruence are computed in *TCT* by procedures **supqc** and **supsqc**, respectively, according to

$$\bar{\mathbf{H}} = \mathbf{supqc}(\mathbf{G}, \mathbf{null}(P)/\mathbf{image}(P))$$

or

$$\bar{\mathbf{H}} = \mathbf{supsqc}(\mathbf{G}, \mathbf{null}(P)/\mathbf{image}(P))$$

Here **G** refers to the standard DES originally given. A silent transition in $\bar{\mathbf{H}}$ is reported with label e. Just as in transition graphs, the *TCT* transition table for $\bar{\mathbf{H}}$ displays direct transitions under ϵ (or e) and $\alpha \in \Sigma_o$, but not compositions. Finally, *TCT* drops μ-selfloops from $\bar{\mathbf{H}}$, simply listing the corresponding states as marked.

Exercise 21: For **G** as shown below, find the canonical forms with respect to quasi-congruence and strong quasi-congruence. Explain the difference between them.

G

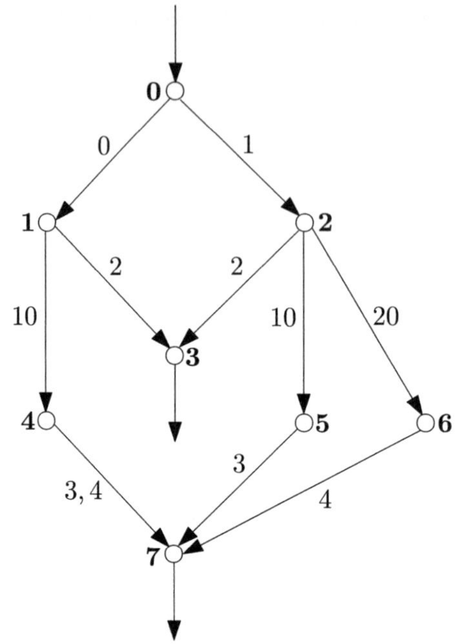

$\mathbf{null}(P) = [10, 20]$

As the most common application of the **supqc** algorithm is to check that a given natural projection is a natural observer, we note here that the prior computation of a natural observer itself can be a significant undertaking. Typically an observer is required whose image event set must contain a specified subset, perhaps the subset of events shared by two DESs (cf. Exercise 6.1.24) or the subset of controllable events listed in a given **condat** table (as in Sect. 6.8 below). We refer to such a listed subset as a Seedlist, used to 'seed' the observer computation. In *TCT* a corresponding full event list can be computed using the procedure **natobs**, according to

$$(\mathbf{OG}, \mathbf{OSEED}) = \mathbf{natobs}(\mathbf{G}, \mathbf{SEED})$$

Here **G** (assumed to be trim) is the DES for which a natural observer is required, while **SEED** and **OSEED**, respectively, represent the given Seedlist and the required natural observer image list in '**allevents**' form (a single state, marked, with all events self-looped). The corresponding projection **OG** is, of course, deterministic provided the result of **natobs** is correct. On that assumption, the ordinary **project** procedure could be used instead. Finally we note that, theoretically, the natural observer for given **G** and Seedlist need not be unique, and (for practical computation) the result of **natobs** need not be of minimal event size; but within these limitations **natobs** is often of practical utility.

Exercise 22: **Complexity of supqc algorithm**

Show that the algorithm of Proposition 4 can be executed in polynomial time, of order $O(|\Sigma| \cdot |Q|^4)$.

Exercise 23: By examining the discussion leading to Theorem 11, confirm that the 'if' statement can be proved without the assumption that **G** is coreachable; namely it remains valid even when **G** is blocking. Also the 'only if' statement is valid if the condition that **G** is nonblocking is replaced by the slightly weaker condition that $\overline{L_m(\mathbf{G})}$ is $(L(\mathbf{G}), P)$-normal.

6.8 Efficient Coordination in Decentralized Control

In this section we introduce a construction to obtain an efficient coordinator for nonblocking in decentralized control. We also propose two practical measures of DES complexity by which the complexity reduction achieved by a given decentralized control architecture can be realistically estimated.

First, consider an arbitrary blocking DES **G**. Let **K** be the (trim) coordinator given by

$$\mathbf{K} := \mathbf{supcon}(\mathbf{G}, \mathbf{ALL})$$

where **ALL = allevents(G)**. We assume that **K** is nonempty.

Now $L_m(\mathbf{K}) = \sup \mathcal{C}(L_m(\mathbf{G}))$, and $\overline{L_m(\mathbf{K})} = L(\mathbf{K})$. We wish to replace **K** by a smaller coordinator having the same (optimal) control action, based on a natural observer **PK**, where $P : \Sigma^* \to \Sigma_o^*$. We assume that Σ_o contains all the controllable events that are disabled in the synthesis (see **condat** table) of **K**. Such Σ_o can be obtained using the *TCT* procedure **natobs** applied to **G**, seeded with these controllable events. Since any controllable events not disabled in the synthesis of **K** could be declared uncontrollable at the outset (as far as concerns the **supcon** operation above), we may safely assume that $\Sigma_o \supseteq \Sigma_c$. Write $L_m(\mathbf{K}) =: K$. The property of P we shall require eventually is that P *normalize* **K**, namely

$$K = L_m(\mathbf{G}) \cap P^{-1}(PK)$$

and

$$\bar{K} = L(\mathbf{G}) \cap P^{-1}(P\bar{K})$$

To this end, an auxiliary condition we now bring in is that (K, P) be *exit-control-consistent (e.c.c.)*. Namely, with $\bar{L} := L(\mathbf{G})$,

$$(\forall s \in \bar{K})(\forall \sigma \in \Sigma_c) s\sigma \in \bar{L} \ \& \ [(\exists s' \in \bar{L}) Ps' = Ps \ \& \ s'\sigma \in \bar{K}] \Rightarrow s\sigma \in \bar{K}$$

Equivalently,

$$(\forall s \in \bar{K})(\forall \sigma \in \Sigma_c) s\sigma \in \bar{L} - \bar{K} \Rightarrow [(\forall s' \in \bar{L}) Ps' = Ps \Rightarrow s'\sigma \notin \bar{K}]$$

Thus if, for synthesis of $\bar{K}$, the controllable event σ is disabled following s (to prevent 'exit' from $\bar{K}$) then σ must ('consistently') be disabled following all look-alike $s' \in \bar{K}$ that satisfy $s'\sigma \in \bar{L}$. Under this condition we shall show that if P is a natural observer for $\bar{L}$, then its restriction to $\bar{K}$ is also a natural observer.

If the e.c.c. property fails for $(\bar{K}, P)$ then we should add suitably chosen P-unobservable events to **image**(P) to obtain a new observer P that provides a richer information base.

Proposition 1

As above, assume $\Sigma_c \subseteq \Sigma_o$, K is controllable, and (K, P) satisfies e.c.c. Suppose P is an $\bar{L}$-observer. Then P is a $\bar{K}$-observer, namely

$$(\forall t \in \bar{K}, w \in \Sigma_o^*)(Pt)w \in \overline{PK} \Rightarrow (\exists x \in \Sigma^*)tx \in \bar{K} \ \&\ Px = w$$

Proof

Let $t \in \bar{K}$, $w \in \Sigma_o^*$, $(Pt)w \in \overline{PK}$. Since $t \in \bar{L}$, $\overline{PK} \subseteq \overline{PL}$, and P is an $\bar{L}$-observer, there is a string $x \in \Sigma^*$ such that $tx \in \bar{L}$ and $Px = w$. We need to show that $tx \in \bar{K}$.

Case (i): $w \in P\Sigma_u^*$

We have $x \in \Sigma_u^*$ (otherwise, as $\Sigma_c \subseteq \Sigma_o$, $Px \notin P\Sigma_u^*$). By controllability of $\bar{K}$, $t \in \bar{K}$ and $tx \in \bar{L}$ imply $tx \in \bar{K}$, as required.

Case (ii): w has a prefix $v\sigma$ with $v \in P\Sigma_u^*$ and $\sigma \in \Sigma_c$

Then x has a prefix $y\sigma$ such that $y \in \Sigma_u^*$ and $Py = v$. Write $ty =: s \in \bar{K}$. Then $s \in \bar{L}$, and $Ps = P(ty) = (Pt)v$, so $(Ps)\sigma \leq (Pt)w$, namely $(Ps)\sigma \in \overline{PK}$. Therefore there is $s' \in \bar{K}$ with $Ps' = Ps$ and $s'\sigma \in \bar{K}$. By e.c.c., $s\sigma \in \bar{K}$, i.e. $ty\sigma \in \bar{K}$. If $y\sigma = x$, we are done.

Case (iii): If in case (ii) $y\sigma < x$, set $ty\sigma =: t_{new}$ and repeat the argument for cases (i) or (ii), obtaining $t_{new}x_{new} \in \bar{K}$, until $t_{new}x_{new} = tx$, as required. □

We now present the main result of this section.

Proposition 2

As in Proposition 1, assume $\Sigma_c \subseteq \Sigma_o$, K is controllable, (K, P) satisfies e.c.c., and P is an $\bar{L}$-observer. Then P normalizes $\mathbf{K}$.

Proof

We first show that $\bar{K} = L(\mathbf{G}) \cap P^{-1}(P\bar{K})$. It suffices to show $(\supseteq)$. For this let $s \in L(\mathbf{G})$ and $P(s) \in P(\bar{K})$. We must show $s \in \bar{K}$. If not, choose $t < s$ to be the longest prefix of s such that $t \in \bar{K}$. Let $t\sigma \leq s$. Since $s \in L(\mathbf{G})$, $t\sigma \notin \bar{K}$, and K is controllable, we have $\sigma \in \Sigma_c$, hence $P(\sigma) = \sigma$. By Proposition 1, P is a natural observer for $\bar{K}$. Since $Pt \leq Ps \in \overline{PK}$, there is $w_o \in \Sigma_o^*$ with $(Pt)w_o = Ps$. By the $(\bar{K}, P)$-observer property, there exists $w \in \Sigma^*$ such that $Pw = w_o$ and $tw \in \bar{K}$. We have $P(\sigma) = \sigma \leq w_o$. Hence there is $x \leq w$ with $x\sigma \leq w$ and $Px = \epsilon$, so $x \in (\Sigma - \Sigma_o)^* \subseteq \Sigma_u^*$. As $tw \in \bar{K}$ implies $tx\sigma \in \bar{K}$, while $t\sigma \notin \bar{K}$, we have that (K, P) fails to be e.c.c, contradicting the hypothesis; so $s \in \bar{K}$ after all.

It remains to show that $K = L_m(\mathbf{G}) \cap P^{-1}(PK)$. By Exercise 3.7.18, K is $L_m(\mathbf{G})$-closed, namely $K = L_m(\mathbf{G}) \cap \bar{K}$. Thus

$$
\begin{aligned}
L_m(\mathbf{G}) \cap P^{-1}(PK) &\subseteq L_m(\mathbf{G}) \cap L(\mathbf{G}) \cap P^{-1}(P\bar{K}) \\
&= L_m(\mathbf{G}) \cap \bar{K} \\
&= K
\end{aligned}
$$

as required. □

The proposition implies that K is also represented by the DES $\mathbf{K}'$ (say) computed by **supscop** with $\mathbf{E} = \mathbf{ALL}$.

To show how the foregoing result yields an efficient coordinator we revisit the example Transfer Line of Sect. 4.6, with particular reference to Exercise 4.6.2 and the monolithic controller **SUPER**(28,65), along with the reduced decentralized controllers for the buffers **B1** and **B2**, say **SIMSUP1**(7,30) and **SIMSUP2**(2,3), respectively. In *TCT* notation these provide the modular supervisor (having minimal-state size)

MODSUPER = sync(SIMSUP1,SIMSUP2,TL) (36,85)

However, the result

nonconflict(MODSUPER,ALLTL) = *false*

indicates that the two decentralized controllers **SIMSUP1, SIMSUP2** jointly conflict with the plant **TL**, and therefore a coordinator is required to achieve nonblocking. For this we let **MODSUPER** play the role of **G** in Proposition 2 above. First compute the global supervisor, say **MODCOORD** (denoted above by **K**) that renders **MODSUPER** nonblocking:

MODCOORD = supcon(MODSUPER,ALLTL) (28,65)

While **MODCOORD** is in fact identical with **SUPER**, its role as coordinator is much more limited, which allows it to be reduced as follows. From the **condat** table for **MODCOORD** we note that only event 1 is ever disabled; we thus use this event to initialize the computation of the required natural observer applied to our blocking modular supervisor **MODSUPER**:

$$
\begin{aligned}
\mathbf{SEED} &= \mathbf{allevents(MODCOORD.DAT)} \;\; (1,1) \\
(\mathbf{OMODSUPER}, \mathbf{OSEED}) &= \mathbf{natobs(MODSUPER, SEED)} \\
&\qquad\qquad\qquad\qquad ((9,18);\,(1,4))
\end{aligned}
$$

Here the extended image set represented by **OSEED** is $\{1, 5, 6, 8\}$. That the resulting natural projection P is indeed an observer for **MODSUPER** is verified by computation of the supremal quasi-congruence[8]

QCMODSUPER = supqc(MODSUPER,OSEED) (9,18)

[8]According to the foregoing theory, the natural observer required is that for the closed behavior of **MODSUPER**, namely the first argument of **natobs** should strictly be **CMODSUPER** (say), obtained from **MODSUPER** by marking all states. For the examples of this section this refinement proves to be unnecessary, but it is recommended to be followed in general.

and confirming by inspection that it is deterministic. To achieve the final reduced coordinator one can work at the abstracted (i.e. projected) level and achieve non-blocking in the usual way:

COORD = **supcon(OMODSUPER,ALLOMODSUPER)** (7,14)
COORD = **condat(OMODSUPER,COORD)** Controllable.

Now reduce to obtain

SIMCOORD = **supreduce(OMODSUPER,COORD,COORD)** (4,9;slb=-1)

It remains to verify that the final result is correct, namely by the sequence

TEST = **sync(TL,SIMSUP1,SIMSUP2,SIMCOORD)** (28,65)
Blocked_events = None
isomorph(SUPER,TEST) = *true*

We observe that the 'optimality' of P could also be confirmed by the sequence

SCOP = **supscop(MODSUPER,ALLTL,image**[1,5,6,8]) (28,65)
isomorph(SUPER,SCOP) = *true*

Note that the e.c.c. property needed for Propositions 1 and 2 was never actually confirmed, it being sufficient to verify the final result itself. In fact this example is small enough to confirm e.c.c. from the transition graph of **MODSUPER** by inspection. Finally, we see that our result **SIMCOORD** is just as 'efficient' (having state size 4) as the result **LOOPCOORD** for **LOOPSPEC** in Exercise 4.6.2. By contrast, the latter result depended on understanding the mechanism of blocking, while the former (our new result) was secured by a straightforward algorithmic procedure.

Exercise 3: **Computational verification of exit-control-consistency**
In the notation of the text, show that K satisfies the e.c.c. condition iff

$$\bar{L} \cap P^{-1}P(\bar{K} \cap \bar{K}\Sigma_c) \subseteq \bar{K}$$

Use this formula and *TCT* to check e.c.c. for Transfer Line, with $L = L_m$(**MODSUPER**) and $K = L_m$(**MODCOORD**) as defined in the text. Suppose in some instance e.c.c. failed. Explore how it might be made to succeed by relabeling certain controllable events. What might be the practical implications of such relabeling?

Before presenting further examples of efficient coordination, we briefly discuss DES complexity as a basis for comparing the efficiencies of different architectures. The 'complexity' of a DES **G** can be defined as the number of bits required to specify it. Consider the family of all DES having state set Q of size $|Q| = n$ and alphabet Σ of size $|\Sigma| = k$. Then there are

$$|Q^{(Q\times\Sigma)}| = |Q|^{|Q\times\Sigma|} = |Q|^{|Q|\cdot|\Sigma|} = n^{kn}$$

possible transition functions, and for each of these $2^{|Q|} = 2^n$ possible marker subsets, so the family contains $n^{kn} \cdot 2^n$ possible DES (fixing the initial state and disregarding state recodings). If each is considered equally likely, then the average complexity is $\log_2(n^{kn} \cdot 2^n)$ bits or

$$C(n, k) := \ln(n^{kn} \cdot 2^n)/\ln(2) = n(1 + k \cdot \ln(n)/\ln(2)) \text{ bits}$$

Instead of $C(n, k)$, however, we shall employ the simple approximation $Z(\text{DES}) =$ state size $= n$ in individual cases. A realistic alternative to Z could be the actual DES file size (not size on disk), say $S(\text{DES})$, reported by the computer in use.

With complexity indices Z and S we can compare monolithic supervisor complexity with complexity of the decentralized architecture including coordinator. For the decentralized case a first choice might be the sum of component complexities; but the disadvantage of a sum rule is that it could produce the same result for 3 small decentralized components (say) as for 2 larger components. Intuitively the former architecture is preferable to the latter, hence should be assigned a more favorable figure. We therefore adopt not the sum but the maximum of the figures for the components. Thus the actual number of components is considered less important than the greatest among their complexities. Applying this rule to Transfer Line we obtain

SUPER (28,65), $Z = 28$, $S = 1490B$
Modular supervisors:
SIMSUP1 (7,30), $Z = 7$, $S = 720B$
SIMSUP2 (2,3), $Z = 2$, $S = 238B$
SIMCOORD (4,9), $Z = 4$, $S = 354B$
Complexity reduction by state size $(Z) = ZR = 28/7 = 4.0$
Complexity reduction by file size $(S) = SR = 1490/720 = 2.1$

A similar approach to efficient coordination can be taken to the AGV system of Sect. 4.7, using as initial decentralized controllers the reduced supervisors **Z1SIM,...,Z4SIM** for the four shared track zones, the reduced supervisors **WS1SIM**, **WS2SIM**, **WS3SIM** for the workstations, and the reduced supervisor **IPSSIM** for the input parts station. First we compute the (blocking) modular behavior

MODSUPER = sync(AGV,Z1SIM,Z2SIM,Z3SIM,Z4SIM,

WS1SIM,WS2SIM,WS3SIM,IPSSIM) (4424,11368)

followed by

MODCOORD = supcon(MODSUPER,ALL) (4406,11338)

From the **condat** file for **MODCOORD** we note that only events 11,21 are ever disabled, and therefore use these to obtain from **natobs** the extended event set; in this case the result (**OSEED**) is again the list [11,21], for the image of the required natural observer P:

$$(\mathbf{OMODSUPER}, \mathbf{OSEED}) = \mathbf{natobs}(\mathbf{MODSUPER}, [11, 21]) \; ((8, 14); (1, 2))$$

As before we confirm that the result of **supscop** using **image**[11,21] is the monolithic supervisor **SUPER** as expected. Finally we compute

COORD = supcon(OMODSUPER,ALLOMODSUPER) (7,12)

for which no further reduction is possible. **COORD** agrees in state size (7) with the global coordinator **CFSIM** (7,78) computed by the more complicated approach (demanding specific insight) of Sect. 4.7. For complexity reduction we obtain the following.

SUPER (4406,11338), $Z = 4406$, $S = 362.kB$
Modular supervisors:
Z1SIM,...,Z4SIM (2,8), $Z = 2$, $S = 318B$
WS1SIM (4,10), $Z = 4$, $S = 370B$
WS2SIM, WS3SIM, IPSSIM (2,4), $Z = 2$, $S = 254B$
COORD (7,12), $Z = 7$, $S = 432B$
Complexity reduction by state size $(Z) = ZR = 4406/7 = 629$
Complexity reduction by file size $(S) = SR = 362/0.432 = 838$

For a third example we revisit the 'manufacturing cell with interacting loops' of Exercise 4.9.7. Leaving details to the reader, we merely recall that 4 coordinators **COORDij** must be designed for the conflicting supervisor pairs **SIMSUPi**, **SIMSUPj** corresponding to the buffer specifications for **BUFi**, **BUFj**, respectively, $((i, j) = (1, 3), (1, 4), (2, 3), (2, 4))$. Results for complexity reduction are the following.

SUPER (225,472), $Z = 225$, $S = 9972B$
Modular supervisors:
SIMSUP1,...,SIMSUP4 (2,2), $Z = 2$, $S = 222B$
COORD13, COORD24 (4,6), $Z = 4$, $S = 306B$
COORD14, COORD23 (4,12), $Z = 4$, $S = 402B$
Complexity reduction by state size $(Z) = ZR = 225/4 = 56.2$
Complexity reduction by file size $(S) = SR = 9972/402 = 24.8$

We surmise from the foregoing results that crude complexity measures of DES can be defined in such a way that the corresponding complexity reduction achieved by decentralized as compared to monolithic supervision can be meaningfully estimated, roughly to within an order of magnitude. In each of the above examples reduction depended on the 'efficient' coordinator design based on Proposition 2.

Exercise 4: Confirm the details for the example of Exercise 4.9.7.

Exercise 5: **Naive coordination**
A one-step design of a coordinator for nonblocking could be simply to apply **supreduce** to the intermediate supervisor denoted above by **MODCOORD**. Test this approach for the three examples in this section, and compare the results for complexity reduction.

Exercise 6: Instead of the 'average' complexity estimate $C(n, k)$ one might base an estimate on the specific triple (n, m, k) where $m =$ #transitions. Assuming the marker state subset is 'small' (as is true in most cases) we consider only encoding the list of m transitions $[q, \sigma, q']$ where $q, q' \in Q$ and $\sigma \in \Sigma$. A possible measure of DES complexity is thus

$$T(n, m, k) = m(\log_2(kn^2)) = m \ln(kn^2))/\ln(2) \text{ bits}$$

Compute both C and T for each DES in the three examples of this section, and appraise the usefulness of these measures in estimating complexity reduction.

6.9 Notes

This chapter is based largely on the doctoral thesis of F. Lin (1987) and related publications Lin and Wonham (1988a, b, 1990), Lin et al. (1989). The result in Exercise 6.2.3 is due to K.-C. Wong (1994, Appendix B). Example 6.3.12 on distributed control by communication is adapted from the thesis (Mannani 2009, Sect. 3.6), although our approach here is different. Further information related to problems of the type of SCOP can be found in Cho and Marcus (1989) and Takai and Ushio (2003).

The definition and results in Sects. 6.2, 6.3 on relative observability are due to K. Cai et al. (2013, 2014, 2015, 2018), Cai and Wonham (2016).

Exercise 6.5.15 is adapted from Cao et al. (1997).

The definition of 'quasi-congruence' in Exercise 1.4.15 and Sect. 6.7 was introduced in Wong and Wonham (2004) as a generalization to nondeterminism of the observer congruence for deterministic systems discussed in Wonham (1976). Application to the computation of 'natural observers' is provided in Feng and Wonham (2006, 2010). As shown in Wong and Wonham (2004), reduction by quasi-congruence is closely related to the notions of bisimulation and observational equivalence due to Milner (1989), see also Arnold (1994). We prefer the 'congruence' terminology as closer to standard algebra.

For Exercise 6.7.19 see e.g. Fig. 17 of Feng and Wonham (2010).

References

Arnold A (1994) Finite transition systems. Prentice-Hall, Upper Saddle River

Brandt RD, Garg V, Kumar R, Lin F, Marcus SI, Wonham WM (1990) Formulas for calculating supremal controllable and normal sublanguages. Syst Control Lett 15:111–117

Cai K, Wonham WM (2016) A new algorithm for computing the supremal relatively observable sublanguage. In: Proceedings thirteenth international workshop on discrete event systems (WODES '16), Xi'an, China, May 30–June 1, 2016, pp 8–13

Cai K, Zhang R-Y, Wonham WM (2013) On relative observability of discrete-event systems. In: Proceedings fifty-second IEEE conference on decision and control, Firenze, Italy, 11–13 Dec 2013, pp 7285–7290

Cai K, Zhang R-Y, Wonham WM (2014) On relative observability of timed discrete-event systems. In: Proceedings of the twelfth IFAC and IEEE international workshop on discrete event systems (WODES '14), Cachan, France, 14–16 May 2014, pp 208–213

Cai K, Zhang R-Y, Wonham WM (2015) On relative coobservability of discrete-event systems. In: Proceedings of the American control conference, Chicago, IL, USA, 1–3 July 2015, pp 371–376

Cai K, Zhang R-Y, Wonham WM (2018) Characterizations and effective computation of supremal relatively observable sublanguages. Discret Event Dyn Syst 28(2):269–287

Cao C, Lin F, Lin Z-H (1997) Why event observation: observability revisited. Discret Event Dyn Syst Theory Appl 7:127–149

Cho H, Marcus SI (1989) Supremal and maximal sublanguages arising in supervisor synthesis problems with partial observations. Math Syst Theory 22:177–211

Feng L, Wonham WM (2006) On the computation of natural observers in discrete-event systems. In: Proceedings of the forty-fifth IEEE conference on decision and control, San Diego, pp 428–433

Feng L, Wonham WM (2010) On the computation of natural observers in discrete-event systems. Discret Event Dyn Syst 20(1):63–102

Lin F (1987) On controllability and observability of discrete event systems. PhD thesis, Department of Electrical Engineering, University of Toronto

Lin F, Wonham WM (1988a) On observability of discrete-event systems. Inf Sci 44(2):173–198

Lin F, Wonham WM (1988b) Supervisory control and observation of discrete-event systems. In: Byrnes CI, Martin CF, Saeks RE (eds) Eighth international symposium on the mathematical theory of networks and systems (MTNS '87), Phoenix (1987). Analysis and control of nonlinear systems, North-Holland, Amsterdam, pp 337–348

Lin F, Wonham WM (1990) Decentralized control and coordination of discrete-event systems with partial observation. IEEE Trans Autom Control 35(12):1330–1337

Lin F, Brandt RD, Wonham WM (1989) A note on supremal controllable and normal sublanguages. In: Proceedings of the twenty-seventh annual Allerton conference on communication, control and computing, University of Illinois, pp 491–500

Mannani A (2009) Synthesis of communicating decentralized supervisors for discrete-event systems with application to communication protocol synthesis. PhD thesis, ECE Department, Concordia University, Montreal

Milner R (1989) Communication and concurrency. Prentice-Hall, Upper Saddle River

Takai S, Ushio T (2003) Effective computation of an L_m(G)-closed, controllable, and observable sublanguage arising in supervisory control. Syst Control Lett 49:191–200

Wong K-C (1994) Discrete-event control architecture: an algebraic approach. PhD thesis, Department of Electrical and Computer Engineering, University of Toronto

Wong KC, Wonham WM (2004) On the computation of observers in discrete-event systems. Discret Event Dyn Syst 14(1):55–107

Wonham WM (1976) Towards an abstract internal model principle. IEEE Trans Syst Man Cybern SMC 6(11):735–740

Chapter 7
State-Based Control of Discrete-Event Systems

In certain contexts it is convenient to employ directly a state space model combined with a simple logic formalism, in preference to the linguistic basis adopted so far. To retain flexibility, we indicate how the two approaches are conceptually dual. This chapter provides the background for vector discrete-event systems, to be introduced in Chap. 8. Two illustrations of discrete state models are adapted from the literature on industrial process control.

7.1 Introduction

In previous chapters our approach to modeling the DES control problem has started from two underlying languages, respectively generated by the plant and accepted by the specification. The system state descriptions were brought in as vehicles for representation and computation rather than as essential to the problem description. In the present chapter we adopt a dual and more conventional viewpoint, in which the underlying state transition structure is assigned a more basic role. Two illustrative examples are provided in the Appendix 7.1. In addition, the state viewpoint will facilitate the treatment of systems—notably the vector discrete-event systems of Chap. 8—where the underlying state set has regular algebraic structure that can be exploited for modeling compactness and computational efficiency.

7.2 Predicates and State Subsets

Let $\mathbf{G} = (Q, \Sigma, \delta, q_0, Q_m)$ be a DES, as defined in previous chapters. In order to place 'conditions' on the state q of $\mathbf{G}$, it will be convenient to use a logic formalism. While not strictly necessary (in fact we could make do with state subsets just as well),

W. M. Wonham and K. Cai, *Supervisory Control of Discrete-Event Systems*, Communications and Control Engineering,
https://doi.org/10.1007/978-3-319-77452-7_7

the logic terminology is sometimes closer to natural language and intuition. Thus, a *predicate* P defined on Q is a function $P : Q \to \{0, 1\}$. P can always be identified with the corresponding state subset

$$Q_P := \{q \in Q | P(q) = 1\} \subseteq Q$$

Thus the complement $Q - Q_P = \{q \in Q | P(q) = 0\}$. We say that P *holds*, or *is satisfied*, precisely on the subset Q_P, and that $q \in Q_P$ *satisfies* P. The satisfaction relation $P(q) = 1$ will often be written $q \models P$ ('q satisfies P'). Write $Pred(Q)$ for the set of all predicates on Q, often identified with the power set $Pwr(Q)$. Boolean expressions will be formed as usual by logical negation, conjunction and disjunction; in standard notation:

$$(\neg P)(q) = 1 \text{ iff } P(q) = 0$$

$$\begin{aligned}(P_1 \wedge P_2)(q) = 1 &\text{ iff } P_1(q) = 1 \text{ and } P_2(q) = 1\\ (P_1 \vee P_2)(q) = 1 &\text{ iff } P_1(q) = 1 \text{ or } \quad P_2(q) = 1\end{aligned}$$

Recall the De Morgan rules

$$\neg(P_1 \wedge P_2) = (\neg P_1) \vee (\neg P_2), \quad \neg(P_1 \vee P_2) = (\neg P_1) \wedge (\neg P_2)$$

Given $P \in Pred(Q)$, we say that $P_1, P_2 \in Pred(Q)$ are *equivalent* with respect to P if

$$P_1 \wedge P = P_2 \wedge P$$

namely P_1 and P_2 coincide when restricted, or 'relativized', to the subset Q_P. As the logic counterpart to subset containment we introduce on $Pred(Q)$ the partial order $\preceq$ defined by

$$\begin{aligned}P_1 \preceq P_2 &\text{ iff } P_1 \wedge P_2 = P_1\\ &\text{ iff } (\forall q \in Q) q \models \neg P_1 \vee P_2\end{aligned}$$

Thus $P_1 \preceq P_2$ (P_1 'precedes' P_2) just when P_1 is stronger than P_2 in the sense that P_1 'implies' P_2; equivalently if $q \models P_1$ then $q \models P_2$; and we also say that P_1 is a *subpredicate* of P_2. Under the identification of $Pred(Q)$ with $Pwr(Q)$ and $\preceq$ with $\subseteq$ it is clear that $(Pred(Q), \preceq)$ is a complete lattice. The top element $\top$ of this lattice, defined by $Q_\top = Q$, will be denoted by *true* (and is the weakest possible predicate), while the bottom element $\bot$, defined by $Q_\bot = \emptyset$, will be written $\bot = false$ (and is the strongest possible predicate).

Exercise 1: Justify the terms 'strong' and 'weak' in this usage.

7.3 Predicate Transformers

First we consider reachability issues for **G**. Let $P \in Pred(Q)$. The *reachability predicate* $R(\mathbf{G}, P)$ is defined to hold precisely on those states that can be reached in **G** from q_0 via states satisfying P, according to the inductive definition:

1. $q_0 \models P \Rightarrow q_0 \models R(\mathbf{G}, P)$
2. $q \models R(\mathbf{G}, P)$ & $\sigma \in \Sigma$ & $\delta(q, \sigma)!$ & $\delta(q, \sigma) \models P \Rightarrow \delta(q, \sigma) \models R(\mathbf{G}, P)$
3. No other states q satisfy $R(\mathbf{G}, P)$

Explicitly, $q \models R(\mathbf{G}, P)$ iff there exist an integer $k \geq 0$, states $q_1, \ldots, q_k \in Q$, and events $\sigma_0, \sigma_1, \ldots, \sigma_{k-1} \in \Sigma$ such that

1. $\delta(q_i, \sigma_i)!$ & $\delta(q_i, \sigma_i) = q_{i+1}, \ i = 0, 1, \ldots, k-1$
2. $q_i \models P, \ i = 0, 1, \ldots, k$
3. $q_k = q$

For fixed **G**, $R(\mathbf{G}, \cdot) : Pred(Q) \to Pred(Q)$ is an example of a *predicate transformer*, i.e. a map transforming predicates to predicates. Clearly $R(\mathbf{G}, P) \preceq P$; also $R(\mathbf{G}, \cdot)$ is monotone with respect to $\preceq$, in the sense that $P \preceq P'$ implies $R(\mathbf{G}, P) \preceq R(\mathbf{G}, P')$. Note that $R(\mathbf{G}, true)$ is the reachable set for **G**, namely the set of states reachable from q_0 without constraint.

Fix $\sigma \in \Sigma$. The *weakest liberal precondition* corresponding to $\sigma \in \Sigma$ is the predicate transformer $M_\sigma : Pred(Q) \to Pred(Q)$ defined as follows:

$$M_\sigma(P)(q) := \begin{cases} 1 & \text{if either } \delta(q, \sigma)! \,\&\, \delta(q, \sigma) \models P, \ \text{or } \textit{not } \delta(q, \sigma)! \\ 0 & \text{otherwise (i.e. } \delta(q, \sigma)! \text{ and } \textit{not } \delta(q, \sigma) \models P) \end{cases}$$

Equivalently and more concisely,

$$q \models M_\sigma(P) \ \text{ iff } \ \delta(q, \sigma) \models P \text{ whenever } \delta(q, \sigma)! \text{ (and by default otherwise)}$$

It is clear that $M_\sigma(\cdot)$ is monotone. The action of M_σ can be visualized in terms of the 'dynamic flow' on Q induced by one-step state transitions under σ, wherever such transitions are defined. Thus $M_\sigma(P)$ is just the condition on (states in) Q that ensures that from a state, say q, there is a one-step transition into Q_P under the occurrence of the event σ (so $M_\sigma(P)$ is indeed a 'precondition' for P), or else that σ cannot occur at q at all, i.e. $\delta(q, \sigma)$ is not defined (in this sense the precondition is 'liberal'). Note that 'weakest' means 'the largest state subset with the asserted property', namely 'the weakest assumption required to establish either that σ was not enabled or that σ occurred and led to P'.

The weakest liberal precondition is retrospective, drawing attention from a present condition (i.e. P) to its one-step antecedent. A dual concept is the predicate transformer *strongest postcondition* N_σ: $Pred(Q) \to Pred(Q)$ defined according to

$$N_\sigma(P)(q) := \begin{cases} 1 & \text{if}(\exists q' \models P)\delta(q',\sigma)!\,\&\,\delta(q',\sigma) = q \\ 0 & \text{otherwise} \\ & (\text{i.e. } (\forall q' \models P) \text{ either } \textit{not } \delta(q',\sigma)! \text{ or } \delta(q',\sigma) \neq q) \end{cases}$$

The strongest postcondition is prospective, drawing attention from a present condition to its one-step consequent; and 'strongest' means 'the smallest state subset with the asserted property', namely 'the strongest inference that can be made solely on the assumption that P held and σ occurred'.

We shall see that weakest liberal precondition plays a role in controllability properties, while it can be shown that strongest postcondition relates to observability.

Later we shall use the connection between state reachability and language behavior, summarized as follows. The *closed behavior* $L(\mathbf{G}, q)$ corresponding to initialization of $\mathbf{G}$ at an arbitrary state $q \in Q$ is defined to be

$$L(\mathbf{G}, q) := \{w \in \Sigma^* | \delta(q, w)!\}$$

while the corresponding *marked behavior* is, of course,

$$L_m(\mathbf{G}, q) := \{w \in L(\mathbf{G}, q) | \delta(q, w) \in Q_m\}$$

Similarly we define the closed (resp. marked) language *induced* by a predicate $P \in Pred(Q)$ to be

$$L(\mathbf{G}, P) := \{w \in L(\mathbf{G}, q_0) | (\forall v \leq w)\delta(q_0, v) \models P\}$$
$$(\text{resp. } L_m(\mathbf{G}, P) := \{w \in L(\mathbf{G}, P) | \delta(q_0, w) \in Q_m\}\ )$$

The *reachable set* $R(\mathbf{G}, q)$ of states reachable from arbitrary $q \in Q$ is then

$$R(\mathbf{G}, q) := \{\delta(q, w) | w \in L(\mathbf{G}, q)\}$$

Exercise 1: Consider an agent observing the behavior of $\mathbf{G}$, who knows only that $\mathbf{G}$ was initialized at some (unspecified) state $q \in Q_P \subseteq Q$ with $P \in Pred(Q)$, and that $\mathbf{G}$ subsequently generated the string $s = \sigma_1\sigma_2 \cdots \sigma_k$. Show that the agent's best estimate of the state at s is the predicate

$$N_{\sigma_k}(\cdots N_{\sigma_2}(N_{\sigma_1}(P)) \cdots)$$

Exercise 2: If P is a predicate and σ is an element of Σ, consider the weakest liberal precondition of the strongest postcondition of P, namely

$$P' := M_\sigma(N_\sigma(P))$$

Similarly let $P'' := N_\sigma(M_\sigma(P))$. Show that

$$P'' \preceq P \preceq P'$$

Illustrate with examples where the inequalities are strict.

7.4 State Feedback and Controllability

We define a *state feedback control (SFBC)* for $\mathbf{G}$ to be a total function

$$f : Q \to \Gamma$$

where

$$\Gamma := \{\Sigma' \subseteq \Sigma | \Sigma' \supseteq \Sigma_u\}$$

Thus f attaches to each state q of $\mathbf{G}$ a subset of events that always contains the uncontrollable events. The event $\sigma \in \Sigma$ is *enabled* at q if $\sigma \in f(q)$, and is *disabled* otherwise; σ is always enabled if $\sigma \in \Sigma_u$. For $\sigma \in \Sigma$ introduce the predicate $f_\sigma : Q \to \{0, 1\}$ defined by

$$f_\sigma(q) := 1 \text{ iff } \sigma \in f(q)$$

Thus a SFBC f is specified by the family of predicates $\{f_\sigma | \sigma \in \Sigma\}$. The closed-loop transition function induced by f will be written δ^f, given by

$$\delta^f(q, \sigma) := \begin{cases} \delta(q, \sigma) & \text{if } \delta(q, \sigma)! \text{ and } f_\sigma(q) = 1 \\ \text{undefined} & \text{otherwise} \end{cases}$$

We write $\mathbf{G}^f := (Q, \Sigma, \delta^f, q_0, Q_m)$ for the closed-loop DES formed from $\mathbf{G}$ and f.

If f is a SFBC for $\mathbf{G}$ then, of course, $R(\mathbf{G}^f, P)$ denotes the reachability predicate for $\mathbf{G}^f$ (initialized at q_0). Since for any q and σ, $\delta^f(q, \sigma)!$ only if $\delta(q, \sigma)!$ it is evident that $R(\mathbf{G}^f, P) \preceq R(\mathbf{G}, P)$.

The following definition is fundamental. We say that $P \in Pred(Q)$ is *controllable* (with respect to $\mathbf{G}$) if

$$P \preceq R(\mathbf{G}, P) \;\&\; (\forall \sigma \in \Sigma_u) P \preceq M_\sigma(P)$$

Thus controllability asserts that if q satisfies P then (i) q is reachable from q_0 via a sequence of states satisfying P, and (ii) if $\sigma \in \Sigma_u$, $\delta(q, \sigma)!$ and $q' = \delta(q, \sigma)$ then q' satisfies P, namely 'P is invariant under the flow induced by uncontrollable events'. Note that the predicate *false* is trivially controllable.

Theorem 1

Let $P \in Pred(Q)$, $P \neq false$. Then P is controllable iff there exists a SFBC f for $\mathbf{G}$ such that $R(\mathbf{G}^f, true) = P$.

Thus a nontrivial predicate P is controllable precisely when it can be 'synthesized' by state feedback.

Proof

(If) Assume $R(\mathbf{G}^f, true) = P$. Since $\delta^f(q, \sigma)!$ implies $\delta(q, \sigma)!$ it is clear that any q such that $q \models P$ can be reached from q_0 along a sequence of states with the same property, so $P \preceq R(\mathbf{G}, P)$. Let $\sigma \in \Sigma_u$ and $q \models P$. If $\delta(q, \sigma)!$ then $\delta^f(q, \sigma)!$ (since f is a SFBC) and then

$$\delta(q, \sigma) = \delta^f(q, \sigma) \models R(\mathbf{G}^f, true) = P$$

This implies that $q \models M_\sigma(P)$, and therefore $P \preceq M_\sigma(P)$.

(Only if) Assume P controllable and define the SFBC $f : Q \to \Gamma$ by

$$(\forall \sigma \in \Sigma_c) f_\sigma := M_\sigma(P)$$

First it will be shown that $R(\mathbf{G}^f, true) \preceq P$. Let $q \models R(\mathbf{G}^f, true)$. Then for some $k \geq 0, q_1, \ldots, q_k (= q) \in Q$ and $\sigma_0, \sigma_1, \ldots, \sigma_{k-1} \in \Sigma$, we have

$$\delta(q_i, \sigma_i)!, \quad \delta(q_i, \sigma_i) = q_{i+1}, \quad f_{\sigma_i}(q_i) = 1, \quad i = 0, 1, \ldots, k-1$$

By hypothesis, $\hat{q} \models P$ for some $\hat{q}$, and by controllability $\hat{q} \models R(\mathbf{G}, P)$, so in particular $q_0 \models P$. We claim that $q_1 \models P$. For if $\sigma_0 \in \Sigma_u$ then controllability implies that $q_0 \models M_{\sigma_0}(P)$, so $\delta(q_0, \sigma_0) = q_1 \models P$; while if $\sigma_0 \in \Sigma_c$ then $f_{\sigma_0}(q_0) = 1$, namely $q_0 \models M_{\sigma_0}(P)$ and again $\delta(q_0, \sigma_0) = q_1 \models P$. By repetition of this argument for $q_2, \ldots, q_k$ we conclude that $q_k = q \models P$. It remains to show that $P \preceq R(\mathbf{G}^f, true)$. Let $q \models P$. By controllability, $q \models R(\mathbf{G}, P)$. For some $k \geq 0, q_1, \ldots, q_k (= q) \in Q$ and $\sigma_0, \sigma_1, \ldots, \sigma_{k-1} \in \Sigma$ we have

$$q_i \models P, \quad i = 0, \ldots, k$$

$$\delta(q_i, \sigma_i)!, \quad \delta(q_i, \sigma_i) = q_{i+1}, \quad i = 0, \ldots, k-1$$

and therefore

$$q_i \models M_{\sigma_i}(P), \quad i = 0, 1, \ldots, k-1$$

If $\sigma_i \in \Sigma_u$ then $f_{\sigma_i}(q_i) = 1$ because f is a SFBC; while if $\sigma_i \in \Sigma_c$ then $f_{\sigma_i}(q_i) = M_{\sigma_i}(P)(q_i) = 1$ (as just shown). Therefore $\delta^f(q_i, \sigma_i)!$ and $\delta^f(q_i, \sigma_i) = q_{i+1}$ $(i = 0, 1, \ldots, k-1)$, namely

$$q = q_k \models R(\mathbf{G}^f, true)$$

as claimed. □

Now suppose $P \in Pred(Q)$ is not controllable. As usual, we seek a controllable predicate that best approximates P from below. Following standard procedure, bring in the family of controllable predicates that are stronger than P, namely

$$\mathcal{CP}(P) := \{K \in Pred(Q) | K \preceq P \,\&\, K \text{ controllable}\}$$

Proposition 2

$\mathcal{CP}(P)$ is nonempty and is closed under arbitrary disjunctions; in particular the supremal element $\sup \mathcal{CP}(P)$ exists in $\mathcal{CP}(P)$.

Proof

We have already seen that $false \in \mathcal{CP}(P)$. Now let $K_\lambda \in \mathcal{CP}(P)$ for $\lambda \in \Lambda$, some index set. It will be shown that

$$K := \bigvee_{\lambda \in \Lambda} K_\lambda \in \mathcal{CP}(P)$$

It is clear that $K \preceq P$, so it remains to test controllability. Let $q \models K$, so $q \models K_\lambda$ for some λ. By controllability of K_λ, $q \models R(\mathbf{G}, K_\lambda)$, and as $K_\lambda \preceq K$ there follows $q \models R(\mathbf{G}, K)$. If $\sigma \in \Sigma_u$ then similarly $q \models M_\sigma(K_\lambda)$; hence $q \models M_\sigma(K)$ as required. Finally, the supremal element of $\mathcal{CP}(P)$ is

$$\sup \mathcal{CP}(P) = \bigvee \{K | K \in \mathcal{CP}(P)\} \qquad \square$$

The following characterization will be useful later. Define the predicate transformer $\langle \cdot \rangle$ according to

$$q \models \langle P \rangle \text{ if } (\forall w \in \Sigma_u^*) \delta(q, w)! \Rightarrow \delta(q, w) \models P$$

Note that $\langle P \rangle \preceq P$ (since $\delta(q, \epsilon) = q$) and in fact $\langle P \rangle$ is the weakest subpredicate of P that is invariant under the flow induced by uncontrollable events.

Proposition 3

$$\sup \mathcal{CP}(P) = R(\mathbf{G}, \langle P \rangle)$$

Proof

Claim 1. $R(\mathbf{G}, \langle P \rangle) \in \mathcal{CP}(P)$

We show that $R(\mathbf{G}, \langle P \rangle)$ is controllable. Let $q \models R(\mathbf{G}, \langle P \rangle)$. Then $q_0 \models \langle P \rangle$ and there are $k \geq 0$, $q_1, \ldots, q_k (= q) \in Q$ and $\sigma_0, \sigma_1, \ldots, \sigma_{k-1} \in \Sigma$ such that

$$q_i \models \langle P \rangle, \quad i = 1, \ldots, k$$

$$\delta(q_i, \sigma_i)!, \quad \delta(q_i, \sigma_i) = q_{i+1}, \quad i = 0, \ldots, k-1$$

We note that $q_i \models R(\mathbf{G}, \langle P \rangle)$ since $q_j \models \langle P \rangle$ for $j = 0, 1, \ldots, i$. In particular $q \models R(\mathbf{G}, R(\mathbf{G}, \langle P \rangle))$, so $R(\mathbf{G}, \langle P \rangle) \preceq R(\mathbf{G}, R(\mathbf{G}, \langle P \rangle))$. Next we choose $\sigma \in \Sigma_u$ with $\delta(q, \sigma)!$ and establish $q \models M_\sigma(R(\mathbf{G}, \langle P \rangle))$, namely $q' := \delta(q, \sigma) \models R(\mathbf{G}, \langle P \rangle)$. For this let $w \in \Sigma_u^*$ with $\delta(q', w)!$. Then $\sigma w \in \Sigma_u^*$, and

$$\delta(q', w) = \delta(\delta(q, \sigma), w) = \delta(q, \sigma w) \models P \quad (\text{since } q \models \langle P \rangle)$$

Thus $q' \models \langle P\rangle$. Then $q \models R(\mathbf{G}, \langle P\rangle)$ together with $q' = \delta(q, \sigma)$ implies $q' \models R(\mathbf{G}, \langle P\rangle)$, so $q \models M_\sigma(R(\mathbf{G}, \langle P\rangle))$, completing the proof of Claim 1.

Claim 2. Let $P' \in \mathcal{CP}(P)$. Then $P' \preceq R(\mathbf{G}, \langle P\rangle)$.

Let $q \models P'$. Then $q_0 \models P'$ and there exist $k \geq 0$, $q_1, \ldots, q_k (= q) \in Q$ and $\sigma_0, \sigma_1, \ldots, \sigma_{k-1} \in \Sigma$ such that

$$q_i \models P', \quad i = 1, \ldots, k$$

$$\delta(q_i, \sigma_i)!, \quad \delta(q_i, \sigma_i) = q_{i+1}, \quad i = 0, \ldots, k-1$$

Fix i and $\sigma \in \Sigma_u$. Then $q_i \models P' \preceq M_\sigma(P')$ and $\delta(q_i, \sigma)!$ imply $\delta(q_i, \sigma) \models P'$. If $w \in \Sigma_u^*$ then by induction on $|w|$ we infer that if $\delta(q_i, w)!$ then $\delta(q_i, w) \models P' \preceq P$. There follows in turn

$$q_0 \models \langle P\rangle, \quad q_1 \models \langle P\rangle, \quad \ldots, \quad q_k \models \langle P\rangle, \quad \text{i.e. } q \models \langle P\rangle$$

namely $q \models R(\mathbf{G}, \langle P\rangle)$, and Claim 2 is proved.

The result follows from Claims 1 and 2. □

Corollary 4

$$\sup \mathcal{CP}(P) \neq \mathit{false} \quad \text{iff} \quad R(\mathbf{G}, \langle P\rangle) \neq \mathit{false} \quad \text{iff} \quad q_0 \models \langle P\rangle$$ □

Under the assumption that $\sup \mathcal{CP}(P) \neq \mathit{false}$, we may define a corresponding 'optimal', i.e. behaviorally least restrictive, SFBC f^* to synthesize $R(\mathbf{G}, \langle P\rangle)$. Imitating the proof ('Only if') of Theorem 1 we may set

$$(\forall \sigma \in \Sigma_c) f_\sigma^*(q) := \begin{cases} 1 & \text{if } \delta(q, \sigma)! \,\&\, \delta(q, \sigma) \models \langle P\rangle \\ 0 & \text{otherwise} \end{cases}$$

Note that $f_\sigma^*(q)$ may be evaluated arbitrarily (in particular $= 0$) if *not* $\delta(q, \sigma)!$. This formula suggests that in practice optimal control can be implemented 'online', namely by testing, at the current state q, the satisfaction relation $\delta(q, \sigma) \models \langle P\rangle$ for each controllable event σ such that $\delta(q, \sigma)!$. Efficient implementation of f_σ^* in an application will thus depend on devising an economical algorithmic representation of $\langle P\rangle$, namely of 'reachability on the uncontrollable subsystem'. While in general this reachability property may be intractable or even undecidable, we shall see in the next chapter that an efficient algorithm is often available for vector discrete-event systems and linear predicates.

Exercise 5: Show that the predicate $P \in \mathit{Pred}(Q)$ is controllable iff $P = R(\mathbf{G}, P)$ and the language $L(\mathbf{G}, P)$ is controllable.

Exercise 6: Show that, for arbitrary $P \in \mathit{Pred}(Q)$,

$$L(\mathbf{G}, \sup \mathcal{CP}(P)) = \sup \mathcal{C}(L(\mathbf{G}, P))$$

7.5 Balanced State Feedback Controls and Modularity

A SFBC $f : Q \to \Gamma$ is *balanced* if

$$(\forall q, q' \in Q)(\forall \sigma \in \Sigma) q, q' \models R(\mathbf{G}^f, true) \;\&\; \delta(q, \sigma)! \;\&\; \delta(q, \sigma) = q' \Rightarrow f_\sigma(q) = 1$$

A balanced SFBC is one which, among all SFBC synthesizing the same (controllable) predicate, enables at every reachable state q the largest possible set of (controllable) events σ for which $\delta(q, \sigma)!$. Thus a balanced SFBC is 'maximally permissive'.

Exercise 1: Show that an arbitrary SFBC can always be replaced by a balanced SFBC without changing the reachable set.

Let $P \in Pred(Q)$, $P \neq false$, be expressible as a conjunction of predicates $P_1, \ldots, P_k$:

$$P = \bigwedge_{i=1}^{k} P_i$$

We shall think of P as a specification for the controlled behavior of the DES $\mathbf{G}$, and the $P_i \in Pred(Q)$ as partial specifications. Our objective will be to implement an optimal (balanced) SFBC for P by means of modular SFBC for the P_i.

Write $R(f/\mathbf{G})$ for $R(\mathbf{G}^f, true)$. For $i = 1, \ldots, k$ let $f_i : Q \to \Gamma$ be an optimal (balanced) SFBC for P_i, i.e.

$$R(f_i/\mathbf{G}) = \sup \mathcal{CP}(P_i)$$

The *modular SFBC* f formed from the f_i is given by

$$f_\sigma := \bigwedge_{i=1}^{k} f_{i,\sigma}, \quad \sigma \in \Sigma$$

i.e. $f_\sigma(q) = 1$ iff $f_{i,\sigma}(q) = 1$ for $i = 1, \ldots, k$: an event is enabled by f iff it is enabled by each f_i. In this case write symbolically

$$f := \bigwedge_{i=1}^{k} f_i$$

Theorem 2

Assume that f_i is balanced ($i = 1, \ldots, k$). Then f is balanced, and

$$R(f/\mathbf{G}) = \sup \mathcal{CP}(P)$$

Proof

Clearly $R(f/\mathbf{G}) \preceq \bigwedge_{i=1}^{k} R(f_i/\mathbf{G})$, from which it easily follows that f is balanced. Next, we have

$$R(f_i/\mathbf{G}) = R(\mathbf{G}, \langle P_i \rangle) \preceq P_i, \quad i = 1, \ldots, k$$

so that $R(f/\mathbf{G}) \preceq P$, hence $R(f/\mathbf{G}) \in \mathcal{CP}(P)$, and therefore $R(f/\mathbf{G}) \preceq \sup \mathcal{CP}(P)$. It remains to check that

$$\sup \mathcal{CP}(P) \preceq R(f/\mathbf{G})$$

Now $\sup \mathcal{CP}(P) = R(\mathbf{G}, \langle P \rangle)$, and

$$\langle P \rangle = \langle \bigwedge_{i=1}^{k} P_i \rangle = \bigwedge_{i=1}^{k} \langle P_i \rangle$$

so it must be shown that

$$R(\mathbf{G}, \bigwedge_{i=1}^{k} \langle P_i \rangle) \preceq R(f/\mathbf{G})$$

Let

$$q \models R(\mathbf{G}, \bigwedge_{i=1}^{k} \langle P_i \rangle)$$

Then there are $m \geq 0$, $q_1, \ldots, q_m (= q)$ and $\sigma_0, \sigma_1, \ldots, \sigma_{m-1}$ such that

$$\delta(q_j, \sigma_j)!, \quad \delta(q_j, \sigma_j) = q_{j+1}, \quad j = 0, 1, \ldots, m-1$$

$$q_j \models \bigwedge_{i=1}^{k} \langle P_i \rangle, \quad j = 0, 1, \ldots, m$$

Thus $q_j \models R(\mathbf{G}, \langle P_i \rangle) = R(f_i/\mathbf{G})$ $(i = 1, \ldots, k;\ j = 0, 1, \ldots, m)$. Since each f_i is balanced,

$$f_{i,\sigma_j}(q_j) = 1, \quad i = 1, \ldots, k; \ j = 0, 1, \ldots, m-1$$

which implies in turn

$$\bigwedge_{i=1}^{k} f_{i,\sigma_j}(q_j) = 1, \quad j = 0, 1, \ldots, m-1$$

$$q_j \models R(f/\mathbf{G})$$

$$q = q_m \models R(f/\mathbf{G})$$

as required. □

Exercise 3: Provide an example to show that the conclusion of Theorem 2 may fail if the assumption that the f_i are balanced is dropped. On the other hand show by another example that the conclusion may hold even though the assumption fails, namely 'balance' is not a necessary condition.

7.6 Dynamic State Feedback Control

The foregoing methods are readily extended to the situation where additional memory elements are included in the state description, permitting the control action to depend not only on the current state of the plant **G**, but also on various properties of past behavior.

Let $\mathbf{G}_i = (Q_i, \Sigma, \delta_i, q_{i,0}, Q_{i,m})$ $(i = 1, 2)$ be DES defined over the same alphabet Σ. We recall that the *product DES* $\mathbf{G} = \mathbf{G}_1 \times \mathbf{G}_2$, $\mathbf{G} = (Q, \Sigma, \delta, q_0, Q_m)$, is defined according to

$$Q = Q_1 \times Q_2$$
$$\delta((q_1, q_2), \sigma) = (\delta_1(q_1, \sigma), \delta_2(q_2, \sigma))$$

(just in case $\delta(q_1, \sigma)!$ and $\delta(q_2, \sigma)!$),

$$q_0 = (q_{1,0}, q_{2,0}), \quad Q_m = Q_{1,m} \times Q_{2,m}$$

Let $\mathbf{H} = (Y, \Sigma, \eta, y_0, Y)$ be a DES with marker subset the entire state set (so marked and closed behaviors coincide). We say that **H** is a *memory* for **G** if $L(\mathbf{H}) \supseteq L(\mathbf{G})$. Such **H** does not constrain **G** under product of DES, namely

$$L(\mathbf{G} \times \mathbf{H}) = L(\mathbf{G}), \quad L_m(\mathbf{G} \times \mathbf{H}) = L_m(\mathbf{G})$$

Memory can provide a medium for the 'recording' of specification languages restricting the behavior of **G**, in the following sense. Let $E \subseteq \Sigma^*$ be a closed language, and let $P \in Pred(Q \times Y)$. We say that the pair (E, P) is *compatible* with $\mathbf{G} \times \mathbf{H}$ if

$$L(\mathbf{G} \times \mathbf{H}, P) = E \cap L(\mathbf{G})$$

Thus enforcement of the predicate P on the state set of the product DES is equivalent to restricting the closed behavior of **G** to the language E. The precise connection is the following, where we write $\mathcal{CP}_{\mathbf{G}\times\mathbf{H}}(\cdot)$ for the family of controllable (with respect to $(\mathbf{G} \times \mathbf{H})$ subpredicates (on $Q \times Y$) of its argument, and $\mathcal{C}_{\mathbf{G}\times\mathbf{H}}(\cdot)$ (resp. $\mathcal{C}_{\mathbf{G}}(\cdot)$)

for the family of controllable (with respect to $\mathbf{G} \times \mathbf{H}$ (resp. $\mathbf{G}$)) sublanguages of its argument.

Theorem 1

Let $\mathbf{H}$ be a memory for $\mathbf{G}$. For $E \subseteq \Sigma^*$ unrestricted,

$$\mathcal{C}_{\mathbf{G}\times\mathbf{H}}(E) = \mathcal{C}_{\mathbf{G}}(E)$$

Also, if $E \subseteq \Sigma^*$ is closed and (E, P) is compatible with $\mathbf{G} \times \mathbf{H}$, then

$$L(\mathbf{G} \times \mathbf{H}, \sup \mathcal{CP}_{\mathbf{G}\times\mathbf{H}}(P)) = \sup \mathcal{C}_{\mathbf{G}}(E \cap L(\mathbf{G}))$$

Proof

Apply the definition of memory, and the result of Exercise 7.4.6 to $\mathbf{G} \times \mathbf{H}$. □

To implement behavior of the type just described, we bring in *dynamic state feedback control (DSFBC)*, defined as SFBC in the sense already employed, but now on the state set of $\mathbf{G} \times \mathbf{H}$. Thus a DSFBC is a map $f : Q \times Y \to \Gamma$, with component predicates $f_\sigma : Q \times Y \to \{0, 1\}$ for $\sigma \in \Sigma$, just as before. For emphasis we may refer to the pair $F := (f, \mathbf{H})$ as a *DSFBC for* $\mathbf{G}$ and write $L(F/\mathbf{G}), L_m(F/\mathbf{G})$ for the corresponding controlled sublanguages of $L(\mathbf{G}), L_m(\mathbf{G})$. Finally we say that the DSFBC F is *balanced* if f is balanced as a SFBC for $\mathbf{G} \times \mathbf{H}$ defined on $Q \times Y$. In this terminology we have the following consequence of Theorems 1 and 7.4.1, and Exercises 3 and 7.5.1.

Corollary 2

Let $E \subseteq \Sigma^*$, E closed, and assume $\sup \mathcal{C}_{\mathbf{G}}(E \cap L(\mathbf{G})) \neq \varnothing$. There exists a balanced DSFBC $F = (f, \mathbf{H})$ for $\mathbf{G}$ such that

$$L(F/\mathbf{G}) = \sup \mathcal{C}_{\mathbf{G}}(E \cap L(\mathbf{G}))$$ □

Exercise 3: Let $E \subseteq L(\mathbf{G})$ be nonempty and closed. Show that there exist a memory $\mathbf{H}$ for $\mathbf{G}$ and a predicate $P \in Pred(Q \times Y)$ such that (E, P) is compatible with $\mathbf{G} \times \mathbf{H}$. **Hint:** Consider the reachability tree for $L(\mathbf{G})$, or alternatively an automaton (recognizer) for E completed by an unmarked dump state so its closed behavior is Σ^*. $\mathbf{H}$ can be constructed by marking all states of this recognizer.

Exercise 4: For a 'distinguished' event $\alpha \in \Sigma$, let $\mathbf{H}$ model a counter that records the number of occurrences of α (mod 10 say). Suggest some pairs (E, P) that could be compatible with $\mathbf{G} \times \mathbf{H}$.

Exercise 5: Prove Theorem 1 and Corollary 2. **Hint:** Use the results of Exercises 3 and 7.4.6.

Finally we indicate how the principle of DSFBC can be adapted to modularity. With a view to later applications we assign to languages the primary role of specifications. Let $E_i \subseteq \Sigma^*$ $(i = 1, \ldots, k)$ be closed languages which we take to be specification languages for the closed behavior of $\mathbf{G}$. Let $E = E_1 \cap \cdots \cap E_k$.

For each i let $\mathbf{H}_i = (Y_i, \Sigma, \eta_i, y_{i,0}, Y_i)$ be a memory for $\mathbf{G}$, and $P_i \in Pred(Q \times Y_i)$ such that (P_i, E_i) is compatible with $\mathbf{G} \times \mathbf{H}_i$. Define $P \in Pred(Q \times Y_1 \times \cdots \times Y_k)$ according to

$$(q, y_1, \ldots, y_k) \models P \quad \text{iff} \quad (q, y_i) \models P_i, \quad i = 1, \ldots, k$$

Let $F_i = (f_i, \mathbf{H}_i)$ be a balanced DSFBC for $\mathbf{G}$ such that $L(F_i/\mathbf{G}) = \sup \mathcal{C}_{\mathbf{G}}(E_i \cap L(\mathbf{G}))$. Now define $F = (f, \mathbf{H}_1 \times \cdots \times \mathbf{H}_k)$ according to

$$f_\sigma(q, y_1, \ldots, y_k) = 1 \quad \text{iff} \quad f_{i,\sigma}(q, y_i) = 1, \quad i = 1, \ldots, k$$

With the foregoing conditions in place, we have

Theorem 6

The DES $\mathbf{H} := \mathbf{H}_1 \times \cdots \times \mathbf{H}_k$ is a memory for $\mathbf{G}$, the pair (E, P) is compatible with $\mathbf{G} \times \mathbf{H}$, $F := (f, \mathbf{H})$ is a balanced DSFBC for $\mathbf{G}$, and

$$L(F/\mathbf{G}) = L(\mathbf{G} \times \mathbf{H}, \sup \mathcal{CP}_{\mathbf{G}\times\mathbf{H}}(P)) = \sup \mathcal{C}_{\mathbf{G}}(E \cap L(\mathbf{G})) \qquad \square$$

Exercise 7: Prove Theorem 6, and provide a concrete illustration.

Exercise 8: Marking and nonblocking

In this chapter so far, the marker states of $\mathbf{G}$ have played no role, and nonblocking has not been explicitly treated. To complete the story in this respect, define a predicate $P \in Pred(Q)$ to be *nonblocking* for $\mathbf{G}$ if

$$(\forall q \models P)(\exists s \in \Sigma^*)\delta(q, s)! \;\&\; \delta(q, s) \in Q_m \;\&\; (\forall w \le s)\delta(q, w) \models P$$

Notice that P = *false* is trivially nonblocking; in any case, if $q \models P$, there is a path in Q leading from q to some state in Q_m along which P is satisfied. Define a balanced SFBC f for $\mathbf{G}$ to be *nonblocking* for $\mathbf{G}$ if $R(\mathbf{G}^f, true)$ is a nonblocking predicate. Assuming $P \neq$ *false*, show that there is a nonblocking SFBC f for $\mathbf{G}$ such that $R(\mathbf{G}^f, true) = P$, iff P is controllable and nonblocking. Next show that the family of nonblocking predicates stronger than a given predicate, say SPEC $\in Pred(Q)$, is closed under arbitrary disjunctions. Then show that the family of predicates that are stronger than SPEC and are both nonblocking and controllable, is closed under arbitrary disjunctions, hence possesses a weakest element, the supremal element of the family. If not *false*, this element will thus determine an optimal nonblocking SFBC that enforces SPEC. Go on to investigate the design requirement of nonblocking when SFBC is modular. As might be expected, the conjunction of nonblocking predicates generally fails to be nonblocking: prove this by example. Say two nonblocking predicates are *nonconflicting* if their conjunction is nonblocking. On this basis, develop corresponding refinements of the previous results of Sects. 7.5 and 7.6. Finally, link the above state-based approach to nonblocking with the linguistic approach of Chap. 3, identifying and proving the relations between dual concepts.

Exercise 9: Partial state observations

Extend the state feedback control theory of this chapter to the case where the state $q \in Q$ is known to the supervisor only modulo some equivalence relation $\rho \in \mathcal{E}(Q)$. Find a necessary and sufficient condition that $P \in Pred(Q)$ can be synthesized by SFBC. **Hint:** Consider (among other things) the 'observability' condition

$$(\forall q, q' \models P)(\forall \sigma \in \Sigma_c) q \equiv q' (\mathrm{mod} \rho) \,\&\, \delta(q,\sigma)! \,\&\, \delta(q',\sigma)! \,\&\, \delta(q,\sigma) \models P \\ \Rightarrow \delta(q',\sigma) \models P$$

Exercise 10: Observable predicate

Explore the definition of observable predicate $P \in Pred(Q)$, along the following lines. Let $\Sigma_o \subseteq \Sigma$ and $R : \Sigma^* \to \Sigma_o^*$ be the natural projection of languages. Say the predicate P is $(\mathbf{G}, R)$-*observable* iff the language $L(\mathbf{G}, P)$ is $(\mathbf{G}, R)$-observable in the sense of Chap. 6. Develop a parallel theory to that of Chap. 6 in the style of 'states and predicates'.

7.7 Notes

Predicate transformers were introduced by Dijkstra (1976) and first applied to DES control theory in Ramadge and Wonham (1987), Wonham and Ramadge (1988). This chapter is based on the latter references together with work of Li (1991), Li and Wonham (1988a, b, 1989, 1993). The two industrial examples in Appendix 7.1 are adapted from Sørensen et al. (1993) and Falcione and Krogh (1993) respectively.

A concrete representation for state-based modeling and supervision, called 'boolean discrete-event systems' (BDES), and including a treatment of memory, nonblocking, and forcing, is provided in the thesis of Wang (2000). In particular the Gas Burner (Exercise 1 in Appendix 7.1 below) is developed in detail.

Appendix 7.1: Two Industrial Examples

In this appendix we sketch two exercises on DES modeling by predicates, specifically in a boolean formalism where the state structure is determined by a list of independent propositional variables. For complementary details the reader may consult the cited references.

Exercise 1: Gas Burner (Sørensen et al. (1993))

The process state is given by the simultaneous truth values of 5 boolean variables; state transitions are triggered by any of 10 distinct events. Some of these are:

predicate HR = 'heat is required'
X = 'an explosion has occurred (system is in an exploded condition)'
event 'on' = 'system on-button is pushed'
'ex' = 'system explodes'

Transitions are determined by rules of two types: enablement rules ENR and next-state rules NSR. An example of ENR is: event 'on' can occur iff predicate HR $=$ *false*; while NSR might be: 'on' takes any state with HR $=$ *false* into the corresponding state with HR $=$ *true*. To each event there may be associated several rules, corresponding to different subsets of states. Multiple rules should be checked for consistency, to ensure transitions are deterministic.

The rule base is easily encoded into a case-statement associating with each event an if-then clause in terms of boolean state components. From this the one-step state transition matrix (32×10) is computed once for all; then the reduced transition structure over the reachability set can be generated by depth-first search. [This naive monolithic approach is exponentially complex in the number of predicates and is not recommended for large systems!]

1. Referring to the cited paper for ENR and NSR, carry out this procedure for the Gas Burner example, and verify the authors' state transition model, writing your own program to calculate the reachability set. You should get a final state size of just 9. In the notation of the paper, label events as follows:

on	off	df1	df2	df3	df4	sr	ff	cr	ex
0	1	2	3	4	5	6	7	8	9

and encode state components according to

HR	D	F	B	X
0	1	2	3	4

Each state component will take values 1 or 0, corresponding to *true* or *false*. States are numbered from 0 to 31 corresponding to all possible boolean combinations of the vector [HR D F B X], evaluated as

$$HR \star 1 + D \star 2 + F \star 4 + B \star 8 + X \star 16$$

where $\star$ means multiplication. For instance, the initial state is 2 = [0 1 0 0 0].

2. Extend the foregoing modeling procedure to include control, e.g. SCT disablement, assuming for instance that $\Sigma_c = \{\text{on, off, sr, cr}\}$. Comment on whether or not the system could then be safely turned on and, if not, suitably modify the design, for instance by using forcible events in the sense of Sect. 3.8. Generalize the approach to take into account that sometimes additional memory (an extended

state space) is required to represent the control specifications, namely when the latter involve conditions on the past behavior of the process. For instance, 'stop the process after 10 sensor failures' could be enforced with the aid of an auxiliary counter.

Exercise 2: Neutralization System (Falcione and Krogh (1993))

A neutralization system in an industrial chemical process could consist of a reaction tank, heater, mixer, and valves for filling and emptying the tank. The plant as seen by the programmable logic controller (PLC) consists of boolean variables to represent the operator setting (process off or on), fluid level, temperature and pressure. The controller variables are boolean variables for the valve states (closed or open), temperature and pH indicator lights (off or on), and heater and mixer states (off or on). Recall that pH is a measure of alkalinity; i.e. if pH is too low, more neutralizer is added. In detail the plant state variables are the following:

$$
\begin{array}{ll}
x_0 = \text{start} = 1 \text{ when process ON} & \\
x_1 = ls1 & \text{level} \geq \text{level1} \\
x_2 = ls2 & \text{level} \geq \text{level2} \\
x_3 = ls3 & \text{level} \geq \text{level3} \\
x_4 = ts & \text{temp} \geq \text{OK} \\
x_5 = as & \text{pH} \geq \text{OK}
\end{array}
$$

Here level1 $\leq$ level2 $\leq$ level3, i.e. the possible level combinations are $(x_1, x_2, x_3) = (0, 0, 0), (1, 0, 0), (1, 1, 0), (1, 1, 1)$.

The controller state variables are:

$$
\begin{aligned}
u_1 &= v1 = 1 \text{ iff valve1 is OPEN (tank fluid feed from reservoir)} \\
u_2 &= m \text{ mixer is ON} \\
u_3 &= h \text{ heater is ON} \\
u_4 &= tl \text{ temp indicator light is ON} \\
u_5 &= v4 \text{ valve4 is OPEN (tank fluid drain back to reservoir)} \\
u_6 &= v2 \text{ valve2 is OPEN (neutralizer feed)} \\
u_7 &= al \text{ pH indicator light is ON} \\
u_8 &= v3 \text{ valve3 is OPEN (tank fluid drain to next tank)}
\end{aligned}
$$

The controller state transition structure is given by assignment (:=) statements:

$$
\begin{aligned}
u_1 &:= (u_1 + x_0).\overline{u_8}.\overline{x_2} \\
u_2 &:= (u_2 + x_2).x_1 \\
u_3 &:= \overline{u_8}.x_2.\overline{x_4} \\
u_4 &:= x_2.x_4 \\
u_5 &:= (u_5 + x_3).x_2
\end{aligned}
$$

$$u_6 := \overline{u_5}.\overline{u_8}.x_2.\overline{x_5}$$

$$u_7 := x_2.x_5$$

$$u_8 := (u_8 + x_2.x_4.x_5).\overline{u_5}.x_1$$

Note that + is boolean disjunction (thus 1+1 = 1), $\overline{\cdot}$ boolean negation and . boolean conjunction.

Initially the process is OFF ($x_0 = 0$), the tank is empty ($x_1 = x_2 = x_3 = 0$), temperature is low ($x_4 = 0$), and pH is unspecified ($x_5 = 0$ or 1). In the controller, all valves are CLOSED ($u_1 = u_5 = u_6 = u_8 = 0$) and the mixer and heater are OFF ($u_2 = u_3 = 0$). From the control logic this implies that the indicator lights are both OFF ($u_4 = u_7 = 0$).

The rules of operation are the following. To start the process, set $x_0 := 1$, opening valve1 ($u_1 = 1$) which stays open until the tank has filled to level2 ($x_2 = 1$). When level2 is reached start the mixer ($u_2 := 1$); if the level drops below level1 ($x_1 = 0$) stop the mixer ($u_2 := 0$). Energize the heater ($u_3 := 1$) if the temperature is too low ($x_4 = 0$) and the tank level is at least level2 ($x_2 = 1$). If pH is too low ($x_5 = 0$) open the neutralizer feed valve2 ($u_6 := 1$). If the tank becomes full (level $\geq$ level3, i.e. $x_3 = 1$) then open valve4 (drain tank fluid back to reservoir, i.e. $u_5 := 1$); this will close valve2 ($u_6 := 0$) to stop the flow of neutralizer. When the fluid level drops just below level2 ($x_2 = 0$), close valve4 ($u_5 := 0$). When both temperature and pH are OK ($x_4 = x_5 = 1$), de-energize the heater ($u_3 := 0$) and open valve3 ($u_8 := 1$) to drain the tank. When the tank is empty ($x_1 = 0$), close valve3 ($u_8 := 0$) and restart the process.

While the description specifies how the controller reacts to the plant, no model has been specified as to how the plant responds to the controller, so the control loop is open, and no analysis of controlled behavior is possible. One could experiment with various DES models for the plant. For instance, take the plant to be the shuffle of 3 processes, LEVEL, TEMP and ALK. LEVEL has the 4 states listed above, with events 'level_rise' and 'level_drop', where 'level_rise' is enabled when valve1 or valve2 is OPEN and valve3 and valve4 are CLOSED (i.e. $(u_1 + u_6).\overline{u_5}.\overline{u_8}$), while 'level_drop' is enabled when valve1 and valve2 are CLOSED and valve3 or valve4 is OPEN (i.e. $\overline{u_1}.\overline{u_6}.(u_5 + u_8)$). TEMP has 2 states ($x_4 = 0$ or 1) with events 'temp_rise', enabled when heater is ON ($u_3 = 1$) and 'temp_drop', enabled when $u_3 = 0$. Similarly ALK has 2 states ($x_5 = 0$ or 1) with events 'alk_rise' enabled when valve2 is OPEN ($u_6 = 1$). Notice that what we have here is a simple qualitative model of the plant physics. The example illustrates why there is current interest in qualitative modeling: this is simply high-level discrete aggregation, and (if the model is valid) is all you need for logic control design.

The plant transition triples are:

[level0,level_drop, level0]

[level0,level_rise, level1]

[level1,level_drop, level0]

[level1,level_rise, level2]

[level2,level_drop, level1]

[level2,level_rise, level3]

[level3,level_drop, level2]

[level3,level_rise, level3]

Note that level_drop (resp. level_rise) leaves level0 (resp. level3) unchanged.

In this formalism one can regard the occurrence of an event in a given time window or sampling interval as a boolean variable. Recalling that level0 = (0, 0, 0), level1 = (1, 0, 0), level2 = (1, 1, 0), level3 = (1, 1, 1), we can bring in new boolean level variables

$$
\begin{aligned}
l0 &= \overline{x_1}.\overline{x_2}.\overline{x_3} \\
l1 &= x_1.\overline{x_2}.\overline{x_3} \\
l2 &= x_1.x_2.\overline{x_3} \\
l3 &= x_1.x_2.x_3
\end{aligned}
$$

with transition assignments

$$
\begin{aligned}
&l0 := l0.\overline{\text{level\_}rise} + l1.\text{level\_drop} \\
&l1 := l1.\overline{\text{level\_rise}}.\overline{\text{level\_drop}} + l0.\text{level\_rise} + l2.\text{level\_drop}
\end{aligned}
$$

and so on. These rules could be expressed in terms of the x_i by the equations

$$
\begin{aligned}
x_1 &= l1 + l2 + l3 \\
x_2 &= l2 + l3 \\
x_3 &= l3
\end{aligned}
$$

The events are enabled or disabled according to the controller and plant states, i.e. events can occur only in control and plant states where they are enabled in accordance with the physical interpretation. Thus the enablement preconditions for rise or drop in level are:

$$
\begin{aligned}
\text{enable(level\_rise)} &= (u_1 + u_6).\overline{u_5}.\overline{u_8} \\
\text{enable(level\_drop)} &= \overline{u_1}.\overline{u_6}.(u_5 + u_8)
\end{aligned}
$$

The transition rules are

$$
\begin{aligned}
\text{level\_rise} &:= \text{enable(level\_rise).random} \\
\text{level\_drop} &:= \text{enable(level\_drop).}\overline{\text{random}}
\end{aligned}
$$

where random (= boolean 0 or 1) represents the mechanism of event selection.

Similarly

$$x_4 := \overline{x_4}.\text{temp\_rise} + x_4.\,\overline{\text{temp\_drop}}$$
$$x_5 := \overline{x_5}.\text{alk\_rise} + x_5$$

with preconditions

$$\begin{aligned}\text{enable(temp\_rise)} &= u_3\\ \text{enable(temp\_drop)} &= \overline{u_3}\\ \text{enable(alk\_rise)} &= u_6\end{aligned}$$

Note that we have embodied a physical assumption that if $u_3 = 0$ (heater is OFF) then temp_drop is enabled, so temperature could drop but need not; whereas if temperature is OK and heater is ON then temp_drop is disabled, and the condition $x_4 = 1$ is stable. A more refined model could include a temperature controller with ON/OFF thresholds bracketing the setpoint.

The equations can be used for simulation or analysis, with an enabled event chosen randomly at each simulation step and the state and control variables updated. The process can now be considered a standard SCT model with the feedback already incorporated: for each state (x, u) certain events will be enabled and exactly one of them will be selected as the next event. Then (x, u) is updated and the process repeats.

In general the model is of the form

x_new := qualitative_physics(x,u,events), u_new := control_logic(x, u)

As always, the total process with state (x, u) is driven by the events constituting the changes in physical state under the action of the plant dynamics and control logic.

The archtypal system of this general kind is a level controller (e.g. water pump or furnace control), where the plant state variables are x_1, x_2 with

$$\begin{aligned}x_1 &= 1 \quad \text{iff level} \geq \text{level1}\\ x_2 &= 1 \quad \text{iff level} \geq \text{level2}\end{aligned}$$

and level2 > level1 as above. The controller variable is u, with $u = 1$ representing that the pump or furnace is ON. The process has a main switch TOGGLE.

Then

$$u := \text{TOGGLE}.(u + \overline{x_1}).\overline{x_2}$$

namely 'keep pumping if (already pumping or level < level1) and (level < level2)'. So the pump starts low, stops high, and tries to keep level between level1 and level2. Once the level reaches level2 the pump shuts off, only turning on again if the level drops below level1. Ignoring TOGGLE we have

$$u := f(u, x_1, x_2), \text{ say}$$

with u_new = 1 iff $x_2 = 0$ and either $u = 1$ or $x_1 = 0$.

The 'correctness' of such a controller means that the plant state subset $\{(x_1, x_2)|x_1 = 1\}$ is globally attractive, like the regulation condition 'tracking error = 0' of standard control theory. Namely the system should eventually reach and remain in this subset from any initialization and in the absence of further disturbances. In the absence of disturbances (e.g. leaks or cooling) a trivial plant model shows the attractive state is (1,1) because, when once $u = 1$, the condition $u = 1$ is maintained until $x_2 = 1$. If $x_2 = 1$ then $x_1 = 1$ (by the plant model) and, again by the plant model, in the absence of disturbances x_2 just remains at 1. So one has to think of the plant model as autonomous except for occasional disturbances, which are then fully corrected, just as in regulation theory. The effect of a disturbance is just to move the plant to some arbitrary initial state. Thus the (undisturbed) plant model has to say that if $x_1 = 0$ and u is kept at 1 then eventually $x_1 = 1$; if $x_1 = 1$ and u is kept at 1 then eventually $x_2 = 1$; and if once $x_2 = 1$ then (regardless of u) $x_2 = 1$ always. One could get u to turn off when $x_2 = 1$ by considering that u is controllable (can be disabled) and introducing an overflow condition that is to be prohibited.

In the neutralization process, the assumptions are that temperature (resp. pH) always rises as long as the heater is ON (resp. the neutralizer valve is OPEN); and that level always rises (resp. falls) under appropriate valve conditions of OPEN or CLOSED. One then needs to prove that from the empty tank condition (or possibly other, disturbed, initial conditions) the target temperature and pH are reached with level $\geq$ level1; and that subsequently the tank is emptied in an appropriate way. This could be done either with brute force calculation or by analysis using logic and/or stability (e.g. Liapunov) arguments. Considerations of this sort are explored in the currently popular area of hybrid (mixed discrete/continuous) systems.

Write a simulation program for the neutralization process, exploring various plant models. Develop standard DES (generator/automaton) representations, and investigate supervisory control and stability.

References

Dijkstra EW (1976) A discipline of programming. Prentice-Hall, Upper Saddle River

Falcione A, Krogh B (1993) Design recovery for relay ladder logic. IEEE Control Syst 13(2):90–98

Li Y (1991) Control of vector discrete-event systems. PhD thesis, Department of Electrical Engineering, University of Toronto

Li Y, Wonham WM (1988a) Controllability and observability in the state-feedback control of discrete-event systems. In: Proceedings of the twenty-seventh IEEE conference on decision and control, New York, pp 203–208

Li Y, Wonham WM (1988b) A state-variable approach to the modeling and control of discrete-event systems. In: Proceedings of the twenty-sixth annual Allerton conference on communication, control, and computing, University of Illinois, pp 1140–1149

Li Y, Wonham WM (1989) Composition and modular state-feedback control of vector discrete-event systems. In: Proceedings of the conference on information sciences and systems, The Johns Hopkins University, Baltimore, pp 103–111

Li Y, Wonham WM (1993) Control of vector discrete-event systems: I - the base model. IEEE Trans Autom Control 38(8):1214–1227; Correction: IEEE Trans Autom Control 39(8):1771, (1994). [Figures 9–11 are reprinted, with permission, from IEEE.]
Ramadge PJ, Wonham WM (1987) Modular feedback logic for discrete event systems. SIAM J Control Optim 25(5):1202–1218
Sørensen EV, Nordahl J, Hansen NH (1993) From CSP models to Markov models. IEEE Trans Softw Eng 19(6):554–570
Wang YW (2000) Supervisory control of boolean discrete-event systems. MASc thesis, Department of Electrical and Computer Engineering, University of Toronto
Wonham WM, Ramadge PJ (1988) Modular supervisory control of discrete event systems. Math Control Signals Syst 1(1):13–30

Chapter 8
Supervision of Vector Discrete-Event Systems

The state-based approach to DES supervisory control of Chap. 7 will be applied to a class of models known as vector addition systems, which when endowed with our standard control technology we call vector discrete-event systems (VDES). VDES offer a modeling option especially for systems incorporating groups of similar entities (e.g. machines in a workcell) which for control purposes need not be individually distinguished; often in practice such systems are equivalent to a synchronous product of buffers. VDES may be represented graphically as Petri nets (PN); results from the extensive PN literature will be exploited to advantage. We further consider problem restrictions under which an optimal supervisor itself admits a VDES representation, providing examples when this is or respectively is not the case.

8.1 Introduction

In this chapter we specialize the control theory of discrete-event systems developed previously, to a setting of vector addition systems. We adopt the state-based approach of Chap. 7. It is natural to enhance the abstract automaton model by exploiting algebraic regularity of internal system structure when it exists. An obvious instance of such structure is arithmetic additivity over the integers. For example, the state of a manufacturing workcell might be the current contents of its buffers and the numbers of machines in various modes of operation: thus, when a machine completes a work cycle, the status vector of machines and the vector of buffer contents would be suitably incremented. Similar examples are furnished by various kinds of traffic systems.

System modeling by vector addition systems is a long-standing technique, especially in the setting of Petri nets. For us, however, Petri nets will serve only as a graphical representation tool, and we make little use of net theory as such. Our

W. M. Wonham and K. Cai, *Supervisory Control of Discrete-Event Systems*, Communications and Control Engineering,
https://doi.org/10.1007/978-3-319-77452-7_8

treatment will be self-contained, using only elementary linear algebra and integer programming.

We first develop the base model, then the feedback synthesis of control-invariant state subsets, along with the appropriate versions of controllability, observability and modularity. These results are illustrated by examples from the literature on manufacturing systems.

8.2 Vector Discrete-Event Systems

Let $\mathbb{Z}$ denote the integers $\{\ldots, -1, 0, 1, \ldots\}$ and $\mathbb{N}$ the natural numbers $\{0, 1, 2, \ldots\}$. Let $\mathbb{Z}^n$ (resp. $\mathbb{N}^n$) denote the space of n-vectors (i.e. ordered n-tuples) with components in $\mathbb{Z}$ (resp. $\mathbb{N}$), along with vector addition, and scalar multiplication by integers (resp. natural numbers). In algebra, $\mathbb{Z}^n$ so equipped is a 'module over the ring of integers', not a vector space; nevertheless we shall loosely speak of its elements as 'vectors' and use vector space terminology on grounds of familiarity. We shall employ the 'direct sum' operation $\oplus$ to form structures like $\mathbb{Z}^n \oplus \mathbb{Z}^m$ ($\simeq \mathbb{Z}^{n+m}$ under the obvious isomorphism) or $\mathbb{N}^n \oplus \mathbb{Z}^m$. In such cases we may write $x = x' \oplus x''$ to denote the decomposition of x into its natural projections onto the direct summands. If v is an arbitrary vector in some $\mathbb{Z}^k$, we write $v \geq 0$ to mean that each component of v is nonnegative, i.e. that in fact $v \in \mathbb{N}^k$, thought of as embedded in $\mathbb{Z}^k$.

A *vector discrete-event system (VDES)* is a DES $\mathbf{G} = (X, \Sigma, \xi, x_0, X_m)$ defined as in previous chapters (although we make a notational change to X from Q and to ξ from δ), but with vector structure, in the foregoing sense, attached to the state set X and transition function ξ. Thus in general $X = \mathbb{N}^n \oplus \mathbb{Z}^m$, while $\xi : X \times \Sigma \to X$ will always have the additive form:

$$\xi(x, \sigma) = x + e_\sigma$$

for some $e_\sigma \in \mathbb{Z}^{n+m}$, the *displacement vector* corresponding to σ. Writing $x = x' \oplus x''$ as above, we note that in general ξ will be a partial function, defined for just those (x, σ) pairs such that $x' \in \mathbb{N}^n$ and $(x + e_\sigma)' = x' + e'_\sigma \in \mathbb{N}^n$, or briefly $x' \geq 0$, $x' + e'_\sigma \geq 0$; however, no such restriction will apply to the second components x'' and $(x + e_\sigma)''$, in $\mathbb{Z}^m$. In particular if the $\mathbb{N}^n$ summand is absent, so $X = \mathbb{Z}^m$, then ξ will be a total function. By the usual inductive definition, ξ is extended to strings $s \in \Sigma^*$, so from now on we consider $\xi : X \times \Sigma^* \to X$ (pfn).

Let Σ be indexed as $\Sigma = \{\sigma_1, \ldots, \sigma_k\}$ and write e_i for e_{σ_i}. With $X = \mathbb{N}^n \oplus \mathbb{Z}^m$ as above, write $p := n + m$ and regard x and the e_i as (column) arrays of size $p \times 1$. Bring in the matrix

$$E := [e_1 \cdots e_k] \in \mathbb{Z}^{p \times k}$$

the *displacement matrix* for $\mathbf{G}$. Now consider strings $s \in \Sigma^*$. It will be useful to count the occurrences of the various event symbols in s. For this define

$$V:\Sigma^* \to \mathbb{N}^k: s \mapsto [v_1(s) \cdots v_k(s)] \in \mathbb{N}^{k\times 1}$$

where $v_j(s)$ is the number of occurrences of σ_j in s. [Note that we may display a column array as a row, to save space: the definition should resolve any ambiguity, and our vector-matrix operations will always be consistent with the array definitions.] $V(s)$ is the *occurrence vector* of s. $V(\cdot)$ can be regarded as a morphism of monoids ($\mathbb{N}^k$ is an *additive* or *commutative*, monoid), with

$$V(\epsilon) = 0 \in \mathbb{N}^k, \quad V(s_1 s_2) = V(s_1) + V(s_2)$$

In this notation we have

$$\xi(x, s)! \Rightarrow \xi(x, s) = x + EV(s)$$

The evaluation of $\xi(x, s)$ depends just on x and $V(s)$, but it makes sense only when $\xi(x, s)$ is defined. With $X \doteq \mathbb{N}^n \oplus \mathbb{Z}^m$ and $x = x' \oplus x''$, $\xi(x, s)!$ iff $(x + EV(w))' \geq 0$ for all $w \leq s$, namely nonnegativity of the $\mathbb{N}^n$-projection is preserved for all prefixes of s. Thus it might well be that for certain x and s one has $x' \geq 0$ and $(x + EV(s))' \geq 0$, but for some w, $\epsilon < w < s$, the nonnegativity condition fails; in that case, $\xi(x, s)$ is not defined.

Remark 1: A more general definition of VDES which might be preferred results on strengthening the enablement conditions as follows. Suppose $X = \mathbb{N}^n$. Given $e_\sigma \in \mathbb{Z}^n$ as before, let $f_\sigma \in \mathbb{N}^n$ be any vector $\geq \max\{0, -e_\sigma\}$ (computed componentwise), and declare that $\xi(x, \sigma)!$ iff $x \geq f_\sigma$. This will guarantee that $x + e_\sigma \geq 0$. Alternatively, one can pick vectors $e_\sigma^+, e_\sigma^- \in \mathbb{N}^n$ independently and define $e_\sigma := e_\sigma^+ - e_\sigma^-$, $f_\sigma := e_\sigma^-$; this is equivalent to the usual transition enablement definition for a Petri net allowing selfloops. In this chapter we use only the restricted definition corresponding to the choice $f_\sigma = -e_\sigma$. See, however, Exercise 8.8.2 for how a selfloop can be simulated when it is needed.

As in Sect. 7.3, denote by $L(\mathbf{G}, x_0)$ the closed behavior generated by $\mathbf{G}$ when initialized at state $x_0 \in X$:

$$L(\mathbf{G}, x_0) := \{s \in \Sigma^* | \xi(x_0, s)!\}$$

The following exercise illustrates a modeling limitation of VDES, namely not all finite-state generators can be so represented, in the sense of closed behavior equivalence.

Exercise 2: Consider the DES $\mathbf{G}$ in Fig. 8.1. Show that there exists no 'generalized' VDES $\mathbf{H}$ (cf. Remark 1) such that $L(\mathbf{H}, x_0) = L(\mathbf{G})$.

Exercise 3: Let $\mathbf{G}$ be a VDES with no selfloops, state set $X = \mathbb{N}^n$, and matrix description

Fig. 8.1 DES **G** (all states considered marked)

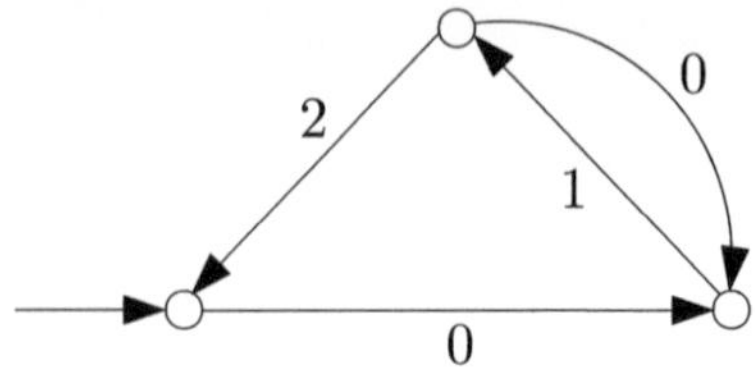

$$x(s) = x_0 + EV(s)$$

for $x_0 \in X, s \in L(\mathbf{G}, x_0)$. Here $E \in \mathbb{Z}^{n\times m}$, $V \in \mathbb{N}^{m\times 1}$. Borrowing the terminology of Petri nets, define **G** to be *structurally bounded* if, for every choice of initial state $x_0 \in X$, there is an integer b such that $|x(s)| \leq b$ for all $s \in L(\mathbf{G}, x_0)$, where e.g. $|x| := \max\{|x_1|, \ldots, |x_n|\}$. Prove that **G** is structurally bounded iff there exists $c = [c_1 \cdots c_n] \in \mathbb{N}^{1\times n}$ with all $c_i > 0$, such that $cE \leq 0$ (i.e. the vector $cE \in \mathbb{Z}^{1\times m}$ has each component ≤ 0). In this case obtain an estimate of b. Extend your result to 'generalized' VDES with selfloops in the sense of Remark 1. **Hint**: Proof of the 'if' statement is straightforward; for the deeper 'only if' statement you will need the following 'alternative' theorem from linear inequalities.

Theorem 4

The dual systems (over $\mathbb{Z}$)

$$t \geq 0, \quad Et \geq 0; \quad cE \leq 0, \quad c \geq 0$$

possess solutions t^* and c^* for which the following alternative holds:

$$\text{either} \quad Et^* \neq 0 \quad \text{or} \quad c^* > 0 \qquad \square$$

Use this theorem to show that if the required vector c fails to exist then **G** is not structurally bounded.

Finally, prove that if **G** is structurally bounded, with vector $c > 0$ as above, then

$$x_i(s) \leq cx_0/c_i\ , \quad s \in L(\mathbf{G}, x_0), \quad i = 1, \ldots, n$$

If $b_i := \lfloor cx_0/c_i \rfloor$ $(i = 1, \ldots, n)$, show how to represent **G** as the synchronous product of n buffers $\mathbf{B_1}, \ldots, \mathbf{B_n}$, where the capacity of $\mathbf{B_i}$ is b_i. [Notation: for $r \in \mathbb{R}$ and $r \geq 0$, $\lfloor r \rfloor := \max\{k \in \mathbb{N} | k \leq r\}$.] Thus a structurally bounded VDES in $\mathbb{N}^n$ is nothing other than the synchronous product of n buffers, one for each state component. This representation can be used to check, for instance, whether the given VDES is nonblocking.

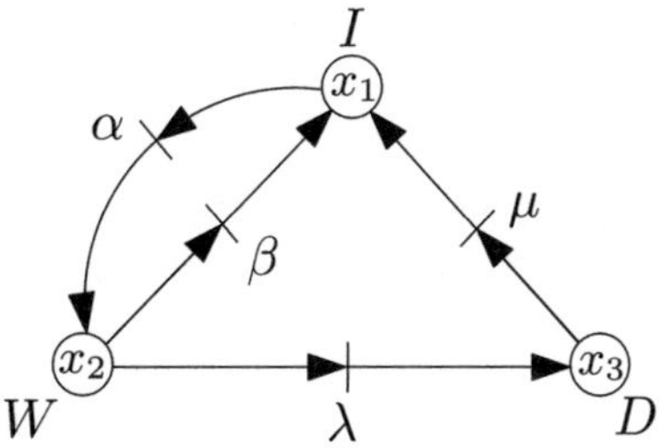

Fig. 8.2 Petri net representation of factory, where ○ denotes a 'place' and $\longrightarrow\!\!\!\!\!\!+\!\!\longrightarrow$ denotes a labeled transition

8.3 VDES Modeling

Example 1: **FACT#1**

Consider a factory consisting of 10 machines, each a DES modeled over the alphabet $\Sigma = \{\alpha, \beta, \lambda, \mu\}$ in the usual way, and displayed as a Petri net in Fig. 8.2. We do not distinguish the machines individually, being interested only in the numbers resident in the three state components indexed $\{1, 2, 3\}$ corresponding respectively to {Idle, Working, Down}. The state of the factory is then $x \in \mathbb{N}^3$, with $x_i \geq 0$ and $x_1 + x_2 + x_3 = 10$. If it is assumed that two or more machines never make a transition simultaneously, then the possible transitions are $x \mapsto x + e_\sigma$ $(\sigma \in \Sigma)$, with

$$E = [e_\alpha \; e_\beta \; e_\lambda \; e_\mu] = \begin{bmatrix} -1 & 1 & 0 & 1 \\ 1 & -1 & -1 & 0 \\ 0 & 0 & 1 & -1 \end{bmatrix}$$

Thus the effect, if any, of each event on a state component is just to increment or decrement it by 1. The initial state could be taken as $x_0 = [10 \; 0 \; 0] \in \mathbb{N}^{3\times 1}$ and a transition is defined iff the condition $x \geq 0$ is preserved.

Note that, at the expense of additional notation, it would be quite straightforward to model the occurrence of compound events, defined as the simultaneous execution of individual events by distinct machines. Just as before, a compound event would be represented by a suitable displacement vector and restricted by the appropriate nonnegativity condition. For example, if either two or three events α could occur together, label these simultaneous events $\alpha 2$, $\alpha 3$ and bring in the corresponding displacements

$$e_{\alpha 2} = [-2 \; 2 \; 0], \qquad e_{\alpha 3} = [-3 \; 3 \; 0]$$

The corresponding transitions are defined just in case, respectively, $x_1 \geq 2$ or $x_1 \geq 3$.

Example 2: **FACT#2**

By augmenting the state of **FACT#1** one can bring in 'memory' to record features of past behavior. For instance, let

$$x_4 := \#\text{workpieces produced}$$
$$= \#\text{occurrences of event } \beta \text{ since initialization}$$

Thus x_4 models a counter for β, with initially $x_4 = 0$. Making this extension to **FACT#1** yields the new state space $X = \mathbb{N}^4$ and displacement matrix for **FACT#2**:

$$E = \begin{bmatrix} -1 & 1 & 0 & 1 \\ 1 & -1 & -1 & 0 \\ 0 & 0 & 1 & -1 \\ 0 & 1 & 0 & 0 \end{bmatrix}$$

Note that x_4 may grow without bound as the process evolves, although in an application one might impose a control specification such as $x_4 \leq 100$, perhaps to respect the capacity of a storage buffer.

Example 3: **FACT#3**

We extend **FACT#2** to allow a count of the excess of successful work cycles over machine breakdowns, defining $x_5 := \#\beta - \#\lambda$ (where $\#\sigma$ denotes the number of occurrences of σ since initialization). Now we must allow $x_5 \in \mathbb{Z}$ and take for the new state space $X = \mathbb{N}^4 \oplus \mathbb{Z}$. Then, for instance,

$$e_\beta = [1\ -1\ 0\ 1\ 1], \quad e_\lambda = [0\ -1\ 1\ 0\ -1] \in \mathbb{Z}^{5\times 1}$$

and the reader will easily construct E.

Remark 4: **Simultaneous events**

We generalize the remark at the end of Example 1 as follows. Regard the event set $\Sigma = \{\sigma_1, \ldots, \sigma_m\}$ as a basis set of simple events for the generation of compound events, defined as arbitrary formal linear combinations $\sigma(r_1, \ldots, r_m) := r_1\sigma_1 + \cdots + r_m\sigma_m$, where the $r_j \in \mathbb{N}$ (thus any simple event is compound). The full event set, including the null event, is now the set of all compound events. The interpretation of a compound event is just that $r_j \geq 0$ instances of σ_j occur simultaneously, for all $j = 1, \ldots, m$. In Example 1, an event $2\alpha + 3\beta + \lambda + 3\mu$ would mean that simultaneously 2 machines start work, 3 end work successfully, 1 breaks down and 3 complete repair. Intuition suggests that this only makes sense if we imagine a small time delay between the 'beginning' of a compound event and its 'termination', so the foregoing event can occur only if the state vector satisfies $x \geq [2\ 3+1\ 3] = [2\ 4\ 3]$. In general for $\sigma \in \Sigma$ write $e_j := e_{\sigma_j}, e_j^- := \max\{-e_j, 0\}$, and declare that $\xi(x, \sigma(r_1, \ldots, r_m))!$ iff $x \geq r_1e_1^- + \cdots + r_me_m^-$, in which case

$$\xi(x, \sigma(r_1, \ldots, r_m)) := x + r_1e_1 + \cdots + r_me_m$$

In this chapter we prohibit (nonsimple) compound events, albeit they are sometimes of interest in applications, especially in problems of mutual exclusion.

Exercise 5: Show that in Example 1, $\sigma(r_1, r_2, r_3, r_4)$ is enabled only if $r_1 + r_2 + r_3 + r_4 \leq 10$, so the set of compound events that can actually occur is finite. Exactly how many are there? Find a general formula for the case $r_1 + \cdots + r_k \leq N$.

Exercise 6: With $\xi(x, s) = x + EV(s)$ whenever $\xi(x, s)!$, and $E \in \mathbb{Z}^{n\times k}$, interpret solutions $t \in \mathbb{Z}^{k\times 1}$ of $Et = 0$, and solutions $p \in \mathbb{Z}^{1\times n}$ of $pE = 0$. In the terminology of Petri nets, such t are 'transition invariants' and p are 'place invariants'. For an application of the latter see Exercise 8.13.20.

8.4 Linear Predicates

It will be appropriate to require of predicates on X that they be compatible with the algebraic setting, in particular that they be generated by basic predicates that are 'linear' in the state. Such predicates occur commonly in the applications: e.g. in conditions like 'either $x_1 \leq 5$ or $x_2 > 10$'. For simplicity of notation consider that $X = \mathbb{Z}^n$ (if X has a component $\mathbb{N}^m$ then embed $\mathbb{N}^m \subseteq \mathbb{Z}^m$). Our basic predicates P will be those that distinguish state subsets $X_P \subseteq X$ of the form

$$X_P := \{x = [x_1 \cdots x_n] \in \mathbb{Z}^n | a_1x_1 + \cdots + a_nx_n \leq b\}$$

where $a_1, \ldots, a_n, b \in \mathbb{Z}$. Representing $a \in \mathbb{Z}^{1\times n}$ (a row array) and $x \in \mathbb{Z}^{n\times 1}$ (a column array), we have succinctly,

$$X_P = \{x \in X | ax \leq b\}$$

or

$$x \models P \text{ iff } ax \leq b$$

Call such P a *linear predicate* on X and denote by $Pred_{\mathbf{lin}}(X)$ the corresponding subset of $Pred(X)$. Finally, let $\overline{Pred_{\mathbf{lin}}(X)}$ be the boolean closure of $Pred_{\mathbf{lin}}(X)$, namely the smallest subset of $Pred(X)$ that contains $Pred_{\mathbf{lin}}(X)$ and is closed under the boolean operations $\neg$, $\wedge$ and $\vee$. We have $P \in \overline{Pred_{\mathbf{lin}}(X)}$ iff P can be built up by applying the boolean operations a finite number of times to a finite number of predicates in $Pred_{\mathbf{lin}}(X)$. Thus $P = (x_1 \leq 5) \vee (\neg(x_2 \leq 10))$ for the example above.

Exercise 1: If $X = \mathbb{N}^n \oplus \mathbb{Z}^m$ and $S \subset X$ is a finite subset, show that the predicate $P = (x \in S)$ belongs to $\overline{Pred_{\mathbf{lin}}(X)}$.

Exercise 2: Show that $\overline{Pred_{\mathbf{lin}}(X)}$ is a proper boolean subalgebra of $Pred(X)$.

8.5 State Feedback and Controllability of VDES

In this section we exemplify for VDES the definitions and results of Sect. 7.4. Recalling **FACT#2** in Sect. 8.3 let us assume that items produced (on occurrence of event β) are placed in a buffer of capacity 100. To prevent possible buffer overflow we must maintain the control specification

$$\#\text{ free slots in buffer } \geq \#\text{ machines working}$$

or $100 - x_4 \geq x_2$, i.e. the linear predicate

$$\text{SPEC} := (x_2 + x_4 \leq 100)$$

Assume $\Sigma_c = \{\alpha\}$. Now SPEC is true under the initial condition $x_0 = [10\ 0\ 0\ 0]$. It is easily seen that SPEC is maintained true provided α is enabled only if SPEC holds with positive slack, i.e. when $x_2 + x_4 < 100$. For this define SFBC $f : X \times \Sigma \to \{0, 1\}$ according to

$$f_\alpha(x) = \begin{cases} 1 \text{ if } x_2 + x_4 < 100 \\ 0 \text{ otherwise} \end{cases}$$

Exercise 1: Draw a graph with vertical axis $x_2 + x_4$, and horizontal axis 'real' time marked with the quasi-random instants of successive events, corresponding to the string $\alpha\alpha\beta\alpha\lambda\alpha\mu\beta\lambda\ldots$. Next, assuming the process has run long enough to approach the 'control boundary' $x_2 + x_4 = 100$, generate a string to display (on a similar graph) the action of the SFBC f.

Exercise 2: Reasoning *ad hoc* show that f enforces SPEC with minimal restriction placed on the behavior of **FACT#2**.

We shall be using the theory of Sect. 7.4 to solve problems of the foregoing type systematically. We begin by illustrating reachability and controllability for VDES.

Example 3: **FACT#4**

We simplify **FACT#2** by eliminating the breakdown/repair feature, taking $\Sigma = \{\alpha, \beta\}$, $\Sigma_c = \{\alpha\}$ as shown in Fig. 8.3. In **FACT#4** $X = \mathbb{N}^3$, with $x = [x_1\ x_2\ x_3] \in \mathbb{N}^{3\times 1}$; x_1, x_2 are the number of machines idle (I) and working (W) respectively; x_3 is the number of occurrences of β since initialization; and the initial state is $x_0 = [10\ 0\ 0]$. Thus

$$E = [e_\alpha\ e_\beta] = \begin{bmatrix} -1 & 1 \\ 1 & -1 \\ 0 & 1 \end{bmatrix}$$

We first consider the predicate $P \in Pred(X)$ given by

Fig. 8.3 FACT#4

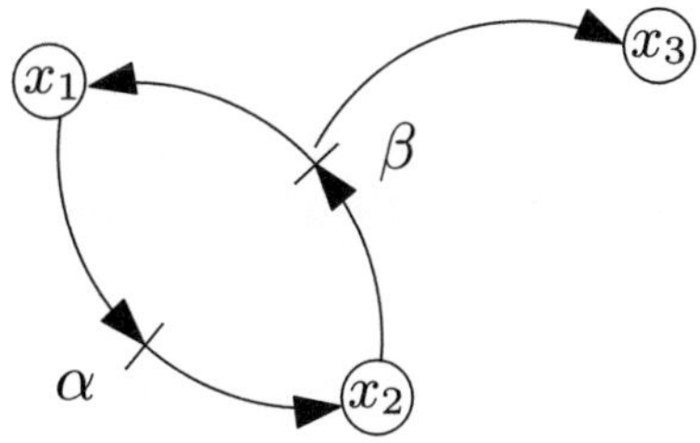

$$x \models P \quad \text{iff} \quad (x_1 + x_2 = 10)\ \&\ (x_3 \leq 100)$$

P amounts to our previous buffer constraint, conjoined with the 'obvious' invariant on the total number of machines in the system, included here for technical convenience.

To investigate whether P is controllable, we start by checking the condition

$$P \preceq R(\mathbf{FACT\#4}, P) \qquad (?)$$

Let $x \models P$, i.e. $x = [10 - j \;\; j \;\; k]$ for some $j, k{:}0 \leq j \leq 10,\ 0 \leq k \leq 100$. We attempt to construct a string that will lead **FACT#4** from x_0 to x while preserving P. For instance, let $s := (\alpha\beta)^k\alpha^j$, corresponding to k successful work cycles followed by j machine transitions from I to W. It is clear that $x = \xi(x_0, s)!$. We claim:

$$(\forall w \leq s) y := \xi(x_0, w)! \quad \& \quad y \models P$$

In fact it is easy to verify $\xi(x_0, w)!$ and $y \models P$ for strings w of each of the prefix types

$$w = (\alpha\beta)^l, \quad 0 \leq l \leq k - 1$$
$$w = (\alpha\beta)^l\alpha, \quad 0 \leq l \leq k - 1$$
$$w = (\alpha\beta)^k\alpha^l, \quad 0 \leq l \leq j$$

This proves the claim, and thus (?) above. Next we check

$$(\forall \sigma \in \Sigma_u) P \preceq M_\sigma(P) \qquad (??)$$

Let $\sigma = \beta$ and $x = [9\ 1\ 100]$. We have $x \models P$, $\xi(x, \beta)!$ and

$$y := \xi(x, \beta) = x + e_\beta = [10\ 0\ 101]$$

Since *not* $y \models P$, (??) fails, so P is not controllable.

Exercise 4: Explain intuitively why P is not controllable, i.e. how P might fail under the occurrence of an uncontrollable event.

Now consider the alternative predicate $Q \in Pred(X)$ given by

$$x \models Q \quad \text{iff} \quad (x_1 + x_2 = 10) \ \& \ (x_3 \le 100 - x_2)$$

In this case $x \models Q$ iff $x = [10 - j \ \ j \ \ k]$ where $0 \le j \le 10, 0 \le k \le 100 - j$. Taking $s = (\alpha\beta)^k \alpha^j$ we verify $x \models R(\mathbf{G}, Q)$ as shown above for P. To check $Q \preceq M_\beta(Q)$ observe that

$$\begin{aligned}\xi(x, \beta)! \ \ &\text{iff} \ \ x + e_\beta \ge 0\\ &\text{iff} \ \ [10 - j + 1 \ \ j - 1 \ \ k + 1] \ge 0\\ &\text{iff} \ \ j \ge 1\end{aligned}$$

i.e. x must satisfy

$$0 \le k \le 100 - j, \qquad 1 \le j \le 10$$

and then

$$y := \xi(x, \beta) = [10 - l \ \ l \ \ m]$$

where $l = j - 1, m = k + 1$. Since $0 \le l \le 9$ and

$$0 \le m \le 1 + (100 - j) = 1 + (100 - (l + 1)) = 100 - l$$

we get that $y \models Q$. It follows that Q is controllable.

According to Theorem 7.4.1 the controllability of Q guarantees the existence of a SFBC f such that $R((\mathbf{FACT\#4})^f, \mathit{true}) = Q$, and in fact the proof showed that f can be defined by

$$f_\sigma := M_\sigma(Q), \qquad \sigma \in \Sigma_c$$

With $\Sigma_c = \{\alpha\}$ we get

$$f_\alpha(x) = 1 \quad \text{iff} \quad \text{either } \mathit{not} \ \xi(x, \alpha)! \text{ or } \xi(x, \alpha)! \ \& \ \xi(x, \alpha) \models Q$$

Now

$$\begin{aligned}\xi(x, \alpha)! \ \ &\text{iff} \ \ x + e_\alpha \ge 0\\ &\text{iff} \ \ [10 - j \ \ j \ \ k] + [-1 \ 1 \ 0] \ge 0\\ &\text{iff} \ \ [10 - j - 1 \ \ j + 1 \ \ k] \ge 0\\ &\text{iff} \ \ (-1 \le j \le 9) \ \& \ (k \ge 0)\end{aligned}$$

For $\xi(x, \alpha) \models Q$ we then require

$$0 \le k \le 100 - (j + 1)$$

or

$$f_\alpha(x) = 1 \quad \text{iff} \quad \text{either } j = 10 \ \mathit{or} \ (j \le 9) \ \& \ (0 \le k \le 100 - (j + 1))$$

For instance, this control would enable α if $j = 0$ and $k \leq 99$, but disable α if $j = 1$ and $k = 99$. Notice that the SFBC synthesizing Q is far from unique: for instance, there is no need to enable α if $j = 10$, as there are then no machines idle. It is clear that the way $f_\sigma(x)$ is defined when *not* $\xi(x, \sigma)!$ may in principle be arbitrary.

From these examples it appears that direct calculation of reachable sets can become rather involved, even for the simplest VDES models. On the other hand, as far as P-invariance is concerned, the calculation need be done only along sequences of uncontrollable events. In the next section we explore this issue in more detail.

8.6 Reachability and Loop-Freeness

Let $\mathbf{G} = (X, \Sigma, \xi, x_0, X_m)$ be a VDES. In later application of the results of this section, $\mathbf{G}$ will be taken as 'the uncontrollable subsystem of the plant', to be defined in due course. For now, we focus generally on how the components of the vector x are successively incremented and decremented under the occurrence of events. Let $X = \mathbb{Z}^n$ and $x = [x_1 \cdots x_n]$. This coordinatization of X will be fixed and the subsequent definitions will depend on it. The corresponding displacement vectors will be written $e_\sigma = [e_\sigma(1) \cdots e_\sigma(n)]$, with the $e_\sigma(i) \in \mathbb{Z}$. Write $I = \{1, \ldots, n\}$ and for $\sigma \in \Sigma$ define

$$\sigma^\uparrow := \{i \in I | e_\sigma(i) < 0\}$$
$$\sigma^\downarrow := \{i \in I | e_\sigma(i) > 0\}$$

Thus for $i \in \sigma^\uparrow$ the components x_i of x are negatively incremented (i.e. positively decremented) by the occurrence of σ, while for $i \in \sigma^\downarrow$ the x_i are positively incremented. The index subsets $\sigma^\uparrow$ (resp. $\sigma^\downarrow$) can be visualized as (labeling) the components of x that are 'upstream' (resp. 'downstream') from σ; the $\sigma^\uparrow$-labeled x-components act as 'source' state variables for σ, while the $\sigma^\downarrow$ act as 'sinks'. Dually, for $i \in I$ define

$$i^\uparrow := \{\sigma \in \Sigma | e_\sigma(i) > 0\}$$
$$i^\downarrow := \{\sigma \in \Sigma | e_\sigma(i) < 0\}$$

The occurrence of an event $\sigma \in i^\uparrow$ positively increments x_i, i.e. σ acts as a source for x_i; while $\sigma \in i^\downarrow$ negatively increments x_i, i.e. σ acts as a sink for x_i. Dually again, $i^\uparrow$ (resp. $i^\downarrow$) represents the subset of events that are immediately upstream (resp. downstream) from the state component x_i. Notice that our notation is necessarily unsymmetrical. With state variables, we need to distinguish between (a) the component index (i), standing for the component 'x_i' as a fixed symbol for a state variable in the ordered n-tuple of state variables, and (b) the (current) value assumed by the state variable x_i, an integer subject to incremental change as the process evolves. For events (labeled) σ no such distinction is applicable.

Observe that $i \in \sigma^\uparrow$ (resp. $\sigma^\downarrow$) iff $\sigma \in i^\downarrow$ (resp. $i^\uparrow$).

Let $\{\tau_1, \ldots, \tau_k\}$ be a list of elements from Σ, possibly with repetitions, and let $[i_1 \cdots i_k]$ be a similar list from I. The interleaved list $L := [\tau_1\ i_1 \cdots \tau_k\ i_k]$ will be called a *loop* in **G** if

$$\tau_1 \in i_1^{\downarrow}, \ldots, \tau_k \in i_k^{\downarrow}$$

and

$$i_1 \in \tau_2^{\downarrow},\ \ i_2 \in \tau_3^{\downarrow},\ \ \ldots,\ \ i_{k-1} \in \tau_k^{\downarrow},\ \ i_k \in \tau_1^{\downarrow}$$

Equivalently the loop relations can be displayed as

$$i_1 \in \tau_2^{\downarrow},\ \ \tau_2 \in i_2^{\downarrow},\ \ \ldots,\ \ i_{k-1} \in \tau_k^{\downarrow},\ \ \tau_k \in i_k^{\downarrow},\ \ i_k \in \tau_1^{\downarrow},\ \ \tau_1 \in i_1^{\downarrow}$$

If no loop in **G** exists, **G** is *loop-free*.

Bring in the *state-variable source subset*

$$I^{\uparrow} := I - \bigcup\{\sigma^{\downarrow} | \sigma \in \Sigma\}$$

Thus for $i \in I^{\uparrow}$, x_i is never positively incremented by the occurrence of an event, namely x_i can only stay constant or decrease in value as the DES **G** evolves.

Lemma 1

Assume $\Sigma \neq \varnothing$. If **G** is loop-free then for some $\sigma \in \Sigma$, $\sigma^{\uparrow} \subseteq I^{\uparrow}$.

Proof

Suppose the contrary, namely

$$(\forall \sigma \in \Sigma)\sigma^{\uparrow} - I^{\uparrow} \neq \varnothing$$

Pick $\tau_1 \in \Sigma$ arbitrarily and let $i_1 \in \tau_1^{\uparrow} - I^{\uparrow}$. Then $i_1 \notin I^{\uparrow}$ implies that $i_1 \in \tau_2^{\downarrow}$ for some $\tau_2 \in \Sigma$. Since $\tau_2^{\uparrow} - I^{\uparrow} \neq \varnothing$, we pick $i_2 \in \tau_2^{\uparrow} - I^{\uparrow}$, and then $i_2 \in \tau_3^{\downarrow}$ for some $\tau_3 \in \Sigma$. Continuing this process we obtain a sequence $\tau_1, i_1, \tau_2, i_2, \ldots, \tau_j, i_j, \ldots$ such that

$$i_j \in \tau_j^{\uparrow} \cap \tau_{j+1}^{\downarrow},\quad j = 1, 2, \ldots$$

or equivalently

$$\tau_j \in i_j^{\downarrow},\ \ i_j \in \tau_{j+1}^{\downarrow},\quad j = 1, 2, \ldots$$

Since the index set I is finite ($|I| = n$) we may select the least $j > 1$, say $j = k + 1$ ($k \geq 1$), such that $i_j = i_l$ for some $l < j$, and without loss of generality assume $l = 1$. Then we have

$$i_1 \in \tau_2^{\downarrow},\ \ \tau_2 \in i_2^{\downarrow},\ \ \ldots,\ \ i_{k-1} \in \tau_k^{\downarrow},\ \ \tau_k \in i_k^{\downarrow},\ \ i_k \in \tau_{k+1}^{\downarrow},\ \ \tau_{k+1} \in i_1^{\downarrow}$$

This states that

$$L := [\tau_{k+1}\ i_1\ \tau_2\ i_2 \cdots \tau_k\ i_k]$$

is a loop in **G**, contrary to hypothesis. □

We shall need the idea of a 'subsystem' of **G** obtained by picking out a subset of the components of the state vector and a subset of events. With I as before, let $\hat{I} \subseteq I$, and let $\hat{\Sigma} \subseteq \Sigma$. The corresponding *subsystem* $\hat{\mathbf{G}}$ of **G** is

$$\hat{\mathbf{G}} := (\hat{X}, \hat{\Sigma}, \hat{\xi}, \hat{x}_0, \hat{X}_m)$$

where $\hat{X}$, $\hat{x}_0$, $\hat{X}_m$ are the natural projections of X, x_0, X_m on the components with indices in $\hat{I}$, and $\hat{\xi}$ is the restriction of ξ:

$$\hat{\xi}: \hat{X} \times \hat{\Sigma} \to \hat{X}: (\hat{x}, \sigma) \mapsto \hat{x} + \hat{e}_\sigma \quad \text{(pfn)}$$

With $\sigma \in \hat{\Sigma}$ we declare $\hat{\xi}(\hat{x}, \sigma)!$ whenever $\xi(x, \sigma)!$ for some x with projection $\hat{x}$. For instance, if $\hat{X} = \mathbb{N}^m$, $\hat{\xi}(\hat{x}, \sigma)!$ provided $\hat{x} \geq 0$ and $\hat{x} + \hat{e}_\sigma \geq 0$. Thus $\hat{\mathbf{G}}$ is indeed a VDES.

Recall the definition of the closed behavior generated by **G** corresponding to initialization at an arbitrary state x:

$$L(\mathbf{G}, x) := \{s \in \Sigma^* | \xi(x, s)!\}$$

Lemma 2

Let $\hat{\mathbf{G}}$ be a subsystem of **G** obtained by removing one or more elements of Σ, but keeping $\hat{X} = X$. Let $s \in L(\mathbf{G}, x)$, $x' := \xi(x, s)$, and $\hat{s} \in \hat{\Sigma}^*$. Then

$$\hat{s} \in L(\hat{\mathbf{G}}, x') \quad \text{iff} \quad s\hat{s} \in L(\mathbf{G}, x)$$ □

Lemma 3

Let $X = \mathbb{N}^n$, $\sigma \in \Sigma$, $k \in \mathbb{N}$. Then

$$x + ke_\sigma \geq 0 \Rightarrow \sigma^k \in L(\mathbf{G}, x)$$

Proof

The statement is true for $k = 0$. Assume inductively that it is true for $k \leq l$, and let $x + (l + 1)e_\sigma \geq 0$. Clearly

$$x' := x + le_\sigma \geq 0$$

so $\sigma^l \in L(\mathbf{G}, x)$ and $x' = \xi(x, \sigma^l)!$. Also $x' + e_\sigma \geq 0$ implies $\xi(x', \sigma)!$, so $\sigma^{l+1} \in L(\mathbf{G}, x)$, as required. □

Lemma 4

Let $X = \mathbb{N}^n$, $x \in X$ and $x + \Sigma\{k_\sigma e_\sigma | \sigma \in \Sigma\} \in X$ for some $k_\sigma \in \mathbb{N}$. For some $\tau \in \Sigma$ assume $\tau^\uparrow \subseteq I^\uparrow$. Then $x + k_\tau e_\tau \in X$.

Proof

If $i \in I^{\uparrow}$ then $\Sigma\{k_\sigma e_\sigma(i)|\sigma \neq \tau\} \leq 0$, so that

$$x_i + k_\tau e_\tau(i) \geq x_i + \Sigma\{k_\sigma e_\sigma(i)|\sigma \in \Sigma\} \geq 0$$

If $i \notin I^{\uparrow}$ then $i \notin \tau^{\uparrow}$, so $k_\tau e_\tau(i) \geq 0$, and again $x_i + k_\tau e_\tau(i) \geq 0$. □

For the remainder of this section we assume $\Sigma = \{\sigma_1, \ldots, \sigma_m\}$ and that $X = \mathbb{N}^n$, namely all states x satisfy $x \geq 0$. Recalling from Sect. 8.2 the definition of the occurrence vector $V(\cdot)$, we know that for $s \in \Sigma^*$, $x \in X$,

$$s \in L(\mathbf{G}, x) \Rightarrow \xi(x, s) = x + EV(s) \geq 0$$

Our main result states that, under the condition of loop-freeness, this implication can be reversed.

Theorem 5

Assume $\mathbf{G}$ is loop-free. Then for every $x \in \mathbb{N}^{n\times 1}$ and $v \in \mathbb{N}^{m\times 1}$,

$$(\exists s \in L(\mathbf{G}, x))V(s) = v \quad \text{iff} \quad x + Ev \geq 0$$

Proof

(Only if) The result is immediate by $\xi(x, s)!$ and $\xi(x, s) = x + EV(s)$.

(If) Let $x \geq 0$, $v \geq 0$, $x + Ev \geq 0$. We may write $v(i)$ for v_i if convenient. By Lemma 1, $\sigma^{\uparrow} \subseteq I^{\uparrow}$ for some $\sigma \in \Sigma$, say σ_1. Lemma 4 (with $k_\sigma = v_\sigma$) yields $x + v_1 e_1 \geq 0$. With $s_1 := \sigma_1^{v(1)}$ Lemma 3 gives $s_1 \in L(\mathbf{G}, x)$, so $\xi(x, s_1)!$ and $x_1 := \xi(x, s_1) \geq 0$. Let $\hat{\mathbf{G}}$ be the subsystem of $\mathbf{G}$ obtained by removing σ_1 (but keeping $\hat{X} = X$), so $\hat{\Sigma} := \Sigma - \{\sigma_1\}$. Let $\hat{I}^{\uparrow}$ be the state-variable source subset for $\hat{\mathbf{G}}$. It is clear that $\mathbf{G}$ loop-free implies $\hat{\mathbf{G}}$ loop-free, so (if $\hat{\Sigma} \neq \emptyset$) we pick $\sigma \in \hat{\Sigma}$, say $\sigma = \sigma_2$, with $\sigma^{\uparrow} \subseteq \hat{I}^{\uparrow}$. Now we have

$$x_1 + \sum_{i=2}^{m} v_i e_{\sigma_i} = x + \sum_{i=1}^{m} v_i e_{\sigma_i} = x + Ev \geq 0$$

By Lemma 4, $x_1 + v_2 e_{\sigma_2} \geq 0$; Lemma 3 gives $s_2 := \sigma_2^{v(2)} \in L(\hat{\mathbf{G}}, x_1)$, and by Lemma 2, $s_1 s_2 \in L(\mathbf{G}, x)$. So $x_2 := \xi(x, s_1 s_2)!$ and $x_2 \geq 0$. Continuing in this way we get finally

$$s := s_1 s_2 \cdots s_m \in L(\mathbf{G}, x)$$

with

$$s_i = \sigma_i^{v(i)}, \qquad i = 1, \ldots, m$$

and $V(s) = v$. □

Exercise 6: Show that the assumption in Theorem 5 that **G** is loop-free cannot be dropped altogether if the stated conclusion is to hold. Also show that the conclusion may hold in some cases where **G** is not loop-free.

8.7 Loop-Freeness and Optimal Control

Let **G** be a VDES as before, with $X = \mathbb{N}^n$. In this section we apply Theorem 8.6.5 to obtain an (often) efficient way to compute SFBC 'online', whenever the specification predicate is linear, and **G** satisfies a condition of loop-freeness. Formally, let $\hat{\Sigma} = \Sigma_u$, $\hat{X} = X$ and define $\mathbf{G}_u$ $(:=\hat{\mathbf{G}})$ to be the *uncontrollable subsystem* of **G**. Let $P \in Pred_{\mathbf{lin}}(X)$, with

$$x \models P \quad \text{iff} \quad ax \leq b$$

for some $a \in \mathbb{Z}^{1\times n}, b \in \mathbb{Z}$. We arrange the indexing so that $\Sigma_u = \{\sigma_1, \ldots, \sigma_m\}$, with

$$E_u = [e_1 \cdots e_m] \in \mathbb{Z}^{n\times m}$$

We write $|s|_i$ to denote the number of occurrences of σ_i in s, and for $s \in \Sigma_u^*$ bring in the occurrence vector

$$V_u(s) := [|s|_1 \cdots |s|_m] \in \mathbb{N}^{m\times 1}$$

Recalling the characterization $\sup \mathcal{CP}(P) = R(\mathbf{G}, \langle P\rangle)$ of Proposition 7.4.3, our first task is to calculate $\langle P\rangle$. Using the fact that $L(\mathbf{G}_u, x)$ is prefix-closed, we have that

$$\begin{aligned} x \models \langle P\rangle \ &\text{iff} \ (\forall s \in L(\mathbf{G}_u, x))\xi(x, s) \models P \\ &\text{iff} \ (\forall s \in L(\mathbf{G}_u, x))x + E_u V_u(s) \models P \\ &\text{iff} \ (\forall s \in L(\mathbf{G}_u, x))ax + aE_u V_u(s) \leq b \\ &\text{iff} \ ax + \max\{aE_u V_u(s) | s \in L(\mathbf{G}_u, x)\} \leq b \end{aligned}$$

In general the indicated maximization problem may be intractable, a fact which makes the following result especially interesting.

Proposition 1

If $\mathbf{G}_u$ is loop-free, then $x \models \langle P\rangle$ iff

$$ax + cv^*(x) \leq b$$

Here $c := aE_u \in \mathbb{Z}^{1\times m}$, and $v^*(x)$ is a solution of the integer linear programming problem:

$$cv = \text{maximum}$$

with respect to $v \in \mathbb{Z}^{m \times 1}$ such that $v \geq 0$ and $E_u v \geq -x$.

Proof

By Theorem 8.6.5 applied to $\mathbf{G}_u$,

$$\{V_u(s) | s \in L(\mathbf{G}_u, x)\} = \{v \in \mathbb{N}^m | x + E_u v \geq 0\}$$

Therefore

$$\max\{a E_u V_u(s) | s \in L(\mathbf{G}_u, x)\} = cv^*(x)$$

and the result for $\langle P \rangle$ follows as claimed. □

We can now provide our final result, on computation of the optimal control.

Theorem 2

Assume $\mathbf{G}_u$ is loop-free. An optimal SFBC f^*, if one exists, enforcing the linear predicate $P := (ax \leq b)$, is given for $\sigma \in \Sigma_c$ by the formula:

$$f^*_\sigma(x) = \begin{cases} 1 & \text{if } \xi(x, \sigma)! \;\&\; ax_{new} + cv^*(x_{new}) \leq b \text{ (where } x_{new} := x + e_\sigma) \\ 0 & \text{otherwise} \end{cases}$$

Furthermore, f^* so defined is balanced.

Proof

The fact that f^* optimally enforces P follows immediately from Sect. 7.4 and Proposition 1. The property that f^* is balanced results by construction: $f^*_\sigma(x) = 1$ whenever both $x \in R(\mathbf{G}, \langle P \rangle)$ and $\xi(x, \sigma)!$ with $\xi(x, \sigma) \in R(\mathbf{G}, \langle P \rangle)$. □

We remark that $\xi(x, \sigma)!$ just when $x_{new} \geq 0$, by VDES dynamics. If *not* $\xi(x, \sigma)!$ then $f^*_\sigma(x)$ can in principle be defined arbitrarily.

Corollary 3

Assume $\mathbf{G}_u$ is loop-free. An optimal SFBC f^* exists iff

$$ax_0 + cv^*(x_0) \leq b$$

If this condition fails, then no SFBC exists for $\mathbf{G}$ that enforces P.

Proof

By the results of Sect. 7.4 an optimal control f^* exists iff $\sup \mathcal{CP}(P) \neq \mathit{false}$, and this is true iff $x_0 \models \sup \mathcal{CP}(P)$. Since $\sup \mathcal{CP}(P) = R(\mathbf{G}, \langle P \rangle)$, f^* exists iff $x_0 \models \langle P \rangle$, and the assertion follows by Proposition 1. □

Exercise 4: For a given VDES $\mathbf{G}$, suppose $\mathbf{G}_u$ is not loop-free, but you decide to use the integer linear programming method of Theorem 2 anyway, because it is computationally convenient. Could such a control design violate the specification? Either prove it could not or find an example to show that it may. If it does not, its only fault might be that it is overly conservative. In that case, create an example to illustrate.

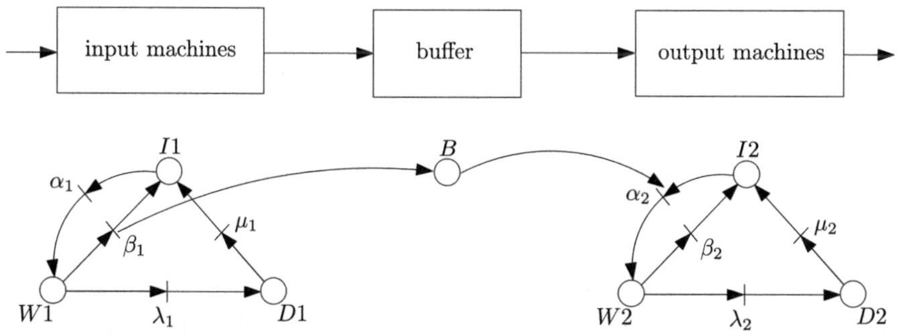

Fig. 8.4 Petri net for **FACT#5**

8.8 Example: FACT#5

We consider the Small Factory with Petri net shown in Fig. 8.4, where a group of 10 input machines feeds a buffer, which in turn supplies a group of 5 machines at the output. Let $I1$, $W1$, $D1$ denote the numbers of input machines in state I, W, D, with a similar notation for output machines, and let B denote the number of items in the buffer. We define the state vector x as

$$x := [I1 \;\; W1 \;\; D1 \;\; I2 \;\; W2 \;\; D2 \;\; B] \in \mathbb{N}^{7\times 1}$$

with initial condition

$$x(0) = [10 \;\; 0 \;\; 0 \;\; 5 \;\; 0 \;\; 0 \;\; 0]$$

Listing the events in order $(\alpha_1, \beta_1, \lambda_1, \mu_1, \alpha_2, \beta_2, \lambda_2, \mu_2)$ we get the displacement matrix $E \in \mathbb{Z}^{7\times 8}$ displayed below.

	α_1	β_1	λ_1	μ_1	α_2	β_2	λ_2	μ_2
$I1$	−1	1	0	1	0	0	0	0
$W1$	1	−1	−1	0	0	0	0	0
$D1$	0	0	1	−1	0	0	0	0
$I2$	0	0	0	0	−1	1	0	1
$W2$	0	0	0	0	1	−1	−1	0
$D2$	0	0	0	0	0	0	1	−1
B	0	1	0	0	−1	0	0	0

Taking $\Sigma_u = \{\beta_1, \lambda_1, \beta_2, \lambda_2\}$ and extracting the corresponding submatrix of E results in

$$E_u = \begin{bmatrix} 1 & 0 & 0 & 0 \\ -1 & -1 & 0 & 0 \\ 0 & 1 & 0 & 0 \\ 0 & 0 & 1 & 0 \\ 0 & 0 & -1 & -1 \\ 0 & 0 & 0 & 1 \\ 1 & 0 & 0 & 0 \end{bmatrix} \in \mathbb{Z}^{7\times 4}$$

It is easy to check—for instance by inspection of its Petri net—that $\mathbf{G}_u$ is loop-free.

Assume that the buffer capacity is 100, and that we undertake to prevent overflow, namely to enforce the predicate $P_{over} := (ax \leq b)$, where

$$a := [0\ 0\ 0\ 0\ 0\ 0\ 1] \in \mathbb{Z}^{1\times 7}, \quad b := 100$$

This gives

$$c := aE_u = [1\ 0\ 0\ 0] \in \mathbb{Z}^{1\times 4}$$

Writing $v := [v_1\ v_2\ v_3\ v_4] \in \mathbb{Z}^{4\times 1}$, we attempt to maximize $cv = v_1$, subject to $v \geq 0$ and $E_u v \geq -x$. With $x = [x_1 \cdots x_7] \in \mathbb{N}^{7\times 1}$ the constraints become

$$\begin{aligned} v_1 &\geq -x_1 \\ -v_1 - v_2 &\geq -x_2 \\ v_2 &\geq -x_3 \\ v_3 &\geq -x_4 \\ -v_3 - v_4 &\geq -x_5 \\ v_4 &\geq -x_6 \\ v_1 &\geq -x_7 \end{aligned}$$

together with $v \geq 0$. All but the second and fifth of these conditions are enforced by VDES dynamics, which maintain $x \geq 0$. Thus the effective constraints reduce to

$$v_1 \geq 0, \quad v_2 \geq 0, \quad v_1 + v_2 \leq x_2, \quad v_3 + v_4 \leq x_5$$

Clearly v_1 is maximized at $v_1 = x_2$, $v_2 = 0$, $v_3 = \omega$, $v_4 = \omega$, where ω denotes 'don't care'. For α_1 the optimal control defined in Sect. 8.7 is therefore

$$\begin{aligned} f^*_{\alpha_1}(x) = 1 \quad &\text{iff} \quad ax_{new} + cv^*(x_{new}) \leq b \ \ (x_{new} := x + e_{\alpha_1}) \\ &\text{iff} \quad x_{7,new} + x_{2,new} \leq 100 \\ &\text{iff} \quad x_7 + (x_2 + 1) \leq 100 \\ &\text{iff} \quad x_2 + x_7 \leq 99 \end{aligned}$$

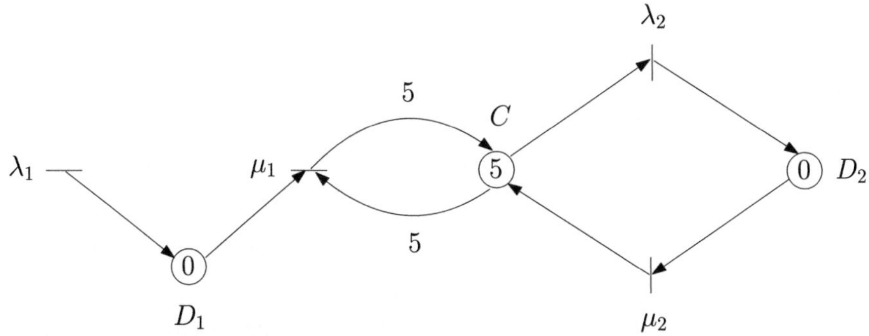

Fig. 8.5 VDES with selfloop

In words, α_1 is enabled iff the number of input machines at work plus the current buffer content is at most one less than the buffer capacity, and this is obviously intuitively correct.

Exercise 1: Under the same assumptions, investigate how to prevent buffer underflow, namely enforce

$$P_{under} := (x_7 \geq 0) = (-x_7 \leq 0)$$

Following a similar procedure for α_2, verify that optimal control enables α_2 in all states, namely the only enablement condition for the occurrence of α_2 is $\xi(x, \alpha_2)!$, or $(x_4 \geq 1)$ & $(x_7 \geq 1)$, and this is enforced automatically by VDES dynamics.

Exercise 2: **Selfloop simulation**

Consider the additional specification that no input machine should be repaired (i.e. μ_1 is disabled) as long as some output machine is broken down. Writing $\#\sigma$ for the number of occurrences of event σ since initialization, we have for the number of output machines broken down,

$$D2 = x_6 = \#\lambda_2 - \#\mu_2$$

Since μ_1 must be disabled if $D2 > 0$, and up to $I2(0) = x_4(0) = 5$ output machines can be down at one time, $D2 > 0$ means $C := 5 - D2 < 5$. This specification can be modeled using a 'VDES with selfloop' as displayed by the Petri net in Fig. 8.5.

To incorporate this specification in our standard model of VDES (from which selfloops are excluded) one may interpose a new coordinate V and uncontrollable transition ν as in Fig. 8.6.

At first glance this device may seem unrealistic, since when $C = 0$ and $V = 1$, following an occurrence of μ_1, the (uncontrollable!) event λ_2 will be disabled pending the occurrence of ν. Recall, however, that no timing constraints are imposed, so ν can be assumed to occur arbitrarily soon after its enablement.

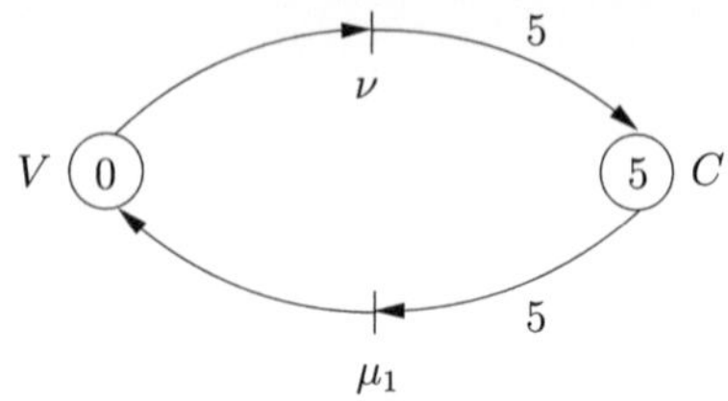

Fig. 8.6 Removal of selfloop using new place and transition

(i) More formally, consider the configuration of Fig. 8.5, defined as a 'generalized' VDES, say **G**, in the sense of Remark 8.2.1. Take for the state vector and initial condition

$$x = [D1\;\; D2\;\; C], \quad x_0 = [0\;\; 0\;\; 5]$$

and for the alphabet $\{\lambda_1, \mu_1, \lambda_2, \mu_2\}$. The enablement conditions for $\lambda_1, \lambda_2, \mu_2$ are just as in a standard VDES, while for μ_1 we define vectors

$$e^+_{\mu_1} = [0\;\; 0\;\; 5],\; e^-_{\mu_1} = [1\;\; 0\;\; 5],\; e_{\mu_1} = [-1\;\; 0\;\; 0]$$

and the enablement condition $x \geq e^-_{\mu_1}$. With the semantics of **G** now well-defined, let $L(\mathbf{G})$ be the corresponding closed behavior, of course over $\{\lambda_1, \mu_1, \lambda_2, \mu_2\}$. Next define a VDES **H** by incorporating Fig. 8.6 into Fig. 8.5. The state vector and initial condition of **H** can be taken as

$$x = [D1\;\; D2\;\; C\;\; V] \in \mathbb{Z}^{4\times 1}, \quad x_0 = [0\;\; 0\;\; 5\;\; 0]$$

and the alphabet as $\{\lambda_1, \mu_1, \lambda_2, \mu_2, \nu\}$. The displacement matrix E is then

$$E = \begin{bmatrix} 1 & -1 & 0 & 0 & 0 \\ 0 & 0 & 1 & -1 & 0 \\ 0 & -5 & -1 & 1 & 5 \\ 0 & 1 & 0 & 0 & -1 \end{bmatrix}$$

With the semantics that of a standard VDES, let the closed behavior be $L(\mathbf{H})$. Finally, if $P{:}\{\lambda_1, \mu_1, \lambda_2, \mu_2, \nu\}^* \rightarrow \{\lambda_1, \mu_1, \lambda_2, \mu_2\}^*$ is the natural projection (erasing ν), show that $L(\mathbf{G}) = PL(\mathbf{H})$.

(ii) With **FACT#5** redefined as a new VDES by incorporating the additional structure of the VDES **H**, in part(i), re-solve for the optimal control to prevent buffer overflow, and compare the result with the control law found in the earlier part of this section.

8.9 Memory and Dynamic State Feedback Control for VDES

We now apply to VDES the constructions of Sect. 7.6. As in Sect. 8.8 assume that the plant VDES **G** has state set $\mathbb{N}^n$. To **G** we shall adjoin a VDES $\mathbf{H} = (Y, \Sigma, \eta, y_0, Y)$, and in view of the vector structures write $\mathbf{G} \oplus \mathbf{H}$ for the 'direct sum' VDES:

$$\mathbf{G} \oplus \mathbf{H} = (X \oplus Y, \Sigma, \xi \oplus \eta, x_0 \oplus y_0, X_m \oplus Y)$$

Typically the state vector $y \in Y$ will play the role of a memory variable (as distinct from a material variable like 'numbers of machines'), and so we shall take $Y = \mathbb{Z}^p$ for some $p > 0$. The displacement vector for **H** corresponding to σ will be denoted by $h_\sigma \in \mathbb{Z}^{p\times 1}$; the corresponding displacement vector in $\mathbf{G} \oplus \mathbf{H}$ is $e_\sigma \oplus h_\sigma \in \mathbb{Z}^{(n+p)\times 1}$, with

$$(\xi \oplus \eta)(x \oplus y, \sigma)! \quad \text{iff} \quad \xi(x, \sigma)! \quad \text{iff} \quad x + e_\sigma \geq 0$$

In other words, the memory **H** places no additional 'physical' constraint on the transitions of $\mathbf{G} \oplus \mathbf{H}$.

As before let $\Sigma = \{\sigma_1, \ldots, \sigma_m\}$. We define a *linear dynamic predicate* to be an element $P_{dyn} \in Pred_{\mathbf{lin}}(\mathbb{N}^m)$:

$$v = [v_1 \cdots v_m] \models P_{dyn} \quad \text{iff} \quad \sum_{i=1}^{m} c_i v_i \leq d$$

where the c_i and $d \in \mathbb{Z}$, i.e.

$$v \models P_{dyn} \quad \text{iff} \quad d - cv \geq 0$$

where $c := [c_1 \cdots c_m] \in \mathbb{Z}^{1\times m}$. For the behavior of **G** subject to P_{dyn}, bring in

$$L(\mathbf{G}, P_{dyn}) := \{s \in L(\mathbf{G}) | (\forall w \leq s) V(w) \models P_{dyn}\}$$

With P_{dyn} we associate the memory **H** as above, where

$$Y := \mathbb{Z}, \quad y_0 := d, \quad \eta(y, \sigma_i) := y - c_i, \quad (i = 1, \ldots, m)$$

It should now be clear that enforcing $s \in L(\mathbf{G}, P_{dyn})$ is tantamount to enforcing the predicate $y \geq 0$ in $\mathbf{G} \oplus \mathbf{H}$. Formally, define $P_{sta} \in Pred_{\mathbf{lin}}(X \oplus Y)$ according to

$$x \oplus y \models P_{sta} \quad \text{iff} \quad -y \leq 0$$

Then we have

Lemma 1

$$L(\mathbf{G} \oplus \mathbf{H}, P_{sta}) = L(\mathbf{G}, P_{dyn}) \qquad \square$$

By Lemma 1 and Theorem 7.6.1, we have the main result, in the notation of Sect. 7.6, as follows.

Theorem 2

Let f^* be an optimal SFBC enforcing P_{sta} in $\mathbf{G} \oplus \mathbf{H}$, and write $F^* = (f^*, \mathbf{H})$. Then F^* is an optimal DSFBC for $\mathbf{G}$ relative to P_{dyn}, namely

$$L(F^*/\mathbf{G}) = \sup \mathcal{C}_{\mathbf{G}}(L(\mathbf{G}, P_{dyn})) \qquad \square$$

With a slight modification of the definition of P_{dyn}, the computational result of Sect. 8.8 will apply in the present situation as well; for an example see Sect. 8.11 below.

8.10 Modular Dynamic State Feedback Control for VDES

By following the scheme outlined at the end of Sect. 7.6 it is quite straightforward to describe modular control of VDES, by adjoining a direct sum of memory VDES of the type $\mathbf{H}$ in the previous section. Again taking $\Sigma = \{\sigma_1, \ldots, \sigma_m\}$, suppose we wish to enforce on the occurrence vector $V(\cdot)$ of $\mathbf{G}$ a predicate $P_{dyn} \in \overline{Pred_{\mathbf{lin}}(\mathbb{N}^m)}$ of conjunctive form:

$$v \models P_{dyn} \quad \text{iff} \quad \bigwedge_{i=1}^{k} (c_i v \leq d_i)$$

where $v \in \mathbb{Z}^{m\times 1}$, $c_i \in \mathbb{Z}^{1\times m}$, $d_i \in \mathbb{Z}$ $(i = 1, \ldots, k)$. For each conjunct $P_{dyn,i} := (c_i v \leq d_i)$, construct memory $\mathbf{H}_i$ as in Sect. 8.9, and let $F_i^* := (f_i^*, \mathbf{H}_i)$ be a corresponding optimal (balanced) DSFBC for $\mathbf{G}$. It follows by Theorem 7.6.6 that

$$F^* := \{f_1^* \wedge \cdots \wedge f_k^*, \mathbf{H}_1 \oplus \cdots \oplus \mathbf{H}_k\}$$

is then an optimal (balanced) DSFBC enforcing P_{dyn} on the behavior of $\mathbf{G}$.

To conclude this section we restate for future reference the correspondence between VDES and conjunctive predicates.

Proposition 1

For any predicate of form P_{dyn} there is a VDES $\mathbf{H}$ with state set $\mathbb{N}^k$ and event set Σ such that

$$L(\mathbf{G}) \cap L(\mathbf{H}) = L(\mathbf{G}, P_{dyn})$$

Dually for any VDES **H** with state set $\mathbb{N}^k$ and event set Σ ($|\Sigma| = m$) there exists a predicate P_{dyn} on $\mathbb{N}^m$ such that the above equality is true.

Proof

For the first statement, apply the construction of Sect. 8.9 to each conjunct of P_{dyn}. For the second statement, reverse the procedure: given $\mathbf{H} = (Y, \Sigma, \eta, y_0, Y)$ with $Y = \mathbb{N}^k$, $y_0 = [y_{01} \cdots y_{0k}]$, $\Sigma = \{\sigma_1, \ldots, \sigma_m\}$ and $\eta(y, \sigma) = y + h_\sigma$ (defined when $y \geq 0,\ y + h_\sigma \geq 0$), write $h_{\sigma_j} =: h_j = [h_{j1} \cdots h_{jk}] \in \mathbb{Z}^{k \times 1}$ $(j = 1, \ldots, m)$ and let

$$d_i := y_{0i}, \quad c_{ij} := -h_{ji}, \quad (i = 1, \ldots, k;\ \ j = 1, \ldots, m)$$

Then define $c_i := [c_{i1} \cdots c_{im}] \in \mathbb{Z}^{1 \times m}$ $(i = 1, \ldots, k)$, and finally

$$P_{dyn} := \bigwedge_{i=1}^{k} (d_i - c_i v \geq 0) \qquad \square$$

Thus each state variable of a VDES can be regarded as a 'memory' variable that records a weighted sum of event occurrence numbers. The initial and occurrence conditions of a VDES with state space $\mathbb{N}^k$ impose the requirement that all k memory state variables be maintained nonnegative. A VDES on $\mathbb{N}^k$ thus expresses the same language as a conjunction of linear dynamic specifications. Thus such a VDES can be used to provide a control specification in the first place.

8.11 Example: FACT#2

Returning to **FACT#2** (Sects. 8.3, 8.5), we attempt to enforce predicates

$$P_1 := (x_4 \leq 100), \qquad P_2 := (10 + \#\beta \geq 3\#\lambda)$$

where $\#\sigma$ means the number of occurrences of σ in the string generated by **G** since initialization. P_2 is supposed to limit the number of breakdowns, relative to the number of workpieces successfully processed; since breakdown (λ) is uncontrollable, this may require that eventually the process be shut down.

Following the procedure of Sect. 8.9, to represent P_2 bring in the memory

$$\mathbf{H} = (Y, \Sigma, \eta, y_0, Y)$$

with

$$Y := \mathbb{Z}, \quad \Sigma = \{\alpha, \beta, \lambda, \mu\}, \quad y_0 := 10$$

$$h_\alpha = 0, \quad h_\beta = 1, \quad h_\lambda = -3, \quad h_\mu = 0$$

where $\eta(y, \sigma) = y + h_\sigma$. It is clear that in $\mathbf{G} \oplus \mathbf{H}$, the state variable y will record the quantity

$$y = 10 + \#\beta - 3\#\lambda$$

and P_2 is tantamount to $(y \geq 0)$. The state vector of $\mathbf{G} \oplus \mathbf{H}$ is

$$x \oplus y = [x_1 \;\; x_2 \;\; x_3 \;\; x_4 \;\; y] \in \mathbb{N}^4 \oplus \mathbb{Z}$$

initialized at $[10 \;\; 0 \;\; 0 \;\; 0 \;\; 10]$. Note that the VDES dynamics for $\mathbf{G} \oplus \mathbf{H}$ automatically enforce only $x \geq 0$, and not $y \geq 0$. The full displacement matrix is

$$E := \begin{bmatrix} -1 & 1 & 0 & 1 \\ 1 & -1 & -1 & 0 \\ 0 & 0 & 1 & -1 \\ 0 & 1 & 0 & 0 \\ 0 & 1 & -3 & 0 \end{bmatrix}$$

For the control design we wish to use integer linear programming as in Sect. 8.7. To do so we must respect the assumption (cf. Theorem 8.6.5) that our VDES state variables remain nonnegative under the defined (and uncontrolled) dynamic transition action. Since the memory variable $y \in \mathbb{Z}$ is not thus constrained we first write it as the difference of two nonnegative variables, say

$$y = y_1 - y_2$$

with $y_1 = 10 + \#\beta$, $y_2 = 3\#\lambda$. We now redefine

$$Y := \mathbb{N}^2, \quad y := \begin{bmatrix} y_1 \\ y_2 \end{bmatrix}, \quad \Sigma = \{\alpha, \beta, \lambda, \mu\}, \quad y_0 := \begin{bmatrix} 10 \\ 0 \end{bmatrix}$$

$$h_\alpha = \begin{bmatrix} 0 \\ 0 \end{bmatrix}, \quad h_\beta = \begin{bmatrix} 1 \\ 0 \end{bmatrix}, \quad h_\lambda = \begin{bmatrix} 0 \\ 3 \end{bmatrix}, \quad h_\mu = \begin{bmatrix} 0 \\ 0 \end{bmatrix}$$

with $\eta(y, \sigma) = y + h_\sigma \in \mathbb{N}^2$ and $P_2 = (y_1 - y_2 \geq 0)$. The new state vector of $\mathbf{G} \oplus \mathbf{H}$ is

$$x \oplus y = [x_1 \;\; x_2 \;\; x_3 \;\; x_4 \;\; y_1 \;\; y_2] \in \mathbb{N}^6$$

initialized at $[10 \;\; 0 \;\; 0 \;\; 0 \;\; 10 \;\; 0]$. Note that the VDES dynamics for $\mathbf{G}$ itself automatically enforce $x \oplus y \geq 0$. The full displacement matrix is

$$E := \begin{bmatrix} -1 & 1 & 0 & 1 \\ 1 & -1 & -1 & 0 \\ 0 & 0 & 1 & -1 \\ 0 & 1 & 0 & 0 \\ 0 & 1 & 0 & 0 \\ 0 & 0 & 3 & 0 \end{bmatrix} \in \mathbb{Z}^{6\times 4}$$

To start the design we note that $(\mathbf{G} \oplus \mathbf{H})_u$ is loop-free. Referring to Sect. 8.7, write E_u for the 6×2 submatrix of E corresponding to events β and λ, i.e. columns 2 and 3. For P_1, let

$$a = [0\ \ 0\ \ 0\ \ 1\ \ 0\ \ 0] \in \mathbb{Z}^{1\times 6}, \quad b = 100$$

Then

$$c := aE_u = [1\ \ 0] \in \mathbb{Z}^{1\times 2}$$

and we take $v = [v_1\ \ v_2] \in \mathbb{N}^{2\times 1}$. We are to maximize $cv = v_1$ subject to $v \geq 0$ and $E_u v \geq -(x \oplus y)$, i.e.

$$v_1 \geq 0, \quad v_2 \geq 0,$$

$$v_1 \geq -x_1, \quad -v_1 - v_2 \geq -x_2, \quad v_2 \geq -x_3,$$

$$v_1 \geq -x_4, \quad v_1 \geq -y_1, \quad 3v_2 \geq -y_2$$

In view of $x_i \geq 0$, $y_j \geq 0$ by the dynamics of $\mathbf{G} \oplus \mathbf{H}$, the effective constraints are

$$v_1 \geq 0, \quad v_2 \geq 0, \quad v_1 + v_2 \leq x_2$$

from which we obtain the solution

$$v_1^* = x_2, \quad v_2^* = 0$$

As expected, the solution for P_1 is independent of the memory element $\mathbf{H}$. From Theorem 8.7.2 the optimal control for P_1 is

$$\begin{aligned} f_\alpha^{(1)*}(x \oplus y) = 1 \ \text{ iff } \ & a(x \oplus y)_{new} + cv^*((x \oplus y)_{new}) \leq b \\ \text{iff } \ & x_{4,new} + x_{2,new} \leq 100 \end{aligned}$$

using $(x \oplus y)_{new} := (x \oplus y) + (e_\alpha \oplus h_\alpha)$.

For P_2 we have $-y_1 + y_2 \leq 0$, or $a(x \oplus y) \leq b$ with

$$a = [0\ \ 0\ \ 0\ \ 0\ \ -1\ \ 1], \quad b = 0$$

Thus $c = aE_u = [-1\ \ 3]$, and our problem is to maximize

$$cv = -v_1 + 3v_2$$

under the same effective conditions as before, namely $v_i \geq 0$, $v_1 + v_2 \leq x_2$. This gives $v_1^* = 0$, $v_2^* = x_2$ and $cv^*(x) = 3x_2$. Thus for $\sigma \in \Sigma_c$ the optimal control for P_2 is

$$f_\sigma^{(2)*}(x \oplus y) = 1 \ \text{ iff } \ -y_{1,new} + y_{2,new} + 3x_{2,new} \leq 0$$

where $(x \oplus y)_{new} := (x \oplus y) + (e_\sigma \oplus h_\sigma)$. Combining the SFBC for P_1 with the DSFBC for P_2 we obtain for the conjunction

$$f_\sigma^*(x \oplus y) = 1 \quad \text{iff} \quad (x_{2,new} + x_{4,new} \le 100) \wedge (3x_{2,new} - y_{1,new} + y_{2,new} \le 0)$$

whenever $\sigma \in \Sigma_c$, in particular for $\sigma = \alpha$.

This example provides an opportunity to implement optimal control by means of a 'control' VDES, say $\mathbf{G_{con}}$, coupled (via $\oplus$) to $\mathbf{G}$. To see how this is done, rearrange the conditions for f_σ^* in the form

$$100 - x_{2,new} - x_{4,new} \ge 0, \qquad -3x_{2,new} + y_{1,new} - y_{2,new} \ge 0$$

Introduce control coordinates z_1, z_2 which we shall try to arrange so that

$$z_1 = 100 - x_2 - x_4, \quad z_2 = -3x_2 + y_1 - y_2$$

as the process evolves; initially $z_1 = 100$, $z_2 = 10$. Heuristically note that

$$z_1 = 100 - (\#\alpha - \#\beta - \#\lambda) - \#\beta = 100 - \#\alpha + \#\lambda$$
$$z_2 = -3(\#\alpha - \#\beta - \#\lambda) + (10 + \#\beta) - 3\#\lambda = 10 - 3\#\alpha + 4\#\beta$$

With the ordering $(\alpha, \beta, \lambda, \mu)$ we therefore take the displacements in z_1, z_2 to be

$$k(z_1) = [-1 \;\; 0 \;\; 1 \;\; 0], \quad k(z_2) = [-3 \;\; 4 \;\; 0 \;\; 0]$$

Thus if we let $\mathbf{G_{con}} := (Z, \Sigma, \zeta, z_0, Z)$ with $Z = \mathbb{N}^2$, and let the foregoing displacements define ζ, then the behavior of $\mathbf{G} \oplus \mathbf{G_{con}}$ (as a VDES with state space $\mathbb{N}^6$) will be exactly the behavior of $\mathbf{G}$ under f^*, inasmuch as σ will be enabled only if $z_{1,new} \ge 0$, $z_{2,new} \ge 0$. Notice that the only negative entries in the k_1, k_2 vectors correspond to controllable events (specifically, α). Thus the requirement of coordinate nonnegativity enforced, by assumption 'physically', by VDES over $\mathbb{N}^6$, captures the control action in a plausible way: no further 'control technology' is needed. This approach will be pursued systematically in Sect. 8.13.

8.12 Modeling and Control of a Production Network

We consider the modeling and control of a production network, adapted from work by Al-Jaar and Desrochers. A Petri net for the system is shown in Fig. 8.7.

The system operates as follows. Machines in Groups 1 and 2 receive parts from a non-depleting inventory and deposit the finished parts in Buffers 1 and 2 respectively. Machines in Group 3 fetch parts from Buffers 1 and 2 for assembling. The assembled workpiece is deposited in Buffer 3 to be further processed by Group 4. The processed

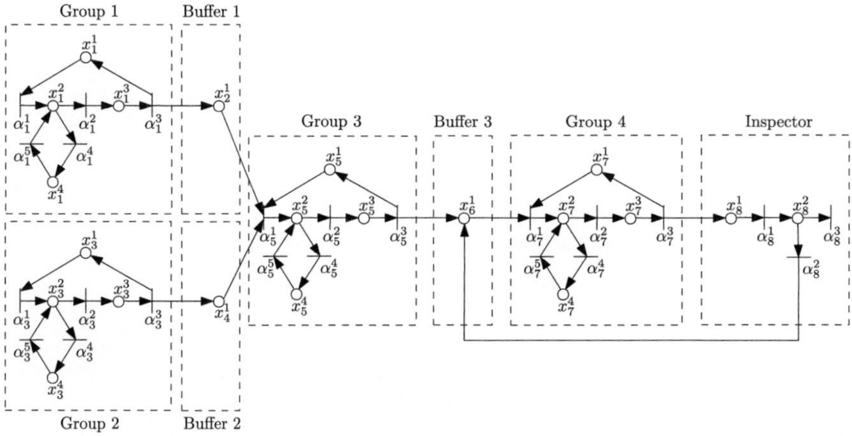

Fig. 8.7 Production network

workpiece is sent to an inspection unit which can either output the workpiece as a finished product or return it for reworking by Group 4.

We use a modular approach to modeling this production network. First we model modules of the network individually and then compose them to form the model of the complete system.

(i) Modeling of the Machines

The state vector of machine group 1, 2, 3, 4 is indexed respectively $i = 1, 3, 5, 7$ and is

$$x_i = [x_i^1 \; x_i^2 \; x_i^3 \; x_i^4] \in \mathbb{N}^{4\times 1}, \qquad i = 1, 3, 5, 7$$

The state components denote, respectively, 'idle', 'processing', 'holding (part prior to sending on)', 'broken down'. With reference to Fig. 8.7, transitions $\alpha_i^1, \alpha_i^2, \alpha_i^3, \alpha_i^4, \alpha_i^5$ ($i = 1, 3, 5, 7$) represent 'start processing', 'finish processing', 'return to idle', 'break down', 'return to processing after repair'.

The corresponding state transition function $\xi_i : X_i \times \Sigma_i \to X_i$ is given by

$$\xi_i(x_i, \alpha_i^j) = x_i + e_{i,\alpha_i^j} \qquad (i = 1, 3, 5, 7; \;\; j = 1, 2, 3, 4, 5)$$

where

$$e_{i,\alpha_i^1} = \begin{bmatrix} -1 \\ 1 \\ 0 \\ 0 \end{bmatrix} \quad e_{i,\alpha_i^2} = \begin{bmatrix} 0 \\ -1 \\ 1 \\ 0 \end{bmatrix}$$

$$e_{i,\alpha_i^3} = \begin{bmatrix} 1 \\ 0 \\ -1 \\ 0 \end{bmatrix} \quad e_{i,\alpha_i^4} = \begin{bmatrix} 0 \\ -1 \\ 0 \\ 1 \end{bmatrix} \quad e_{i,\alpha_i^5} = \begin{bmatrix} 0 \\ 1 \\ 0 \\ -1 \end{bmatrix}$$

Initially, all machines are in the 'idle' state:

$$x_{i,0} = \begin{bmatrix} g_i \\ 0 \\ 0 \\ 0 \end{bmatrix}$$

The vector models of these machines are then

$$\mathbf{G_i} = (X_i, \Sigma_i, \xi_i, x_{i,0})$$

where

$$\Sigma_i = \{\alpha_i^1, \alpha_i^2, \alpha_i^3, \alpha_i^4, \alpha_i^5\}$$

with $\Sigma_{i,c} = \{\alpha_i^3\}$, i.e. we assume that only the event 'return to idle' is controllable.

(ii) Modeling of the Inspection Unit

The model for[1] the inspector is

$$\mathbf{G_8} = (X_8, \Sigma_8, \xi_8, x_{8,0})$$

where

$$\Sigma_8 = \{\alpha_7^3, \alpha_8^1, \alpha_8^2, \alpha_8^3\}$$

$$\Sigma_{8,c} = \{\alpha_8^1\}$$

and

$$x_8 = [x_8^1\ x_8^2] \in \mathbb{N}^{2\times 1}$$

$$x_{8,0} = [0\ 0]$$

$$\xi_8(x_8, \alpha_8^i) = x_8 + e_{8,\alpha_8^i}$$

with

$$e_{8,\alpha_7^3} = \begin{bmatrix} 1 \\ 0 \end{bmatrix} \quad e_{8,\alpha_8^1} = \begin{bmatrix} -1 \\ 1 \end{bmatrix} \quad e_{8,\alpha_8^2} = \begin{bmatrix} 0 \\ -1 \end{bmatrix} \quad e_{8,\alpha_8^3} = \begin{bmatrix} 0 \\ -1 \end{bmatrix}$$

(iii) Modeling of Buffers

The three buffers can be modeled as scalar systems. For Buffer 1, we have

$$\mathbf{G_2} = (X_2, \Sigma_2, \xi_2, x_{2,0})$$

with

[1]The 'coupling event' α_7^3 is already controllable in $\mathbf{G_7}$.

$$\Sigma_2 = \{\alpha_1^3, \alpha_5^1\}, \quad \Sigma_{2,c} = \{\alpha_1^3\}$$

$$X_2 = \mathbb{N} \quad x_{2,0}^1 = 0$$

$$\xi_2(x_2^1, \alpha_1^3) = x_2^1 + 1$$
$$\xi_2(x_2^1, \alpha_5^1) = x_2^1 - 1$$

Buffer 2 is modeled similarly. For Buffer 3, we have

$$\mathbf{G_6} = (X_6, \Sigma_6, \xi_6, x_{6,0})$$

with

$$\Sigma_6 = \{\alpha_5^3, \alpha_7^1, \alpha_8^2\}, \quad \Sigma_{6,c} = \{\alpha_5^3\}$$

$$X_6 = \mathbb{N}, \quad x_{6,0}^1 = 0$$

$$\xi_6(x_6^1, \alpha_5^3) = \xi_6(x_6^1, \alpha_8^2) = x_6^1 + 1$$

$$\xi(x_6^1, \alpha_7^1) = x_6^1 - 1$$

(iv) Composition

Finally we compose the above components to obtain the VDES model of the production network:

$$\mathbf{G} = (X, \Sigma, \xi, x_0) = \bigoplus_{i=1}^{8} \mathbf{G_i}$$

where

$$\Sigma = \bigcup_{i=1}^{8} \Sigma_i, \quad \Sigma_c = \bigcup_{i=1}^{8} \Sigma_{i,c}$$

$$X = \bigoplus_{i=1}^{8} X_i, \quad x_0 = \bigoplus_{i=1}^{8} x_{i,0}$$

$$\xi(x, \alpha_i^j) = x + e_{\alpha_i^j} \quad \text{with} \quad e_{\alpha_i^j} = \bigoplus_{k=1}^{8} e_{k,\alpha_i^j}$$

where we define $e_{k,\alpha_i^j} = 0$ if α_i^j is not in Σ_k. The connections of the system modules are displayed in Fig. 8.8.

Note that the connection of $\mathbf{G_1}$ and $\mathbf{G_2}$ is serial, that of $\mathbf{G_2}$ and $\mathbf{G_4}$ parallel, and that of $\mathbf{G_6}$ and $\mathbf{G_8}$ feedback.

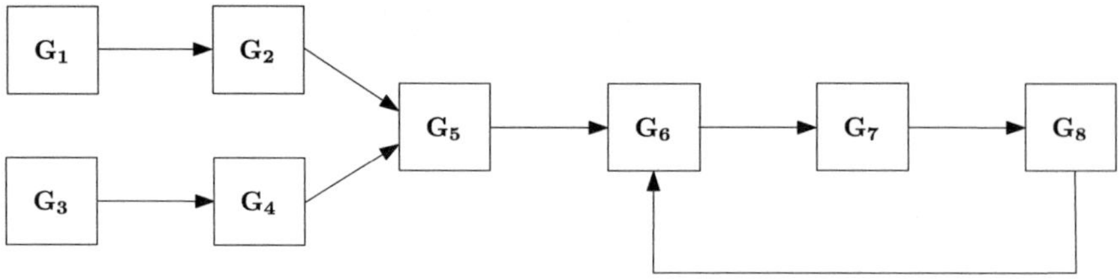

Fig. 8.8 Production network connection

(v) Control of the Production Network

We now discuss how to synthesize a modular controller to satisfy the performance specification of the system. The specification is that no buffer overflow and that at most one part be inspected at a given time. We assume that Buffer i has capacity k_i, and the buffer inside the inspection unit (x_8^1) has capacity 1. This specification can be formalized as a predicate on the state space X:

$$P = \bigwedge_{i=1}^{4} P_i$$

with

$$\begin{aligned} P_1 &= (x_2^1 \leq k_1) \\ P_2 &= (x_4^1 \leq k_2) \\ P_3 &= (x_6^1 \leq k_3) \\ P_4 &= (x_8^1 \leq 1) \end{aligned}$$

We list the optimal[2] subcontrollers for the above linear specifications.

$$\begin{aligned} f_{1,\alpha}(x) = 0 \quad &\Leftrightarrow \quad \alpha = \alpha_1^3 \text{ and } x_2^1 \geq k_1 \\ f_{2,\alpha}(x) = 0 \quad &\Leftrightarrow \quad \alpha = \alpha_3^3 \text{ and } x_4^1 \geq k_2 \\ f_{3,\alpha}(x) = 0 \quad &\Leftrightarrow \quad (\alpha = \alpha_5^3 \text{ or } \alpha = \alpha_8^1) \text{ and } x_6^1 + x_8^2 \geq k_3 \\ f_{4,\alpha}(x) = 0 \quad &\Leftrightarrow \quad \alpha = \alpha_7^3 \text{ and } x_8^1 \geq 1 \end{aligned}$$

The conjunction of these subcontrollers is

$$f := \bigwedge_{i=1}^{5} f_i$$

[2]While $\mathbf{G}_u$ is not loop-free, and therefore Theorem 8.7.2 not strictly applicable, the asserted optimality is obvious by inspection.

It is easy to check that all subcontrollers in f are balanced. Therefore, this modular controller is optimal in the sense that it synthesizes a largest reachable state set among all controllers which enforce the specification

$$P = \bigwedge_{i=1}^{4} P_i$$

as asserted by Theorem 7.5.2.

The above modular controller can lead to deadlock of the controlled system. To see this, consider the state at which $x_2^1 = k_1$, $x_4^1 = k_2$, $x_6^1 = k_3$, $x_8^1 = 1$ and $x_i^3 = g_i$ ($i = 1, 3, 5, 7$), with all other state variables being 0. At this state all controllable events are disabled and no uncontrollable event can occur. One way to remove the deadlock in the system is to add another subspecification which ensures that the deadlock state cannot be reached.

For this it is sufficient to ensure that the number of empty slots in Buffer 3 ($k_3 - x_6^1$) is maintained at least as great as the number of workpieces that could potentially be returned to Buffer 3 on being tested defective. In the worst case this is the number of workpieces being processed by the machines in Group 4 together with the Inspector, namely

$$x_7^2 + x_7^3 + x_7^4 + x_8^1 + x_8^2$$

So our new subspecification can be taken to be

$$P_5 = (x_6^1 + x_7^2 + x_7^3 + x_7^4 + x_8^1 + x_8^2 \leq k_3)$$

Notice that P_5 implies P_3, so the latter may now be discarded and the controls redesigned on the basis of P_1, P_2, P_4 and P_5.

Exercise 1: Redesign the controls as just specified. By detailed reasoning from your control design, prove that the controlled system is maximally permissive and nonblocking with respect to the prescribed initial state as marker state. For a plausible numerical assignment of the k_i (not too large!) verify this conclusion computationally using the DES representation described in Exercise 8.2.3.

To illustrate dynamic control, let us consider the following linear dynamic specification:

$$|\alpha_5^3| - |\alpha_8^3| \leq k$$

which specifies that the number of parts in the inspection loop never exceeds an integer k. Here $|\alpha_i^3|$ ($i = 5, 8$) denotes the number of occurrences of α_i^3. A one-dimensional memory **H** can be easily constructed from this specification and is shown in Fig. 8.9.

The dynamic specification is then equivalent to a static specification

$$y \leq k$$

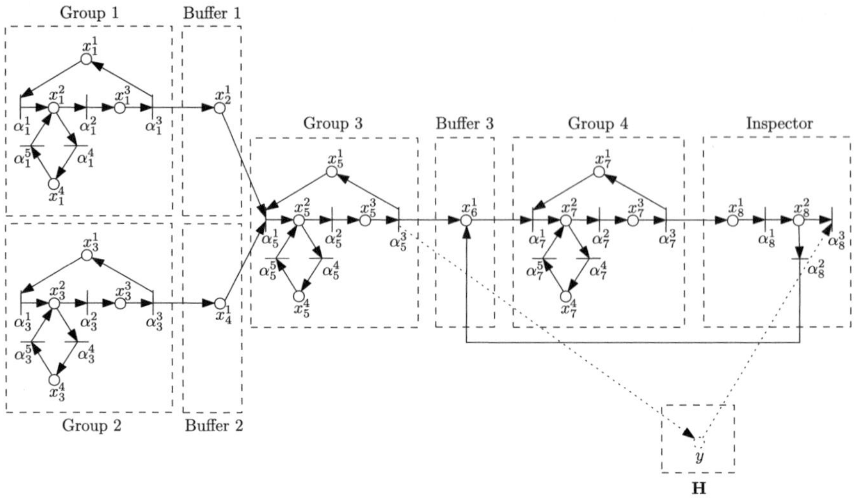

Fig. 8.9 Production network with dynamic control

on the extended state space $X \oplus Y$ with Y being the one-dimensional state space of **H**. By Theorem 8.9.2, the optimal controller f enforcing this static specification can be defined as

$$f_\alpha = \begin{cases} 0 & \text{if } \alpha = \alpha_5^3 \text{ and } y \geq k \\ 1 & \text{otherwise} \end{cases}$$

Exercise 2: Verify this result by calculating f^* as in Theorem 8.9.2.

8.13 Representation of Optimal Control by a Control VDES

In this section we return to the problem, illustrated in Sect. 8.11, of representing the optimal control by a VDES. This kind of result has the appeal that control does not require departure from the basic model class, a feature that offers convenience in control implementation, and ease of analysis and simulation of controlled behavior. On the other hand, insisting on a VDES implementation of the control does impose further restrictions on the structure of the plant **G** (although not on the control specification).

As usual let $\mathbf{G} = (X, \Sigma, \xi, x_0, X_m)$, with $X = \mathbb{N}^n$ and where ξ is defined by displacement vectors $e_\sigma \in \mathbb{Z}^{n\times 1}$. Assume that the control specification is provided in the form of a VDES $\mathbf{S} = (Y, \Sigma, \eta, y_0, Y)$, where $Y = \mathbb{N}^p$, and with displacement vectors $h_\sigma \in \mathbb{Z}^{p\times 1}$. One may think of **S** as tracking the behavior of **G**, with the specification expressed as the predicate $(y \geq 0)$ on $X \oplus Y$. Write $S := L(\mathbf{S})$, the closed behavior of **S**. We shall assume that $\sup \mathcal{C}_{\mathbf{G}}(L(\mathbf{G}) \cap S) \neq \varnothing$, so an optimal DSFBC F^* for **G** exists (as a SFBC on $X \oplus Y$), such that

$$L(F^*/\mathbf{G}) = \sup \mathcal{C}_{\mathbf{G}}(L(\mathbf{G}) \cap S)$$

Let $\mathbf{G}_{\mathbf{con}} := (Z, \Sigma, \zeta, z_0, Z)$ be a VDES with $Z = \mathbb{N}^r$. We say that $\mathbf{G}_{\mathbf{con}}$ is a *VDES implementation (VDESI)* of F^* provided

$$L(F^*/\mathbf{G}) = L(\mathbf{G} \oplus \mathbf{G}_{\mathbf{con}})$$

We shall provide a constructive sufficient condition (jointly due to Huang 1991, and Chen 1992) under which $\mathbf{G}_{\mathbf{con}}$ exists; it will then turn out that $r = p$.

Let $A \subseteq \Sigma$ and $\alpha \in A$. Define the event subset $\Sigma(\alpha, A)$ and coordinate (index) set $I(\alpha, A)$ inductively by the rules:

1. $\alpha \in \Sigma(\alpha, A)$
2. $\sigma \in \Sigma(\alpha, A)$ & $i \in \sigma^\uparrow \Rightarrow i \in I(\alpha, A)$
3. $i \in I(\alpha, A)$ & $\sigma \in i^\uparrow \cap A \Rightarrow \sigma \in \Sigma(\alpha, A)$
4. No other elements belong to $\Sigma(\alpha, A)$ or $I(\alpha, A)$

Note that Rule 2 says that i is placed in $I(\alpha, A)$ if $e_\sigma(i) < 0$. The restriction of **G** to $I(\alpha, A)$ and $\Sigma(\alpha, A)$ is the subsystem of **G** that is upstream from α, taking into account only the flow due to transitions in A.

Next take $A := \Sigma_u$ and consider the (one-dimensional) **S** with $p = 1$, $Y = \mathbb{N}$. Define

$$\Sigma_u^- := \{\sigma \in \Sigma_u | h_\sigma < 0\}$$
$$\Sigma^\nabla := \bigcup \{\Sigma(\sigma, \Sigma_u) | \sigma \in \Sigma_u^-\}$$
$$I^\nabla := \bigcup \{I(\sigma, \Sigma_u) | \sigma \in \Sigma_u^-\}$$

Finally, denote by $\mathbf{G}^\nabla$ the restriction of **G** to I^∇, Σ^∇. Thus $\mathbf{G}^\nabla$ is just the subsystem of **G** of which the flow is uncontrollable and effect is to decrement the (scalar) specification coordinate $y \in Y$ via events in Σ_u^-. Since the specification is precisely that y be maintained nonnegative, it is the structure of $\mathbf{G}^\nabla$ that is crucial for that of the optimal control.

Example 1: For **FACT#2** we had $P_2 = (10 + \#\beta \geq 3\#\lambda)$, which may be converted to a VDES **S** with $Y = \mathbb{N}$, $y_0 = 10$, and $[h_\alpha\ h_\beta\ h_\lambda\ h_\mu] = [0\ 1\ -3\ 0]$. We have

$$\Sigma_u^- = \{\lambda\}, \quad \Sigma^\nabla = \{\lambda\}, \quad I^\nabla = \lambda^\uparrow = \{2\}$$

Now we can state

Theorem 2 (Nianqing Huang, Shu-Lin Chen)

Given a VDES **G** and a specification language S represented by a 1-dimensional VDES **S** as above. Assume that an optimal DSFBC F^* for **G** exists (i.e. $\sup \mathcal{C}_{\mathbf{G}}(L(\mathbf{G}) \cap S) \neq \varnothing$). In addition assume the conditions

1. $\mathbf{G}^{\nabla}$ is loop-free.
2. For all $\sigma \in \Sigma^{\nabla}$, the displacement vector e_σ in **G** has at most one negative component, i.e. $|\sigma^{\uparrow}| \leq 1$; furthermore if for some i, $e_\sigma(i) < 0$, then $e_\sigma(i) = -1$ (by Rule 2 above, $\sigma^{\uparrow} = \{i\}$ and $i \in I^{\nabla}$).

Then a VDESI $\mathbf{G_{con}}$ for F^* exists. □

Example 3: For **FACT#2** the conditions of Theorem 2 are clearly satisfied; a VDESI was constructed *ad hoc* in Sect. 8.11.

Our proof will be constructive, and include a test for the existence of F^*. Roughly, the procedure is to successively transform the specification VDES **S**, moving upstream in $\mathbf{G}^{\nabla}$ against the flow of (uncontrollable) events, until the controllability condition on the transformed version of **S** is satisfied. The loop-freeness Condition 1 guarantees termination, while Condition 2 serves to rule out 'disjunctive' control logic (cf. Exercise 16 below).

We begin by defining a family of transformations T_α ($\alpha \in \Sigma_u$) on 1-dimensional VDES. As above let $\mathbf{S} = (Y, \Sigma, \eta, y_0, Y)$ with $y_0 \geq 0$ and $\eta(y, \sigma) = y + h_\sigma$. **NULL** will stand for the 'empty' VDES with $L(\mathbf{NULL}) := \varnothing$. Under Condition 2 of Theorem 2, define

$$\mathbf{S}_{new} := T_\alpha \mathbf{S}$$

as follows.

1. If $h_\alpha \geq 0$ then $T_\alpha = id$ (identity operator), i.e. $\mathbf{S}_{new} := \mathbf{S}$.
2. If $h_\alpha < 0$ and $\alpha^{\uparrow} = \{i\}$ (thus $e_\alpha(i) < 0$) then

$$y_{new,0} := y_0 + x_0(i) h_\alpha$$

$$h_{new,\sigma} := \begin{cases} h_\sigma + e_\sigma(i) h_\alpha & \text{if } \sigma \in i^{\uparrow} \cup i^{\downarrow} \\ h_\sigma & \text{otherwise} \end{cases}$$

   If $y_{new,0} \geq 0$ then accept the VDES $\mathbf{S}_{new}$ with $y_{new,0}$, $h_{new,\sigma}$ ($\sigma \in \Sigma$): otherwise $\mathbf{S}_{new} := \mathbf{NULL}$.

3. If $h_\alpha < 0$ and $\alpha^{\uparrow} = \varnothing$ then $\mathbf{S}_{new} := \mathbf{NULL}$.
4. For all $\sigma \in \Sigma$, $T_\sigma(\mathbf{NULL}) := \mathbf{NULL}$.

In part 2 of the definition of T_α, clearly $y_{new,0} \leq y_0$. Also, Condition 2 of Theorem 2 implies $e_\alpha(i) = -1$, so

$$h_{new,\alpha} = h_\alpha + e_\alpha(i) h_\alpha = 0$$

In general $h_{new,\sigma}$ is made up of the direct contribution h_σ to y on the occurrence of σ, plus a contribution to y of $e_\sigma(i)h_\alpha$ due to occurrences of α (see the proof of Lemma 9 below).

Note that activation of part 3 of the definition of T_α is not ruled out by Condition 2 of Theorem 2.

Example 4: In **FACT#2** recall that $y = 10 + \#\beta - 3\#\lambda$, $h = [0\ 1\ \ -3\ 0]$, $\lambda \in \Sigma_u$, $h_\lambda = -3 < 0$, $\lambda^\uparrow = \{2\}$. Thus, $\mathbf{S}_{new} = T_\lambda \mathbf{S}$ is calculated according to

$$y_{new,0} = y_0 + x_0(2)h_\lambda = y_0 = 10$$

$$2^\uparrow \cup 2^\downarrow = \{\alpha, \beta, \lambda\}$$

$$h_{new,\alpha} = h_\alpha + e_\alpha(2)h_\lambda = 0 + (+1)(-3) = -3$$

$$h_{new,\beta} = h_\beta + e_\beta(2)h_\lambda = (+1) + (-1)(-3) = 4$$

$$h_{new,\lambda} = h_\lambda + e_\lambda(2)h_\lambda = 0$$

$$h_{new,\mu} = h_\mu$$

Thus

$$h_{new} = [-3\ 4\ 0\ 0]$$

For $\mathbf{S}_{new}$,

$$\{\sigma \in \Sigma_u | h_\sigma < 0\} = \varnothing$$

so that now all $T_\sigma = id$.

Let $\hat{I} \subseteq I := \{1, \ldots, n\}$, $\hat{\Sigma} \subseteq \Sigma$ and let $\hat{\mathbf{G}}$ be the restriction of $\mathbf{G}$ to $\hat{I}$, $\hat{\Sigma}$. An event $\sigma \in \hat{\Sigma}$ is a *leaf event of* $\hat{\mathbf{G}}$ if, for all $i \in \sigma^\downarrow \cap \hat{I}$, $i^\downarrow \cap \hat{\Sigma} = \varnothing$, or briefly $(\sigma^\downarrow \cap \hat{I})^\downarrow \cap \hat{\Sigma} = \varnothing$ (in particular this is true if $\sigma^\downarrow \cap \hat{I} = \varnothing$). Evidently a leaf event of $\hat{\mathbf{G}}$ cannot contribute to the occurrence in $\hat{\mathbf{G}}$ of an immediately following event in $\hat{\Sigma}$.

Let LeafEvent($\hat{\mathbf{G}}$, α) denote a procedure that selects an arbitrary leaf event α of $\hat{\mathbf{G}}$ (or returns *error* if no leaf event exists). To compute $\mathbf{G_{con}}$ we start with

Procedure (1) (index Σ^∇ by leaf property):

```
Gvar := G∇;
Σvar := Σ∇;
k := |Σ∇|;
index := 1;
if k = 0 then LEList := [ ] else
   while index ≤ k do
      begin
          [] LeafEvent(Gvar, α);
          LEList[index] := α;
```

$\Sigma_{var} := \Sigma_{var} - \{\alpha\}$;
$\mathbf{G_{var}} :=$ restriction of $\mathbf{G_{var}}$ to I^{∇}, Σ_{var};
index := index + 1
end.

Proposition 5

Under the conditions of Theorem 2, Procedure (1) is well-defined.

Proof

It suffices to check that $\mathbf{G_{var}}$ always has a leaf event if index $\leq k$. But this follows by Condition 1 of Theorem 2 that $\mathbf{G}^{\nabla}$ is loop-free. □

Exercise 6: Supply the details.

Procedure (1) returns a listing of Σ^{∇}, LEList = [ ] or $[\alpha_1, \ldots, \alpha_k]$. Procedure (2) will then compute the final result as follows.

Procedure (2):

If LEList = [ ] then $\mathbf{S_{fin}} := \mathbf{S}$ else

$$\mathbf{S_{fin}} := T_{\alpha_k} T_{\alpha_{k-1}} \cdots T_{\alpha_1}(\mathbf{S})$$

It will be shown that, under the Conditions 1 and 2 of Theorem 2, either $\mathbf{S_{fin}} = \mathbf{NULL}$, in which case $\sup \mathcal{C}_{\mathbf{G}}(L(\mathbf{G}) \cap S) = \varnothing$ and F^* does not exist, or else $\mathbf{S_{fin}} = \mathbf{G_{con}}$ is a VDESI for F^*.

Exercise 7: Let $\mathbf{G}$ be a 1-dimensional VDES with $\Sigma_u = \Sigma = \{\alpha, \beta\}$ and displacement matrix $E = [-1 \ \ -1]$. Define $\mathbf{S}$ by $h = [-5 \ \ -7]$. Check that α, β are both leaf events and that

$$\mathbf{S_{fin}} = T_{\alpha} T_{\beta}(\mathbf{S}) = T_{\beta} T_{\alpha}(\mathbf{S})$$

with $y_{fin,0} = y_0 - 7x_0$ and $h_{fin} = [2 \ 0]$. Thus $\mathbf{S_{fin}} \neq \mathbf{NULL}$ iff $y_0 \geq 7x_0$. Explain intuitively.

Exercise 8: Consider the 6-dimensional VDES $\mathbf{G}$ with $\Sigma_u = \Sigma = \{\sigma_i | i = 1, \ldots, 7\}$, and

$$E := \begin{bmatrix} 0 & -1 & -1 & 0 & 0 & 0 & 0 \\ 0 & 0 & 0 & 4 & -1 & 0 & 0 \\ 2 & 1 & 0 & 0 & 0 & -1 & 0 \\ 0 & 0 & 3 & 0 & 2 & 0 & -1 \\ -1 & 0 & 0 & 0 & 0 & 0 & 0 \\ 0 & 0 & 0 & -1 & 0 & 0 & 0 \end{bmatrix}$$

For $\mathbf{S}$ let

$$h := [0 \ 0 \ 0 \ 0 \ 0 \ -2 \ -3]$$

Thus $\Sigma_u^- = \{\sigma_6, \sigma_7\}$. Show that a possible LEList is $[\sigma_6, \sigma_2, \sigma_7, \sigma_1, \sigma_5, \sigma_3, \sigma_4]$. Calculate h_{fin} and

$$y_{fin,0} = y_0 - cx_0$$

for suitable $c \in \mathbb{N}^{1\times 6}$. Interpret the result in terms of a 'worst case' event sequence that maximally decrements y.

Write $S = L(\mathbf{S})$, $S_{new} = L(\mathbf{S_{new}})$.

Lemma 9

Let $\mathbf{S_{new}} = T_\alpha \mathbf{S}$ (with $\alpha \in \Sigma_u$). Then

$$\sup \mathcal{C}_{\mathbf{G}}(L(\mathbf{G}) \cap S_{new}) = \sup \mathcal{C}_{\mathbf{G}}(L(\mathbf{G}) \cap S) \qquad (*)$$

Proof

If $h_\alpha \geq 0$ then $T_\alpha = id$, $\mathbf{S_{new}} = \mathbf{S}$, and there is nothing to prove.

If $h_\alpha < 0$ and $\alpha^\uparrow = \varnothing$ then $\mathbf{S_{new}} = \mathbf{NULL}$. Also, as $\alpha^j \in L(\mathbf{G})$ for all $j \geq 0$, and $y_0 + jh_\alpha < 0$ for j sufficiently large, we have $\sup \mathcal{C}_{\mathbf{G}}(L(\mathbf{G}) \cap S) = \varnothing$, establishing $(*)$ for this case.

It remains to assume $h_\alpha < 0$ and $\alpha^\uparrow \neq \varnothing$, namely $\alpha^\uparrow = \{i\}$, with $e_\alpha(i) = -1$. Let $s \in L(\mathbf{G})$, $n_\sigma = |s|_\sigma$ $(\sigma \in \Sigma)$. Then

$$0 \leq k := \xi(x_0, s)(i) = x_0(i) + \sum_\sigma n_\sigma e_\sigma(i)$$

Note that it suffices to sum over $\sigma \in i^\uparrow \cup i^\downarrow$. First suppose $s \in \sup \mathcal{C}_{\mathbf{G}}(L(\mathbf{G}) \cap S_{new})$. Then

$$\begin{aligned}
0 &\leq \eta_{new}(y_{new,0}, s) \\
&= y_{new,0} + \sum_\sigma n_\sigma h_{new,\sigma} \\
&= y_0 + x_0(i)h_\alpha + \sum_{\sigma \in i^\uparrow \cup i^\downarrow} n_\sigma [h_\sigma + e_\sigma(i)h_\alpha] + \sum_{\sigma \notin i^\uparrow \cup i^\downarrow} n_\sigma h_\sigma \\
&= y_0 + \sum_\sigma n_\sigma h_\sigma + \left[x_0(i) + \sum_{\sigma \in i^\uparrow \cup i^\downarrow} n_\sigma e_\sigma(i) \right] h_\alpha \\
&= y_0 + \sum_\sigma n_\sigma h_\sigma + k h_\alpha
\end{aligned}$$

and so, as $kh_\alpha \leq 0$,

$$y_0 + \sum_\sigma n_\sigma h_\sigma \geq 0$$

The same argument applies to each prefix $s' \leq s$, showing that $\eta(y_0, s)!$, namely $s \in L(\mathbf{G}) \cap S$. Therefore

$$\sup \mathcal{C}_{\mathbf{G}}(L(\mathbf{G}) \cap S_{new}) \subseteq L(\mathbf{G}) \cap S$$

so that

$$\sup \mathcal{C}_{\mathbf{G}}(L(\mathbf{G}) \cap S_{new}) \subseteq \sup \mathcal{C}_{\mathbf{G}}(L(\mathbf{G}) \cap S)$$

For the reverse inclusion, take $s \in \sup \mathcal{C}_{\mathbf{G}}(L(\mathbf{G}) \cap S)$. With k as before we have, as $e_\alpha(i) = -1$,

$$x_0(i) + \sum_{\sigma \neq \alpha} n_\sigma e_\sigma(i) + (n_\alpha + k) e_\alpha(i) = 0 \qquad (\dagger)$$

so that $\xi(x_0, s\alpha^j)!$ for $0 \leq j \leq k$, with $\xi(x_0, s\alpha^k)(i) = 0$. By controllability

$$s\alpha^k \in \sup \mathcal{C}_{\mathbf{G}}(L(\mathbf{G}) \cap S)$$

In particular $\eta(y_0, s\alpha^k)!$, namely

$$y_0 + \sum_{\sigma \neq \alpha} n_\sigma h_\sigma + (n_\alpha + k) h_\alpha \geq 0 \qquad (\dagger\dagger)$$

Calculating as before,

$$\begin{aligned} y_{new,0} + \sum_\sigma n_\sigma h_{new,\sigma} &= y_0 + \sum_\sigma n_\sigma h_\sigma + \left[x_0(i) + \sum_{\sigma \in i^\uparrow \cup i^\downarrow} n_\sigma e_\sigma(i) \right] h_\alpha \\ &= y_0 + \sum_{\sigma \neq \alpha} n_\sigma h_\sigma + n_\alpha h_\alpha - k e_\alpha(i) h_\alpha \\ &\quad + \left[x_0(i) + \sum_{\sigma \neq \alpha} n_\sigma e_\sigma(i) + (n_\alpha + k) e_\alpha(i) \right] h_\alpha \\ &= y_0 + \sum_{\sigma \neq \alpha} n_\sigma h_\sigma + (n_\alpha + k) h_\alpha \\ &\geq 0 \end{aligned}$$

using (†) and (††). By the same argument applied to each prefix of s, we conclude that $\eta_{new}(y_{new,0}, s)!$, namely

$$s \in L(\mathbf{G}) \cap S_{new}$$

and therefore

$$\sup \mathcal{C}_{\mathbf{G}}(L(\mathbf{G}) \cap S) \subseteq \sup \mathcal{C}_{\mathbf{G}}(L(\mathbf{G}) \cap S_{new})$$

as required. □

Lemma 10

Let $\mathbf{G}, \mathbf{S}$ satisfy Condition 2 of Theorem 2 and assume $\alpha \in \Sigma^\nabla$ with $\alpha^\uparrow = \{i\}$. Let $\mathbf{S_{new}} = T_\alpha \mathbf{S}$. Then

$$(\forall \sigma \notin i^\uparrow) h_{new,\sigma} \geq h_\sigma$$ □

Corollary 11

Under the conditions of Theorem 2, let the result of Procedure (1) be

$$\text{LEList} = [\alpha_1, \ldots, \alpha_k]$$

Then for $j = 2, \ldots, k$, T_{α_j} does not decrease the components of h_{α_i} for $i = 1, \ldots, j-1$.

Proof

Let $\alpha_j^{\uparrow} = \{l\}$ and in turn set $i = 1, \ldots, j-1$. Note that $\alpha_i \notin l^{\uparrow}$, since otherwise α_i cannot be a leaf event in the restriction of $\mathbf{G}^{\nabla}$ to I^{∇}, $[\alpha_i, \ldots, \alpha_j, \ldots, \alpha_k]$, contrary to Procedure (1). The result follows by Lemma 10 with $\alpha = \alpha_j$ and putting $\sigma = \alpha_1, \ldots, \alpha_{j-1}$ in turn. □

Exercise 12: In the example of Exercise 8, check that T_{σ_2} does not decrease h_6; T_{σ_7} does not decrease $h_6, h_2; \ldots; T_{\sigma_4}$ does not decrease h_i for $i = 6, 2, 7, 1, 5, 3$.

Proof of Theorem 2

Assume first that Procedure (2) yields $\mathbf{S_{fin}} \neq \mathbf{NULL}$. It will be shown that $\mathbf{S_{fin}}$ is a VDESI for F^*. By construction, LEList contains all $\alpha \in \Sigma_u$ such that $h_\alpha < 0$. Also, if at some stage in Procedure (2) we have $\mathbf{S}'' = T_\alpha \mathbf{S}'$, say, then $h_\alpha(\mathbf{S}'') \geq 0$. By Corollary 11 it follows that for all $\sigma \in \Sigma_u$, $h_{fin,\sigma} \geq 0$.

Write $S_{fin} = L(\mathbf{S_{fin}})$. We claim that S_{fin} is controllable with respect to $\mathbf{G}$. Indeed if $s \in S_{fin} \cap L(\mathbf{G})$ and $\sigma \in \Sigma_u$ with $s\sigma \in L(\mathbf{G})$, let $\eta_{fin}(y_{fin,0}, s) = y$, so

$$\eta_{fin}(y_{fin,0}, s\sigma) = y + h_{fin,\sigma} \geq y$$

namely $s\sigma \in S_{fin}$.

It follows that

$$\sup \mathcal{C}_{\mathbf{G}}(L(\mathbf{G}) \cap S_{fin}) = L(\mathbf{G}) \cap S_{fin}$$

By Lemma 9,

$$\sup \mathcal{C}_{\mathbf{G}}(L(\mathbf{G}) \cap S_{fin}) = \sup \mathcal{C}_{\mathbf{G}}(L(\mathbf{G}) \cap S)$$

and thus

$$L(\mathbf{G}) \cap S_{fin} = \sup \mathcal{C}_{\mathbf{G}}(L(\mathbf{G}) \cap S)$$

namely $\mathbf{S_{fin}}$ is a VDESI for F^*.

Finally we note that $\mathbf{S_{fin}} \neq \mathbf{NULL}$ iff, at each stage of Procedure (2), we have both $\alpha^{\uparrow} \neq \varnothing$ and $y_{new,0} \geq 0$. But if $\alpha^{\uparrow} = \varnothing$ (i.e. $\alpha \in \Sigma^{\nabla}$ is permanently enabled) there must exist a string $s \in \Sigma_u^* \cap L(\mathbf{G}^{\nabla})$ such that $\eta(y_0, s) < 0$, i.e. $s \notin L(\mathbf{S})$, hence $\sup \mathcal{C}_{\mathbf{G}}(L(\mathbf{G}) \cap S) = \varnothing$. The same conclusion follows if $y_{new,0} < 0$ at some stage. Thus (under the conditions of the theorem) the requirement $\mathbf{S_{fin}} \neq \mathbf{NULL}$ is necessary and sufficient for the existence of F^*. □

Remark 13: Procedure (2) could be modified by dropping the condition that $y_{new,0} \geq 0$ at each stage, and simply reporting whatever value of $y_{fin,0}$ is calculated at termination. A result $y_{fin,0} < 0$ would then represent the least amount by which the original value of y_0 should be raised to yield an acceptable (nonnegative) result.

It is straightforward to extend Theorem 2 to the case of a p-dimensional specification, merely by treating each component as an independent scalar in modular fashion.

Exercise 14: Justify the last statement in detail, and illustrate with a modular VDESI of dimension 2.

Exercise 15: Let **G** be a 5-dimensional VDES over $\Sigma = \{\sigma_1, \sigma_2, \sigma_3, \sigma_4, \sigma_5\}$ with $\Sigma_c = \{\sigma_4\}$ and

$$E = \begin{bmatrix} -1 & -1 & 2 & 0 & 0 \\ 0 & -2 & 0 & 0 & 0 \\ 0 & 0 & -1 & 1 & -2 \\ 0 & 0 & 0 & -1 & 0 \\ 0 & 0 & 0 & 0 & -1 \end{bmatrix}$$

$$h = [-1\ 0\ 0\ 0\ 0]$$

Let $x_0 = [0\,3\,0\,4\,1]$, $y_0 = 3$. Apply Theorem 2 with Procedures (1) and (2) to obtain a VDESI for F^*.

Exercise 16: Show that Condition 2 of Theorem 2 cannot be dropped altogether. **Hint**: In the following example, verify that Condition 2 is violated, and F^* exists but cannot be implemented as a VDES. Let **G** be 2-dimensional with $\Sigma = \{\alpha, \beta, \gamma\}$, $\Sigma_u = \{\alpha\}$ and

$$E = \begin{bmatrix} -1 & 0 & 1 \\ -1 & 1 & 0 \end{bmatrix}$$

$$h = [-1\ 0\ 0]$$

Let $x_0 = [0\ 0]$, $y_0 = 2$. Show that optimal control is given by

$$F^*_\beta(x_1, x_2, y) = 1 \ \text{ iff } \min\{x_1, x_2 + 1\} \le y$$

$$F^*_\gamma(x_1, x_2, y) = 1 \ \text{ iff } \min\{x_1 + 1, x_2\} \le y$$

Show that neither of these enablement conditions can be written as an inequality (or conjunction of inequalities) linear in the occurrence vector, so there can be no VDESI for either F^*_β or F^*_γ. In particular, show that the enablement region for β in the (x_1, x_2)-plane is the non-convex L-shaped region $\{0 \le x_1 \le y,\ x_2 \ge 0\} \cup \{x_1 \ge y,\ 0 \le x_2 \le y - 1\}$.

Exercise 17: Show that Condition 2 of Theorem 2 is not necessary. **Hint**: In the following example, verify that Condition 2 is violated, but F^* exists and does have a VDES implementation. Let **G** be 2-dimensional with $\Sigma = \{\alpha, \beta, \gamma\}$, $\Sigma_c = \{\gamma\}$ and

$$E = \begin{bmatrix} -1 & 0 & 1 \\ -1 & 1 & 0 \end{bmatrix}$$
$$h = [-1 \ \ 0 \ \ 0]$$

Let $x_0 = [0 \ 0]$, $y_0 = 2$. Check that F_γ^* exists and has the VDESI **S** with $h(\mathbf{S}) = [0 \ 0 \ -1]$ and initial value 2.

Exercise 18: Let **G** be 1-dimensional with $\Sigma = \{\alpha, \beta, \gamma\}$, $\Sigma_c = \{\gamma\}$,

$$E = [-2 \ \ -1 \ \ 1]$$
$$h = [-3 \ \ 1 \ \ 0]$$

Investigate the existence of F^* and a VDES implementation. What conclusion can be drawn about Theorem 2?

Exercise 19: Repeat Exercise 18 for the following. Let **G** be 4-dimensional with $\Sigma = \{\alpha, \beta, \lambda, \mu\}$, $\Sigma_u = \{\lambda\}$,

$$E = \begin{bmatrix} -1 & 0 & 0 & 1 \\ 0 & -1 & 0 & 2 \\ 2 & 1 & -2 & 0 \\ 0 & 0 & 1 & -2 \end{bmatrix}$$
$$x(0) = [1 \ \ 2 \ \ 0 \ \ 2] \in \mathbb{N}^{4\times 1}$$

and specification $x_4 \leq 2$.

Exercise 20: Continuing Exercise 19, find a 'place invariant' (cf. Exercise 6) $[c_1 \ c_2 \ c_3 \ c_4]$ with the $c_i > 0$. From this derive *a priori* bounds $x_1, x_4 \leq 4; x_2, x_3 \leq 8$. Using these bounds construct state models $\mathbf{X_i}$ for the VDES components x_i, and form the plant model **X** as their synchronous product. Use *TCT* to obtain the optimal supervisor enforcing ($x_4 \leq 2$). Verify consistency with your result in Exercise 19.

8.14 Notes

Supervisory control theory in the sense of Ramadge (1983), Ramadge and Wonham (1987) was subsequently formulated in Petri nets by Krogh (1987), who employed a mechanism of 'control places' and 'control arcs'. This chapter is based mainly on work of Li (1991), Li and Wonham (1988a, b, 1989a, b), Li and Wonham (1993), Brandin and Wonham (1994), N.-Q. Huang (Huang et al. 1989; Huang 1991), and S.-L. Chen (Chen and Wonham 1992; Chen 1992). Further developments can be found in Chen (1996). The structural boundedness criterion of Exercise 3 seems to be due to Memmi and Roucairol (1980, p. 215, Theorem 1); see also Murata (1989, p. 567, Theorem 29). The 'alternative theorem' we suggest using here is an easy extension from $\mathbb{R}$ to $\mathbb{Z}$ of Tucker (1956, p. 11, Corollary 3A(i)). In the Petri net literature the result of Theorem 8.6.5 has been attributed to Hiraishi and Ichikawa (1988).

Exercise 8.13.19 is adapted from Moody and Antsaklis (1998), Sect. 4.5, and the Production Network of Sect. 8.12 from Al-Jaar and Desrochers (1988).

Further information on linking supervisory control theory to VDES or Petri net models can be found in Wonham (2003), Uzam and Wonham (2006).

An extensive recent treatment of controlled Petri nets which addresses the nonblocking specification is available in the monograph by Chen and Li (2013).

Appendix 8.1: Three Examples from Petri Nets

We provide three examples to show how supervisory control problems described in terms of Petri nets can be treated in the automaton framework of this monograph, Chaps. 3, 4, and 6. Commented output from the *TCT* MAKEIT.TXT files is provided for the reader's convenience. Details of problem formulation can be found in the cited literature.

Example 1: **Manufacturing Workcell**

Reference: K. Barkaoui, I. Ben Abdallah. A deadlock prevention method for a class of FMS. Proc. IEEE Int. Conf. on Systems, Man, and Cybernetics. Vancouver, Canada, October 1995, pp. 4119–4124.

The system is a manufacturing workcell consisting of two input bins I1, I2, four machines M1,...,M4, two robots R1, R2, and two output bins O1, O2. Two production sequences, for RED and GRN workpieces, run concurrently; these are:

GRN: I1 $\rightarrow$ R1 $\rightarrow$ (M1 or M2) $\rightarrow$ R1 $\rightarrow$ M3 $\rightarrow$ R2 $\rightarrow$ O1

RED: I2 $\rightarrow$ R2 $\rightarrow$ M4 $\rightarrow$ R1 $\rightarrow$ M2 $\rightarrow$ R1 $\rightarrow$ O2

In the simplest case, treated here, at most one workpiece of each type (red, green) is allowed in the system at any one time.

Since the machines and production sequences share the robots as resources, there is the *a priori* possibility of deadlock. In fact, without control there is exactly one deadlock state (20) in **TEST** = **meet**(**CELL**,**SPEC**). At this state the components are in states:

R1	R2	M1	M2	M3	M4	RED	GRN
1	0	0	1	0	0	2	3

One deadlock sequence is (writing ev for event):

R1 takes in green part (ev 11 in **CELL**)
R2 takes in red part (ev 91)
R1 loads M2 with green part (ev 21)
R2 loads M4 with red part (ev 101)
R1 unloads red part from M4 (ev 111)

At this point, R1 holds a red, having just unloaded it from M4 (ev 111), while M2 holds a green, having finished processing it. R1 must load M2 with the red it's holding (ev 121) but cannot do so because M2 holds the green, which only R1 can unload (ev 41). The deadlock occurs because both R1 and M2 are 'full' (with a red, green respectively), and there is no mechanism for making the required swap. The cure is easy: simply eliminate state 20 from **TEST**; the result is then exactly the optimal controlled behavior **SUP**. So instead of the old blocking sequence 11.91.21.101.111 we now have 11.91.21.101.41; in other words GRN is allowed to progress to its state 4 (ev 41) before RED is allowed to progress to its state 3 (ev 111). The problem arises because a single robot (R1) is shared by 2 machines M2, M4; and M2 is shared by 2 processes (RED,GRN). At the deadlock state RED and GRN have conflicting requirements for M2 and therefore on R1, which is deadlocked because of its previous action on M4. Conclusion: careful sequencing of the interleaved processes RED, GRN is needed to avoid deadlock due to conflicting demands on the shared resources R1 and M2. Of course, this is achieved 'automatically' by **supcon**.

R1 = **create(R1**,[mark 0],[tran [0,11,1],[0,41,1],[0,51,1],
[0,111,1],[0,131,1],[1,21,0],[1,31,0],[1,61,0],[1,121,0],
[1,141,0]]) (2,10)
R2 = **create(R2**,[mark 0],[tran [0,71,1],[0,91,1],[1,81,0],
[1,101,0]]) (2,4)
M1 = **create(M1**,[mark 0],[tran [0,31,1],[1,51,0]]) (2,2)
M2 = **create(M2**,[mark 0],[tran [0,21,1],[0,121,1],[1,41,0],
[1,131,0]]) (2,4)
M3 = **create(M3**,[mark 0],[tran [0,61,1],[1,71,0]]) (2,2)
M4 = **create(M4**,[mark 0],[tran [0,101,1],[1,111,0]]) (2,2)
CELL = **sync(R1,R2)** (4,28) Blocked_events = None
CELL = **sync(CELL,M1)** (8,52) Blocked_events = None
CELL = **sync(CELL,M2)** (16,88) Blocked_events = None
CELL = **sync(CELL,M3)** (32,160) Blocked_events = None
CELL = **sync(CELL,M4)** (64,288) Blocked_events = None
ALL = **allevents(CELL)** (1,14)
GRN = **create(GRN**,[mark 0],[tran [0,11,1],[1,21,2],[1,31,3],
[2,41,4],[3,51,4],[4,61,5],[5,71,6],[6,81,0]]) (7,8)
RED = **create(RED**,[mark 0],[tran [0,91,1],[1,101,2],[2,111,3],
[3,121,4],[4,131,5],[5,141,0]]) (6,6)
SPEC = **sync(RED,GRN)** (42,90) Blocked_events = None
nonconflict(CELL,SPEC) = *false*
TEST = **meet(CELL,SPEC)** (35,61)
SUP = **supcon(CELL,SPEC)** (34,60)
SUP = **condat(CELL,SUP)** Controllable.

[Only events 51,71 do not appear in the **condat** table; therefore they could be replaced by uncontrollable counterparts (say 50,70) without changing the controlled behavior.]

SIMSUP = **supreduce(CELL,SUP,SUP)** (11,38)
STEST = **meet(CELL,SIMSUP)** (34,60)
isomorph(STEST,SUP) = *true*
SIMSUP = **condat(CELL,SIMSUP)** Controllable.

The following shows that removing the single blocking state 20 is enough to obtain the optimal control from the naive behavior **TEST**.

ETEST = **edit(TEST**,[states -[20]]) (34,60)
isomorph(ETEST,SUP) = *true*

Example 2: **Piston Rod Robotic Assembly Cell**

Reference: J.O. Moody, P.J. Antsaklis. Supervisory Control of Discrete Event Systems Using Petri Nets. Kluwer, 1998; Sect. 8.4.

With reference to Fig. 8.11 of the cited text, the system consists of an M-1 robot performing various tasks (Petri net places p4, p5, p6, p7), and similarly an S-380 robot (p2, p3); p1 is used for initialization.

M-1 ROBOT

To model this we replace p4 by a generator capable of holding up to two piston pulling tools in a two-slot buffer MR1; the tools are generated by event 40 and selected for use by event 41. The event sequence 41.51.60.70.80 tracks the installation of a cap on a piston rod and the conveyance of its engine block out of the work space. In our model, up to four operations in this sequence could be progressing simultaneously, although a specification (below) will limit this number to one.

MR1 = **create(MR1**,[mark 0],[tran [0,40,1],[1,40,2],[1,41,0],
[2,41,1]]) (3, 4)
MR2 = **create(MR2**,[mark 0],[tran [0,41,1],[1,51,0]]) (2,2)
MR3 = **create(MR3**,[mark 0],[tran [0,51,1],[1,60,0]]) (2,2)
MR4 = **create(MR4**,[mark 0],[tran [0,60,1],[1,70,0]]) (2,2)
MR5 = **create(MR5**,[mark 0],[tran [0,70,1],[1,80,0]]) (2,2)
MROB = **sync(MR1,MR2)** (6,9)
MROB = **sync(MROB,MR3)** (12,21)
MROB = **sync(MROB,MR4)** (24,48)
MROB = **sync(MROB,MR5)** (48,108)

S-380 ROBOT

Starting from the ready-to-work condition, this robot performs the event sequence 10.20.30 corresponding to readying parts for assembly; its work cycle is closed by event 80.

SROB = **create**(**SROB**,[mark 0],[tran [0,10,1],[1,20,2],[2,30,3], [3,80,0]]) (4,4)
PLANT = **sync**(**MROB**,**SROB**) (192,504)

Note that the only controllable events are 41,51, which more than satisfies the authors' requirement that events 60,70,80 be uncontrollable.

There are three specifications, as detailed in the authors' equations (8.11)–(8.13) in (Moody and Antsaklis 1998). These are linear inequalities on markings, which are easily converted (by inspection of the Petri net in (Moody and Antsaklis 1998)) into counting constraints on suitable event pairs. For instance, (8.12) requires that m4+m5+m6+m7 $\leq$ 1, where mi is the marking of place pi; by inspection, this is equivalent to

$$(|41| - |51|) + (|51| - |60|) + (|60| - |70|) + (|70| - |80|) \leq 1$$

or simply $|41| - |80| \leq 1$; here $|k|$ is the number of firings of transition k since the start of the process. By inspection of the Petri net in (Moody and Antsaklis 1998) it is clear that the inequality forces events 41,80 to alternate, with 41 occurring first; hence **SPEC2**, below.

SPEC1 = **create**(**SPEC1**,[mark 0],[tran [0,10,1],[1,30,0]]) (2,2)
SPEC2 = **create**(**SPEC2**,[mark 0],[tran [0,41,1],[1,80,0]]) (2,2)
SPEC3 = **create**(**SPEC3**,[mark 0],[tran [0,30,1],[1,51,0]]) (2,2)
SPEC = **sync**(**SPEC1**,**SPEC2**) (4,8)
SPEC = **sync**(**SPEC**,**SPEC3**) (8,18)
PLANTALL = **allevents**(**PLANT**) (1,9)
SPEC = **sync**(**SPEC**,**PLANTALL**) (8,50)

The supremal supervisor can now be computed, then simplified by the control-congruence reduction procedure.

SUPER = **supcon**(**PLANT**,**SPEC**) (33,60)
SUPER = **condat**(**PLANT**,**SUPER**) Controllable.
SIMSUP = **supreduce**(**PLANT**,**SUPER**,**SUPER**) (3,12)

Thus **SIMSUP** is strictly minimal.

The authors specify four auxiliary constraints (8.14)–(8.17) (see (Moody and Antsaklis 1998)), of the form already discussed; we model these as follows, and create the auxiliary specification **ASPEC**. We test these constraints against the existing controlled behavior **SUPER**, and confirm that they are already satisfied.

ASP1 = **create**(**ASP1**,[mark 0],[tran [0,20,1],[1,30,0]]) (2,2)
ASP2 = **create**(**ASP2**,[mark 0],[tran [0,41,1],[1,60,0]]) (2,2)
ASP3 = **create**(**ASP3**,[mark 0],[tran [0,60,1],[1,70,0]]) (2,2)
ASP4 = **create**(**ASP4**,[mark 0],[tran [0,70,1],[1,80,0]]) (2,2)
ASPEC = **sync**(**ASP1**,**ASP2**) (4,8)
ASPEC = **sync**(**ASPEC**,**ASP3**) (8,18)
ASPEC = **sync**(**ASPEC**,**ASP4**) (16,40)
ASPEC = **sync**(**ASPEC**,**PLANTALL**) (16,88)
COASPEC = **complement**(**ASPEC**,[]) (17,153)
X = **meet**(**SUPER**,**COASPEC**) (33,60)
TX = **trim**(**X**) (0,0)

Unobservable events: we assume with (Moody and Antsaklis 1998) that events 51,60,70 have become unobservable. As a simplifying assumption on supervisor design, we consider that controllable event 51 will now not be subject to disablement. Thus we could (but will not) relabel event 51 as 50 throughout. Our new assumption allows us to treat the problem as an instance of SCOP (Sect. 6.5). We therefore compute as follows.

N = **supnorm**(**PLANT**,**SPEC**,**null**[51,60,70]) (24,39)
NO = **project**(**N**,**null**[51,60,70]) (15,24)
PLANTO = **project**(**PLANT**,**null**[51,60,70]) (60,129)
SUPERO = **supcon**(**PLANTO**,**NO**) (15,24)
['Observer's supervisor']
SUPERO = **condat**(**PLANTO**,**SUPERO**) Controllable.
['Observer's supervisor']
OSUPER = **selfloop**(**SUPERO**,[51,60,70]) (15,69)
[Feasible supervisor]
nonconflict(**PLANT**,**OSUPER**) = *true*
K = **meet**(**PLANT**,**OSUPER**) (24,39)
[Controlled behavior using feasible supervisor]
SIMSUPO = **supreduce**(**PLANTO**,**SUPERO**,**SUPERO**) (2,7)

Thus **SIMSUPO** is strictly minimal.

SIMSUPO = **condat(PLANTO,SIMSUPO)** Controllable.
TESTO = **meet(PLANTO,SIMSUPO)** (15,24)
isomorph(TESTO,SUPERO) = *true*

We will check that, as expected, controlled behavior **K** using the feasible supervisor is more restricted than the original controlled behavior **SUPER** (which of course was computed without assuming any observational constraint). Nevertheless, **K** is adequate for performance of the assembly process: for instance the **K**-string 10.20.30.40.41.51.60.70.80 is a full assembly cycle.

COSUPER = **complement(SUPER,[])** (34,306)
X = **meet(K,COSUPER)** (24,39)
TX = **trim(X)** (0,0)

Some routine checks, in principle redundant, are:

nonconflict(PLANT,OSUPER) = *true*
OSUPER = **condat(PLANT,OSUPER)** Controllable.

As expected, **OSUPER** never disables unobservable event 51.

Example 3: **Unreliable Machine (Deadlock Avoidance)**

Reference: J.O. Moody, P.J. Antsaklis. Deadlock avoidance in Petri nets with uncontrollable transitions. Proc. 1998 American Automatic Control Conference. Reproduced in J.O. Moody, P.J. Antsaklis. Supervisory Control of Discrete Event Systems Using Petri Nets. Kluwer, 1998; Sect. 8.3 (pp. 122–129).

This is a problem which, in the authors' Petri net formulation, requires finding the system's two 'uncontrolled siphons'. By contrast, the *TCT* solution is fast and immediate, requiring no special analysis.

The system model consists of a machine M1 containing two 1-slot output buffers M1C (for completed workpieces) and M1B (for damaged workpieces, which result when M1 breaks down), together with two dedicated AGVs to clear them. M1 is the SCT machine model (e.g. Sect. 3.2). Event 10 (successful completion) increments M1C, which must be cleared (event 14) by AGV1 before M1 can restart; event 12 (breakdown) increments M1B, which must be cleared (event 16) by AGV2 before M1 can be repaired after breakdown (event 13); these requirements are enforced by **SPEC1C**, **SPEC1B**, respectively. The workspace near the buffers can be occupied by only one AGV at a time: this is enforced by **SPEC1**; the final **SPEC** model is **sync(SPEC1,SPEC1C,SPEC1B)**. Blocking would occur if, for instance, AGV1 moved into position to clear its buffer M1C, but M1B rather than M1C was filled; or AGV2 moved into position to clear its buffer M1B, but M1C rather than M1B was filled; in each case the positioned AGV would lock out the other.

Modeling the plant

M1 = **create**(**M1**,[mark 0],[tran [0,11,1],[1,10,0],[1,12,2],
[2,13,0]]) (3,4)
M1C = **create**(**M1C**,[mark 0],[tran [0,10,1],[1,14,0]]) (2,2)
M1B = **create**(**M1B**,[mark 0],[tran [0,12,1],[1,16,0]]) (2,2)
AGV1 = **create**(**AGV1**,[mark 0],[tran [0,101,1],[1,14,2],
[2,100,0]]) (3,3)
AGV2 = **create**(**AGV2**,[mark 0],[tran [0,201,1],[1,16,2],
[2,200,0]]) (3,3)
P = **sync**(**M1,M1C**) (6,10) Blocked_events = None
P = **sync**(**P,M1B**) (12,24) Blocked_events = None
P = **sync**(**P,AGV1**) (36,84) Blocked_events = None
P = **sync**(**P,AGV2**) (108,288) Blocked_events = None
PALL = **allevents**(**P**) (1,10)

Modeling the specification

SPEC1 = **create**(**SPEC1**,[mark 0],[tran [0,101,1],[0,201,2],[1,100,0],
[2,200,0]]) (3,4)
SPEC1C = **create**(**SPEC1C**,[mark 0],[tran [0,10,1],[0,11,0],[1,14,0]])
(2,3)
SPEC1B = **create**(**SPEC1B**,[mark 0],[tran [0,12,1],[0,13,0],[1,16,0]])
(2,3)
SPEC = **sync**(**SPEC1,SPEC1C**) (6,17) Blocked_events = None
SPEC = **sync**(**SPEC,SPEC1B**) (12,52) Blocked_events = None
SPEC = **sync**(**SPEC,PALL**) (12,52) Blocked_events = None
nonconflict(**P,SPEC**) = *false*

Blocking could occur in the absence of supervisory control. Some blocking strings are 11.10.201, 201.11.10, 11.12.101, 101.11.12. These result in the situations described earlier, where an AGV in the workspace locks out the other, required AGV.

PSPEC = **meet**(**P,SPEC**) (24,40)
nonconflict(**PSPEC,PALL**) = *false*
MPSPEC = **minstate**(**PSPEC**) (23,40)

Computing the supremal supervisor

SUP = **supcon(P,SPEC)** (16,24)
SUP = **condat(P,SUP)** Controllable.

Computing a simplified supervisor

SIMSUP = **supreduce(P,SUP,SUP)** (5,23)
X = **meet(P,SIMSUP)** (16,24)
isomorph(SUP,X) = *true*
SIMSUP = **condat(P,SIMSUP)** Controllable.

It is easy to check by inspection that **SIMSUP** prohibits the blocking sequences listed above.

References

Al-Jaar RY, Desrochers AA (1988) A modular approach for the performance analysis of automated manufacturing systems using generalized stochastic Petri nets. Rpt. RAL #116, Robotics and Automation Laboratory, Rensselaer Polytechnic Institute, Troy, New York

Chen S-L (1992) Existence and design of supervisors for vector discrete event systems. MASc thesis, Department of Electrical and Computer Engineering, University of Toronto

Chen S-L (1996) Control of discrete-event systems of vector and mixed structural type. PhD thesis, Department of Electrical and Computer Engineering, University of Toronto

Chen S-L, Wonham WM (1992) Existence and design of supervisors for vector discrete-event systems. In: Proceedings of thirtieth annual Allerton conference on communication, control and computing, University of Illinois, pp 604–613

Chen Y, Li Z (2013) Optimal supervisory control of automated manufacturing systems. CRC Press, Boca Raton

Hiraishi K, Ichikawa A (1988) A class of Petri nets that [sic] a necessary and sufficient condition for reachability is obtainable. Trans Soc Instrum Control Eng (SICE) 24(6):91–96 (in Japanese)

Huang NQ, Li Y, Wonham WM (1989) Supervisory control of vector discrete-event systems. In: Proceedings of twenty-seventh annual Allerton conference on communication, control and computing, University of Illinois, pp 925–934

Huang N-Q (1991) Supervisory control of vector discrete-event processes. MASc thesis, Department of Electrical and Computer Engineering, University of Toronto

Krogh BH (1987) Controlled Petri nets and maximally permissive feedback logic. In: Proceedings of twenty-fifth annual Allerton conference on communication control and computing, Allerton, pp 317–326

Li Y (1991) Control of vector discrete-event systems. PhD thesis, Department of Electrical and Computer Engineering, University of Toronto

Li Y, Wonham WM (1988a) A state-variable approach to the modeling and control of discrete-event systems. In: Proceedings of twenty-sixth annual Allerton conference on communication, control, and computing, University of Illinois, pp 1140–1149

Li Y, Wonham WM (1988b) Controllability and observability in the state-feedback control of discrete-event systems. In: Proceedings of twenty-seventh IEEE conference on decision and control, New York, pp 203–208

Li Y, Wonham WM (1989a) Composition and modular state-feedback control of vector discrete-event systems. In: Proceedings of the conference on information sciences and systems, The Johns Hopkins University, Baltimore, pp 103–111

Li Y, Wonham WM (1989b) Linear integer programming techniques in the control of vector discrete-event systems. In: Proceedings of twenty-seventh annual Allerton conference on communication, control and computing, University of Illinois, pp 528–537

Li Y, Wonham WM (1993) Control of vector discrete-event systems: I - the base model. IEEE Trans Autom Control 38(8):1214–1227. Correction: IEEE Trans Autom Control 39(8):1771 (1994). [Figures 9–11 are reprinted, with permission, from IEEE.]

Memmi G, Roucairol G (1980) Linear algebra in net theory. In: Brauer W (ed) Net theory and applications. Lecture notes on computer science (LNCS), vol 84. Springer, Berlin, pp 213–223

Merlin PM, Farber DJ (1976) Recoverability of communication protocols - implications of a theoretical study. IEEE Trans Commun 24:1036–1043

Moody J, Antsaklis P (1998) Supervisory control of discrete event systems using Petri nets. Kluwer

Murata T (1989) Petri nets: properties, analysis and applications. Proc IEEE 77(4):541–580

Ramadge PJ (1983) Control and supervision of discrete event processes. PhD thesis, Department of Electrical and Computer Engineering, University of Toronto

Ramadge PJ, Wonham WM (1987) Supervisory control of a class of discrete event processes. SIAM J Control Optim 25(1):206–230

Tucker AW (1956) Dual systems of homogeneous linear relations. In: Kuhn HW, Tucker AW (eds) Linear inequalities and related systems. Annals of mathematics studies, vol 38. Princeton University Press, Princeton, pp 3–18

Uzam M, Wonham WM (2006) A hybrid approach to supervisory control of discrete event systems coupling RW supervisors to Petri nets. Int J Adv Manuf Technol 28(7–8):747–760

Wonham WM (2003) Supervisory control theory: models and methods. In: Proceedings of ATPN - workshop on discrete event systems control, twenty-fourth international conference on application theory of Petri nets (ATPN '03), Eindhoven, The Netherlands, pp 1–14

Chapter 9
Supervisory Control of Timed Discrete-Event Systems

This chapter provides a framework for the study of theoretical issues in the design of supervisory controls for timed discrete-event systems. The model incorporates both time delays and hard deadlines, and admits both forcing and disablement as means of control. In addition it supports composition of modular subsystems and systematic synthesis. In particular, the model retains the concept of design that is qualitatively optimal in the sense of minimally restricting the behavior of the underlying controlled system, subject to constraints imposed by formal specifications of performance.

Because the theory is expressed in the elementary framework of regular languages and finite automata, only rudimentary control scenarios can be treated directly. For instance, no explicit provision is made for program variables, or such constructs of a real-time programming language as interrupts and logically conditioned events and procedures. In higher-level approaches where such constructs are supported, design approaches tend to be heuristic. With the introduction of suitable architecture the present framework may supply a basis for rendering such approaches more formal and systematic.

9.1 Introduction

In this chapter we augment the framework of Chaps. 3 and 4 with a timing feature. The occurrence of an event, relative to the instant of its enablement, will be constrained to lie between a lower and upper time bound, synchronized with a postulated global digital clock. In this way we are able to capture timing issues in a useful range of control problems. Timing introduces a new dimension of DES modeling and control, of considerable power and applied interest, but also of significant complexity. Nevertheless, it will turn out that our previous concept of controllability, and the existence of maximally permissive supervisory controls, can be suitably generalized. The enhanced

W. M. Wonham and K. Cai, *Supervisory Control of Discrete-Event Systems*, Communications and Control Engineering,
https://doi.org/10.1007/978-3-319-77452-7_9

setting admits subsystem composition (analogous to synchronous product), and the concept of forcible event as an event that preempts the tick of the clock. An example of a manufacturing cell illustrates how the timed framework can be used to solve control synthesis problems which may include logic-based, temporal and quantitative optimality specifications.

Our timed framework is amenable to computation in the style of *TCT*; the enhanced package, designed to be used with this text, is *TTCT*.

9.2 Timed Discrete-Event Systems

To develop the base model we begin with the usual 5-tuple of form

$$\mathbf{G_{act}} = (A, \Sigma_{act}, \delta_{act}, a_0, A_m)$$

except that the 'state set' often designated Q has been replaced with an *activity set* A whose elements are *activities* a. While in principle the activity set need not be finite, in applications it nearly always is; here we shall restrict A to be finite for technical simplicity. Σ_{act} is a finite alphabet of *event labels* (or simply, *events*). We stress that, in the interpretation, activities have duration in time, while events are instantaneous. The *activity transition function* is, as expected, a partial function $\delta_{act} : A \times \Sigma_{act} \to A$. An *activity transition* is a triple $[a, \sigma, a']$, with $a' = \delta_{act}(a, \sigma)$. In line with standard terminology, a_0 is the *initial activity* and $A_m \subseteq A$ is the subset of *marker activities*. Let $\mathbb{N}$ denote the natural numbers $\{0, 1, 2, \ldots\}$. In Σ_{act}, each transition (label) σ will be equipped with a *lower time bound* $l_\sigma \in \mathbb{N}$ and an *upper time bound* $u_\sigma \in \mathbb{N} \cup \{\infty\}$. To reflect two distinct possibilities of basic interest we partition Σ_{act} according to

$$\Sigma_{act} = \Sigma_{spe} \dot\cup \Sigma_{rem}$$

where '*spe*' denotes 'prospective' and '*rem*' denotes 'remote'. If an event σ is *prospective*, its upper time bound u_σ is finite ($0 \leq u_\sigma < \infty$) and $0 \leq l_\sigma \leq u_\sigma$; while if σ is *remote*, we set $u_\sigma = \infty$ and require $0 \leq l_\sigma < \infty$. The modeling function of time bounds is straightforward: l_σ would typically represent a delay, in communication or in control enforcement; u_σ a hard deadline, imposed by legal specification or physical necessity. The formal role of time bounds will be treated in detail below. The triples $(\sigma, l_\sigma, u_\sigma)$ will be called *timed events*, and for these we write

$$\Sigma_{tim} := \{(\sigma, l_\sigma, u_\sigma) | \sigma \in \Sigma_{act}\}$$

For $j, k \in \mathbb{N}$ write $[j, k]$ for the set of integers i with $j \leq i \leq k$, and let

$$T_\sigma = \begin{cases} [0, u_\sigma] & \text{if } \sigma \in \Sigma_{spe} \\ [0, l_\sigma] & \text{if } \sigma \in \Sigma_{rem} \end{cases}$$

T_σ will be called the *timer interval* for σ. We can now define the *state set*

$$Q := A \times \prod\{T_\sigma | \sigma \in \Sigma_{act}\}$$

Thus a *state* is an element of form

$$q = (a, \{t_\sigma | \sigma \in \Sigma_{act}\})$$

where $a \in A$ and the $t_\sigma \in T_\sigma$; namely q consists of an activity a together with a tuple assigning to each event $\sigma \in \Sigma_{act}$ an integer in its timer interval T_σ. The component t_σ of q will be called the *timer* of σ in q. If $\sigma \in \Sigma_{spe}$, the *current deadline* for σ is t_σ, while the *current delay* is $\max\{t_\sigma + l_\sigma - u_\sigma, 0\}$. If $\sigma \in \Sigma_{rem}$, the *current delay* is t_σ (while the current deadline may be regarded as infinite). The value u_σ (resp. l_σ) for a prospective (resp. remote) event σ will be called the *default value* of t_σ. The *initial state* is

$$q_0 := (a_0, \{t_{\sigma 0} | \sigma \in \Sigma_{act}\})$$

where the t_σ are set to their default values

$$t_{\sigma 0} := \begin{cases} u_\sigma & \text{if } \sigma \in \Sigma_{spe} \\ l_\sigma & \text{if } \sigma \in \Sigma_{rem} \end{cases}$$

The *marker state subset* will be taken to be of the form

$$Q_m \subseteq A_m \times \prod\{T_\sigma | \sigma \in \Sigma_{act}\}$$

namely a marker state comprises a marker activity together with a suitable assignment of the timers.

We introduce one additional event, written *tick*, to represent 'tick of the global clock', and take for our total set of events

$$\Sigma := \Sigma_{act} \dot\cup \{tick\}$$

The *state transition function* will be defined in detail below; as expected it will be a partial function

$$\delta : Q \times \Sigma \to Q$$

We now write

$$\mathbf{G} = (Q, \Sigma, \delta, q_0, Q_m)$$

With the above definitions, including the partition of Σ as

$$\Sigma = \Sigma_{spe} \dot\cup \Sigma_{rem} \dot\cup \{tick\}$$

and an assignment of time bounds, **G** will be called a *timed discrete-event system (TDES)*. For the purpose of display we may employ the *activity transition graph (ATG)* of **G**, namely the ordinary transition graph of $\mathbf{G_{act}}$, and the *timed transition graph (TTG)* of **G**, namely the ordinary transition graph of **G**, incorporating the *tick* transition explicitly. In addition, by projecting out *tick* from $L(\mathbf{G})$ we can derive the *timed activity* DES $\mathbf{G_{tact}}$ over Σ_{act}, and display its transition structure as the *timed activity transition graph (TATG)* of **G**. While the TATG suppresses *tick*, it does incorporate the constraints on ordering of activities induced by time bounds. As illustrated by Examples 1 and 2 below, $\mathbf{G_{tact}}$ may be much more complex than $\mathbf{G_{act}}$.

Before defining the behavioral semantics of the TDES **G** in detail, we provide an informal summary. As is customary with DES, events are thought of as instantaneous and occurring at quasi-random moments of real time $\mathbb{R}^+ = \{t|0 \leq t < \infty\}$. However, we imagine measuring time only with a global digital clock with output *tickcount*: $\mathbb{R}^+ \to \mathbb{N}$, where

$$tickcount(t) := n, \quad n \leq t < n + 1$$

Temporal conditions will always be specified in terms of this digital clock time; real-valued time as such, and the clock function *tickcount*, will play no formal role in the model. The temporal resolution available for modeling purposes is thus just one unit of clock time. The event *tick* occurs exactly at the real-time moments $t = n$ ($n \in \mathbb{N}$). As usual, **G** is thought of as a generator of strings in Σ^*; intuitively **G** incorporates the digital clock, and thus its 'generating action' extends to the event *tick*.

Events are generated as follows. **G** starts from q_0 at $t = 0$ and executes state transitions in accordance with its transition function δ, i.e. by following its TTG. $\delta(q, \sigma)$ is defined at a pair (q, σ), written $\delta(q, \sigma)!$, provided (i) $\sigma =$ *tick*, and no deadline of a prospective event in q is zero (i.e. no prospective event is *imminent*); or (ii) σ is prospective, $q = (a, \_)$, $\delta_{act}(a, \sigma)!$, and $0 \leq t_\sigma \leq u_\sigma - l_\sigma$; or (iii) σ is remote, $q = (a, \_)$, $\delta_{act}(a, \sigma)!$, and $t_\sigma = 0$. An event $\sigma \in \Sigma_{act}$ is said to be *enabled* at $q = (a, \_)$ if $\delta_{act}(a, \sigma)!$, and to be *eligible* if, in addition, its timer evaluation is such that $\delta(q, \sigma)!$. Only an eligible event 'can actually occur'. If σ is not enabled, it is said to be *disabled*; if σ is not eligible, it is *ineligible*; an enabled but ineligible event will be called *pending*.

The occurrence of *tick* at q causes no change in the activity component a of q; however, the timer components t_σ are altered in accordance with the detailed rules given below. The occurrence of $\sigma \in \Sigma_{act}$ at q always resets t_σ to its default value; again, the effect on other timers will be described below. After $\sigma \in \Sigma_{act}$ first becomes enabled, its timer t_σ is decremented by 1 at each subsequent *tick* of the clock, until either t_σ reaches zero, or σ occurs, or σ is disabled as a result of the occurrence of some eligible transition (possibly σ itself). If σ occurs, or becomes disabled owing to some transition to a new activity, t_σ is reset to its default value, where it is held until σ next becomes re-enabled, when the foregoing process repeats.

An event $\sigma \in \Sigma_{act}$ cannot become eligible (and so, occur) prior to l_σ *ticks* of the clock after it last became enabled. A prospective event σ cannot be delayed longer than $u_\sigma - l_\sigma$ *ticks* after t_σ has 'ticked down' to $u_\sigma - l_\sigma$; thus when t_σ 'times out' to

0, σ cannot be delayed except by preemptive occurrence of some other eligible event in Σ_{act}. A remote event σ can occur any time (although it need not occur at all), as long as it remains enabled, and provided l_σ *ticks* have elapsed after it last became enabled. It is important to note that, because of its possible non-occurrence (even if continuously enabled) a remote event is not just a 'limiting case' of a prospective event.

With the above as guidelines we now provide the formal definition of δ. Write $\delta(q, \sigma) = q'$, where

$$q = (a, \{t_\tau | \tau \in \Sigma_{act}\}), \quad q' = (a', \{t'_\tau | \tau \in \Sigma_{act}\})$$

Then $\delta(q, \sigma)!$ iff

(i) $\sigma = tick$ and $(\forall \tau \in \Sigma_{spe})\delta_{act}(a, \tau)! \Rightarrow t_\tau > 0$; or
(ii) $\sigma \in \Sigma_{spe}$, $\delta_{act}(a, \sigma)!$, and $0 \leq t_\sigma \leq u_\sigma - l_\sigma$; or
(iii) $\sigma \in \Sigma_{rem}$, $\delta_{act}(a, \sigma)!$, and $t_\sigma = 0$

The entrance state q' is defined as follows.

(i) Let $\sigma = tick$. Then $a' := a$, and

$$\text{if } \tau \text{ is prospective,} \quad t'_\tau := \begin{cases} u_\tau & \text{if } \textit{not } \delta_{act}(a, \tau)! \\ t_\tau - 1 & \text{if } \delta_{act}(a, \tau)! \text{ and } t_\tau > 0 \end{cases}$$

(Recall that if τ is prospective, $\delta_{act}(a, \tau)!$ and $t_\tau = 0$ then *not* $\delta(q, tick)!$)

$$\text{if } \tau \text{ is remote,} \quad t'_\tau := \begin{cases} l_\tau & \text{if } \textit{not } \delta_{act}(a, \tau)! \\ t_\tau - 1 & \text{if } \delta_{act}(a, \tau)! \text{ and } t_\tau > 0 \\ 0 & \text{if } \delta_{act}(a, \tau)! \text{ and } t_\tau = 0 \end{cases}$$

(ii) Let $\sigma \in \Sigma_{act}$. Then $a' := \delta_{act}(a, \sigma)$, and

$$\begin{array}{ll} \text{if } \tau \neq \sigma \text{ and } \tau \text{ is prospective,} & t'_\tau := \begin{cases} u_\tau & \text{if } \textit{not } \delta_{act}(a', \tau)! \\ t_\tau & \text{if } \delta_{act}(a', \tau)! \end{cases} \\ \text{if } \tau = \sigma \text{ and } \sigma \text{ is prospective,} & t'_\tau := u_\sigma \\ \text{if } \tau \neq \sigma \text{ and } \tau \text{ is remote,} & t'_\tau := \begin{cases} l_\tau & \text{if } \textit{not } \delta_{act}(a', \tau)! \\ t_\tau & \text{if } \delta_{act}(a', \tau)! \end{cases} \\ \text{if } \tau = \sigma \text{ and } \sigma \text{ is remote,} & t'_\tau := l_\sigma \end{array}$$

To complete the general definition of TDES we impose a final technical condition, to exclude the physically unrealistic possibility that a *tick* transition might be preempted indefinitely by repeated execution of an activity loop within a fixed unit time interval. A TDES is said to have an *activity loop* if

$$(\exists q \in Q)(\exists s \in \Sigma_{act}^{+})\delta(q, s) = q$$

We rule this out, and declare that all TDES must be *activity-loop-free (a.l.f.)*, namely

$$(\forall q \in Q)(\forall s \in \Sigma_{act}^{+})\delta(q, s) \neq q$$

It should be stressed that the a.l.f. condition refers to the timed transition structure, not to the activity transition structure. The latter may quite safely contain loops provided the time bounds associated with the relevant events in Σ_{act} are appropriate.

With the definition of TDES transition structure now complete, the Σ^* behavioral semantics of **G** is defined in the usual way: the *closed behavior* $L(\mathbf{G})$ of **G** is the subset of all strings in Σ^* that can be generated by iteration of δ starting from q_0 (i.e. the strings s such that $\delta(q_0, s)!$); while the *marked behavior* $L_m(\mathbf{G})$ is the subset of all strings in $L(\mathbf{G})$ for which the terminal state belongs to Q_m (i.e. the strings s such that $\delta(q_0, s) \in Q_m$). Note that a TDES never 'stops the clock': at any state either some transition $[\_, \sigma, \_]$ with $\sigma \in \Sigma_{act}$ is eligible, or at least the *tick* transition is defined. By activity-loop-freeness, no infinite (Σ^ω-)string generated by the transition structure of a TDES can be *tick*-free; indeed in any infinite string *tick* must occur infinitely often.[1]

Exercise 1: Verify the foregoing remark in detail. That is, if Q is finite and the activity-loop-freeness (a.l.f.) condition holds for **G**, then every string in $L(\mathbf{G})$ can be extended in $L(\mathbf{G})$ (i.e. by use of the TDES transition structure) to include an additional occurrence of *tick*, and no string can be extended indefinitely without an occurrence of *tick*. This shows that in every infinite string generated by TDES, *tick* must occur infinitely often.

9.3 Example 1

The following example illustrates how timing constraints can strongly influence complexity of the language generated by a TDES. Let

$$\mathbf{G_{act}} = (A, \Sigma_{act}, \delta_{act}, a_0, A_m)$$

with

$$\Sigma_{act} = \{\alpha, \beta\}, \quad A = A_m = \{0\}, \quad a_0 = 0$$

$$\delta_{act}(0, \alpha) = \delta_{act}(0, \beta) = 0$$

and timed events $(\alpha, 1, 1)$, $(\beta, 2, 3)$, both prospective. The ATG for $\mathbf{G_{act}}$ is simply:

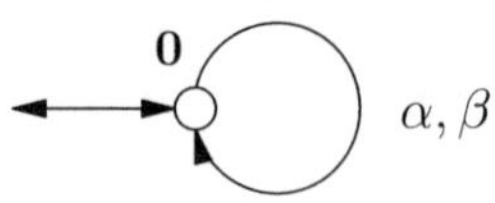

[1]Here the fact that A, and so Q, are finite sets is crucial.

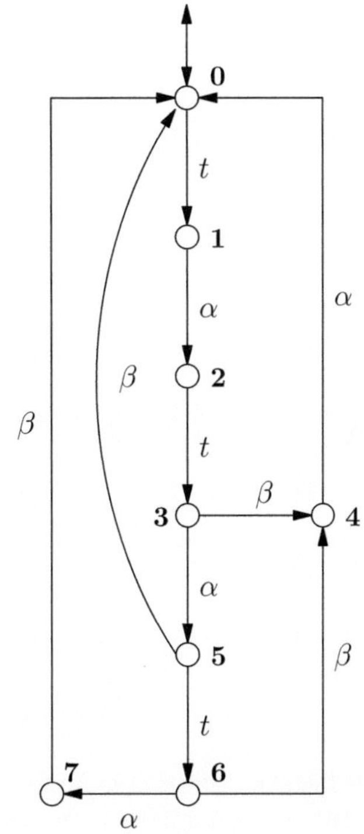

Fig. 9.1 Timed transition graph, Example 1

Thus α, β are always enabled. The state set for **G** is

$$Q = \{0\} \times T_\alpha \times T_\beta = \{0\} \times [0, 1] \times [0, 3]$$

and has size $|Q| = 8$. We take $Q_m = \{(0, [1, 3])\}$. The TTG for **G** is easily constructed and is displayed in Fig. 9.1; it has 11 transitions, over the event set $\{\alpha, \beta, tick\}$; the pairs $[t_\alpha, t_\beta]$ corresponding to the states $(0, \{t_\alpha, t_\beta\})$ of **G** are tabled below. The event α is pending at states 0, 2, 5, 7 and eligible at states 1, 3, 4, 6, while β is pending at 0, 1, 2, 4 and eligible at 3, 5, 6, 7. Notice that *tick* is preempted by α or β if either of these events has deadline 0 (namely is imminent).

State (node of TTG):	0	1	2	3	4	5	6	7
Components $[t_\alpha, t_\beta]$:	[1,3]	[0,2]	[1,2]	[0,1]	[0,3]	[1,1]	[0,0]	[1,0]

To obtain the TATG of **G** we require a projection operation on TDES defined (in outline) as follows. Let **G** be an arbitrary TDES, over the alphabet Σ, with closed and marked behaviors $L(\mathbf{G})$, $L_m(\mathbf{G})$, respectively. Let $\Sigma_{pro} \subseteq \Sigma$ and write $\Sigma_{nul} := \Sigma - \Sigma_{pro}$. Let $P : \Sigma^* \to \Sigma_{pro}^*$ be the natural projection whose action on a string $s \in \Sigma^*$ is just to erase any symbols of Σ_{nul} that appear in s. Now let $\mathbf{G_{pro}}$

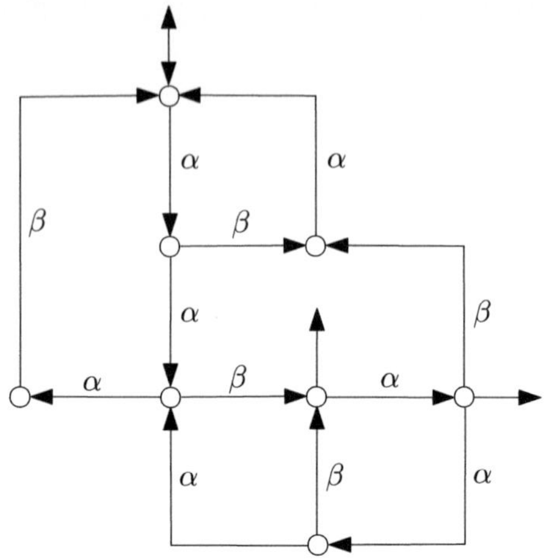

Fig. 9.2 Timed activity transition graph, Example 1

be any TDES over Σ_{pro} with closed and marked behaviors

$$L(\mathbf{G_{pro}}) := PL(\mathbf{G}), \quad L_m(\mathbf{G_{pro}}) := PL_m(\mathbf{G})$$

If the state set of **G** is finite, it is convenient in practice to select $\mathbf{G_{pro}}$ such that its state set is of minimal size. In any case, for a suitable determination of $\mathbf{G_{pro}}$ we can define an operation **project** according to

$$\mathbf{G_{pro}} = \mathbf{project}(\mathbf{G}, \mathbf{null}[\Sigma_{nul}])$$

In examples Σ_{nul} will be written as a list. We can now specify the TATG of **G** as

$$\mathbf{G_{tact}} = \mathbf{project}(\mathbf{G}, \mathbf{null}[tick])$$

For the example of this section the result is displayed in Fig. 9.2. Notice that $\mathbf{G_{tact}}$ happens to have just as many states (8) as **G**, illustrating the logical complexity that may be induced on the ordering of events by time bounds. This ordering, which could also be thought of as a set of phase relationships, is exhibited in the TATG (Fig. 9.2) but not in the ATG (above).

9.4 Example 2

Let $\Sigma = \{\alpha, \beta, \gamma\}$ with the timed events α, β as in Example 1. Adjoin to the structure of Example 1 the timed remote event $(\gamma, 2, \infty)$ with activity transition $[0, \gamma, 0]$; $\mathbf{G_{act}}$ is otherwise unchanged. The state size of **G** turns out to be $|Q| = 24$, with 41 transitions. It is found that $\mathbf{G_{tact}} := \mathbf{project}(\mathbf{G}, \mathbf{null}[tick])$ has 52 states and 108 transitions, thus being even more complex than **G** itself! While at first glance surprising, inasmuch as the occurrence or non-occurrence of γ does not appear to constrain that of α or β, this result can be thought of as a consequence of the nondeterminism created in the transition structure of **G** when *tick* is first projected out: an increase in complexity is the penalty exacted (by **project**) for replacing this nondeterministic description by a behaviorally equivalent deterministic one.

9.5 Time Bounds as Specifications

The imposition of time bounds on an event $\sigma \in \Sigma_{act}$ can be thought of as a specification over the alphabet $\{\sigma, tick\}$. If $\sigma \in \Sigma_{spe}$, with bounds $0 \leq l_\sigma \leq u_\sigma$, then the corresponding DES, say **SPEC**σ, will have state set $\{0, \ldots, u_\sigma\}$, with transitions $[i, tick, i+1]$ for $0 \leq i \leq u_\sigma - 1$, together with $[i, \sigma, 0]$ for $l_\sigma \leq i \leq u_\sigma$. To state i corresponds the evaluation $t_\sigma = u_\sigma - i$. Similarly if $\sigma \in \Sigma_{rem}$, with bound l_σ, $0 \leq l_\sigma < \infty$, then **SPEC**σ has state set $\{0, \ldots, l_\sigma\}$, with transitions $[i, tick, i+1]$ for $i = 0, \ldots, l_\sigma - 1$, $[l_\sigma, tick, l_\sigma]$, and $[l_\sigma, \sigma, 0]$. To state i corresponds the evaluation $t_\sigma = l_\sigma - i$. The specifications for the events α, β, γ of Examples 1 and 2 are displayed in Fig. 9.3.

It should be observed that in **SPEC**σ all events other than σ are ignored, in particular events whose occurrence may reinitialize **SPEC**σ by transition to an activity

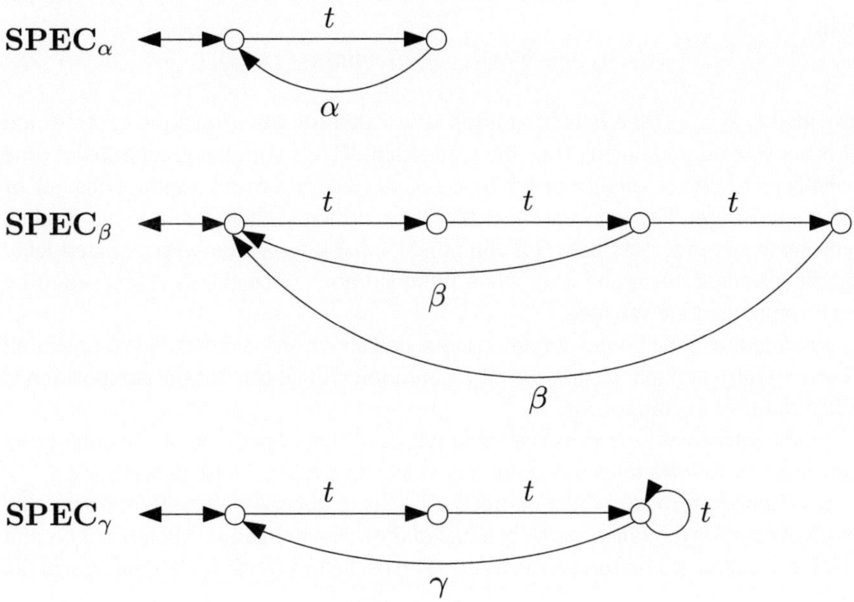

Fig. 9.3 Time bound specifications, Examples 1 and 2

where σ is disabled. Unfortunately there seems to be no simple way to obtain **G** by straightforward combination of $\mathbf{G_{act}}$ with the **SPEC**σ. In general, to obtain **G** (or its reachable subgraph) one must compute the reachable subset of Q, starting from q_0 and systematically examining the timed transition rules (for δ) in conjunction with the transition structure (δ_{act}) of $\mathbf{G_{act}}$.

9.6 Composition of TDES

Complex TDES can be built up from simpler ones by a composition operator **comp**. Let $\mathbf{G_1}, \mathbf{G_2}$ be TDES, over alphabets Σ_1, Σ_2 respectively, where $\Sigma_i = \Sigma_{i,act} \cup \{tick\}$. In general $\Sigma_{1,act}, \Sigma_{2,act}$ need not be disjoint. To form the *composition* $\mathbf{G} = \mathbf{comp}(\mathbf{G_1}, \mathbf{G_2})$ we start by defining the alphabet of **G** as $\Sigma_1 \cup \Sigma_2$, and the activity transition structure of **G** as the synchronous product of the component activity structures:

$$\mathbf{G_{act}} = \mathbf{sync}(\mathbf{G_{1,act}}, \mathbf{G_{2,act}})$$

The time bounds (l_σ, u_σ) in **G** of an event

$$\sigma \in (\Sigma_{1,act} - \Sigma_{2,act}) \cup (\Sigma_{2,act} - \Sigma_{1,act})$$

remain unchanged from their definition in the corresponding component structure, while if $\sigma \in \Sigma_{1,act} \cap \Sigma_{2,act}$ then its time bounds in **G** are defined in obvious notation to be

$$(l_\sigma, u_\sigma) = (\max\{l_{1,\sigma}, l_{2,\sigma}\}, \min\{u_{1,\sigma}, u_{2,\sigma}\})$$

provided $l_\sigma \leq u_\sigma$. If the latter condition is violated for any σ then the composition **G** is considered undefined. Thus the component TDES with the greater lower time bound (respectively smaller upper time bound) determines the timing behavior of the composition. This convention extends the principle that synchronous product represents an agreement between components that a transition with a shared label can be executed when and only when the conditions imposed on its execution by each component are satisfied.[2]

Provided the time bound conditions as stated above are satisfied, the composition **G** is now fully defined; clearly the a.l.f. condition will be true for the composition if it is true for each component.

Since synchronous product is associative, as a binary operation on the underlying languages, it follows that composition is associative in this sense as well.

It is important to stress that $\mathbf{comp}(\mathbf{G_1}, \mathbf{G_2})$ is in general quite different from the result of forming the synchronous product of the *timed* transition structures of $\mathbf{G_1}$ and $\mathbf{G_2}$, for the latter would force synchronization of the *tick* transition as it occurs in the

[2]While this convention respects physical behavior in many applications it need not be considered sacrosanct for all future modeling exercises.

component TTGs. Such a rule of composition places a constraint on the interaction of component TDES that proves unrealistic for the modeling requirements in many applications; it may even lead to temporal deadlock ('stopping the clock') as in the example of Sect. 9.7.

Exercise 1: Let **G1**, **G2** be TDES with $\Sigma_{1,act} \cap \Sigma_{2,act} = \varnothing$. Show that, in this special case,

$$\mathbf{comp(G1, G2)} \approx \mathbf{sync(G1, G2)}$$

where $\approx$ denotes that the closed and marked behaviors respectively of the two TDES on either side coincide.

9.7 Example 3

Consider the TDES **G1,G2** with ATGs displayed in Fig. 9.4; $\Sigma_{1,act} = \{\alpha, \beta\}$, $\Sigma_{2,act} = \{\beta, \gamma\}$; time bounds are as in Example 2. Let **G = comp(G1,G2)** (17,21). The ATG of **G** is also shown in Fig. 9.4; as the time bounds for the shared event label β are the same in **G1** and **G2**, the time bounds for $\Sigma_{act} = \{\alpha, \beta, \gamma\}$ are as specified already. While the structure of **G** is fairly rich, and admits strings of arbitrary length, the synchronous product of the *timed* transition graphs of **G1** and **G2** turns out to have a closed behavior which terminates with deadlock (i.e. no subsequent transition is defined) after just 9 transitions.[3] Thus, it does not even represent a TDES.

Exercise 1: Verify this example using *TTCT*. Explain why the timed behavior fails to be nonblocking.

9.8 Controllability of TDES

To use TDES as models for supervisory control, it is necessary to specify the ways in which TDES transitions can be controlled by an external agent or supervisor. From a theoretical viewpoint it is natural and convenient to impose two criteria on our 'control technology': (i) control should at most restrict uncontrolled behavior, never enlarge it; and (ii) controlled behavior subject to a specification constraint should admit optimization in the sense of maximal permissiveness.

By analogy with our theory of untimed DES, we first seek the counterpart of 'controllable events', namely transitions that can be disabled. Intuitively, if an event can be 'disabled', then it can be prevented indefinitely from occurring. In view of (i) this suggests that only remote events may belong to this category, for if a prospective event were disabled then it might be prohibited from occurring even when imminent and when no competing event is eligible to preempt it. This situation would result in behavior that could never be realized in the absence of control. On this basis we bring

[3]The language generated is the closure of the string pair $\{tick\ \alpha\ tick^2\ \beta\ tick^2\ (\gamma\ tick \mid tick\ \gamma)\}$.

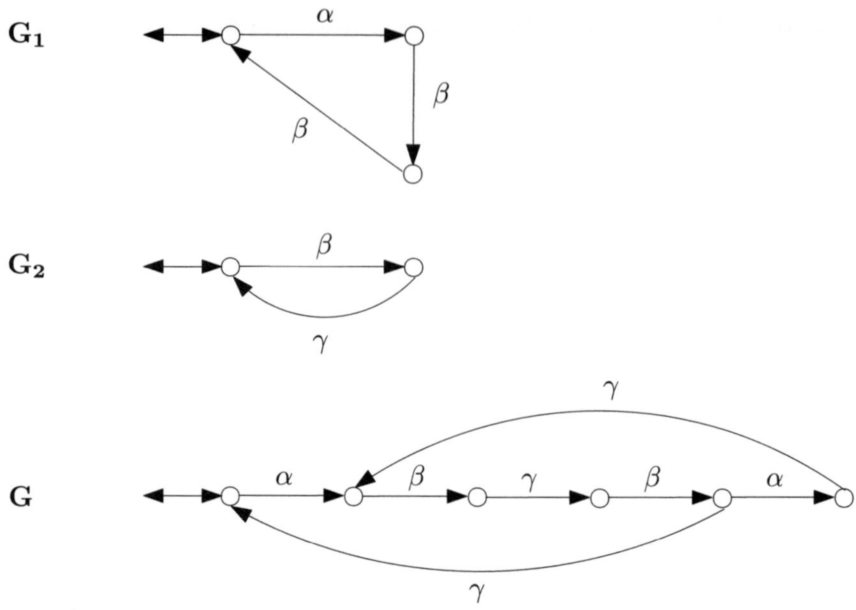

Fig. 9.4 Activity transition graphs, Example 3

in a new subset $\Sigma_{hib} \subseteq \Sigma_{rem}$ to label the *prohibitible* events. Our 'technology' will permit the supervisor to erase a prohibitible event from the current list of eligible transitions at a given state q of the supervised TDES. Of course, just as in the original model, the erased event may be reinstated if and when **G** revisits q on a subsequent occasion.

The following example may render the notion of 'prohibitible' clearer. Suppose σ is a remote event with lower time bound $l_\sigma = b$, and whose occurrence causes an ATG transition [0, σ, 1], after which σ is disabled. The corresponding TTG, say **G**, is displayed below, where t means *tick*.

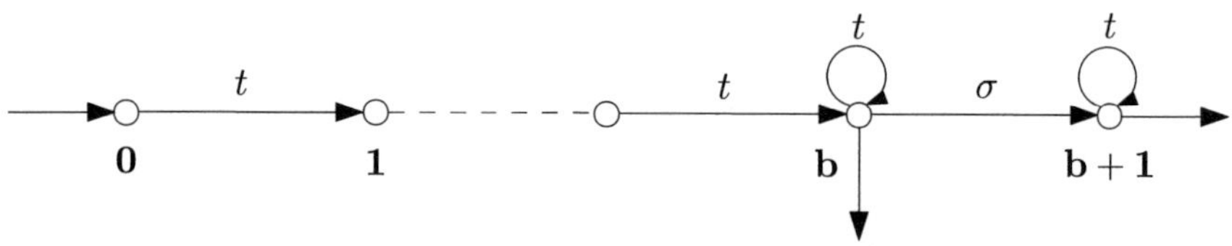

With states **b**, **b + 1** marked, **G** can be thought of as generating the language

$$t^b t^* \cup t^b t^* \sigma t^*$$

The first of these sublanguages can be interpreted as the behavior 'event σ never occurs at all', while the second says that 'after at least b *ticks*, σ occurs eventually, but with no hard deadline'. Notice that in Σ^*-semantics, 'never' means 'not in any finite time'; but as Σ^* contains only finite strings we can provide no 'witness' string (t^ω) of infinite length and free of σ. Now suppose σ is prohibitible; then disabling σ at state b will guarantee its nonoccurrence (ever); and doing so will respect the principle that controlled behavior should be a sublanguage of uncontrolled behavior.

Next we consider a new category of events that arises naturally in the presence of timing: the *forcible* events, or elements of a new subset $\Sigma_{for} \subseteq \Sigma_{act}$. A forcible event is one that can preempt a *tick* of the clock. If at a given state of **G**, *tick* is defined and one or more elements of Σ_{for} are eligible, then our supervisory control technology permits the effective erasure of *tick* from the current list of defined events, namely the guaranteed preemptive occurrence of some one of the eligible events in Σ_{act}, whether a member of Σ_{for} or otherwise. Thus forcible events are 'events of last resort to beat the clock'. There is no particular relation postulated *a priori* between Σ_{for} and any of Σ_{hib}, Σ_{rem} or Σ_{spe}. In particular an event in Σ_{rem} might be both forcible and prohibitible.

It is convenient to define the *uncontrollable* event set

$$\begin{aligned}\Sigma_{unc} &:= \Sigma_{act} - \Sigma_{hib}\\ &= \Sigma_{spe} \cup (\Sigma_{rem} - \Sigma_{hib})\end{aligned}$$

Eligible events in Σ_{unc} can never be erased by control action. Finally, we define the (complementary) *controllable* event set

$$\begin{aligned}\Sigma_{con} &:= \Sigma - \Sigma_{unc}\\ &= \Sigma_{hib} \cup \{tick\}\end{aligned}$$

Note that a forcible event may be controllable or uncontrollable; a forcible event that is uncontrollable cannot be directly prevented from occurring by disablement.[4] Also, while formally designated 'controllable' to simplify terminology, the status of *tick* lies intuitively between 'controllable' and 'uncontrollable': no technology could 'prohibit' *tick* in the sense of 'stopping the clock', although a forcible event, if eligible, may preempt it.

Exercise 1: 'Delayable' events

Consider an event α that is both prohibitible and forcible, with the requirement that α occur no earlier than 2 *ticks* (from enablement) and no later than 4 *ticks*. Provide the corresponding specification (as a language over the full alphabet). More generally, suppose the requirement is that α be delayed until some other event β occurs, but not for longer than 4 *ticks*, and that when α does occur, β ceases to be enabled. Assume, for instance, that β is uncontrollable and has time bounds $(0, \infty)$. Show that this specification can be modeled on a structure with 11 states.

[4] An instance: air defense could force a plane to land within 20 min (say) but not prevent it from landing eventually; the landing is forcible but not controllable.

The simplest way to visualize the behavior of a TDES **G** under supervision is first to consider the (infinite) reachability tree of **G** before any control is operative. Each node n of the tree corresponds to a unique string s of $L(\mathbf{G})$ (of course, over the full alphabet Σ including *tick*). At n we may define the subset of eligible events, say $\text{Elig}_{\mathbf{G}}(s) \subseteq \Sigma$. Thus

$$\text{Elig}_{\mathbf{G}}(s) := \{\sigma \in \Sigma | s\sigma \in L(\mathbf{G})\}$$

Henceforth we shall use the term 'eligible' in this extended sense, to apply to *tick* as well as to events in Σ_{act}. By our assumptions on **G**, $\text{Elig}_{\mathbf{G}}(s) \neq \varnothing$ for all $s \in L(\mathbf{G})$. A supervisor will be considered a decision-maker that, at n, selects a nonempty subset of $\text{Elig}_{\mathbf{G}}(s)$ in accordance with the rules stated above. It is now clear that, under these rules, our criterion (i) is satisfied, and it will later be shown that criterion (ii) is satisfied as well.

To formalize the rules we proceed as follows. Define a *supervisory control* to be any map $V : L(\mathbf{G}) \to Pwr(\Sigma)$ such that, for all $s \in L(\mathbf{G})$,

$$V(s) \supseteq \begin{cases} \Sigma_{unc} \cup (\{tick\} \cap \text{Elig}_{\mathbf{G}}(s)) \\ \qquad \text{if } V(s) \cap \text{Elig}_{\mathbf{G}}(s) \cap \Sigma_{for} = \varnothing \\ \Sigma_{unc} \ \text{ if } V(s) \cap \text{Elig}_{\mathbf{G}}(s) \cap \Sigma_{for} \neq \varnothing \end{cases}$$

Notice that if V' and V'' are both supervisory controls, then so is $V := V' \vee V''$, defined by $V(s) := V'(s) \cup V''(s)$. This property will imply the satisfaction of criterion (ii).

Fix a supervisory control V. The remainder of the discussion proceeds by analogy with Chap. 3. Write $V/\mathbf{G}$ to denote the pair $(\mathbf{G}, V)$ ('**G** under the supervision of V'). The *closed behavior* of $V/\mathbf{G}$ is the language $L(V/\mathbf{G}) \subseteq L(\mathbf{G})$ defined inductively according to:

(i) $\epsilon \in L(V/\mathbf{G})$
(ii) If $s \in L(V/\mathbf{G})$, $\sigma \in V(s)$, and $s\sigma \in L(\mathbf{G})$ then $s\sigma \in L(V/\mathbf{G})$
(iii) No other strings belong to $L(V/\mathbf{G})$

Thus $\{\epsilon\} \subseteq L(V/\mathbf{G}) \subseteq L(\mathbf{G})$, and $L(V/\mathbf{G})$ is nonempty and closed. The *marked behavior* of $V/\mathbf{G}$ is

$$L_m(V/\mathbf{G}) = L(V/\mathbf{G}) \cap L_m(\mathbf{G})$$

and thus $\varnothing \subseteq L_m(V/\mathbf{G}) \subseteq L_m(\mathbf{G})$. As usual we say V is *nonblocking for* **G** provided

$$\overline{L_m(V/\mathbf{G})} = L(V/\mathbf{G})$$

Exercise 2: Show that, for all $s \in L(\mathbf{G})$,

$$V(s) \cap \text{Elig}_{\mathbf{G}}(s) \neq \varnothing$$

We shall characterize those sublanguages of $L_m(\mathbf{G})$ that qualify as the marked behavior of some supervisory control V. First let $K \subseteq L(\mathbf{G})$ be arbitrary, and write

$$\text{Elig}_K(s) := \{\sigma \in \Sigma | s\sigma \in \bar{K}\}, \qquad s \in \Sigma^*$$

We define K to be *controllable (with respect to* **G***)* if, for all $s \in \bar{K}$,

$$\text{Elig}_K(s) \supseteq \begin{cases} \text{Elig}_{\mathbf{G}}(s) \cap (\Sigma_{unc} \cup \{tick\}) & \text{if } \text{Elig}_K(s) \cap \Sigma_{for} = \varnothing \\ \text{Elig}_{\mathbf{G}}(s) \cap \Sigma_{unc} & \text{if } \text{Elig}_K(s) \cap \Sigma_{for} \neq \varnothing \end{cases}$$

Thus K controllable means that an event σ (in the full alphabet Σ including *tick*) may occur in K if σ is currently eligible in **G** and either (i) σ is uncontrollable, or (ii) $\sigma =$ *tick* and no forcible event is currently eligible in K. The effect of the definition is to allow the occurrence of *tick* (when it is eligible in **G**) to be ruled out of K only when a forcible event is eligible in K and could thus (perhaps among other events in Σ_{act}) be relied on to preempt it. Notice, however, that a forcible event need not preempt the occurrence of competing non-*tick* events that are eligible simultaneously. In general our model will leave the choice of *tick*-preemptive transition nondeterministic.

In one form or another, the notion of forcing as preemption is inherent in control. Our notion of forcing is 'weakly preemptive', in that only the clock *tick* is assuredly preempted if forcing is invoked; however, the preemptive occurrence of a competing non-forcible but eligible event is not ruled out. A more conventional notion of forcing would require 'strong preemption', namely that a forcible event actually preempt any competing eligible event. If the control technology to be modeled actually admits 'forcing' in the strongly preemptive sense just indicated, then that feature would be modeled in our setup by suitably defining the activity transition structure.[5]

Notice finally that the controllability of K is a property only of $\bar{K}$, and that the languages $\varnothing$, $L(\mathbf{G})$, and Σ^* are all trivially controllable.

Our first main result is the analog of Theorem 3.4.3. Since the *tick* transition needs special treatment, the proof will be given in full.

Theorem 3

Let $K \subseteq L_m(\mathbf{G})$, $K \neq \varnothing$. There exists a nonblocking supervisory control V for **G** such that $L_m(V/\mathbf{G}) = K$ if and only if

(i) K is controllable with respect to **G**, and
(ii) K is $L_m(\mathbf{G})$-closed

Proof

(If) For $s \in \bar{K}$ define

$$V(s) := \Sigma_{unc} \cup (\Sigma_{con} \cap \text{Elig}_K(s))$$

while if $s \in L(\mathbf{G}) - \bar{K}$ assign the 'don't care' value

$$V(s) := \Sigma$$

[5]For instance, if a forcible event σ = 'stop' is to strictly preempt κ = 'collision', our model requires interposing at least one *tick* between σ and κ, and a structure in which σ causes transition to an activity where κ ceases to be enabled. This seems quite intuitive on physical grounds.

First of all, V really is a supervisory control. Indeed, $V(s) \supseteq \Sigma_{unc}$ always. Next, if

$$V(s) \cap \mathrm{Elig}_{\mathbf{G}}(s) \cap \Sigma_{for} = \varnothing$$

then

$$\Sigma_{unc} \cap \mathrm{Elig}_{\mathbf{G}}(s) \cap \Sigma_{for} = \varnothing$$

and

$$\Sigma_{con} \cap \mathrm{Elig}_{K}(s) \cap \Sigma_{for} = \varnothing$$

Therefore

$$(\Sigma_{unc} \cup \Sigma_{con}) \cap \mathrm{Elig}_{K}(s) \cap \Sigma_{for} = \varnothing$$

i.e.

$$\mathrm{Elig}_{K}(s) \cap \Sigma_{for} = \varnothing$$

By controllability of K,

$$\{tick\} \cap \mathrm{Elig}_{\mathbf{G}}(s) \subseteq \mathrm{Elig}_{K}(s)$$

and so

$$\{tick\} \cap \mathrm{Elig}_{\mathbf{G}}(s) \subseteq V(s)$$

as required.

Next we show that $L(V/\mathbf{G}) = \bar{K}$ and begin with $L(V/\mathbf{G}) \subseteq \bar{K}$. We have $\epsilon \in L(V/\mathbf{G})$ by definition, and $\epsilon \in \bar{K}$ since $K \neq \varnothing$. Arguing by induction, let $s \in L(V/\mathbf{G})$, $s \in \bar{K}$, $s\sigma \in L(V/\mathbf{G})$. Then $\sigma \in V(s) \cap \mathrm{Elig}_{\mathbf{G}}(s)$. If $\sigma \in \Sigma_{unc}$ then $\sigma \in \mathrm{Elig}_{\mathbf{G}}(s) \cap \Sigma_{unc}$, so $\sigma \in \mathrm{Elig}_{K}(s)$ since K is controllable. If $\sigma \in \Sigma_{con} \cap \mathrm{Elig}_{K}(s)$ then again $\sigma \in \mathrm{Elig}_{K}(s)$. In either case $\sigma \in \mathrm{Elig}_{K}(s)$, so $s\sigma \in \bar{K}$. For $\bar{K} \subseteq L(V/\mathbf{G})$, we proceed similarly, letting $s \in \bar{K}$, $s \in L(V/\mathbf{G})$, $s\sigma \in \bar{K}$. If $\sigma \in \Sigma_{unc}$ then $\sigma \in V(s)$. Since $s\sigma \in \bar{K}$ we have $s\sigma \in L(\mathbf{G})$ and so $s\sigma \in L(V/\mathbf{G})$. If $\sigma \in \Sigma_{con}$ then $\sigma \in \Sigma_{con} \cap \mathrm{Elig}_{K}(s)$, i.e. $\sigma \in V(s)$, and again $s\sigma \in L(V/\mathbf{G})$. We have now proved $L(V/\mathbf{G}) = \bar{K}$.

Finally

$$\begin{aligned} L_m(V/\mathbf{G}) &= L(V/\mathbf{G}) \cap L_m(\mathbf{G}) \qquad \text{(by definition)} \\ &= \bar{K} \cap L_m(\mathbf{G}) \\ &= K \qquad \text{(since } K \text{ is } L_m(\mathbf{G})\text{-closed)} \end{aligned}$$

and $\overline{L_m(V/\mathbf{G})} = \bar{K} = L(V/\mathbf{G})$, so V is nonblocking for $\mathbf{G}$.

(Only if) Let V be a nonblocking supervisory control for $\mathbf{G}$ with $L_m(V/\mathbf{G}) = K$. Since V is nonblocking, we have $L(V/\mathbf{G}) = \bar{K}$, so

$$K = L(V/\mathbf{G}) \cap L_m(\mathbf{G}) = \bar{K} \cap L_m(\mathbf{G})$$

i.e. K is $L_m(\mathbf{G})$-closed. To show that K is controllable let $s \in \bar{K}$, so $s \in L(V/\mathbf{G})$, and by definition of $L(V/\mathbf{G})$,

$$\text{Elig}_K(s) = V(s) \cap \text{Elig}_{\mathbf{G}}(s)$$

Thus

$$\text{Elig}_K(s) \supseteq \Sigma_{unc} \cap \text{Elig}_{\mathbf{G}}(s)$$

always. Also if $\text{Elig}_K(s) \cap \Sigma_{for} = \varnothing$ then

$$V(s) \cap \text{Elig}_{\mathbf{G}}(s) \cap \Sigma_{for} = \varnothing$$

Because V is a supervisory control,

$$V(s) \supseteq \{tick\} \cap \text{Elig}_{\mathbf{G}}(s)$$

hence

$$\text{Elig}_K(s) \supseteq \{tick\} \cap \text{Elig}_{\mathbf{G}}(s)$$

as required. So K is controllable, as claimed. □

For brevity we refer to a nonblocking supervisory control (for $\mathbf{G}$, understood) as an NSC. It is useful to introduce a slight generalization of NSC in which the supervisory action includes marking as well as control. For this, let $M \subseteq L_m(\mathbf{G})$. Define a *marking nonblocking supervisory control for the pair* $(M, \mathbf{G})$, or *MNSC*, as a map $V : L(\mathbf{G}) \rightarrow Pwr(\Sigma)$ exactly as before; but now for the marked behavior of $V/\mathbf{G}$ we define

$$L_m(V/\mathbf{G}) = L(V/\mathbf{G}) \cap M$$

One may think of the marking action of the MNSC V as carried out by a recognizer for M that monitors the closed behavior of $V/\mathbf{G}$, sounding a beep exactly when a string in M has been generated. As a sublanguage of $L_m(\mathbf{G})$, these strings could be thought of as representing a subset of the 'tasks' that $\mathbf{G}$ (or its underlying physical referent) is supposed to accomplish. For instance in a manufacturing problem, one might define a 'batch' to consist of 10 fully processed workpieces. M might then be taken as the set of strings that represent the successful processing of N integral batches, $N \geq 0$, with all machines returned to the idle state and all buffers empty.

The counterpart result to Theorem 3 actually represents a simplification, as the condition of $L_m(\mathbf{G})$-closedness can now be dropped.

Theorem 4

Let $K \subseteq L_m(\mathbf{G})$, $K \neq \varnothing$. There exists an MNSC V for $(K, \mathbf{G})$ such that

$$L_m(V/\mathbf{G}) = K$$

if and only if K is controllable with respect to $\mathbf{G}$.

Proof

(If) With V defined as in the proof of Theorem 3, it may be shown as before that $L(V/\mathbf{G}) = \bar{K}$. Then

$$L_m(V/\mathbf{G}) = L(V/\mathbf{G}) \cap K = \bar{K} \cap K = K$$

so that $\overline{L_m(V/\mathbf{G})} = \bar{K} = L(V/\mathbf{G})$, namely V is nonblocking for $\mathbf{G}$.

(Only if) We have $\bar{K} = \overline{L_m(V/\mathbf{G})} = L(V/\mathbf{G})$. Then the proof that K is controllable is unchanged from that of Theorem 3. □

9.9 Supremal Controllable Sublanguages and Optimal Supervision

Let $\mathbf{G} = (\_, \Sigma, \_, \_, \_)$ be a controlled TDES with Σ partitioned as in the previous section. Let $E \subseteq \Sigma^*$. As in Chap. 3, we introduce the set of all sublanguages of E that are controllable with respect to $\mathbf{G}$:

$$\mathcal{C}(E) = \{K \subseteq E | K \text{ is controllable with respect to } \mathbf{G}\}$$

Proposition 1

$\mathcal{C}(E)$ is nonempty and is closed under arbitrary unions.

Proof

Since the empty language is trivially controllable, $\mathcal{C}(E) \neq \varnothing$. Suppose $K_1, K_2 \in \mathcal{C}(E)$. Let $K = K_1 \cup K_2$; then $\bar{K} = \bar{K}_1 \cup \bar{K}_2$. For any $s \in \Sigma^*$, clearly

$$\text{Elig}_K(s) = \text{Elig}_{K_1}(s) \cup \text{Elig}_{K_2}(s)$$

Let $s \in \bar{K}$. Since at least one of the two subsets on the right satisfies the inclusion condition appearing in the definition of controllability, so does $\text{Elig}_K(s)$, and therefore K is controllable. Extension of the argument to an arbitrary union is obvious. □

We may now assert the existence of a unique supremal element $\sup \mathcal{C}(E)$ in E.

Let $E, L \subseteq \Sigma^*$. We say that E is *L-marked* if $E \supseteq \bar{E} \cap L$, namely any prefix of E that belongs to L must also belong to E.

Proposition 2

Let $E \subseteq \Sigma^*$ be $L_m(\mathbf{G})$-marked. Then $\sup \mathcal{C}(E \cap L_m(\mathbf{G}))$ is $L_m(\mathbf{G})$-closed.

Proof

We have $E \supseteq \bar{E} \cap L_m(\mathbf{G})$, from which there follows in turn

$$\bar{E} \cap L_m(\mathbf{G}) \subseteq E \cap L_m(\mathbf{G})$$
$$\bar{E} \cap \overline{L_m(\mathbf{G})} \cap L_m(\mathbf{G}) \subseteq E \cap L_m(\mathbf{G})$$
$$\overline{E \cap L_m(\mathbf{G})} \cap L_m(\mathbf{G}) \subseteq E \cap L_m(\mathbf{G})$$

so that $F := E \cap L_m(\mathbf{G})$ is $L_m(\mathbf{G})$-closed. Let $K = \sup \mathcal{C}(F)$. If K is not $L_m(\mathbf{G})$-closed, i.e. $K \subsetneqq \bar{K} \cap L_m(\mathbf{G})$, there is a string $s \in \bar{K} \cap L_m(\mathbf{G})$ with $s \notin K$. Let $J = K \cup \{s\}$. Since $\bar{J} = \bar{K}$ we have that J is controllable. Also $K \subseteq F$ implies that

$$\bar{K} \cap L_m(\mathbf{G}) \subseteq \bar{F} \cap L_m(\mathbf{G}) = F$$

so that $s \in F$ and thus $J \subseteq F$. Therefore $J \in \mathcal{C}(F)$ and $J \supsetneqq K$, contradicting the fact that K is supremal. □

Now we can present the main result of this section.

Theorem 3

Let $E \subseteq \Sigma^*$ be $L_m(\mathbf{G})$-marked, and let $K = \sup \mathcal{C}(E \cap L_m(\mathbf{G}))$. If $K \neq \varnothing$, there exists a nonblocking supervisory control (NSC) V for $\mathbf{G}$ such that $L_m(V/\mathbf{G}) = K$.

Proof

K is controllable and, by Proposition 2, $L_m(\mathbf{G})$-closed. The result follows by Theorem 9.8.3. □

As in Chap. 3, the result may be paraphrased by saying that K is (if nonempty) the maximally permissive (or minimally restrictive) solution of the problem of supervising $\mathbf{G}$ in such a way that its behavior belongs to E and control is nonblocking. In this sense the supervisory control provided by Theorem 3 is (qualitatively) optimal.

As might be expected, if we place part of the burden of 'marking action' on the supervisory control itself we may relax the prior requirement on E. By an application of Theorem 9.8.4 the reader may easily obtain the following analog of Theorem 3.4.5.

Theorem 4

Let $E \subseteq \Sigma^*$ and let $K = \sup \mathcal{C}(E \cap L_m(\mathbf{G}))$. If $K \neq \varnothing$ there exists a marking nonblocking supervisory control (MNSC) for $\mathbf{G}$ such that $L_m(V/\mathbf{G}) = K$. □

Finally, the implementation of supervisory controls by automata is formally no different from procedures already described in Sect. 3.6.

Exercise 5: Specialization of timed to untimed supervisory control

Investigate how the TDES control theory can be specialized to the untimed theory. Provide an example where you run an untimed problem using *TTCT*. Note that the specialization should be legal within the framework of TDES. So, while $tick$ must remain in the model, the timers should place no logical constraints on the occurrence of activity events. Finally the corresponding *TCT* solution of the untimed problem should be obtainable by projecting $tick$ out of the solution in *TTCT*.

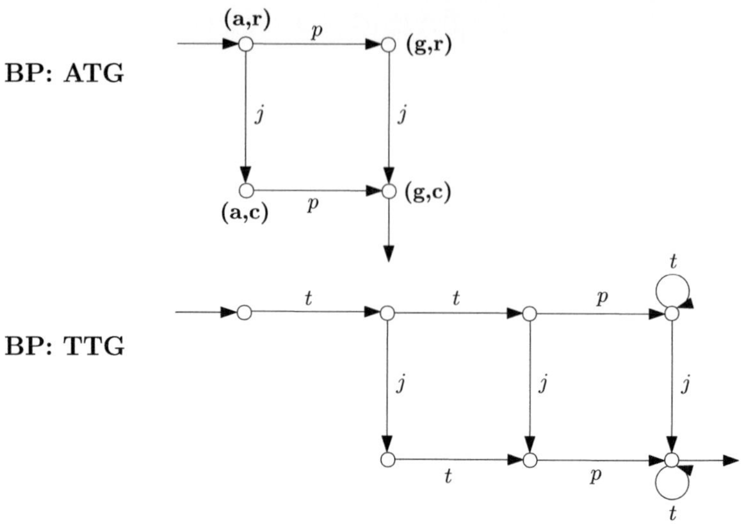

Fig. 9.5 Activity and timed transition graphs for **BP**

9.10 Example 4: Endangered Pedestrian

Consider two TDES

$$\mathbf{BUS} = (\{a, g\}, \{pass\}, \{[a, pass, g]\}, a, \{g\}), \quad (pass, 2, 2)$$
$$\mathbf{PED} = (\{r, c\}, \{jump\}, \{[r, jump, c]\}, r, \{c\}), \quad (jump, 1, \infty)$$

(where in place of δ we merely list the one activity transition). These model respectively a bus that makes a single transition *pass* between the activities 'approaching' and 'gone by', and a pedestrian who may make a single transition *jump* from 'road' to 'curb'. These entities are combined in the TDES

$$\mathbf{BP} = \mathbf{comp}(\mathbf{BUS}, \mathbf{PED})$$

The ATG and TTG of **BP** are displayed in Fig. 9.5.

We bring in 'control technology', with the assumption

$$\Sigma_{hib} = \Sigma_{for} = \{jump\}$$

However, nothing can stop the bus from passing in the interval between 2 and 3 *ticks* of the clock.

Suppose it is required that the pedestrian be saved. As a first scenario, we specify a TDES that imposes no *a priori* timing constraints on events, but merely requires the pedestrian to jump before the bus passes:

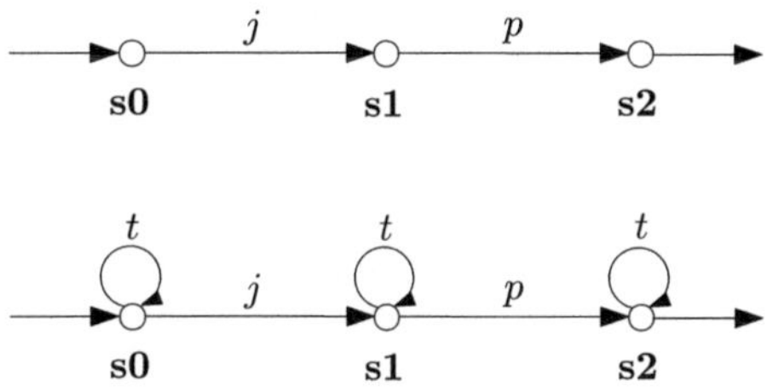

Fig. 9.6 Activity and timed transition graphs for **SAVE**

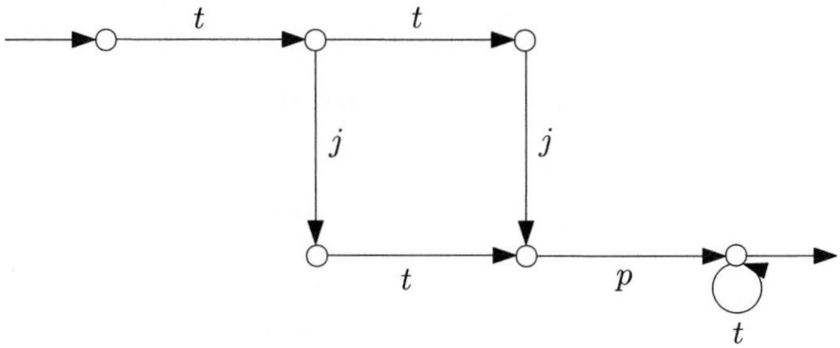

Fig. 9.7 Timed transition graph: **BPSAVE**

$$\mathbf{SAVE} = (\{s0, s1, s2\}, \{jump, pass\}, \{[s0, jump, s1], [s1, pass, s2]\}, s0, \{s2\})$$

with set of timed events $\Sigma_{tim} = \{(jump, 0, \infty), (pass, 0, \infty)\}$. The ATG and TTG are displayed in Fig. 9.6; the TTG is obtained from the ATG by self-looping *tick*.

Here and later we use the operation meet, defined on TDES according to

$$\mathbf{G3} = \mathbf{meet}(\mathbf{G1}, \mathbf{G2})$$

where $L(\mathbf{G3}) := L(\mathbf{G1}) \cap L(\mathbf{G2})$, $L_m(\mathbf{G3}) := L_m(\mathbf{G1}) \cap L_m(\mathbf{G2})$. As usual with such operations it is understood that **meet** is uniquely defined at implementation.

Now we can bring in the relevant ('physically possible') strings of $L(\mathbf{SAVE})$ as those shared with **BP**, namely the behavior of

$$\mathbf{BPSAVE} = \mathbf{meet}(\mathbf{BP}, \mathbf{SAVE})$$

as displayed in the TTG of Fig. 9.7. The closed behavior $L(\mathbf{BPSAVE})$ is not controllable, since

$$\text{Elig}_{\mathbf{BP}}(tick^2) \cap \Sigma_{for} = \{jump\} \neq \varnothing, \qquad \text{Elig}_{\mathbf{BP}}(tick^2) \cap \Sigma_{unc} = \{pass\}$$

but

$$\text{Elig}_{\mathbf{BPSAVE}}(tick^2) = \{jump\}$$

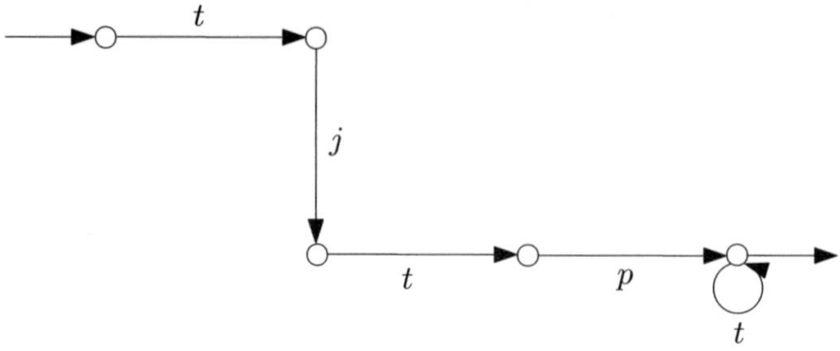

Fig. 9.8 Timed transition graph: **BPSAFE**

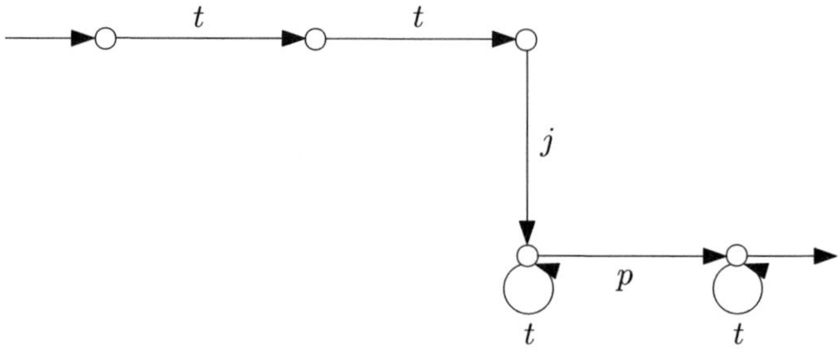

Fig. 9.9 Timed transition graph: **PRISK**

Evidently, after $tick^2$ nothing can prevent the bus from passing before the pedestrian jumps. But all is not lost: L_m(**BPSAVE**) has the supremal controllable sublanguage L_m(**BPSAFE**) as in Fig. 9.8. Note that, while $tick \in \mathrm{Elig}_{\mathbf{BP}}(tick)$, nonetheless

$$\mathrm{Elig}_{\mathbf{BP}}(tick) \cap \Sigma_{for} = \{jump\} \neq \emptyset, \qquad \mathrm{Elig}_{\mathbf{BP}}(tick) \cap \Sigma_{unc} = \emptyset$$

and thus the second *tick* can be reliably preempted by the forcible event *jump* (i.e. the pedestrian can be 'forced' to jump between the first *tick* and the second).

In a less optimistic scenario the pedestrian is again supposed to be saved, but at least 2 *ticks* must elapse from initialization before a *jump* (perhaps the pedestrian is handicapped); since $jump \in \Sigma_{hib}$, the string *tick.jump* can surely be prohibited as a prefix of controlled behavior. The resulting specification **PRISK** is shown in Fig. 9.9. Just as for **BPSAVE**, it is uncontrollable; but now the supremal controllable sublanguage of L_m(**PRISK**) is empty, and the control problem is unsolvable.

As a more complicated variation, suppose the bus can be stopped by a traffic officer and held up for 1 *tick*. Introduce **NEWBUS**, like **BUS** but with new timed events $(stop, 0, \infty)$ with $stop \in \Sigma_{for} \cap \Sigma_{hib}$, and $(wait, 1, 1)$, having ATG in Fig. 9.10. With the string $delay1 := stop.tick.wait$ the TTG of **NEWBUS** can be displayed as in Fig. 9.11. In effect, *delay1* can be used to preempt *tick* at any state of the TTG where both are defined, although it cannot be relied on to preempt *pass* if *pass* is

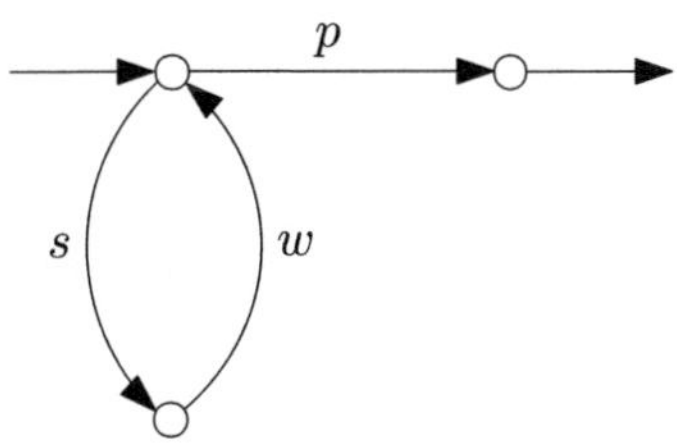

Fig. 9.10 Activity transition graph: **NEWBUS**

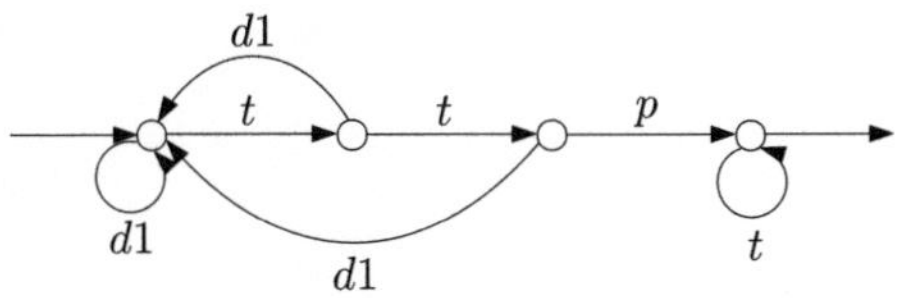

Fig. 9.11 Timed transition graph: **NEWBUS**

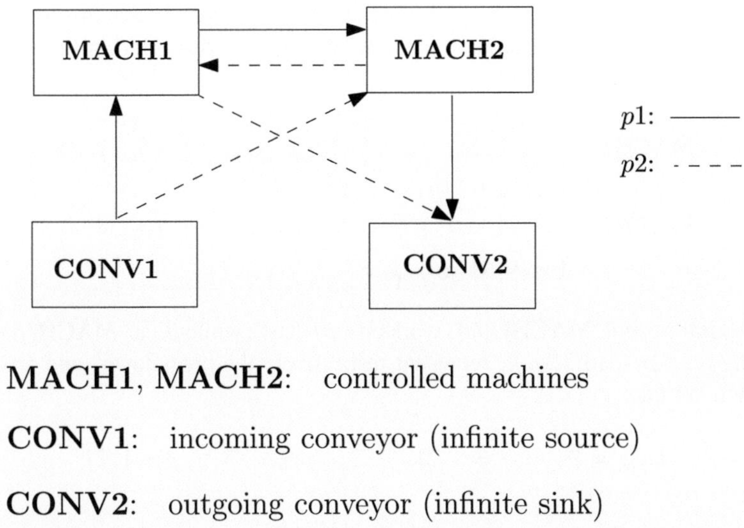

Fig. 9.12 Manufacturing cell

imminent. By use of *delay1* safety of the handicapped pedestrian can be guaranteed, for instance by forcing *stop* initially but disabling *stop* thereafter.

Exercise 1: Work out the details for **NEWBUS** using *TTCT*.

9.11 Example 5: Timed Manufacturing Cell

The manufacturing cell of Fig. 9.12 consists of machines **MACH1**, **MACH2**, with an input conveyor **CONV1** as an infinite source of workpieces and output conveyor

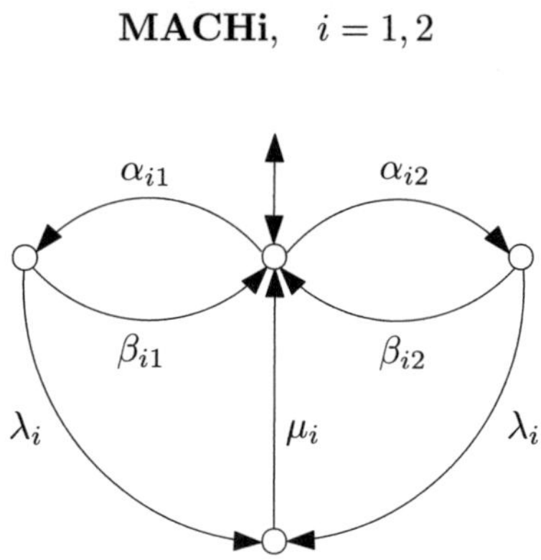

Fig. 9.13 Controlled machines

CONV2 as an infinite sink. Each machine may process two types of parts, $p1$ and $p2$; and each machine is liable to break down, but then may be repaired. For simplicity, the transfer of parts between machines will be absorbed as a step in machine operation. The machine ATGs (identical up to event labeling) are displayed in Fig. 9.13 and the timed events listed below.

$$\begin{aligned}\textbf{MACH1}: \quad & (\alpha_{11}, 1, \infty) \quad (\beta_{11}, 3, 3) \quad (\alpha_{12}, 1, \infty) \quad (\beta_{12}, 2, 2) \\ & (\lambda_1, 0, 3) \quad (\mu_1, 1, \infty) \\ \textbf{MACH2}: \quad & (\alpha_{21}, 1, \infty) \quad (\beta_{21}, 1, 1) \quad (\alpha_{22}, 1, \infty) \quad (\beta_{22}, 4, 4) \\ & (\lambda_2, 0, 4) \quad (\mu_2, 1, \infty)\end{aligned}$$

Here α_{ij} is the event '**MACHi** starts work on a pj-part', while β_{ij} is '**MACHi** finishes working on a pj-part'; λ_i, μ_i represent respectively the breakdown[6] and repair of **MACHi**. We take

$$\Sigma_{for} = \{\alpha_{ij} | i, j = 1, 2\}, \qquad \Sigma_{unc} = \{\lambda_i, \beta_{ij} | i, j = 1, 2\}$$

$$\Sigma_{hib} = \Sigma_{for} \cup \{\mu_1, \mu_2\}$$

The TTGs of **MACH1** and **MACH2** are shown in Figs. 9.14, 9.15.

We shall impose (i) *logic-based* specifications, (ii) a *temporal* specification, and (iii) a *quantitative optimality* specification as follows:

(i) • a given part can be processed by just one machine at a time
- a $p1$-part must be processed first by **MACH1** and then by **MACH2**
- a $p2$-part must be processed first by **MACH2** and then by **MACH1**

[6]Since breakdown can occur only when a machine is working, the upper time bound u_λ assigned to a breakdown event need not exceed the (finite) upper time bound u_β for completion of the corresponding work cycle. The u_λ could be replaced by anything larger, including ∞, without affecting behavior.

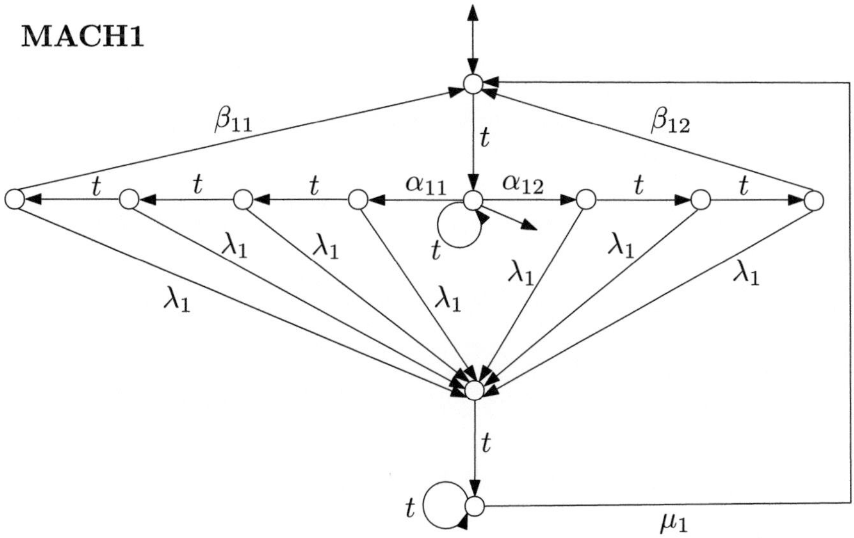

Fig. 9.14 TTG of **MACH1**

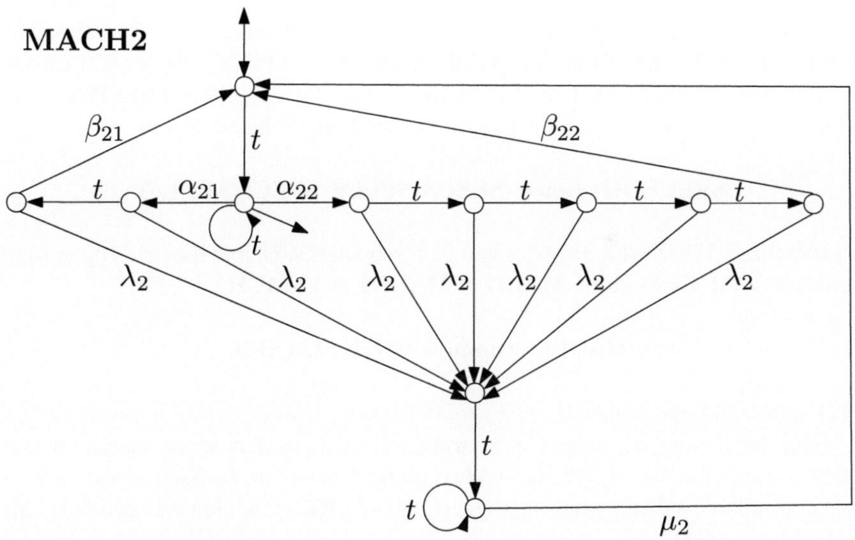

Fig. 9.15 TTG of **MACH2**

- one $p1$-part and one $p2$-part must be processed in each production cycle
- if both machines are down, **MACH2** is always repaired before **MACH1**

(ii) in the absence of breakdown/repair events a production cycle must be completed in at most 10 time units

(iii) subject to (ii), production cycle time is to be minimized

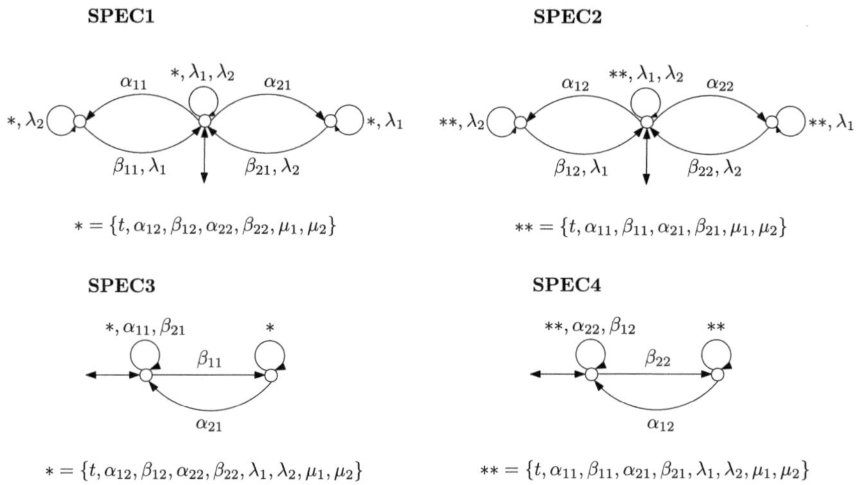

Fig. 9.16 SPEC1 - SPEC4

The first three specifications (i) are formalized as generalized TDES[7] **SPEC1-SPEC4**, displayed in Fig. 9.16, while the fourth specification (i) is formalized as **SPEC5**, Fig. 9.17, and the fifth (breakdown/repair) as **SPEC6**, Fig. 9.18. It can be verified that, in fact, **SPEC1** and **SPEC2** are automatically enforced by **SPEC3** and **SPEC4** together. We therefore define the complete logic-based specification

$$\mathbf{SPECLOG} = \mathbf{meet}(\mathbf{SPEC3}, \mathbf{SPEC4}, \mathbf{SPEC5}, \mathbf{SPEC6})$$

a generalized TDES with 32 states and 224 transitions. Define the cell's open-loop behavior as the composition **MACH** of **MACH1** and **MACH2**:

$$\mathbf{MACH} = \mathbf{comp}(\mathbf{MACH1}, \mathbf{MACH2})$$

(121 states, 345 transitions).

Here and below we write **G3 = supcon(G1,G2)** to denote the operation that returns a generalized TDES **G3** whose marked behavior $L_m(\mathbf{G3})$ is the supremal controllable sublanguage $\sup \mathcal{C}(L_m(\mathbf{G1}) \cap L_m(\mathbf{G2}))$; while its closed behavior $L(\mathbf{G3}) = \overline{L_m(\mathbf{G3})}$.

The maximally permissive proper supervisor for **MACH** that enforces **SPECLOG** can now be computed as

$$\mathbf{SUPLOG} = \mathbf{supcon}(\mathbf{MACH}, \mathbf{SPECLOG})$$

[7] We refer to any DES over an alphabet which includes *tick* as a *generalized* TDES; it need not be a (strict) TDES constructed according to the rules in Sect. 9.2. Generalized TDES are needed to model specifications and supervisors.

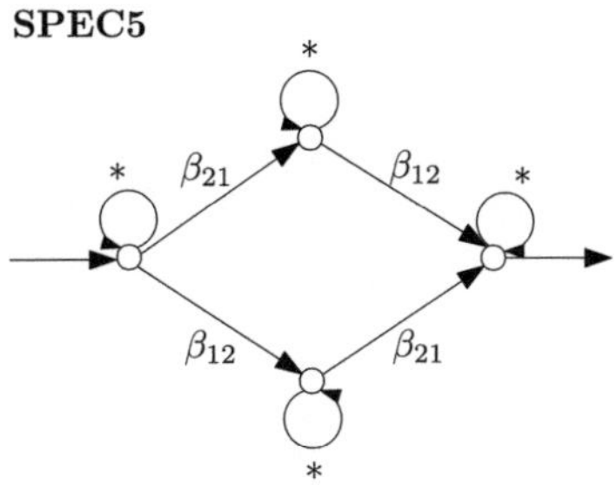

$* = \{t, \alpha_{11}, \beta_{11}, \alpha_{12}, \alpha_{21}, \alpha_{22}, \beta_{22}, \lambda_1, \lambda_2, \mu_1, \mu_2\}$

Fig. 9.17 SPEC5

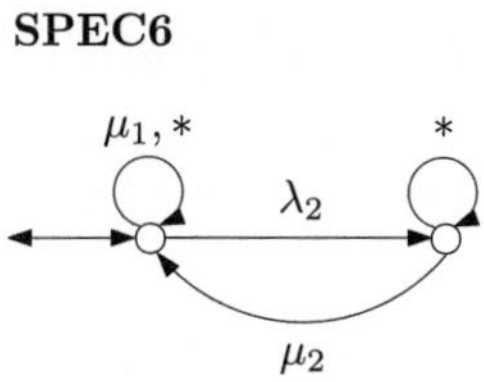

$* = \{t, \alpha_{11}, \beta_{11}, \alpha_{12}, \beta_{12}, \alpha_{21}, \beta_{21}, \alpha_{22}, \beta_{22}, \lambda_1\}$

Fig. 9.18 SPEC6

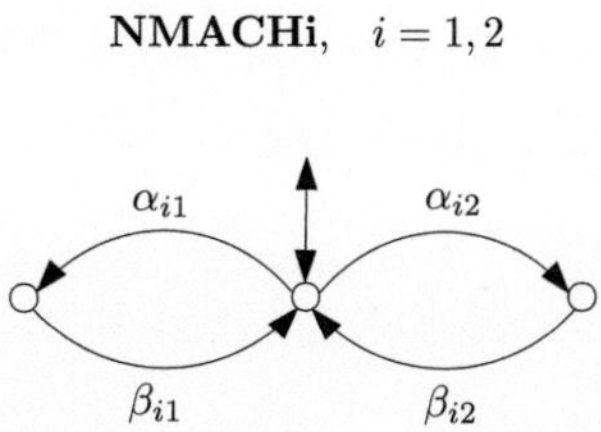

Fig. 9.19 New machine activity transition graphs

(264 states, 584 transitions). In this controlled behavior, forcing plays no role.

To address the temporal specification (ii) we first recompute the results for (i), under the stated assumption that breakdowns are absent. For this we define new machines **NMACH1**, **NMACH2** with simplified ATGs, Fig. 9.19. The new logic-based specification selfloop is

$$\mathbf{NSPECLOG} = \mathbf{meet}(\mathbf{SPEC3}, \mathbf{SPEC4}, \mathbf{SPEC5})$$

(16 states, 72 transitions). The open-loop behavior of the simplified cell is

$$\mathbf{NMACH} = \mathbf{comp}(\mathbf{NMACH1}, \mathbf{NMACH2})$$

(81 states, 121 transitions). These definitions yield the new supervisor

$$\mathbf{NSUPLOG} = \mathbf{supcon}(\mathbf{NMACH}, \mathbf{NSPECLOG})$$

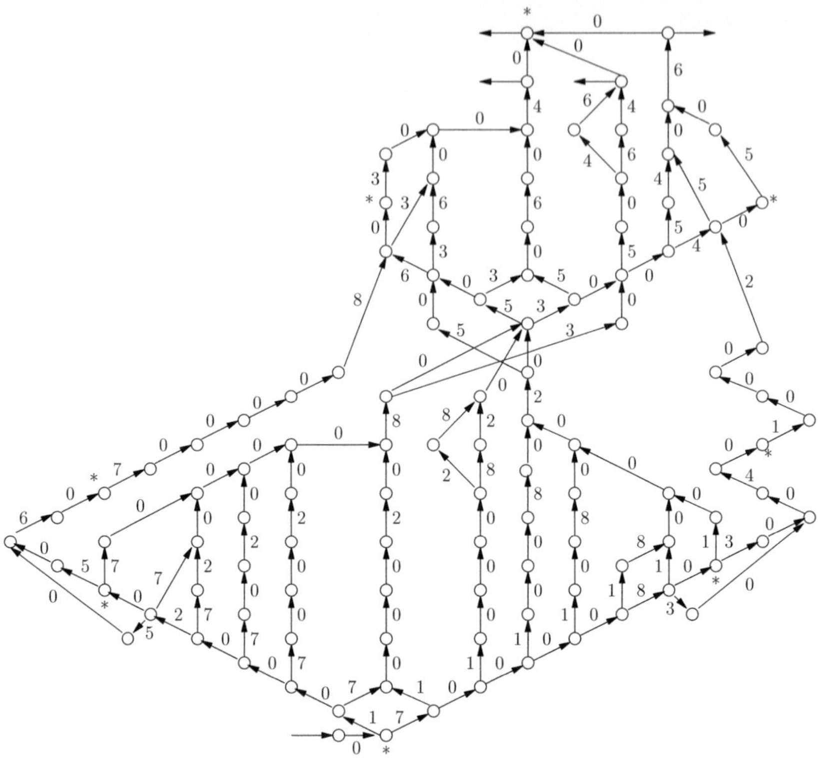

$1{:}\alpha_{11}$, $2{:}\beta_{11}$, $3{:}\alpha_{12}$, $4{:}\beta_{12}$, $5{:}\alpha_{21}$, $6{:}\beta_{21}$, $7{:}\alpha_{22}$, $8{:}\beta_{22}$, $0{:}tick$, $*$:selfloop $tick$

Fig. 9.20 NSUPLOG

(108 states, 144 transitions), displayed in Fig. 9.20. Now we consider the temporal specification itself. Bring in the generalized TDES **TIMER10** displayed in Fig. 9.21. **TIMER10** is simply an 11-*tick* sequence all of whose states are marked. **TIMER10** forces any TDES with which it is synchronized by **meet** to halt after at most 10 *ticks*, i.e. after 11 *ticks* to execute no further event whatever except the *tick* event. Thus it extracts the marked strings (if any) which satisfy this constraint, namely the 'tasks' of TDES which can be accomplished in at most 10 *ticks*. Of course, the designer must guarantee that the 10-*tick* deadline is actually met, if necessary by suitable forcing action. To determine whether such a guarantee is feasible, it suffices to check that the corresponding supremal controllable sublanguage is nonempty. The result is

$$\textbf{TNSUPLOG} = \textbf{supcon}(\textbf{NSUPLOG}, \textbf{TIMER10})$$

TIMER10

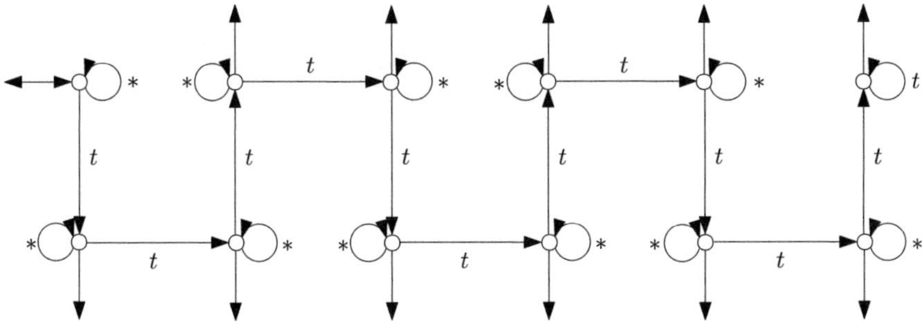

$* = \{\alpha_{11}, \beta_{11}, \alpha_{12}, \beta_{12}, \alpha_{21}, \beta_{21}, \alpha_{22}, \beta_{22}\}$

Fig. 9.21 **TIMER10** (12 states, 100 transitions)

(209 states, 263 transitions), so the check succeeds. We conclude that, in the absence of breakdowns, a production cycle can indeed be forced to complete in 10 *ticks* or less. Here the use of forcible events is, of course, essential.

Finally, to address specification (iii), we proceed as in (ii) with successive timer sequences of *tick*-length 9,8,..., until **supcon** returns an empty result. For this example the minimum enforcible production time turns out to be 7 *ticks*, with behavior

$$\mathbf{OTNSUPLOG} = \mathbf{supcon}(\mathbf{NSUPLOG}, \mathbf{TIMER7})$$

(19 states, 21 transitions) shown in Fig. 9.22. Initially both **NMACH1** and **NMACH2** are forced to start work on a $p1$-part and $p2$-part, respectively (events α_{11}, α_{22}). Forcing occurs as soon as these events become eligible, thus preempting a *tick* which would take the system along a suboptimal (slower) path (see Fig. 9.20). **NMACH1** (**NMACH2**) finishes work on its $p1$-part ($p2$-part) within 3 (resp. 4) time units. As soon as **NMACH2** has finished with its $p2$-part (event β_{22}), **NMACH1** is forced to start working on it (α_{12}), again preempting a *tick* that would take the system along a suboptimal path. Finally, **NMACH2** is forced to work on the $p1$-part, enabling the system to finish its production cycle in minimum time.

9.12 Modular Supervision of Generalized TDES

Let

$$\mathbf{G} = (Q, \Sigma, \delta, q_0, Q_m), \quad \mathbf{S} = (X, \Sigma, \xi, x_0, X_m)$$

be TDES, with $\Sigma = \Sigma_{act} \cup \{tick\}$. We assume that **G** is equipped with control structure, as in Sect. 9.8, and consider when **S** can be used as a supervisor for **G**. As in

OTNSUPLOG

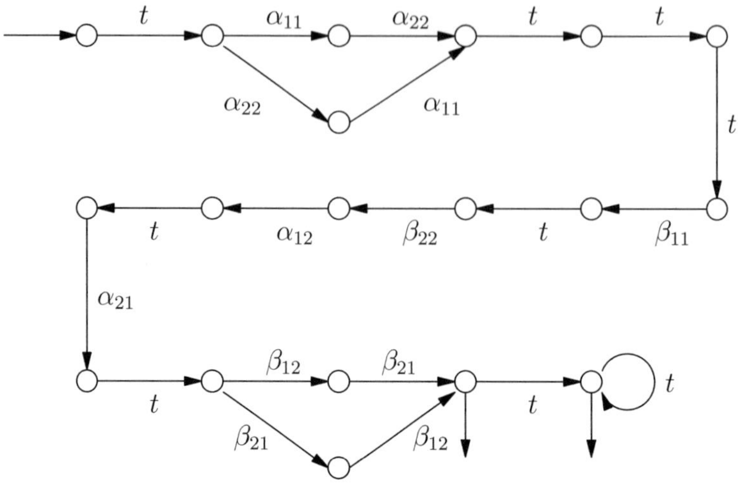

Fig. 9.22 OTNSUPLOG

Chap. 4 write $\mathbf{S} \wedge \mathbf{G}$ for the conjunction of $\mathbf{S}$ and $\mathbf{G}$ (implemented by *TCT* **meet**), so

$$L_m(\mathbf{S} \wedge \mathbf{G}) = L_m(\mathbf{S}) \cap L_m(\mathbf{G}), \quad L(\mathbf{S} \wedge \mathbf{G}) = L(\mathbf{S}) \cap L(\mathbf{G})$$

As in Sect. 3.6 we say that $\mathbf{S}$ is a *proper supervisor* for $\mathbf{G}$ if

(i) $\mathbf{S}$ is trim
(ii) $\mathbf{S}$ is controllable with respect to $\mathbf{G}$ (i.e. $L_m(\mathbf{S} \wedge \mathbf{G})$ is controllable)
(iii) $\mathbf{S} \wedge \mathbf{G}$ is nonblocking

Since by (iii), $\overline{L_m(\mathbf{S} \wedge \mathbf{G})} = L(\mathbf{S} \wedge \mathbf{G})$, (ii) means that

$$\mathrm{Elig}_{L(\mathbf{S}\wedge\mathbf{G})}(s) \supseteq \begin{cases} \mathrm{Elig}_{\mathbf{G}}(s) \cap (\Sigma_{unc} \cup \{tick\}) & \text{if } \mathrm{Elig}_{L(\mathbf{S}\wedge\mathbf{G})}(s) \cap \Sigma_{for} = \varnothing \\ \mathrm{Elig}_{\mathbf{G}}(s) \cap \Sigma_{unc} & \text{if } \mathrm{Elig}_{L(\mathbf{S}\wedge\mathbf{G})}(s) \cap \Sigma_{for} \neq \varnothing \end{cases}$$

We remark that if $L_m(\mathbf{S} \wedge \mathbf{G}) \neq \varnothing$ then

$$(\forall s \in L(\mathbf{S} \wedge \mathbf{G}))\mathrm{Elig}_{L(\mathbf{S}\wedge\mathbf{G})}(s) \neq \varnothing$$

Exercise 1: Justify this statement.

Let $K \subseteq L(\mathbf{G})$. The following definition extracts the feature of controllability that expresses the preemption of *tick* by a forcible event. We say that K is *coercive* with respect to $\mathbf{G}$ if

$$(\forall s \in \bar{K})tick \in \mathrm{Elig}_{\mathbf{G}}(s) - \mathrm{Elig}_K(s) \Rightarrow \mathrm{Elig}_K(s) \cap \Sigma_{for} \neq \varnothing$$

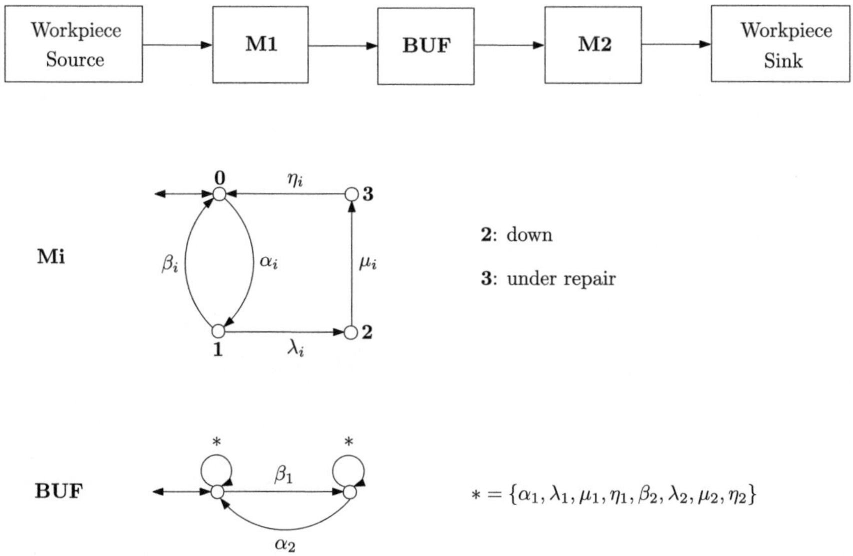

Fig. 9.23 Timed workcell

i.e.

$$(\forall s \in \bar{K})\text{Elig}_K(s) \cap \Sigma_{for} = \varnothing \; and \; tick \in \text{Elig}_{\mathbf{G}}(s) \Rightarrow tick \in \text{Elig}_K(s)$$

We say that languages $K_1, K_2 \subseteq L(\mathbf{G})$ are *jointly coercive* with respect to **G** if $K_1 \cap K_2$ is coercive with respect to **G**. Now let **S1**, **S2** be proper supervisors for **G**.

Theorem 2

S1 $\wedge$ **S2** is a proper supervisor for **G** if

(i) **S1** $\wedge$ **S2** is trim,
(ii) $L_m(\mathbf{S1} \wedge \mathbf{G})$, $L_m(\mathbf{S2} \wedge \mathbf{G})$ are nonconflicting, and
(iii) $L_m(\mathbf{S1} \wedge \mathbf{G})$, $L_m(\mathbf{S2} \wedge \mathbf{G})$ are jointly coercive with respect to **G** □

Exercise 3: Prove Theorem 2.

Example 4: Timed Workcell

Consider a workcell consisting of two machines **M1, M2** linked by a one-slot buffer **BUF** as shown in Fig. 9.23.

Let

$$\Sigma_{unc} = \{\beta_i, \lambda_i, \eta_i | i = 1, 2\}, \quad \Sigma_{for} = \Sigma_{hib} = \{\alpha_i, \mu_i | i = 1, 2\}$$

with corresponding timed events

$$(\alpha_1, 0, \infty), \; (\beta_1, 1, 2), \; (\lambda_1, 0, 2), \; (\mu_1, 0, \infty), \; (\eta_1, 1, \infty)$$

$$(\alpha_2, 0, \infty), \; (\beta_2, 1, 1), \; (\lambda_2, 0, 1), \; (\mu_2, 0, \infty), \; (\eta_2, 2, \infty)$$

Fig. 9.24 Specification **BR**

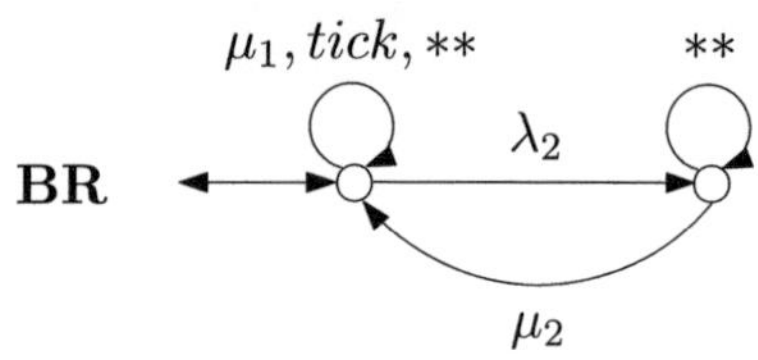

$$** = \{\alpha_1, \beta_1, \lambda_1, \eta_1, \alpha_2, \beta_2, \eta_2\}$$

Fig. 9.25 Modular supervisor **XBUF**

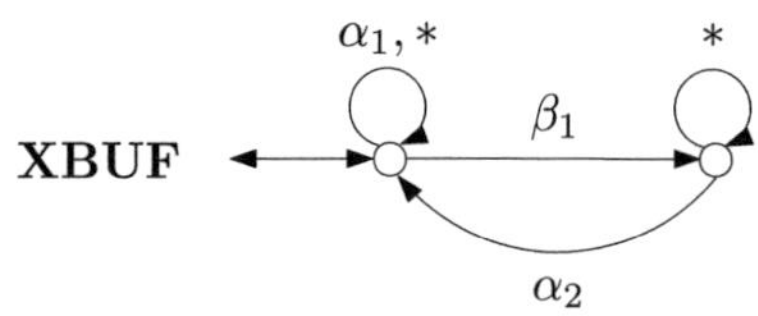

$$* = \{\lambda_1, \mu_1, \eta_1, \beta_2, \lambda_2, \mu_2, \eta_2, tick\}$$

For the TDES to be controlled we take **WORKCELL = comp(M1, M2)**, under the informal specifications

1. **BUF** must not overflow or underflow
2. If **M2** goes down, its repair must be started 'immediately', and prior to starting repair of **M1** if **M1** is currently down

BUF itself will serve as the first of these specifications. The second is formalized as **BR**, in Fig. 9.24.

For the monolithic supervisor we compute

$$\begin{aligned}\mathbf{SPEC} &= \mathbf{meet(BUF, BR)} \quad (4, 32)\\ \mathbf{SUPER} &= \mathbf{supcon(WORKCELL, SPEC)} \quad (54, 124)\\ \mathbf{SUPER} &= \mathbf{condat(WORKCELL, SUPER)}\end{aligned}$$

Now let us attempt to find modular supervisors. It can be verified that **BR** is already controllable and nonconflicting with respect to **WORKCELL. BUF** is not controllable but can be made so by disabling α_1 at state 1 to form **XBUF**, as in Fig. 9.25.

Now Theorem 2 can be used to confirm that **XBUF** $\wedge$ **BR** is proper and yields the controlled behavior representation

$$\mathbf{WCXBUFBR} = \mathbf{meet(WORKCELL, meet(XBUF, BR))} \quad (49, 110)$$

It can be verified that the marked behavior of **WCXBUFBR** is strictly smaller than that of **SUPER**; namely the behavior under the above modular control, while nontrivial and correct, is suboptimal. The reason is that, in the timed structure of **WORKCELL**, disablement of α_1 is not always necessary when the buffer is full, for if **M2** is idle it is sufficient immediately to force α_2, thus clearing the buffer and preventing overflow. Our modular supervisor **XBUF** is overly conservative.

Exercise 5: Check the conditions of Theorem 2, taking **S1** = **XBUF**, **S2** = **BR**, **G** = **WORKCELL**. Use *TTCT* **condat** to check joint coerciveness.

Exercise 6: In the example above, show that the jointly coercive property of **S1**, **S2** comes into play on the occurrence of λ_2: the repair of **M2** must be initialized without delay, thus preempting the occurrence of *tick* (by the forcible event μ_2) in the transition structure of **S2** and consequently of **S1**.

Exercise 7: Consider the simplified problem with **M1**, **M2**, **BUF** replaced with **NM1**, **NM2**, **NBUF** on deleting the events λ_i, μ_i, η_i $(i = 1, 2)$ and with **BR** omitted. Compute

$$\mathbf{NM} = \mathbf{comp}(\mathbf{NM1}, \mathbf{NM2}) \quad (12, 23)$$
$$\mathbf{NBUFSUP} = \mathbf{supcon}(\mathbf{NM}, \mathbf{NBUF}) \quad (13, 21)$$

Show that there exists a reduced (timed) supervisor **NBUFSIM** (3,10) that is control-equivalent to **NBUFSUP** with respect to **NM** and explain its operation. Also show that a 'naive' extension **XBUFSIM** formed by self-looping λ_i, μ_i, η_i is blocking for **WORKCELL** and explain in detail why this is so. Can you find a small, proper buffer supervisor for **WORKCELL**, say **BUFSIM**, that is compatible with **BR** (in the sense of Theorem 2)?

9.13 Notes

The timed DES framework of this chapter is a simplified version of that for 'timed transition models' treated by Ostroff (1987, 1989, 1990), Ostroff and Wonham (1990); the new controllability features, including forcing, are due to Brandin (1989, 1993), Brandin et al. (1993), Brandin and Wonham (1992, 1994). In Petri nets, time bounds on events were previously employed by Merlin and Farber (1976) and Berthomieu and Diaz (1991), while forcing in DES was investigated by Golaszewski and Ramadge (1987). From a different perspective, timed automata are described in Alur and Dill (1990) and applied to supervisory control in Wong-Toi and Hoffmann (1991).

References

Alur R, Dill D (1990) Automata for modeling real-time systems. In: Proceedings of the seventeenth international colloquium on automata, languages and programming. Lecture notes on computer science (LNCS), vol 443. Springer, Berlin, pp 322–335

Berthomieu B, Diaz M (1991) Modeling and verification of time dependent systems using time Petri nets. IEEE Trans Softw Eng 17(3):259–273

Brandin BA (1989) The supervisory control of discrete event systems with forcible events. MASc thesis, Department of Electrical and Engineering, University of Toronto

Brandin BA (1993) Real-time supervisory control of automated manufacturing systems. PhD thesis, Department of Electrical and Computer Engineering, University of Toronto

Brandin BA, Wonham WM (1992) Supervisory control of timed discrete-event systems. In: Proceedings of the thirty-first IEEE conference on decision and control, Tucson, Arizona, pp 3357–3362

Brandin BA, Wonham WM (1994) Supervisory control of timed discrete-event systems. IEEE Trans Autom Control 39(2):329–342. [Figures 1–14, 17–21 are reprinted, with permission, from IEEE.]

Brandin BA, Wonham WM, Benhabib B (1993) Manufacturing cell supervisory control - a modular timed discrete-event system approach. In: Proceedings of the IEEE international conference on robotics and automation, Atlanta, Georgia, pp 931–936

Golaszewski CH, Ramadge PJ (1987) Control of discrete event processes with forced events. In: Proceedings of the twenty-sixth IEEE conference on decision and control, Los Angeles, CA, Dec 1987, pp 247–251

Merlin PM, Farber DJ (1976) Recoverability of communication protocols - implications of a theoretical study. IEEE Trans Commun 24:1036–1043

Ostroff J (1987) Real-time computer control of discrete systems modelled by extended state machines: a temporal logic approach. PhD thesis, Department of Electrical and Engineering, University of Toronto

Ostroff JS (1989) Temporal logic for real-time systems. Research Studies Press

Ostroff JS (1990) Deciding properties of timed transition models. IEEE Trans Parallel Distrib Syst 1(2):170–183

Ostroff JS, Wonham WM (1990) A framework for real-time discrete event control. IEEE Trans Autom Control 35(4):386–397

Wong-Toi H, Hoffmann G (1991) The control of dense real-time discrete event systems. In: Proceedings of the thirtieth IEEE conference on decision and control, Brighton, UK, pp 1527–1528

Appendix A
Supervisory Control of a Mine Pump

Untimed Mine Problem in *TCT*

This exercise is adapted from Chap. 16 of the book: A. Burns, A. Wellings. *Real-Time Systems and Their Programming Languages*. Addison-Wesley, 1990. To follow the steps of the exercise it is recommended that the reader be familiar with the main synthesis procedures of Chaps. 3, 4, and 5.

The plant consists of a water pump (PUMP), water-level sensor (WLSENSOR), methane sensor (MESENSOR), human operator (OPERATOR), and water-flow sensor (WFSENSOR). Each sensor reports either a high or a low value for its variable, with corresponding higher- or lower-indexed event. There are two modes, automatic and manual. In automatic mode, the pump (motor) is controlled by the methane and water-level sensors; it is turned on iff the water level is high and the methane level is low. The water-flow sensor monitors pump operation; it reports either 'ok' (low) or 'problem' (high); in case of 'problem' (i.e. the pump is turned on but there is no water flow) it should sound an alarm. Manual mode overrides automatic; in manual, the sensors are disconnected from the pump control, and the operator is free to turn the pump on or off at will.

The DES model is built up in stages, so the designer has a clear picture of control action as more features are added. The *TCT* **project** command can be used to examine various facets of system behavior.

STAGE 0: Create the individual plant units.

PUMP = **create**(PUMP,[mark 0],[tran [0,11,0],[0,13,0]]) (1,2)
11 = TurnOff; 13 = TurnOn
WLSENSOR = **create**(WLSENSOR,[mark 0],[tran [0,21,1],[1,22,0],
[1,24,0]]) (2,3)
21 = Initialize, 22 = ReportWaterLevelLo, 24 = ReportWaterLevelHi
MESENSOR = **create**(MESENSOR,[mark 0],[tran [0,31,1],[1,32,0],
[1,34,0]]) (2,3)

W. M. Wonham and K. Cai, *Supervisory Control of Discrete-Event Systems*, Communications and Control Engineering,
https://doi.org/10.1007/978-3-319-77452-7

31 = Initialize, 32 = ReportMethaneLevelLo, 34 = ReportMethaneLevelHi
OPERATOR = **create**(OPERATOR,[mark 0],[tran [0,41,0],[0,43,0]]) (1,2)
41 = EnterAutomaticMode, 43 = EnterManualMode
WFSENSOR = **create**(WFSENSOR,[mark 0],[tran [0,51,1],[0,53,0],
[1,52,0],[1,54,0]]) (2,4)
51 = Initialize, 52 = ReportOK, 53 = SoundAlarm, 54 = ReportProblem

---

STAGE 1: Consider just the pump and the two most critical sensors, for water-level and methane.

PLANT1 = **sync**(PUMP,WLSENSOR) (2,7)
PLANT1 = **sync**(PLANT1,MESENSOR) (4,20)
ALL1 = **create**(ALL1,[mark 0]) (1,0)
ALL1 = **selfloop**(ALL1,[11,13,21,22,24,31,32,34]) (1,8)

Bring in two specifications. CYCSPEC1 forces WLSENSOR and MESENSOR to operate in turn, re-initializing after commanding the pump to turn off or on, as appropriate, and then recycling.

CYCSPEC1 = **create**(CYCSPEC1,[mark 0],[tran [0,21,1],[1,22,2],
[1,24,2],[2,31 ,3],[3,32,4],[3,34,4], [4,11,0],[4,13,0]]) (5,8)
CYCSPEC1 = **sync**(CYCSPEC1,ALL1) (5,8)

PMPSPEC1 allows the pump to be forced (or kept) on only when water-level is high and methane level is low. This is a boolean condition expressing the enablement or disablement of pump events 11 and 13 as a function of the most recent sensor reports.

PMPSPEC1 = **create**(PMPSPEC1,[mark 0],[tran [0,11,0],[0,22,0],
[0,24,1],[0,32,0],[0,34,2],[1,13,1],[1,22,0],[1,24,1],[1,32,1],
[1,34,3],[2,11,2],[2,22,2],[2,24,3],[2,32,0],[2,34,2],[3,11,3],
[3,22,2],[3,24,3],[3,32,1],[3,34,3]]) (4,20)
PMPSPEC1 = **sync**(PMPSPEC1,ALL1) (4,28)

Construct the first-stage specification and compute the first-stage supervisor.

SPEC1 = **meet**(CYCSPEC1,PMPSPEC1) (20,28)
MSPEC1 = **minstate**(SPEC1) (10,13)
SUPER1 = **supcon**(PLANT1,SPEC1) (20,28)
SUPER1 = **condat**(PLANT1,SUPER1) Controllable.
MSUPER1 = **minstate**(SUPER1) (10,13)

The transition graph of the (state-minimized) controlled behavior MSUPER1 is as expected, and quite transparent. Investigate modular control.

nonconflict(PLANT1,CYCSPEC1) = *true*
nonconflict(PLANT1,PMPSPEC1) = *true*
MODSUP1 = **meet**(CYCSPEC1,PMPSPEC1) (20,28)
isomorph(SUPER1,MODSUP1) = *true*

Modular control with separate supervisors for the sensor cycle and the pump control has been validated for the Stage 1 system.

STAGE 2: Incorporate operator's override. This is to mean that the operator can force the system from automatic to manual mode at any time (ev43); when returning to automatic mode (ev41) the automatic system is re-initialized to state 0. Thus SPEC2 is created by adding one new state (10) to MSPEC1, with ev43 taking the system from any state of MSPEC1 to the new manual-operation state (10); then ev41 takes 10 back to 0. Note the convenience of using a wildcard ([∗, 43, 10] while editing.

SPEC2 = **edit**(MSPEC1,[trans +[0,43,10],+[1,43,10],+[2,43,10],
+[3,43,10],+[4,43,10],+[5,43,10],+[6,43,10],+[7,43,10],
+[8,43,10],+[9,43,10],+[10,11,10],+[10,13,10],+[10,21,10],
+[10,22,10],+[10,24,10],+[10,31,10],+[10,32,10],+[10,34,10],
+[10,41,0]]) (11,32)

ALL2 = **selfloop**(ALL1,[41,43]) (1,10)

PLANT2 = **sync**(PLANT1,OPERATOR) (4,28)

SUPER2 = **supcon**(PLANT2,SPEC2) (14,44)

SUPER2 = **condat**(PLANT2,SUPER2) Controllable.

Investigate modular control. From CYCSPEC1 and PMPSPEC1 respectively construct CYCSPC12 and PMPSPC12 to incorporate manual override.

CYCSPC12 = **edit**(CYCSPEC1,[trans +[0,43,5],+[1,43,5],+[2,43,5],
+[3,43,5],+[4,43,5],+[5,41,0]]) (6,14)

CYCSPC12 = **edit**(CYCSPC12,[trans +[5,11,5],+[5,13,5],+[5,21,5],
+[5,22,5],+[5,24,5],+[5,31,5],+[5,32,5],+[5,34,5]]) (6,22)

CYCSPC12 = **condat**(PLANT2,CYCSPC12) Uncontrollable.

CYCSUP12 = **supcon**(PLANT2,CYCSPC12) (9,34)

PMPSPC12 = **edit**(PMPSPEC1,[trans +[0,43,4],+[1,43,4],+[2,43,4],
+[3,43,4],+[4,11,4],+[4,13,4],+[4,21,4],+[4,22,4],
+[4,24,4],+[4,31,4],+[4,3 2,4],+[4,34,4],+[4,41,0]]) (5,41)

nonconflict(PLANT2,PMPSPC12) = *true*

PMPSPC12 = **condat**(PLANT2,PMPSPC12) Controllable.

Thus PMPSPC12 is a proper supervisor for PLANT2.

MODSUP2 = **meet**(CYCSUP12,PMPSPC12) (24,69)

nonconflict(PLANT2,MODSUP2) = *true*

MODSUP2 = **condat**(PLANT2,MODSUP2) Controllable.

MMODSUP2 = **minstate**(MODSUP2) (14,44)

isomorph(SUPER2,MMODSUP2) = *true*

The two modular supervisors are together equivalent to the centralized supervisor. Check to see if they are of minimal state size.

MCYCSP12 = **minstate**(CYCSUP12) (9,34)

MPMSPC12 = **minstate**(PMPSPC12) (5,41)

They are both minimal. Make an additional check on behavior.

MODBEH2 = **meet**(PLANT2,MODSUP2) (24,69)

MMODBEH2 = **minstate**(MODBEH2) (14,44)

isomorph(SUPER2,MMODBEH2) = *true*

Modular supervision for the Stage 2 system has been validated.

STAGE 3: Bring in WFSENSOR, specified by WFSPEC to sound an alarm (ev53) if it detects a problem with water flow (ev54); it should be initialized every time the pump is turned on (ev13); and the WFSENSOR cycle should run to completion before the pump is reactivated (turned off or on again).

ALL3= **selfloop**(ALL2,[51,52,53,54]) (1,14)
PLANT3 = **sync**(PLANT2,WFSENSOR) (8,72)
SPEC23 = **sync**(SPEC2,ALL3) (11,76)
WFSPEC = **create**(WFSPEC,[mark 0],[tran [0,11,0],[0,13,1],[1,51,2],
[2,52,0], [2,54,3],[3,53,0]]) (4,6)
WFSPEC = **sync**(WFSPEC,ALL3) (4,38)
SPEC3 = **meet**(SPEC23,WFSPEC) (44,157)
SPEC3 = **condat**(PLANT3,SPEC3) Uncontrollable.
SUPER3 = **supcon**(PLANT3,SPEC3) (56,199)
MSUPER3 =**minstate**(SUPER3) (56,199)
isomorph(MSUPER3,SUPER3) = *true*
SUPER3 = **condat**(PLANT3,SUPER3) Controllable.

Investigate modular supervision. We already know that modularity at Stage 2 is equivalent to using SUPER2, so first upgrade SUPER2 to SUPER23 to make it compatible with PLANT3.

SUPER23 = **sync**(SUPER2,ALL3) (14,100)
nonconflict(PLANT3,WFSPEC) = *true*
WFSPEC = **condat**(PLANT3,WFSPEC) Controllable.

WFSPEC is a proper supervisor for PLANT3; rename it WFSUP3. Try SUPER23 and WFSUP3 as modular supervisors.

WFSUP3 = **edit**(WFSPEC) (4,38)
MODSUP3 =**meet**(SUPER23,WFSUP3) (56,199)
isomorph(SUPER3,MODSUP3) = *true*

As a check, go back and adapt the modular components previously obtained for SUPER2, to obtain three modular supervisors for the Stage 3 system.

CYCSUP13 = **sync**(CYCSUP12,ALL3) (9,70)
PMPSPC13 = **sync**(PMPSPC12,ALL3) (5,61)

Combine the three modular supervisors and check the behavior under modular control.

TEST3 = **sync**(WFSUP3,CYCSUP13) (36,142)
TEST3 = **sync**(TEST3,PMPSPC13) (72,244)
MODBEH3 =**meet**(PLANT3,TEST3) (72,244)
MMODBEH3 = **minstate**(MODBEH3) (56,199)
isomorph(SUPER3,MMODBEH3) = *true*

Modular behavior for the Stage 3 system has been validated.

STAGE 4: Reduce SUPER3 to SUPER4, the behavior of the automatic mode alone, by removing ev41 and ev43.

ALL4 = **edit**(ALL3,[trans -[0,41,0],-[0,43,0]]) (1,12)
SUPER4 = **meet**(SUPER3,ALL4) (37,76)
MSUPER4 = **minstate**(SUPER4) (37,76)
isomorph(MSUPER4,SUPER4) = *true*

As a check, exhibit the ordering in automatic mode of evs11,13,21,31,51,53.

PSUPER4 = **project**(SUPER4,**null**[22,24,32,34,52,54]) (10,17)

As another check, recover the Stage 1 supervisor by projecting out WFSENSOR from SUPER4.

QSUPER4 =**project**(SUPER4,**null**[51,52,53,54]) (10,13)
MSUPER1 =**minstate**(SUPER1) (10,13)
isomorph(QSUPER4,MSUPER1) = *true*

STAGE 5: Make some enhancements to PLANT3 to incorporate a counter which raises a flag (ev90) if there are at least 3 consecutive readings of both water-level high (ev24) and of methane-level high (ev34). We also add a general alarm (ev91) that can be used to order that the mine be evacuated.

WLEMERG = **create**(WLEMERG,[mark 0,3],[tran [0,22,0],[0,24,1],
[1,22,0],[1,24,2],[2,22,0],[2,24,3],[3,22,3],[3,24,3],
[3,90,3]]) (4,9)
MEEMERG = **create**(MEEMERG,[mark 0,3],[tran [0,32,0],[0,34,1],
[1,32,0],[1,34,2],[2,32,0],[2,34,3],[3,32,3],[3,34,3],
[3,90,3]]) (4,9)
EMERG = **sync**(WLEMERG,MEEMERG) (16,65)
EVACMINE = **create**(EVACMINE,[mark 0],[tran [0,91,0]]) (1,1)
PLANT5 = **sync**(PLANT3,EMERG) (128,1160)
PLANT5 = **sync**(PLANT5,EVACMINE) (128,1288)
ALL5= **selfloop**(ALL3,[90,91]) (1,16)

Add SPEC5 to make it impossible for the operator to turn on the pump when the methane level is high.

MESPEC = **create**(MESPEC,[mark 0],[tran [0,13,0],[0,32,0],[0,34,1],
[1,32,0],[1,34,1]]) (2,5)
MESPEC = **sync**(MESPEC,ALL5) (2,31)
SPEC35 = **sync**(SPEC3,ALL5) (44,245)
SPEC5 = **meet**(SPEC35,MESPEC) (73,422)
nonconflict(PLANT5,SPEC5) = *true*
SUPER5 = **supcon**(PLANT5,SPEC5) (804,3927)
MSUPER5 = **minstate**(SUPER5) (776,3826)

STAGE 6: Provide a crisis scenario to force the operating mode to manual (ev43) and order that the mine be evacuated, when the methane/water-level flag is raised (ev90). Note that the sensors are disabled at the terminal state 2.

CRISIS = **create**(CRISIS,[mark 0,2],[tran [0,21,0],[0,31,0],[0,51,0],
[0,90,1],[1,21,1],[1,31,1],[1,43,2],[1,51,1],[1,90,1],[2,90,2],
[2,91,2]]) (3,11)
CRISIS = **sync**(CRISIS,ALL5) (3,41)
SPEC6 = **meet**(SPEC5,CRISIS) (102,386)
SUPER6 = **supcon**(PLANT5,SPEC6) (292,681)
MSUPER6 = **minstate**(SUPER6) (278,652)

[STAGE 7: has been deleted]

STAGE 8: Using vocalization, develop a simple abstract model of SUPER1, taking into account just the methane readings, with the objective of letting a high-level manager execute a system shutdown (ev81).

SUPER8 = **edit**(MSUPER1) (10,13)
SHUTDOWN = **create**(SHUTDOWN,[mark 0],[tran [0,81,0]]) (1,1)
ALL8= **selfloop**(ALL1,[81]) (1,9)
PLANT8 = **sync**(PLANT1,SHUTDOWN) (4,24)

The shutdown specification SDNSPEC will require that both pump transitions, along with the sensor specifications, be disabled.

SDNSPEC = **create**(SDNSPEC,[mark 0,1],[tran [0,11,0],[0,13,0],[0,21,0],
[0,31,0],[0,81,1]]) (2,5)
SDNSPEC = **sync**(SDNSPEC,ALL8) (2,13)

Adapt the Stage 1 sensor cycle to Stage 8.

CYCSPEC8 = **sync**(CYCSPEC1,ALL8) (5,13)

For shutdown to occur without supervisor blocking, the former restriction that only the 'normal' state be marked must be relaxed, inasmuch as shutdown will occur only after *abnormal* behavior! So now all four of the Stage 1 PMPSPEC1 states must be marked.

PMPSPEC8 = **edit**(PMPSPEC1,[mark +[1],+[2],+[3]]) (4,28)
PMPSPEC8 = **sync**(PMPSPEC8,ALL8) (4,32)
SPEC8 = **meet**(PMPSPEC8,CYCSPEC8) (20,48)
SPEC8 = **meet**(SPEC8,SDNSPEC) (40,64)
SUPER8 = **supcon**(PLANT8,SPEC8) (24,32)
MSUPER8 = **minstate**(SUPER8) (9,12)
SUPER8 = **edit**(MSUPER8) (9,12)

Next vocalize the significant events corresponding to methane-level readings (evs32,34) and shutdown (ev81).

VSUPER8 = **vocalize**(SUPER8,[[0,81,81],[5,34,34],[6,34,34],[5,32,32],
[6,32,32]]) (10,13)

VSUPER8 = **BFS**(VSUPER8) (10,13)
VSUP8HI = **higen**(VSUPER8) (2,3)
OCVSUP8 = **outconsis**(VSUPER8) (10,13)
HCVSUP8 = **hiconsis**(VSUPER8) (10,13)
isomorph(HCVSUP8,OCVSUP8) = *false*

OCVSUP8 has 4 vocalized states and 3 distinct vocal outputs, whereas HCVSUP8 has 5 vocalized states with 5 vocal outputs.

X= **hiconsis**(OCVSUP8) (10,13)
isomorph(X,HCVSUP8) = *true*
HCVSP8HI = **higen**(HCVSUP8) (3,5)

Reasonably small high-level abstraction of the low-level controlled behavior: 3 states versus 10. Bring in SPEC8HI to shut the system down after 3 consecutive high methane readings.

SPEC8HI = **create**(SPEC8HI,[mark 0,4],[tran [0,320,0],[0,340,1],
[1,320,0],[1,340,2],[2,320,0],[2,340,3],[3,811,4]]) (5,7)
ALL8HI = **create**(ALL8HI,[mark 0]) (1,0)
ALL8HI = **selfloop**(ALL8HI,[320,340,811,821,831]) (1,5)
SPEC8HI = **sync**(SPEC8HI,ALL8HI) (5,17)
SUPER8HI = **supcon**(HCVSP8HI,SPEC8HI) (8,13)
Y= **minstate**(SUPER8HI) (8,13)
isomorph(Y,SUPER8HI;identity) = *true*

By inspection of SPEC8HI an obvious simplified supervisor can be created.

SMSUP8HI = **edit**(SPEC8HI,[trans -[3,821,3],-[3,831,3]]) (5,15)
TEST = **meet**(HCVSP8HI,SMSUP8HI) (8,13)
isomorph(TEST,SUPER8HI;identity) = *true*

Hierarchical treatment of SUPER1 is successful. However, it was not quite straightforward, as the subtle effects of marking had to be dealt with.

STAGES 9,10:	Adapt the model of Stage 8 to incorporate the operator, as in Stage 2, so we have both automatic and manual modes.

PLANT9 = **sync**(PLANT8,OPERATOR) (4,32)
MSPEC8 = **minstate**(SPEC8) (12,23)

Modify MSPEC8 to allow transition from automatic to manual (ev43) and back (ev41) to occur only between the initial state and the designated entrance state [12] to manual. Note that the latter state [12] is now marked. Shutdown (ev81) is restricted to occur only at the initial state [0] or the entrance state to manual [2]; all action ceases after shutdown and the system remains at the shutdown state [3]. The modeling intent is to cut back to a small number of instances of significant events in order to produce a small high-level model.

SPEC10 = **edit**(MSPEC8,[trans +[0,43,12],+[12,11,12],+[12,13,12],
+[12,21,12],+[12,22,12],+[12,24,12],+[12,31,12],+[12,32,12],
+[12,34,12],+[12,41,0],+[12,81,12]]) (13,34)
SPEC10 = **edit**(SPEC10,[trans -[1,81,5],-[3,81,7],-[4,81,7],-[6,81,10],
-[8,81,10],-[9,81,7],-[11,81,7]]) (13,27)

SPEC10 = **edit**(SPEC10,[mark +[12]],[trans +[12,81,2],-[12,81,12]]) (13,27)
PLANT10 = **edit**(PLANT9) (4,32)
SUPER10 = **supcon**(PLANT10,SPEC10) (13,35)
MSUPER10 = **minstate**(SUPER10) (13,35)
isomorph(MSUPER10,SUPER10) = *true*

Now we create the high-level model, first vocalizing SUPER10 with respect to methane readings (evs32,34), automatic/manual transitions (evs41,43), and shutdown (ev81).

VSUPER10 = **vocalize**(SUPER10,[[0,43,43],[0,81,81],[2,41,41],[2,81,81], [7,32,32],[7,34,34],[8,32,32],[8,34,34],[9,32,32],[10,34,34], [9,34,34],[13,81,81],[10,32,32],[13,43,43],[14,81,81], [15,81,81],[16,81,81]]) (20,67)
X= **minstate**(VSUPER10) (20,67)
isomorph(X,VSUPER10) = *true*
VSP10HI = **higen**(VSUPER10) (3,8)
OCVSP10 = **outconsis**(VSUPER10) (20,67)
isomorph(OCVSP10,VSUPER10) = *false*

In fact, VSUPER10 and OCVSP10 are actually isomorphic, up to the recoding of state outputs done by **outconsis**.

HCVSP10 = **hiconsis**(VSUPER10) (21,72)
X= **hiconsis**(OCVSP10) (21,72)
isomorph(X,HCVSP10) = *true*
HCVS10HI = **higen**(HCVSP10) (6,18)
X= **minstate**(HCVS10HI) (6,18)
isomorph(X,HCVS10HI) = *true*

The high-level model HCVS10HI, of size 6, does not represent a large reduction over the low-level controlled model SUPER10 of size 13, but is some improvement.

PHCVS10H = **project**(HCVS10HI,**null**[820,821,831,841,851]) (3,8)

Notice that the manager may only be interested in the original events that he declared to be significant, of course after any splitting that is required to resolve their control status. Here we have projected out the additional events introduced by **hiconsis**. This small model incorporates only the original events selected by the manager, and in terms of which he would be most likely to express his high-level specification. This can be thought of as the third (and highest) hierarchical level, the level of HCVS10HI being a middle level that the high-level manager need not see (because the additional events due to **hiconsis** may well be confusing and difficult to interpret). This middle level is, of course, the one used by *TCT* in computing the high-level control.

Now create a high-level specification to apply to HCVS10HI.

ALL10HI = **create**(ALL10HI,[mark 0]) (1,0)
ALL10HI = **selfloop**(ALL10HI,[320,340,411,431,811,820,821,831,841, 851]) (1,10)

Check that ALL10HI includes all the events that occur in HCVS10HI.

X= **meet**(HCVS10HI,ALL10HI) (6,18)
isomorph(X,HCVS10HI) = *true*

For the high-level specification we decide to force an initial switch to manual (ev431). When this is followed by 3 consecutive readings of methane-high (ev340), there is a forced return to automatic (ev411) followed by shutdown (ev811).

SPEC10HI = **create**(SPEC10HI,[mark 0,5],[tran [0,320,0],[0,340,1],
[0,431,0],[1,320,0],[1,340,2],[2,320,0],[2,340,3],[3,411,4],
[4,811,5]]) (6,9)

This specification must be augmented by self-looping the auxiliary high-level events that are of no direct interest to the manager.

SPEC10HI = **sync**(SPEC10HI,ALL10HI) (6,39)
SUP10HI = **supcon**(HCVS10HI,SPEC10HI) (13,33)
X= **minstate**(SUP10HI) (13,33)
isomorph(X,SUP10HI) = *true*

Now project from this 'middle-level' behavior the undesired auxiliary events.

PSUP10HI = **project**(SUP10HI,**null**[820,821,831,841,851]) (7,9)

This is reasonably small, and easy to comprehend. Here we stress an important theme: how we achieve a comprehensible state model at any particular stage of (human) design. Indeed this is the basic rationale for the entire approach.

SPC10LO = **create**(SPC10LO,[mark 0,5],[tran [0,32,0],[0,34,1],[0,43,0],
[1,32,0],[1,34,2],[2,32,0],[2,34,3],[3,32,0],[3,41,4],
[4,81,5]]) (6,10)

SPC10LO = **sync**(SPC10LO,ALL9) (6,46)
SPEC10LO = **meet**(SPEC10,SPC10LO) (40,78)
SUP10LO = **supcon**(PLANT10,SPEC10LO) (17,70)
X= **minstate**(SUP10LO) (17,70)
isomorph(X,SUP10LO) = *true*

This displays a conceptual advantage in hierarchical control. The low-level controlled behavior has size (17,70) compared with the high-level controlled behavior of (7,9). The latter is far more comprehensible.

PSUP10LO = **project**(SUP10LO,**null**[11,13,21,22,24,31]) (7,9)

However, when we project out the events not of direct relevance to the high-level specification, we end up with a result of exactly the same size as PSUP10HI.

For interest, change the high-level specification by requiring an initial forcing to automatic, and a final forcing to manual.

ASPC10HI = **edit**(SPC10HI,[trans +[0,411,0],+[3,431,4],-[0,431,0],
-[3,411,4]]) (6,33)

ASUP10HI = **supcon**(HCVS10HI,ASPC10HI) (9,14)
X= **minstate**(ASUP10HI) (9,14)
isomorph(X,ASUP10HI) = *true*

A significant reduction in state size, from (13,33) for SUP10HI, to (9,14) for ASUP10HI.

ASUP10HI = **condat**(HCVS10HI,ASUP10HI) Controllable.
PASP10HI = **project**(ASUP10HI,**null**[831,841]) (6,8)

Compare with the size (7,9) of PSUP10HI.

STAGE 11: Bring in WFSENSOR and repeat the previous exercise with the enlarged plant.

ALL11 = **selfloop**(ALL9,[51,52,53,54]) (1,15)

Modify WFSPEC by marking state 1, after pump has been turned on (ev13) but before WFSENSOR is initialized (ev51).

WFSPEC = **edit**(WFSPEC,[mark +[1]]) (4,38)
WFSPEC11 = **sync**(WFSPEC,ALL11) (4,42)
PLANT11 = **sync**(PLANT10,WFSENSOR) (8,80)
SP1011= **sync**(SPEC10,ALL11) (10,63)
SPEC11 = **meet**(SP1011,WFSPEC11) (40,120)
SUPER11 = **supcon**(PLANT11,SPEC11) (52,162)
MSUPER11 = **minstate**(SUPER11) (52,162)
isomorph(MSUPER11,SUPER11) = *true*

For comparison, try using the previous supervisor SUPER10, synched with WFSENSOR, as the new plant, and the augmented version of WFSPEC, viz. WFSPEC11, as the specification.

BSPEC11 = **edit**(WFSPEC11) (4,42)
BPLANT11 = **sync**(SUPER10,WFSENSOR) (26,122)
BSUP11 = **supcon**(BPLANT11,BSPEC11) (52,162)
isomorph(BSUP11,SUPER11) = *true*

This shows consistency of the two ways of incrementally enlarging the system at this stage. Next we create the high-level model, again regarding only the events 32,34,41,43,81 as significant. The wild-card feature in *TCT* **vocalize** is almost essential to ensure complete data entry.

VSUPER11 = **vocalize**(SUPER11,[[0,43,43],[2,81,81],[0,81,81],[2,41,41],
[6,41,41],[6,81,81],[8,32,32],[9,34,34],[8,34,34],[10,34,34],
[9,32,32],[10,32,32],[12,32,32],[14,41,41],[12,34,34],[13,43,43],
[13,81,81],[14,81,81],[16,34,34],[16,32,32],[19,32,32],[19,34,34],
[23,81,81],[21,34,34],[23,43,43],[24,81,81],[21,32,32],[24,41,41],
[26,32,32],[26,34,34],[28,34,34],[32,81,81],[32,43,43],[28,32,32],
[34,32,32],[35,34,34],[34,34,34],[35,32,32],[37,34,34],[37,32,32],
[41,32,32],[44,34,34],[41,34,34],[44,32,32],[47,32,32],[49,34,34],
[47,34,34],[52,81,81],[49,32,32],[52,43,43],[53,81,81],[54,81,81],
[57,81,81],[55,81,81],[58,81,81],[57,41,41],[59,81,81],[58,41,41],
[59,41,41],[65,81,81],[64,81,81],[66,81,81],[65,41,41],[66,41,41],
[67,81,81],[67,41,41],[74,81,81],[75,81,81],[75,41,41],[76,81,81],
[76,41,41],[77,81,81],[77,41,41],[81,81,81]]) (89,296)

X= **minstate**(VSUPER11) (89,296)
isomorph(X,VSUPER11) = *true*
HCVSPR11 = **hiconsis**(VSUPER11) (99,335)
X= **minstate**(HCVSPR11) (99,335)
isomorph(X,HCVSPR11) = *true*
PLNT11HI = **higen**(HCVSPR11) (20,84)
X= **minstate**(PLNT11HI) (20,84)

isomorph(X,PLNT11HI) = *true*
OCVSPR11 = **outconsis**(VSUPER11) (89,296)
X= **hiconsis**(OCVSPR11) (99,335)
isomorph(HCVSPR11,X) = *true*
T= **trim**(PLNT11HI) (20,84)
isomorph(T,PLNT11HI) = *true*
ALL11HI = **allevents**(PLNT11HI) (1,31)
SPC11HI = **sync**(SPC10HI,ALL11HI) (6,165)
SUP11HI = **supcon**(PLNT11HI,SPC11HI) (31,135)
X= **minstate**(SUP11HI) (31,135)
isomorph(X,SUP11HI) = *true*
PSUP11H = **project**(SUP11HI,**image**[320,340,411,431,811]) (7,9)
isomorph(PSUP11HI,PSUP10HI) = *true*

We've done nothing new at the high level, and get the same final result as before. Of course the vocalized structure is now much larger (cf. VSUPER11 with size (89,296), VSUPER10 with size (20,67)).

Also the projection of PLNT11HI is as expected, with shutdown correctly handled.
PPLN11HI = **project**(PLNT11HI,Image[320,340,411,431,811]) (3,8)

Timed Mine Problem in *TTCT*

We continue with the mine pump model based on the book by Burns and Wellings. To follow the steps of the exercise it is recommended that the reader be familiar with the main synthesis procedures of Chap. 9.

Note: the previous abbreviation 'ME' for 'methane' is here replaced by 'CH4'.

First version of timed model: PLANT8 = WL || CH4 || SHUTDOWN.

Our modeling will follow the lines of stages 0,1,8 of the untimed case, but with a subtlety brought out at the end.

STAGE 0: Create the five basic components, and combine to form plant1.

---

pump = **create**(pump,[mark 0],[timebounds [11,0,1000],[13,0,1000]],
[forcible 11,],[tran [0,11,0],[0,13,0]]) (1,2)
wlsensor = **create**(wlsensor,[mark 0],[timebounds [21,0,1000],[22,0,2],
[24,0,2]],[forcible 21],[tran [0,21,1],[1,22,0],[1,24,0]]) (2,3)
ch4sensor = **create**(ch4sensor,[mark 0],[timebounds [31,0,1000],[32,0,1],
[34,0,1]],[forcible 31],[tran [0,31,1],[1,32,0],[1,34,0]]) (2,3)
wfsensor = **create**(wfsensor,[mark 0],[timebounds [51,0,1000],[52,0,3],
[53,0,1000],[54,0,3]],[forcible 51,],[tran [0,51,1],[0,53,0],[1,52,0],
[1,54,0]]) (2,4)
operator = **create**(operator,[mark 0],[timebounds [41,0,1000],[43,0,1000]],
[forcible 41,],[tran [0,41,0],[0,43,0]]) (1,2)
wlsensor = **timed_graph**(wlsensor) (4,10)

ch4sensor = **timed_graph**(ch4sensor) (3,7)
wfsensor = **timed_graph**(wfsensor) (5,14)
operator = **timed_graph**(operator) (1,3)
pump = **timed_graph**(pump) (1,3)
plant1= **sync**(pump,wlsensor) (4,18)
plant1= **sync**(plant1,ch4sensor) (12,71)
mplant1 = **minstate**(plant1) (12,71)
isomorph(mplant1,plant1) = *true*

STAGE 1: Create the two basic specifications. Then compute centralized supervisor and develop modular supervisors.

---

cycspec1 = **create**(cycspec1,[mark 0],[tran [0,21,1],[1,0,1],[1,22,2],[1,24,2],
[2,31,3],[3,0,3],[3,32,4],[3,34,4],[4,0,5],[5,11,0],[5,13,0]],
[forcible 11,13,21,31]) (6,11)
pmpspec1 = **create**(pmpspec1,[mark 0],[tran [0,11,0],[0,22,0],[0,24,1],
[0,32,0],[0,34,2],[1,13,1],[1,22,0],[1,24,1],[1,32,1],[1,34,3],
[2,11,2],[2,22,2],[2,24,3],[2,32,0],[2,34,2],[3,11,3],[3,22,2],
[3,24,3],[3,32,1],[3,34,3]],[forcible 11,13]) (4,20)
all= **create**(all1,[mark 0]) (1,0)
all= **selfloop**(all1,[0,11,13,21,22,24,31,32,34],
[new forcible 11,13,21,31]) (1,9)
pmpspec1 = **sync**(pmpspec1,all1) (4,32)
spec1 = **meet**(cycspec1,pmpspec1) (24,40)
mspec1 = **minstate**(spec1) (13,19)
isomorph(mspec1,spec1) = *false*
super1= **supcon**(plant1,mspec1) (17,28)
super1= **condat**(plant1,super1)
msuper1 = **minstate**(super1) (17,28)
isomorph(msuper1,super1) = *true*
psuper1 = **project**(super1,[0]) (10,13)

Develop modular control.

nonconflict(plant1,cycspec1) = *true*
cycspec1 = **condat**(plant1,cycspec1)
modsup1 = **meet**(cycspec1,pmpspec1) (24,40)
isomorph(modsup1,spec1) = *true*
nonconflict(modsup1,plant1) = *true*
modsup1 = **condat**(plant1,modsup1)
mmodsup1 = **minstate**(modsup1) (13,19)
isomorph(modsup1,spec1) = *true*
x =**meet**(plant1,modsup1) (36,68)
isomorph(super1,x) = *false*
mx= **minstate**(x) (17,28)
isomorph(msuper1,mx) = *true*

isomorph(msuper1,super1) = *true*
This verifies that modular supervision is optimal.

STAGE 2: Bring in new process shutdown with ev81.

---

all8= **selfloop**(all1,[81],[new forcible 81]) (1,10)
super8 = **edit**(msuper1) (17,28)
shutdown = **create**(shutdown,[mark 0],[timebounds [81,0,1000]], [forcible 81],[tran [0,81,0]]) (1,1)
shutdown = **timed_graph**(shutdown) (1,2)
plant8= **sync**(plant1,shutdown) (12,83)
sdnspec = **create**(sdnspec,[mark 0,1],[tran [0,0,0],[0,11,0],[0,13,0],[0,21,0], [0,31,0],[0,81,1],[1,0,1]],[forcible 11,13,21,31,81]) (2,7)
sdnspec = **sync**(sdnspec,all8) (2,15)
cycspec8 = **sync**(cycspec1,all8) (6,17)
pmpspec8 = **edit**(pmpspec1,[mark +[1],+[2],+[3]]) (4,32)
pmpspec8 = **sync**(pmpspec8,all8) (4,36)
spec8 = **meet**(cycspec8,pmpspec8) (24,64)
spec8 = **meet**(spec8,sdnspec) (48,92)
mspec8 = **minstate**(spec8) (15,33)
isomorph(mspec8,spec8) = *false*
super8 = **supcon**(plant8,spec8) (36,68)
msuper8 = **minstate**(super8) (14,25)
isomorph(msuper8,super8) = *false*
pmsuper8 =**project**(msuper8,Null[0]) (8,11)

This supervisor has eliminated ev81 because it would lead to blocking. This of course was not the intention!

nonconflict(plant8,spec8) = *false*

Revise SPEC8 so it releases CYCSPEC8 when ev81 occurs, namely allows ev0 (*tick*) to occur at every target state for ev81. This will prevent the blocking that would have occurred before. Notice that adding one spec after another can have unintended side effects. Here CYCSPEC8 disabled *tick* (ev0) by forcing initialization of the sensor cycle (ev21), so when in turn ev21 was disabled after ev81, the occurrence of ev81 would have brought the system to a halt. For this reason, **supcon** naturally disabled ev81 everywhere, to prevent blocking. Thus 'change of mode' must be treated with care: it cannot in general be created just by using **meet**.

spec8 = **edit**(spec8,[trans +[2,0,2],+[7,0,7],+[9,0,9],+[24,0,24],+[26,0,26], +[28,0,28],+[30,0,30],+[32,0,32],+[34,0,34],+[36,0,36], +[43,0,43],+[45,0,45]]) (48,104)
super8 = **supcon**(plant8,spec8) (40,76)
msuper8 = **minstate**(super8) (15,27)
pmsuper8 =**project**(msuper8,**null**[0]) (9,12)

Now it looks fine. Shutdown (ev81) can occur without stopping the clock.

STAGE 3: Bring in the operator events ev41, ev43, and enhance SPEC8 to allow for switching from automatic mode to manual mode (ev43) and back (ev41). The shutdown event (ev81) should work from manual mode as well.

---

all9 = **selfloop**(all8,[41,43],[new forcible 41,43]) (1,12)
plant9 = **sync**(plant8,operator) (12,107)
mspec8= **minstate**(spec8) (14,34)
spec9 = **edit**(mspec8,[mark +[14]],[forcible +[41],+[43]],[trans +[0,43,14],
+[14,0,14],+[14,11,14],+[14,13,14],+[14,21,14],+[14,22,14],
+[14,24,14],+[14,31,14],+[14,32,14],+[14,34,14],+[14,41,0],
[14,81,2]]) (15,46)

STAGE 4: Next we create a CH4 emergency specification, to result in a forced change to automatic mode (ev41) followed by shutdown (ev81), after 3 consecutive high CH4 readings (ev34).

---

emerg= **create**(emerg,[mark 0,5],[tran [0,0,0],[0,32,0],[0,34,1],[1,0,1],
[1,32,0],[1,34,2],[2,0,2],[2,32,0],[2,34,3],[3,0,3],[3,41,4],[4,81,5],
[5,0,5]],[forcible 41,81]) (6,13)
semerg = **sync**(emerg,all9) (6,55)
spec10 = **meet**(spec9,semerg) (49,118)
super10= **supcon**(plant9,spec10) (85,320)
super10= **condat**(plant9,super10)
psuper10 =**project**(super10,**null**[0,21,22,24,31]) (15,35)

PSUPER10 displays how manual mode overrides the previous specifications, which apply in automatic mode, allowing the pump to be turned on (ev13) even after a CH4-high reading (ev34).

msuper10 =**minstate**(super10) (85,320)
p0super10 =**project**(super10,Null[0]) (42,114)

STAGE 5: Now we bring in WFSENSOR and a suitable specification. Developing a specification caused some difficulty with blocking: the design is very sensitive to the timing requirements of WFSPEC.

---

all13= **selfloop**(all9,[51,52,53,54],[new forcible 51,53]) (1,16)
spec1013 = **sync**(spec10,all13) (49,314)
wfsensor = **edit**(wfsensor,[trans +[0,0,0]]) (5,14)
plant13 = **sync**(plant9,wfsensor) (60,649)
wfspec1= **create**(wfspec1,[mark 0,2],[tran [0,0,0],[0,11,0],[0,13,1],[1,51,2],
[2,0,2],[2,11,0],[2,13,2],[2,51,2],[2,53,2]],[forcible 11,13,51,53])
(3,9)

Getting WFSPEC1 right was troublesome. Initially the attempt was made to force initialization of WFSENSOR (ev51) exactly 2 ticks after turning on the pump (ev13), but this caused blocking. When several timed objects are synchronized, blocking due to clock-stopping can become a serious problem. It was found that ev51 had to be forced immediately after ev13.

wfspec1= **sync**(wfspec1,all13) (3,42)

WFSPEC2 guarantees that if WFSENSOR detects a problem with water flow (ev54), no OK will be reported (ev52) and an alarm must be sounded (ev53).

wfspec2 = **create**(wfspec2,[mark 0],[tran [0,0,0],[0,52,0],[0,54,1],[1,53,0]], [forcible 53]) (2,4)

wfspec2= **sync**(wfspec2,all13) (2,28)

wfspec = **meet**(wfspec1,wfspec2) (6,75)

spec13 = **meet**(spec1013,wfspec) (200,788)

super13 = **supcon**(plant13,spec13) (425,2214)

super13′ = **edit**(super13) (425,2214)

msuper13 = **minstate**(super13) (418,2204)

super13 = **edit**(msuper13) (418,2204)

super13 = **condat**(plant13,super13)

Because the specification constraints are removed in manual mode, it is important to examine the automatic operation in isolation to make sure that there is no removal of function by the supcon operation, to prevent unintended blocking. To this end, form AALL13 (ALL13 reduced for automatic mode by eliminating ev43), and **meet** with SUPER13 to form ASUPER13.

aall13= **edit**(all13,[trans -[0,43,0]]) (1,15)

asuper13 = **meet**(super13′,aall13) (73,137)

ASUPER13 is correct, inasmuch as none of the evs51,52,53,54 was eliminated in automatic mode by the supcon operation.

pasuper13 = **project**(asuper13,**null**[21,22,24,31,32,34,52,53,54]) (13,35)

masuper13 = **minstate**(asuper13) (64,122)

Bibliography

Chapter References

Akesson K, Flordal H, Fabian M (2002) Exploiting modularity for synthesis and verification of supervisors. In: Proceedings of the fifteenth IFAC world congress on automatic control, pp 175–180. Barcelona, Spain

Al-Jaar RY, Desrochers AA (1988) A modular approach for the performance analysis of automated manufacturing systems using generalized stochastic Petri nets. Rpt. RAL #116, Robotics and Automation Laboratory, Rensselaer Polytechnic Institute, Troy, New York

Alur R, Dill D (1990) Automata for modeling real-time systems. In: Proceedings of the seventeenth international colloquium on automata, languages and programming. Lecture notes on computer science (LNCS), vol 443. Springer, Berlin, pp 322–335

Arnold A (1994) Finite transition systems. Prentice-Hall, Upper Saddle River

Berthomieu B, Diaz M (1991) Modeling and verification of time dependent systems using time Petri nets. IEEE Trans Softw Eng 17(3):259–273

Brandin BA (1989) The supervisory control of discrete event systems with forcible events. MASc thesis, Department of Electrical and Engineering, University of Toronto

Brandin BA (1993) Real-time supervisory control of automated manufacturing systems. PhD thesis, Department of Electrical and Computer Engineering, University of Toronto

Brandin BA, Wonham WM (1992) Supervisory control of timed discrete-event systems. In: Proceedings of the thirty-first IEEE conference on decision and control, Tucson, Arizona, pp 3357–3362

Brandin BA, Wonham WM (1994) Supervisory control of timed discrete-event systems. IEEE Trans Autom Control 39(2):329–342. [Figures 1–14, 17–21 are reprinted, with permission, from IEEE.]

Brandin BA, Wonham WM, Benhabib B (1993) Manufacturing cell supervisory control - a modular timed discrete-event system approach. In: Proceedings of the IEEE international conference on robotics and automation, Atlanta, Georgia, pp 931–936

Brandt RD, Garg V, Kumar R, Lin F, Marcus SI, Wonham WM (1990) Formulas for calculating supremal controllable and normal sublanguages. Syst Control Lett 15:111–117

Brave Y, Heymann M (1990) Stabilization of discrete-event processes. Int J Control 51:1101–1117

Cai K (2008) Supervisor localization: a top-down approach to distributed control of discrete-event systems. MASc thesis, Department of Electrical and Computer Engineering, University of Toronto

Cai K, Wonham WM (2009a) Supervisor localization: a top-down approach to distributed control of discrete-event systems. In: Beji L, Otmane S, Abichou A (eds) Proceedings of the second Mediterranean conference on intelligent systems and automation (CISA '09), Zarzis, Tunisia, Mar 23–25, pp 302–308

W. M. Wonham and K. Cai, *Supervisory Control of Discrete-Event Systems*, Communications and Control Engineering,
https://doi.org/10.1007/978-3-319-77452-7

Cai K, Wonham WM (2009b) Supervisor localization for large-scale discrete-event systems. In: Joint forty-eighth IEEE conference on decision and control and twenty-eighth Chinese control conference, Shanghai, China, Dec 16–18, pp 3099–3105

Cai K, Wonham WM (2010a) Supervisor localization: a top-down approach to distributed control of discrete-event systems. IEEE Trans Autom Control 55(3):605–618

Cai K, Wonham WM (2010b) Supervisor localization for large discrete-event systems: case study, production cell. Int J Adv Manufact Technol 50(9–12):1189–1202

Cai K, Wonham WM (2016a) A new algorithm for computing the supremal relatively observable sublanguage. In: Proceedings thirteenth international workshop on discrete event systems (WODES ’16), Xi’an, China, May 30–June 1 2016, pp 8–13

Cai K, Wonham WM (2016b) Supervisor localization: a top-down approach to distributed control of discrete-event systems. Lecture notes in control and information sciences (LNCIS), vol 459. Springer International Publishing, Switzerland

Cai K, Zhang R-Y, Wonham WM (2013) On relative observability of discrete-event systems. In: Proceedings fifty-second IEEE conference on decision and control, Firenze, Italy, 11–13 Dec 2013, pp 7285–7290

Cai K, Zhang R-Y, Wonham WM (2014) On relative observability of timed discrete-event systems. In: Proceedings of the twelfth IFAC and IEEE international workshop on discrete event systems (WODES ’14), Cachan, France, 14–16 May 2014, pp 208–213

Cai K, Zhang R-Y, Wonham WM (2015) On relative coobservability of discrete-event systems. In: Proceedings of the American control conference, Chicago, IL, USA, 1–3 July 2015, pp 371–376

Cai K, Zhang R-Y, Wonham WM (2018) Characterizations and effective computation of supremal relatively observable sublanguages. Discret Event Dyn Syst 28(2):269–287

Cao C, Lin F, Lin Z-H (1997) Why event observation: observability revisited. Discret Event Dyn Syst Theory Appl 7:127–149

Carroll J, Long D (1989) Theory of finite automata. Prentice-Hall, Englewood Cliffs

Cassandras CG (1993) Discrete event systems. Irwin

Chen S-L (1992) Existence and design of supervisors for vector discrete event systems. MASc thesis, Department of Electrical and Computer Engineering, University of Toronto

Chen S-L (1996) Control of discrete-event systems of vector and mixed structural type. PhD thesis, Department of Electrical and Computer Engineering, University of Toronto

Chen S-L, Wonham WM (1992) Existence and design of supervisors for vector discrete-event systems. In: Proceedings of thirtieth annual Allerton conference on communication, control and computing, University of Illinois, pp 604–613

Chen Y, Li Z (2013) Optimal supervisory control of automated manufacturing systems. CRC Press, Boca Raton

Cho H, Marcus SI (1989) Supremal and maximal sublanguages arising in supervisor synthesis problems with partial observations. Math Syst Theory 22:177–211

Davey BA, Priestley HA (1990) Introduction to lattices and order. Cambridge University Press, Cambridge

de Queiroz MH, Cury JER (2000) Modular control of composed systems. In: Proceedings of the American control conference, Chicago, pp 4051–4055

Desrochers AA, Al-Jaar RY (1995) Applications of Petri nets in manufacturing systems. IEEE Press, New York

Dijkstra EW (1971) Hierarchical ordering of sequential processes. Acta Informatica 1:115–138

Dijkstra EW (1976) A discipline of programming. Prentice-Hall, Upper Saddle River

Eilenberg S (1974) Automata, languages, and machines, vol A. Academic Press, Orlando

Falcione A, Krogh B (1993) Design recovery for relay ladder logic. IEEE Control Syst 13(2):90–98

Feng L (2007) Computationally efficient supervisor design for discrete-event systems. PhD thesis, Department of Electrical and Computer Engineering, University of Toronto

Feng L, Cai K, Wonham WM (2009) A structural approach to the nonblocking supervisory control of discrete-event systems. Int J Adv Manufact Technol 41(11):1152–1167

Feng L, Wonham WM (2006a) Computationally efficient supervisor design: control-flow decomposition. In: Proceedings of the 8th international workshop on discrete-event systems (WODES '06), University of Michigan, Ann Arbor, pp 9–14

Feng L, Wonham WM (2006b) Computationally efficient supervisor design: abstraction and modularity. In: Proceedings of the eighth international workshop on discrete-event systems (WODES '06), University of Michigan, Ann Arbor, pp 3–8

Feng L, Wonham WM (2006c) On the computation of natural observers in discrete-event systems. In: Proceedings of the forty-fifth IEEE conference on decision and control, San Diego, pp 428–433

Feng L, Wonham WM (2008) Supervisory control architecture for discrete-event systems. IEEE Trans Autom Control 53(6):1449–1461

Feng L, Wonham WM (2010) On the computation of natural observers in discrete-event systems. Discret Event Dyn Syst 20(1):63–102

Germundsson R (1995) Symbolic systems - theory, computation and applications. PhD thesis no. 389, Department of Electrical Engineering, Linköping University

Gunnarsson J (1997) On modeling of discrete event dynamic systems, Using symbolic algebraic methods. Thesis no. 502, Division of Automatic Control, Lund University

Heymann M (1990) Concurrency and discrete event control. IEEE Control Syst 10(4):103–112

Hiraishi K, Ichikawa A (1988) A class of Petri nets that [sic] a necessary and sufficient condition for reachability is obtainable. Trans Soc Instrum Control Eng (SICE) 24(6):91–96 (in Japanese)

Hoare, CAR (1985) Communicating sequential processes. Prentice-Hall

Holloway LE, Krogh BH (1990) Synthesis of feedback logic control for a class of controlled Petri nets. IEEE Trans Autom Control 35(5):514–523

Hopcroft JE, Ullman JD (1979) Introduction to automata theory, languages and computation. Addison-Wesley, Boston

Huang N-Q (1991) Supervisory control of vector discrete-event processes. MASc thesis, Department of Electrical and Computer Engineering, University of Toronto

Huang NQ, Li Y, Wonham WM (1989) Supervisory control of vector discrete-event systems. In: Proceedings of twenty-seventh annual Allerton conference on communication, control and computing, University of Illinois, pp 925–934

Hubbard P, Caines PE (2002) Dynamical consistency in hierarchical supervisory control. IEEE Trans Autom Control 47(1):37–52

Jiao T, Gan Y, Xiao G, Wonham WM (2017) Exploiting symmetry of state tree structures for discrete-event systems with parallel components. Int J Control 90(8):1639–1651

Krogh BH (1987) Controlled Petri nets and maximally permissive feedback logic. In: Proceedings of twenty-fifth annual Allerton conference on communication control and computing, Allerton, pp 317–326

Leduc RJ (2002) Hierarchical interface-based supervisory control. PhD thesis, Department of Electrical and Computer Engineering, University of Toronto

Leduc RJ, Brandin BA, Lawford M, Wonham WM (2005) Hierarchical interface-based supervisory control, part I: serial case. IEEE Trans Autom Control 50(9):1322–1335

Leduc RJ, Brandin BA, Wonham WM, Lawford M (2001) Hierarchical interface-based supervisory control: serial case. In: Proceedings of the fortieth IEEE conference on decision and control, Orlando, pp 4116–4121

Leduc RJ, Lawford M, Wonham WM (2001) Hierarchical interface-based supervisory control: AIP example. In: Proceedings of the thirty-ninth annual Allerton conference on communications, control, and computing, Allerton, pp 396–405

Leduc RJ, Lawford M, Wonham WM (2005) Hierarchical interface-based supervisory control, part II: parallel case. IEEE Trans Autom Control 50(9):1336–1348

Leduc RJ, Wonham WM, Lawford M (2001) Hierarchical interface-based supervisory control: parallel case. In: Proceedngs of the thirty-ninth annual Allerton conference on communications, control, and computing, Allerton, pp 386–395

Li Y (1991) Control of vector discrete-event systems. PhD thesis, Department of Electrical Engineering, University of Toronto

Li Y, Wonham WM (1988a) A state-variable approach to the modeling and control of discrete-event systems. In: Proceedings of twenty-sixth annual Allerton conference on communication, control, and computing, University of Illinois, pp 1140–1149

Li Y, Wonham WM (1988b) Controllability and observability in the state-feedback control of discrete-event systems. In: Proceedings of the twenty-seventh IEEE conference on decision and control, New York, pp 203–208

Li Y, Wonham WM (1989a) Composition and modular state-feedback control of vector discrete-event systems. In: Proceedings of the conference on information sciences and systems, The Johns Hopkins University, Baltimore, pp 103–111

Li Y, Wonham WM (1989b) Linear integer programming techniques in the control of vector discrete-event systems. In: Proceedings of twenty-seventh annual Allerton conference on communication, control and computing, University of Illinois, pp 528–537

Li Y, Wonham WM (1993) Control of vector discrete-event systems: I - the base model. IEEE Trans Autom Control 38(8):1214–1227; Correction: IEEE Trans Autom Control 39(8):1771, (1994). [Figures 9–11 are reprinted, with permission, from IEEE.]

Lin F (1987) On controllability and observability of discrete event systems. PhD thesis, Department of Electrical Engineering, University of Toronto

Lin F, Brandt RD, Wonham WM (1989) A note on supremal controllable and normal sublanguages. In: Proceedings of the twenty-seventh annual Allerton conference on communication, control and computing, University of Illinois, pp 491–500

Lin F, Wonham WM (1988a) Decentralized supervisory control of discrete-event systems. Inf Sci 44(2):199–224

Lin F, Wonham WM (1988b) On observability of discrete-event systems. Inf Sci 44(2):173–198

Lin F, Wonham WM (1988c) Supervisory control and observation of discrete-event systems. In: Eighth international symposium on the mathematical theory of networks and systems (MTNS '87), Phoenix (1987); Byrnes CI, Martin CF, Saeks RE (eds) Analysis and control of nonlinear systems. North-Holland, Amsterdam, pp 337–348

Lin F, Wonham WM (1990) Decentralized control and coordination of discrete-event systems with partial observation. IEEE Trans Autom Control 35(12):1330–1337

Ma C (2004) Nonblocking supervisory control of state tree structures. PhD thesis, Department of Electrical and Computer Engineering. University of Toronto

Ma C, Wonham WM (2003) Control of state tree structures. In: Proceedings of the eleventh IEEE Mediterranean conference on control and automation, Rhodes, Greece, paper T4–005 (6 pp)

Ma C, Wonham WM (2005a) A symbolic approach to the supervision of state tree structures. In: Proceedings thirteenth IEEE Mediterranean conference on control and automation (MED '05), Limassol, Cyprus, pp 908–913

Ma C, Wonham WM (2005b) Nonblocking supervisory control of state tree structures. Lecture notes in control and information sciences (LNCIS), vol 317. Springer

Mac Lane S, Birkhoff G (1993) Algebra, 3rd edn. Chelsea, New York

Mannani A (2009) Synthesis of communicating decentralized supervisors for discrete-event systems with application to communication protocol synthesis. PhD thesis, ECE Department, Concordia University, Montreal

Memmi G, Roucairol G (1980) Linear algebra in net theory. In: Brauer W (ed) Net theory and applications. Lecture notes on computer science (LNCS), vol 84. Springer, Berlin, pp 213–223

Merlin PM, Farber DJ (1976) Recoverability of communication protocols - implications of a theoretical study. IEEE Trans Commun 24:1036–1043

Mesarovic MD, Macko D, Takahara Y (1970) Theory of hierarchical, multilevel, systems. Academic Press

Milner R (1989) Communication and concurrency. Prentice-Hall, Upper Saddle River

Moody J, Antsaklis P (1998) Supervisory control of discrete event systems using Petri nets. Kluwer

Murata T (1989) Petri nets: properties, analysis and applications. Proc IEEE 77(4):541–580

Nooruldeen A, Schmidt KW (2015) State attraction under language specification for the reconfiguration of discrete event systems. IEEE Trans Autom Control 60(6):1630–1634

Ostroff J (1987) Real-time computer control of discrete systems modelled by extended state machines: a temporal logic approach. PhD thesis, Department of Electrical and Engineering, University of Toronto

Ostroff JS (1989) Temporal logic for real-time systems. Research Studies Press

Ostroff JS (1990) Deciding properties of timed transition models. IEEE Trans Parallel Distrib Syst 1(2):170–183

Ostroff JS, Wonham WM (1990) A framework for real-time discrete event control. IEEE Trans Autom Control 35(4):386–397

Pu KQ (2000) Modeling and control of discrete-event systems with hierarchical abstraction. MASc thesis, Department of Electrical and Computer Engineering, University of Toronto

Ramadge PJ (1983) Control and supervision of discrete event processes. PhD thesis, Department of Electrical Engineering, University of Toronto

Ramadge PJ, Wonham WM (1981) Algebraic decomposition of controlled sequential machines. In: Eighth Triennial world congress, International federation automatic control (IFAC), vol 3, Kyoto. Preprints, pp 37–41

Ramadge PJ, Wonham WM (1982a) Supervision of discrete event processes. In: Proceedings of the twenty-first IEEE conference on decision and control, New York, pp 1228–1229

Ramadge PJ, Wonham WM (1982b) Supervisory control of discrete event processes. Joint workshop on feedback and synthesis of linear and nonlinear systems, Istituto di Automatica, University di Roma. In: Hinrichsen D, Isidori A (eds) Feedback control of linear and nonlinear systems. Lecture notes on control and information sciences, vol 39. Springer, Berlin, pp 202–214

Ramadge PJ, Wonham WM (1984) Supervisory control of a class of discrete event processes. In: Bensoussan A, Lions JL (eds) Proceedings of the sixth international conference on analysis and optimization of systems, Nice, June 1984. Analysis and optimization of systems. Lecture notes on control and information sciences, Part 2, vol 63. Springer, Berlin, pp 477–498

Ramadge PJ, Wonham WM (1987a) Modular feedback logic for discrete event systems. SIAM J Control Optim 25(5):1202–1218

Ramadge PJ, Wonham WM (1987b) Supervisory control of a class of discrete event processes. SIAM J Control Optim 25(1):206–230

Rudie K (1992) Decentralized control of discrete-event systems. PhD thesis, Department of Electrical Engineering, University of Toronto

Rudie K, Wonham WM (1992) Think globally, act locally: decentralized supervisory control. IEEE Trans Autom Control 37(11):1692–1708. (Reprinted in Sadjadi FA (eds) Selected papers on sensor and data fusion, 1996, ISBN 0-8194-2265-7)

Sørensen EV, Nordahl J, Hansen NH (1993) From CSP models to Markov models. IEEE Trans Softw Eng 19(6):554–570

Schwartz B (1992) State aggregation of controlled discrete-event systems, MASc thesis, Department of Electrical Engineering, University of Toronto

Simon H (1962) The architectecture of complexity. In: Proceeding of the American philosophical society, vol 106, Dec 1962, pp 467–482. Reprinted in: Simon HA, The sciences of the artificial (Chap. 7, pp 193–229), 2nd edn. The MIT Press, Cambridge (1981)

Su R (2004) Distributed diagnosis for discrete-event systems. PhD thesis, Department of Electrical and Computer Engineering, University of Toronto

Su R, Wonham WM (2001) Supervisor reduction for discrete-event systems. Conference on information sciences and systems, The Johns Hopkins University, 21–23 Mar 2001, [6 pp]

Su R, Wonham WM (2004) Supervisor reduction for discrete-event systems. Discret Event Dyn Syst 14(1):31–53

Takai S, Ushio T (2003) Effective computation of an L_m(G)-closed, controllable, and observable sublanguage arising in supervisory control. Syst Control Lett 49:191–200

Theunissen RJM, Petreczky M, Schiffelers RRH, van Beek DA, Rooda JE (2010) Application of supervisory control synthesis to MRI scanners: improving evolvability. SE Report: Nr. 2010-06, Eindhoven, Apr 2010

Tucker AW (1956) Dual systems of homogeneous linear relations. In: Kuhn HW, Tucker AW (eds) Linear inequalities and related systems. Annals of mathematics studies, vol 38. Princeton University Press, Princeton, pp 3–18

Uzam M, Wonham WM (2006) A hybrid approach to supervisory control of discrete event systems coupling RW supervisors to Petri nets. Int J Adv Manuf Technol 28(7–8):747–760

Vaz A, Wonham WM (1985) On supervisor reduction in discrete event systems. In: Proceedings of the twenty-third annual Allerton conference on communication, control and computing, University of Illinois, Urbana, pp 933–939

Vaz AF, Wonham WM (1986) On supervisor reduction in discrete-event systems. Int J Control 44(2):475–491

Viswanadham N, Narahari Y (1992) Performance modeling of automated manufacturing systems. Prentice-Hall

Wang B (1995) Top-down design for RW supervisory control theory, MASc thesis, Department of Electrical and Computer Engineering, University of Toronto

Wang YW (2000) Supervisory control of Boolean discrete-event systems. MASc thesis, Department of Electrical and Computer Engineering, University of Toronto

Willner Y, Heymann M (1991) On supervisory control of concurrent discrete-event systems. Int J Control 54(5):1143–1169

Wong KC (1990) An algebraic description of hierarchical control in discrete-event systems, MASc thesis, Department of Electrical Engineering, University of Toronto

Wong K-C (1994) Discrete-event control architecture: an algebraic approach. PhD thesis, Department of Electrical and Computer Engineering, University of Toronto

Wong KC (1998) On the complexity of projections of discrete-event systems. In: Proceedings of the international workshop on discrete event systems (WODES '98). Cagliari, pp 201–206

Wong KC, Wonham WM (1996a) Hierarchical control of discrete-event systems. Discret Event Dyn Syst 6(3):241–273

Wong KC, Wonham WM (1996b) Hierarchical control of timed discrete-event systems. Discret Event Dyn Syst 6(3):274–306

Wong KC, Wonham WM (2004) On the computation of observers in discrete-event systems. Discret Event Dyn Syst 14(1):55–107

Wong-Toi H, Hoffmann G (1991) The control of dense real-time discrete event systems. In: Proceedings of the thirtieth IEEE conference on decision and control, Brighton, UK, pp 1527–1528

Wonham WM (1976) Towards an abstract internal model principle. IEEE Trans Syst Man Cybern SMC 6(11):735–740

Wonham WM (1985) Linear multivariable control: a geometric approach, 3rd edn. Springer, Berlin

Wonham WM (2003) Supervisory control theory: models and methods. In: Proceedings of ATPN - workshop on discrete event systems control, twenty-fourth international conference on application theory of Petri nets (ATPN '03), Eindhoven, The Netherlands, pp 1–14

Wonham WM, Ramadge PJ (1984) On the supremal controllable sublanguage of a given language. In: Proceedings of the twenty-third IEEE conference on decision and control, New York, pp 1073–1080

Wonham WM, Ramadge PJ (1987) On the supremal controllable sublanguage of a given language. SIAM J Control Optim 25(3):637–659

Wonham WM, Ramadge PJ (1988) Modular supervisory control of discrete event systems. Math Control Signals Syst 1(1):13–30

Zhang ZH (2001) Smart TCT: An efficient algorithm for supervisory control design. MASc thesis, Department of Electrical and Computer Engineering, University of Toronto
Zhang ZH, Wonham WM (2001) STCT: an efficient algorithm for supervisory control design. In Caillaud B, Xie X (eds) Proceedings of the symposium on the supervisory control of discrete event systems (SCODES '01), INRIA, Paris, pp 82–93. See also: Caillaud B et al (eds) (2002) Synthesis and control of discrete event systems, Kluwer, pp 77–100
Zhong H (1987) Control of discrete-event systems: decentralized and hierarchical control. MASc thesis, Department of Electrical Engineering, University of Toronto
Zhong H (1992) Hierarchical control of discrete-event systems. PhD thesis, Department of Electrical Engineering, University of Toronto
Zhong H, Wonham WM (1989) Hierarchical control of discrete-event systems: computation and examples. In: Proceedings of the twenty-seventh annual Allerton conference on communication, control and computing, University of Illinois, pp 511–519
Zhong H, Wonham WM (1990a) Hierarchical coordination. In: Proceeding of the fifth IEEE international symposium on intelligent control, Philadelphia, 5–7 Sept 1990, pp 8–14
Zhong H, Wonham WM (1990b) On consistency of hierarchical supervision in discrete-event systems. IEEE Trans Autom Control 35(10):1125–1134. [Figures 1–5, 10–12, 14–19 are reprinted, with permission, from IEEE.]

For Further Reading

Textbooks, Monographs and Proceedings

Discrete-Event Systems

Baccelli F, Cohen G, Olsder GJ, Quadrat J-P (1992) Synchronization and linearity. Wiley, New York
Balemi S, Kozak P, Smedinga R (eds) (1993) Discrete event systems: modeling and control. Birkhaüser
Boel, R., G. Stremersch (eds) (2000) Discrete event systems: analysis and control [Proceedings of WODES '00, Ghent, August 21–23, 2000]. Kluwer
Caillaud B, Darondeau P, Lavagno L, Xie X (eds) (2002) Synthesis and control of discrete event systems. [Proceedings of SCODES '01, INRIA, Paris, July 2001]. Kluwer
Cassandras CG, Lafortune S (2008) Introduction to discrete event systems. 2nd edn. Springer, Berlin
Cohen G, Quadrat J-P (eds) (1994) Proceedings of the 11th international conference on analysis and optimization of systems: discrete event systems. Sophia-Antipolis, June 1994. Lecture notes in control and information sciences, vol 119. Springer, Berlin
Darondeau P, Kumagai S (eds) Proceedings of the ATPN-workshop on discrete event systems control. In: 24th international conference on application theory of Petri nets (ATPN 2003), Eindhoven, The Netherlands, June 23–27, 2003
Ho Y-C (eds) (1989) Special issue on dynamics of discrete-event systems. Proc IEEE 77(1),
Ho Y-C, Zheng Y-P (eds) (1991) Proceedings of the 1991 IFAC workshop on discrete event system theory and applications in manufacturing and social phenomena. International Academic
Ho Y-C, Zheng Y-P (eds) (1996) International workshop on discrete event systems - WODES '96. Institute of Electrical Engineers, London, UK
Ho Y-C, Zheng Y-P (eds) International workshop on discrete event systems - WODES '98. Institute of Electrical Engineers, London, UK
Lafortune S, Lin F, Tilbury D (eds) (2006) Proceedings of the eighth international workshop on discrete event systems (WODES '06), Ann Arbor, MI, July 10–12. IEEE, New York

Lennartson B et al (eds) (2008) Proceedings of the ninth international workshop on discrete event systems (WODES '08), Göteborg, Sweden, May 28–30, 2008. IEEE, New Yok
Lesage J-J, Faure J-M, Cury JER, Lennartson B (eds) Proceedings of the twelfth IFAC and IEEE international workshop on discrete event systems (WODES '14), Cachan, France, May 14–16
Ma C, Wonham WM (2005) Nonblocking supervisory control of state tree structures. Lecture notes in control and information sciences (LNCIS), vol 317. Springer, Berlin
Sanchez A (1996) Formal specification and synthesis of procedural controllers for process systems. Lecture notes in control and information sciences (LNCIS), vol 212. Springer, Berlin
Seatzu C, Silva M, van Schuppen JH (eds) (2013) Control of discrete-event systems. Lecture notes in control and information sciences (LNCIS), vol 433. Springer, Berlin
Silva M, Giua A, Colom J-M (eds) (2002) Proceedings of the sixth international workshop on discrete event systems (WODES '02), Zaragoza, Spain
Varaiya P, Kurzhanski AB (eds) (1987) discrete event systems: models and applications. Lecture notes in control and information sciences (LNCIS), vol 103. Springer, Berlin
Zaytoon J, Carré-Ménétrier V, Cassandras C, Cao X (eds) (2004) Proceedings of the seventh international workshop on discrete event systems (IFAC workshop on discrete event systems, WODES '04), Reims, France, pp 22–24

Automata and Languages

Lewis HR, Papadimitriou CH (1981) Elements of the theory of computation. Prentice-Hall, Upper Saddle River

Algebra

Burris S, Sankappanavar HP (1981) A course in universal algebra. Springer, Berlin
Szasz G (1963) Introduction to lattice theory. Academic, New York
Wechler W (1992) Universal algebra for computer scientists. Springer, Berlin

Temporal Logic

Manna Z, Pnueli A (1992) The temporal logic of reactive and concurrent systems. Springer, Berlin

Petri Nets

David R, Alla H (1992) Petri nets and grafcet. Prentice-Hall, Upper Saddle River
Iordache MV, Antsaklis PJ (2006) Supervisory control of concurrent systems: a Petri net structural approach. Birkhäuser, Basel
Peterson JL (1981) Petri net theory and modeling of systems. Prentice-Hall, Upper Saddle River
Reisig W (1985) Petri nets. Springer, Berlin
Zhou MC, DiCesare F (1993) Petri net synthesis for discrete event control of manufacturing systems. Kluwer
Zhou MC, Venkatesh K (1999) Modeling, simulation and control of flexible manufacturing systems: a Petri net approach. World Scientific, Singapore
Zhou MC (eds) (1995) Petri nets in flexible and agile automation. Kluwer

Discrete-Event Simulation

Birtwistle GM (1987) DEMOS - a system for discrete event modelling on simula. Springer, Berlin
Fishman GS (1978) Principles of discrete event simulation. Wiley, New York
Law AM, Kelton WD (1991) Modeling and analysis simulation. McGraw-Hill, New York
Zeigler BP (1984) Multifacetted modeling and discrete event simulation. Academic, New York
Zeigler BP (1990) Object-oriented simulation with hierarchical, modular models. Academic, New York

Concurrency: Modeling, Programming, and Computation

Ben-Ari M (1982) Principles of concurrent programming. Prentice-Hall International, Upper Saddle River
Holt RC, Graham GS, Lazowska ED, Scott MA (1978) Structured concurrent programming with operating systems applications. Addison-Wesley, Boston
Lynch N (1996) Distributed algorithms. Morgan Kaufmann, Burlington
Magee J, Kramer J (1999) Concurrency: state models and Java programs. Wiley, New York

Implementation and Engineering

Bennett S (1988) Real-time computer control: an introduction. Prentice-Hall, Upper Saddle River
Burns A, Wellings A (1997) Real-time systems and their programming languages. 2nd edn. Addison-Wesley, Boston
Fleming PI (1988) Parallel processing in control: the transputer and other architectures. Peregrinus
Gray D (1999) Introduction to the formal design of real-time systems. Springer, Berlin
Krishna CM, Shin KG (1997) Real-time systems. McGraw-Hill, New York
Raynal M (1986) Algorithms for mutual exclusion. The MIT Press, Cambridge
Wagner F, Schmuki R, Wagner T, Wolstenholme P (2006) Modeling software with finite state machines: a practical approach. Auerbach Publications, Boca Raton

Supplementary Reports and Articles

Kay A, Reed JN (1993) A relay and guarantee method for timed CSP: a specification and design of a telephone exchange. IEEE Trans Softw Eng 19(6):625–639
Lyngbaek K (2011) Nonblocking distributed state-tree-structures. Master of science thesis, Department of Software Technology, Technical University of Delft
Theunissen RJM (2015) Supervisory control in health care systems. PhD thesis, Mechanical Engineering Department, Eindhoven University of Technology

Systems Control Group Publications and Theses on Discrete-Event Systems

Encyclopedia Article

Wonham WM (2015) Supervisory control of discrete-event systems. In Baillieul J, Samad T (eds) Encyclopedia of systems and control. Springer, Berlin. https://www.springer.com/us/book/9781447150572, under Engineering, Encyclopedia of Systems and Control

Journal Publications

Afzalian A, Saadatpoor A, Wonham WM (2008) Systematic supervisory control solutions for under-load tap-changing transformers. Control Eng Pract 16(9):1035–1054

Bourdon SE, Lawford M, Wonham WM (2005) Robust nonblocking supervisory control of discrete-event systems. IEEE Trans Autom Control 50(12):2015–2021

Cai K, Wonham WM (2014) Supervisor localization of discrete-event systems based on state tree structures. IEEE Trans Autom Control 59(5):1329–1335

Cai K, Wonham WM (2015) New results on supervisor localization, with case studies. Discret Event Dyn Syst 25(1–2):203–226

Cai K, Zhang R-Y, Wonham WM (2015) Relative observability of discrete-event systems and its supremal sublanguages. IEEE Trans Autom Control 60(3):659–670

Cai K, Zhang R-Y, Wonham WM (2016) Relative observability and coobservability of timed discrete-event systems. IEEE Trans Autom Control 61(11):3382–3395

Cai K, Zhang R-Y, Wonham WM (2017) Correction to "Relative observability of discrete-event systems and its supremal sublanguages". IEEE Trans Autom Control 62(1):511

Cai K, Zhang R-Y, Wonham WM (2017) Relative coobservability in decentralized supervisory control of discrete-event systems. Int J Control. http://dx.doi.org/10.1080/00207179.2017.1397754

Cai K, Zhang R-Y, Wonham WM (2018) Characterizations and effective computation of supremal relatively observable sublanguages. Discret Event Dyn Syst 28(2):269–287

Chao W-J, Gan Y-M, Wang Z-O, Wonham WM (2013) Modular supervisory control and coordination of state tree structures. Int J Control 86(1):9–21

Chen PCY, Wonham WM (2001) Stable supervisory control of flexible manufacturing systems with fixed supply and demand rates. Int J Product Res Spec Issue Model Specific Anal Manuf Syst 39(2):347–368

Chen PCY, Wonham WM (2002) Real-time supervisory control of a processor for non-preemptive execution of periodic tasks. Real Time Syst J 23(3):183–208

Chen Y-Z, Wonham WM (2007) Supervisory control based on vector synchronous product of automata. Stud Inf Control 16(1):7–18

Feng L, Wonham WM, Thiagarajan PS (2007) Designing communicating transaction processes by supervisory control theory. Formal Methods Syst Des 30(2):117–141

Gohari P, Wonham WM (2000) On the complexity of supervisory control design in the RW framework. IEEE Trans Syst Man Cybern Part B Cybern Spec Issue Discret Syst Control 30(5):643–652

Gohari P, Wonham WM (2003) Reduced supervisors for timed discrete-event systems. IEEE Trans Autom Control 48(7):1187–1198

Gohari P, Wonham WM (2005) Efficient implementation of fairness in discrete-event systems using queues. IEEE Trans Autom Control 50(11):1845–1849

Hashtrudi Zad S, Kwong RH, Wonham WM (1999) Supremum operators and computation of supremal elements in system theory. SIAM J Control Optim 37(3):695–709
Hashtrudi Zad S, Kwong RH, Wonham WM (1999) Fault diagnosis in finite-state automata and timed discrete-event systems. In: D. Miller, L. Qiu (eds) Topics in control and its applications, pp 81–105. Springer, Berlin
Hashtrudi Zad S, Kwong RH, Wonham WM (2003) Fault diagnosis in discrete-event systems: framework and model reduction. IEEE Trans Autom Control 48(7):1199–1212
Hashtrudi Zad S, Kwong RH, Wonham WM (2005) Fault diagnosis in discrete-event systems: incorporating timing information. IEEE Trans Autom Control 50(7):1010–1015
Hashtrudi Zad S, Moosaei M, Wonham WM (2005) On computation of supremal controllable, normal sublanguages. Syst Control Lett 54(9):871–876
Kozak P, Wonham WM (1995) Fully decentralized solutions of supervisory control problems. IEEE Trans Autom Control 40(12):2094–2097
Kozak P, Wonham WM (1996) Design of transaction management protocols. IEEE Trans Autom Control 41(9):1330–1335
Lawford M, Wonham WM (1995) Equivalence preserving transformations for timed transition models. IEEE Trans Autom Control 40(7):1167–1179
Li Y, Wonham WM (1988) On supervisory control of real-time discrete-event systems. Inf Sci 46(3):159–183
Li Y, Wonham WM (1994) Control of vector discrete-event systems: II - controller synthesis. IEEE Trans Autom Control 39(3):512–531
Li Y, Wonham WM (1995) Concurrent vector discrete-event systems. IEEE Trans Autom Control 40(4):628–638
Lin F, Vaz A, Wonham WM (1988) Supervisor specification and synthesis for discrete event systems. Int J Control 48(1):321–332
Lin F, Wonham WM (1991) Verification of nonblocking in decentralized supervision. Control Theory Adv Technol 7(1):19–29
Lin F, Wonham WM (1995) Supervisory control of timed discrete event systems under partial observation. IEEE Trans Autom Control 40(3):558–562
Ma C, Wonham WM (2006) Nonblocking supervisory control of state tree structures. IEEE Trans Autom Control 51(5):782–793
Nabavi A, Iravani R, Afzalian A, Wonham WM (2009) Discrete-event systems supervisory control for a dynamic flow controller. IEEE Trans Power Delivery 24(1):219–230
Queiroz Max H, Cury José E, Wonham WM (2005) Multitasking supervisory control of discrete-event systems. Discret Event Dyn Syst 15(4):375–395
Ramadge PJ, Wonham WM (1989) The control of discrete event systems. Proc IEEE Special Issue Discret Event Dyn Syst 77(1):81–98
Rudie K, Wonham WM (1990) The infimal prefix-closed and observable superlanguage of a given language. Syst Control Lett 15:361–371
Saadatpoor A, Wonham WM (2007) State-based control of timed discrete-event systems using binary decision diagrams. Syst Control Lett 56(1):62–74
Su R, Wonham WM (2005) Global and local consistencies in distributed fault diagnosis for discrete-event systems. IEEE Trans Autom Control 50(12):1923–1935
Su R, Wonham WM (2006) Hierarchical fault diagnosis for discrete-event systems under global consistency. Discret Event Dyn Syst 16(1):39–70
Thistle JG, Wonham WM (1986) Control problems in a temporal logic framework. Int J Control 44(4):943–976
Thistle JG, Wonham WM (1994) Control of infinite behavior of finite automata. SIAM J Control Optim 32(4):1075–1097
Thistle JG, Wonham WM (1994) Supervision of infinite behavior of discrete-event systems. SIAM J Control Optim 32(4):1098–1113
Ushio T, Li Y, Wonham WM (1992) Concurrency and state feedback in discrete-event systems. IEEE Trans Autom Control 37(8):1180–1184

Wong KC, Wonham WM (1998) Modular control and coordination of discrete event systems. Discret Event Dyn Syst 8(3):247–297

Zhang R-Y, Cai K, Gan Y, Wang Z, Wonham WM (2013) Supervision localization of timed discrete-event systems. Automatica 49(9):2786–2794

Zhang R-Y, Cai K, Gan Y, Wang Z, Wonham WM (2016) Distributed supervisory control of discrete-event systems with communication delay. Discret Event Dyn Syst 26(2):263–293

Zhang R-Y, Cai K, Wonham WM (2016) Delay-robustness in distributed control of timed discrete-event systems based on supervisor localization. Int J Control 89(10):2055–2072

Zhang R-Y, Cai K, Wonham WM (2017) Supervisor localization of discrete-event systems under partial observation. Automatica 81(7):142–147

Conference Papers

Abdelwahed S, Wonham WM (1999) Interacting discrete event systems. In: Proceedings of the thirty-seventh annual Allerton conference on communication, control and computing, Allerton IL, pp 85–92

Abdelwahed S, Wonham WM (2002) Supervisory control of interacting discrete event systems. In: Proceedings of the forty-first IEEE conference on decision and control, Las Vegas, pp 1175–1180

Abdelwahed S, Wonham WM (2003) Blocking detection in discrete-event systems. In: Proceedings of the American control conference, Denver, pp 1673–1678

Afzalian AA, Saadatpoor A, Wonham WM (2006) Discrete-event system modeling and supervisory control for under-load tap-changing transformers. In: IEEE international conference on control applications and IEEE international symposium on computer-aided control systems design and IEEE international symposium on intelligent control (CCA/CACSD/ISIC '06), Tech University München, Munich, Germany, pp 1867–1872

Afzalian AA, Wonham WM (2006) Discrete-event system supervisory controller design for an electrical power transmission network. In: Fourteenth Iranian conference on electrical engineering (ICEE '06), Tehran

Bourdon SE, Lawford M, Wonham WM (2002) Robust nonblocking supervisory control of discrete-event systems. In: Proceedings of the American control conference, Anchorage, pp 730–735

Bourdon SE, Wonham WM, Lawford M (2000) Invariance under scaling of time bounds in timed discrete-event systems. In: Proceedings of the thirty-eighth annual Allerton conference on communications, control, and computing, Allerton, IL, October 2000, pp 1145–1154

Brandin BA, Benhabib B, Wonham WM (1991) Discrete event system supervisory control applied to the management of manufacturing workcells. In: Proceedings of the seventh international conference on computer-aided production engineering, Cookeville TN USA, August 1991. Elsevier, Amsterdam, pp 527–536

Brandin BA, Wonham WM (1993) Modular supervisory control of timed discrete-event systems. In: Proceedings of the thirty-second IEEE conference on decision and control, New York, pp 2230–2235

Cai K, Wonham WM (2012) New results on supervisor localization, with application to multi-agent formations. In Treviño AR, Mellado EL, Lesage J-J, Silva M (eds) Preprints, eleventh international conference on discrete event systems (WODES '12), Guadalajara, Mexico, 3–5 Oct 2012, pp 233–238

Cai K, Wonham WM (2012) Supervisor localization of discrete-event systems based on state tree structures. In: Proceedings of the fifty-first IEEE conference on decision and control, Maui, Hawaii, USA, 10–13 Dec 2012, pp 5822–5827

Cai K, Zhang R-Y, Wonham WM (2013) Supervision localization of timed discrete-event systems. In: Proceedings of the American control conference, Washington, DC, USA, 17–19 June 2013, pp 5686–5691

Chao W, Gan Y, Wonham WM, Wang Z (2013) Nonblocking supervisory control of flexible manufacturing systems based on state tree structures. In: Workshop on discrete-event control systems,

Xidian University, Xi'an, China, 16 February 2011 (6 pp). See also full version. In: Li Z, Al-Ahmari AM (eds) Formal methods in manufacturing systems: recent advances. IGI Global, Engineering Science Reference, pp 1–19

Chao W-J, Gan Y-M, Wang Z-A, Wonham WM (2012) Decentralized supervisory control of discrete-event systems based on state tree structures. In: Proceedings of the thirty-first Chinese control conference (CCC '12), Hefei, China, 25–27 July 2012, pp 2184–2190

Chao W-J, Gan Y-M, Wang Z-A, Wonham WM (2012) Representation of supervisory controls using state tree structures, binary decision diagrams, automata, and supervisor reduction. In: Proceedings of the twenty-fourth IEEE Chinese control and decision conference (CCDC '12), Taiyuan, China, 23–25 May 2012, pp 45–50

Chen CYP, Wonham WM (1998) Non-preemptive scheduling of periodic tasks: a discrete-event control approach. In: Proceedings of the fifth international conference on control, automation, robotics and vision, Nanyang Technological University, Singapore, 8–11 Dec 1998, pp 1674–1678

Chen CYP, Wonham WM (1999) Real-time supervisory control of a processor for non-preemptive execution of periodic tasks. In: Proceedings of the IFAC world congress, Beijing, vol J, pp 13–18

Chen PCY, Wonham WM (2002) Computation of complete schedules for execution of periodic tasks on a processor with preemption. In: Proceedings of the fourth Asia control conference, Singapore, 25–27 Sept 2002, pp 1904–1909

Chen SL, Wonham WM (1995) Existence and design of supervisors for vector discrete-event systems. In: Proceedings of the Canadian conference on electrical and computer engineering, Montreal, pp 805–808

Chen SL, Wonham WM (1995) Supervisory control of finite automata under dynamic predicate specifications. In: Proceedings of the thirty-third annual Allerton conference on communication, control and computing, University of Illinois, pp 501–509

Chen YZ, Wonham WM (2006) Supervisory control based on vector synchronous product of automata. Multi-conference on computational engineering in systems applications, Beijing, pp 4–6

de Queiroz MH, Cury JER, Wonham WM (2004) Multi-tasking supervisory control of discrete-event systems. In: Proceedings of the 7th international workshop on discrete event systems (WODES '04), Reims, pp 175–180

Dietrich P, Malik R, Wonham WM, Brandin BA (2001) Implementation considerations in supervisory control. In: Caillaud B, Xie X (eds) Proceedings of the symposium on the supervisory control of discrete event systems (SCODES '01), INRIA, Paris, July 2001, pp 27–38

Feng L, Wonham WM, Thiagarajan PS (2004) Designing communicating transaction processes by supervisory control theory. In: Proceedings of the forty-second annual Allerton conference on communications, control, and computing, Allerton, pp 1060–1069

Gohari P, Wonham WM (2000) Reduced supervisors for timed discrete-event systems. In: Boel R, Stremersch G (eds) discrete event systems: analysis and control [Proceedings of the WODES '00], Kluwer, pp 119–130

Hashtrudi Zad S, Kwong RH, Wonham WM (1997) Supremum operators and computation of supremal elements in system theory. In: Proceedings of the thirty-sixth IEEE conference on decision and control, San Diego, CA, pp 2946–2951

Hashtrudi Zad S, Kwong RH, Wonham WM (1998) Fault diagnosis in discrete-event systems: framework and model reduction. In: Proceedings of the thirty-seventh IEEE conference on decision and control, Tampa FL, pp 3769–3774

Hashtrudi Zad S, Kwong RH, Wonham WM (1999) Fault diagnosis in timed discrete-event systems. In: Proceedings of the thirty-eighth IEEE conference on decision and control, Phoenix, AZ, pp 1756–1761

Hashtrudi Zad S, Kwong RH, Wonham WM (2000) Fault diagnosis and consistency in hybrid systems. In: Proceedings of the thirty-eighth annual Allerton conference on communications, control, and computing, Allerton, IL, Oct 2000, pp 1135–1144

Ho T-J, Wonham WM (1992) A framework for timed discrete-event systems. In: Proceedings of the thirtieth annual Allerton conference on communication, control and computing, University of Illinois, 1992, pp 650–651

Jiao T, Gan Y, Xiao G, Wonham WM (2016) Exploiting symmetry of state tree structures for discrete-event systems with parallel components. In: Proceedings of the thirteenth international workshop on discrete event systems (WODES '16), Xi'an, China, May 30–June 1, 2016, pp 97–102

Jiao T, Gan Y, Yang X, Wonham WM (2015) Exploiting symmetry of discrete-event systems with parallel components by relabeling. In: Proceedings of the IEEE region 10 conference (TENCON '15), Macau, 1–4 Nov, 2015, pp 1–4

Lawford M, Ostroff JS, Wonham WM (1996) Model reduction of modules for state-event temporal logics. Workshop on application of formal methods to system development: telecommunications, VLSI and real-time computerized control systems, Jacques Cartier centre, Montreal, pp 281–287

Lawford M, Ostroff JS, Wonham WM (1996) Model reduction of modules for state-event temporal logics. In R. Gotzhein, J. Bredereke (eds) Formal description techniques IX, Chapman and Hall, London, 1996, pp 263–278. [IFIP TC6/6.1 international conference on formal description techniques IX/protocol specification, testing and verification XVI, Kaiserslautern, Germany, 8–11 Oct 1996]

Lawford M, Wonham WM (1992) Equivalence preserving transformations for timed transition models. In: Proceedings of the thirty-first IEEE conference on decision and control, Tucson, Arizona, pp 3350–3356

Lawford M, Wonham WM (1993) Supervisory control of probabilistic discrete event systems. In: Proceedings of the thirty-sixth Midwest symposium on circuits and systems, Detroit, MI, pp 327–331

Lawford M, Wonham WM, Ostroff JS (1994) State-event observers for labeled transition systems. In: Proceedings of the thirty-third IEEE conference on decision and control, Lake Buena Vista, FL, pp 3642–3648

Leduc RJ, Brandin BA, Wonham WM (2000) Hierarchical interface-based non-blocking verification. In: Canadian conference on electrical and computer engineering (CCECE '00), Halifax, 7–10 May 2000, pp 1–6

Leduc RJ, Wonham WM (1995) Discrete event systems modeling and control of a manufacturing testbed. In: Proceedings of the Canadian conference on electrical and computer engineering, Montreal, pp 793–796

Leduc RJ, Wonham WM (1995) PLC implementation of a DES supervisor for a manufacturing testbed. In: Proceedings of the thirty-third annual Allerton conference on communication, control and computing, University of Illinois, pp 519–528

Li Y, Wonham WM (1987) Supervisory control of real-time discrete-event systems. In: Proceedings of the American control conference, Minneapolis, pp 1715–1720

Li Y, Wonham WM (1988) Deadlock issues in supervisory control of discrete-event systems. In: Proceedings of the conference on information sciences and systems, Department of Electrical Engineering, Princeton University, 1988, pp 57–63

Li Y, Wonham WM (1989) Strict concurrency and nondeterministic control of discrete-event systems. In: Proceedings of the twenty-eighth IEEE conference on decision and control, New York, pp 2731–2736

Lin F, Wonham WM (1985) On the computation of supremal controllable sublanguages. In: Proceedings of the twenty-third annual Allerton conference on communication, control and computing, University of Illinois, Urbana, Oct 1985, pp 942–950

Lin F, Wonham WM (1988) Decentralized control and coordination of discrete-event systems. In: Proceedings of the twenty-seventh IEEE conference on decision and control, New York, pp 1125–1130

Lawford M, Wonham WM (1992) An application of real-time transformational equivalence. In: Proceedings of the conference on information sciences and systems, vol 1. Princeton University, Princeton, pp 233–238

Ma C, Wonham WM (2008) STSLib and its application to two benchmarks. In: Proceedings of the eighth international workshop on discrete event systems (WODES '08), Göteborg, Sweden, May 28–30, 2008, pp 119–124

Minhas R, Wonham WM (1999) Modelling of timed discrete event systems. In: Proceedings of the thirty-seventh annual Allerton conference on communication, control and computing, Allerton, IL, pp 75–84

Minhas R, Wonham WM (2003) Online supervision of discrete event systems. In: Proceedings of the American control conference, Denver, pp 1685–1690

Ostroff JS, Wonham WM (1985) A temporal logic approach to real time control. In: Proceedings of the twenty-fourth IEEE conference on decision and control, New York, pp 656–657

Ostroff JS, Wonham WM (1987) Modelling, specifying and verifying real-time embedded computer systems. In: Proceedings of the eighth real-time systems symposium, New York, pp 124–132

Ostroff JS, Wonham WM (1987) State machines, temporal logic and control: a framework for discrete event systems. In: Proceedings of the twenty-sixth IEEE conference on decision and control, New York, pp 681–686

O'Young SD, Wonham WM (1989) Object-oriented computation and simulation of large-scale discrete event systems. In: Proceedings of the twenty-seventh annual Allerton conference on communication, control and computing, University of Illinois, pp 945–954

Ramadge PJ, Wonham WM (1986) Modular feedback logic for discrete event systems. In: Fourth IFAC/IFORS symposium, large scale systems: theory and applications, Zurich, August 1986. In: Geering HP, Mansour M (eds) Large scale systems: theory and applications, vol 1. Pergamon Press, Oxford, pp 83–88

Ramadge PJ, Wonham WM (1986) Modular supervisory control of discrete event systems. In: Seventh international conference on analysis and optimization of systems, Antibes, June 1986. In: Bensoussan A, Lions JL (eds) Analysis and optimization of systems, vol 83. Lecture notes on control and information sciences (LNCIS). Springer, New York, pp 202–214

Rudie K, Wonham WM (1990) Supervisory control of communicating processes. In: Tenth international IFIP WG 6.1 symposium on protocol specification, testing and verification, Ottawa. In: Logrippo L, Probert RL, Ural H (eds) Protocol specification, testing and verification, X. Elsevier (North-Holland), 1990, pp 243–257

Rudie K, Wonham WM (1991) Think globally, act locally: decentralized supervisory control. In: Proceedings of the American control conference, Boston, June 1991, pp 898–903

Rudie K, Wonham WM (1992) Protocol verification using discrete-event systems. In: Proceedings of the thirty-first IEEE conference on decision and control, Tucson, Arizona, pp 3770–3777

Saadatpoor A, Afzalian AA, Wonham WM (2006) Hierarchical control of under-load tap-changing transformers using DES. Third international conference on informatics in control, automation and robotics (ICINCO '06), Setúbal, Portugal

Saadatpoor A, Afzalian AA, Wonham WM (2006) Timed discrete-event systems supervisory control for under-load tap-changing transformers. In: Third international conference on informatics in control, automation and robotics (ICINCO '06), Setúbal, Portugal

Saadatpoor A, Ma C, Wonham WM (2008) Supervisory control of timed state tree structures. In: Proceedings of the American control conference, Seattle, Washington, pp 11–13

Saadatpoor A, Wonham WM (2004) State based control of timed discrete event systems using binary decision diagrams. In: Proceedings of the forty-second annual Allerton conference on communications, control, and computing, Allerton, pp 1070–1079

Saadatpoor A, Wonham WM (2007) Supervisor state size reduction for timed discrete-event systems. In: Proceedings of the American control conference, New York, pp 4280–4284

Su R, Wonham WM (2000) Decentralized fault diagnosis for discrete-event systems. In: Conference on information sciences and systems, Princeton University, 15–17 Mar 2000, pp TP1.1–TP1.6

Su R, Wonham WM (2002) Distributed diagnosis for qualitative systems. In: Proceedings of the international workshop on discrete event systems (WODES '02), Zaragoza, Spain, pp 169–174

Su R, Wonham WM (2002) Probabilistic reasoning in distributed diagnosis for qualitative systems. In: Proceedings of the forty-first IEEE conference on decision and control, Las Vegas, pp 429–434

Su R, Wonham WM (2004) A model of component consistency in distributed diagnosis. In: Proceedings of the seventh international workshop on discrete event systems (WODES '04), Reims, pp 427–432

Su R, Wonham WM (2004) An algorithm to achieve local consistency in distributed systems. In: Proceedings of the forty-third IEEE conference on decision and control, Paradise Islands, The Bahamas, pp 998–1003

Su R, Wonham WM (2004) Distributed diagnosis under global consistency. In: Proceedings of the forth-third IEEE conference on decision and control, Paradise Islands, The Bahamas, pp 525–530

Su R, Wonham WM (2004) Hierarchical distributed diagnosis under global consistency. In: Proceedings of the seventh international workshop on discrete event systems (WODES '04), Reims, pp 157–162

Su R, Wonham WM (2004) Undecidability of termination of CPLC for computing supremal local support. In: Proceedings of the forty-second annual Allerton conference on communications, control, and computing, Allerton, pp 604–613

Thistle JG, Wonham WM (1985) On the use of temporal logic in control theory. In: Proceedings of the twenty-third annual Allerton conference on communication, control and computing, University of Illinois, pp 961–970

Thistle JG, Wonham WM (1987) Supervisory control with infinite-string specifications. In: Proceedings of the twenty-fifth annual Allerton conference on communication, control and computing, vil 1, University of Illinois, pp 327–334

Thistle JG, Wonham WM (1988) On the synthesis of supervisors subject to ω-language specifications. In: Proceedings of the conference on information sciences and systems, Department of Electrical Engineering, Princeton University, pp 440–444

Thistle JG, Wonham WM (1992) Control of ω-automata, Church's problem, and the emptiness problem for tree ω-automata. In: Proceedings of the computer science logic (CSL '91), Institut fur Informatik und angewandte Mathematik, Universitat Bern, Bern, Switzerland, Oct 1991. In: Börger E, Jäger G, K Büning H, Richter MM (eds) Lecture notes in computer science (LNCS), vol 626. Springer, Berlin, pp 367–381

Ushio T, Li Y, Wonham WM (1989) Basis feedback, weak interaction, and concurrent well-posedness in concurrent discrete-event systems. In: Proceedings of the twenty-eighth IEEE conference on decision and control, New York, pp 127–131

Wong KC, Wonham WM (1992) Hierarchical and modular control of discrete-event systems. In: Proceedings of the thirtieth annual Allerton conference on communication, control and computing, University of Illinois, pp 614–623

Wong KC, Wonham WM (1993) Hierarchical control of timed discrete-event systems. In: Proceedings of the second European control conference, Groningen, The Netherlands, pp 509–512

Wong KC, Wonham WM (1996) Modular control and coordination of discrete-event systems. In: Proceedings of the fourth IEEE Mediterranean symposium on new directions in control and automation, Chania, Crete, Greece, pp 595–599

Wong KC, Wonham WM (1996) Modular control and coordination of a transfer line: a tutorial example. In: Pre-Proceedings of the workshop on theoretical problems on manufacturing systems design and control, University of Minho and University of Parma, Lisbon, pp 21–29

Wong KC, Wonham WM (2000) On the computation of observers in discrete-event systems. In: Conference on information sciences and systems, Princeton University, 15–17 Mar 2000, pp TP1.7–TP1.12

Wonham WM (1986) On control of discrete event systems. In: Seventh international symposium on the mathematical theory of networks and systems (MTNS '85), Stockholm, (1985) In: Byrnes CI, Lindquist A (eds) Computational and combinatorial methods in systems theory. North-Holland, Amsterdam, pp 159–174

Wonham WM (1986) Some remarks on control and computer science. In: Workshop on future directions in system theory and applications, University of Santa Clara

Wonham WM (1987) Logic and language in control theory. In: Proceedings of the twenty-fifth annual Allerton conference on communication, control and computing, vol 1, University of Illinois, pp 1–3

Wonham WM (1988) A control theory for discrete-event systems. In: Denham MJ, Laub AJ (eds) Advanced computing concepts and techniques in control engineering, NATO ASI series, vol F47. Springer, Berlin, pp 129–169

Wonham WM (1991) Some current research directions in control of discrete-event systems. In: Presented at European control conference (ECC '91), Grenoble, July 1991. [Preprint available separately from Proc.]

Wonham WM (1995) Supervisory control automata for discrete event systems. In: Preprints, ADEDOPS workshop (Analysis and design of event-driven operations in process systems), Centre for Process Systems Engineering, Imperial College of Science Technology and Medicine, London UK, (34 pp)

Wonham WM (2000) Supervisory control of discrete-event systems: an introduction. In: Proceedings of the IEEE international conference on industrial technology (ICIT '00), Goa, India, 19–22 Jan 2000, pp 474–479

Wonham WM (2001) Supervisory control of discrete-event systems. In Datta AK, Deb A, Sengupta S (eds) Proceedings of the international conference on control, instrumentation and information communication (CIIC '01), Department of Applied Physics, University of Calcutta, 13–15 Dec 2001, pp 321–330

Wonham WM (1989) A language-based control theory of discrete-event systems. Preprints, shell conference on logistics, Appeldoorn, The Netherlands (1988). In: van Rijn CFH (ed) Logistics - where ends have to meet. Pergamon, Oxford (UK), pp 158–169

Wonham WM, Gohari-Moghadam P (1998) A linguistic framework for controlled hierarchical DES. In: International workshop on discrete event systems (WODES '98), IEE, London, pp 207–212

Wonham WM, Ramadge PJ (1984) On modular synthesis of supervisory controls for discrete event processes. In: Proceedings of the international conference on computers, systems and signal processing, IEEE and I.I.Sc., Bangalore, Dec 1984, pp 500–504

Wu W, Su H, Chu J, Wonham WM (2002) Colored Petri net control of OR-logic. In: Proceedings of the IEEE international symposium on intelligent control, Vancouver, 27–30 Oct 2002, pp 795–800

Zhang R-Y, Cai K, Gan Y-M, Wang Z, Wonham WM (2012) Checking delay-robustness of distributed supervisors of discrete-event systems. In: Proceedings of the IET international conference on information science and control engineering (ICISCE '12), Shenzhen, China, 7–9 Dec 2012, pp 350–355

Zhang R-Y, Cai K, Wonham WM (2014) Delay-robustness in distributed control of timed discrete-event systems based on supervisor localization. In: Proceedings of the fifty-third IEEE conference on decision and control, Los Angeles, CA, 15–17 Dec 2014, pp 6719–6724

Zhang R-Y, Gan Y, Wonham WM, Wang Z (2011) Distributed maximally permissive nonblocking control of flexible manufacturing systems. In: Workshop on discrete-event control systems, Xidian University, Xi'an, China, 16 Feb 2011 (6 pp)

Zhang R-Y, Gan Y-M, Wang Z-A, Wonham WM (2012) On computation of distributed supervisory controllers in discrete-event systems. In: Proceedings of the twenty-fourth IEEE Chinese control and decision conference (CCDC '12), Taiyuan, China, 23–25 May 2012, pp 781–786

Zhong H, Wonham WM (1988) On hierarchical control of discrete-event systems. In: Proceedings of the conference on information sciences and systems, Department of Electrical Engineering, Princeton University, 1988, pp 64–70

Graduate Theses

Abdelwahed S (2002) Interacting discrete-event systems: modeling, verification, and supervisory control. PhD thesis, Department of Electrical and Computer Engineering, University of Toronto

Caulder IEH (1991) Applications of decentralized hierarchical des control to telecommunications network protocols. MASc thesis, Department of Electrical Engineering, University of Toronto
Chan RSW (1984) Simulation and modelling of a supervisory control system using concurrent euclid. MASc thesis, Department of Electrical Engineering, University of Toronto
Cheung B (2008) Computing hierarchical nonblocking discrete-event systems. MASc thesis, Department of Electrical and Computer Engineering, University of Toronto
Chun TYL (1996) Diagnostic supervisory control: a DES approach. MASc thesis, Department of Electrical and Computer Engineering, University of Toronto
Gohari-Moghadam P (1998) A linguistic framework for controlled hierarchical DES. MASc thesis, Department of Electrical and Computer Engineering, University of Toronto
Gohari-Moghadam P (2002) Fair supervisory control of discrete-event systems. PhD thesis, Department of Electrical and Computer Engineering, University of Toronto
Guan Y-C (1997) Implementation of hierarchical observer theory. MASc thesis, Department of Electrical and Computer Engineering, University of Toronto
Hashtrudi Zad S (1999) Fault diagnosis in discrete-event and hybrid systems. PhD thesis, Department of Electrical and Computer Engineering, University of Toronto
Ho T-J (1997) The control of real-time discrete-event systems subject to predicate-based constraints. PhD thesis, Department of Electrical and Computer Engineering, University of Toronto
Izadian S (2013) Supply chain operation modeling and automation using untimed and timed state tree structures. MASc thesis, Department of Electrical and Computer Engineering, University of Toronto
Lawford MS (1992) Transformational equivalence of timed transition models. MASc thesis, Department of Electrical Engineering, University of Toronto
Lawford MS (1997) Model reduction of discrete real-time systems. PhD thesis, Department of Electrical and Computer Engineering, University of Toronto
Leduc RJ (1996) PLC implementation of a DES supervisor for a manufacturing testbed: an implementation perspective. MASc thesis, Department of Electrical and Computer Engineering, University of Toronto
Lin F (1984) Supervisor synthesis for discrete event processes. MASc thesis, Department of Electrical Engineering, University of Toronto
Li Y (1986) Supervisory control of real-time discrete event systems. MASc thesis, Department of Electrical Engineering, University of Toronto
Liao J (1992) Hierarchical control in vector discrete-event systems. MASc thesis, Department of Electrical Engineering, University of Toronto
Ma C (1999) A computational approach to top-down hierarchical supervisory control of DES. MASc thesis, Department of Electrical and Computer Engineering, University of Toronto
Minhas RS (2002) Complexity reduction in discrete event systems. PhD thesis, Department of Electrical and Computer Engineering, University of Toronto
Pai T-M (1990) Real time implementation of discrete-event control systems. MASc thesis, Department of Electrical Engineering, University of Toronto
Qu X-Q (2013) Efficient abstractions in hierarchical supervisory control of discrete-event systems. MASc thesis, Department of Electrical and Computer Engineering, University of Toronto
Ravichandran VJ (2002) Distributed diagnosis for state-based discrete-event systems. MASc thesis, Department of Electrical and Computer Engineering, University of Toronto
Rudie K (1988) Software for the control of discrete-event systems: a complexity study. MASc thesis, Department of Electrical Engineering, University of Toronto
Saadatpoor A (2004) State-based control of timed discrete-event systems using binary decision diagrams. MASc thesis, Department of Electrical and Computer Engineering, University of Toronto
Saadatpoor A (2009) Timed state-tree structures: supervisory control and fault diagnosis. PhD thesis, Department of Electrical and Computer Engineering, University of Toronto
Shifman J (1987) A hierarchical approach to path planning for robots. MASc thesis, Department of Electrical Engineering, University of Toronto

Su R (2000) Decentralized fault diagnosis for discrete-event systems. MASc thesis, Department of Electrical and Computer Engineering, University of Toronto

Thistle JG (1985) Control problems in a temporal logic setting. MASc thesis, Department of Electrical Engineering, University of Toronto

Thistle JG (1991) Control of infinite behaviour of discrete-event systems. PhD thesis, Department of Electrical Engineering, University of Toronto

Wong RK (1990) State-based discrete-event modeling and control of concurrent systems. MASc thesis, Department of Electrical Engineering, University of Toronto

Yan L-Y (2011) State-based control of discrete-event systems with observational abstractions. MASc thesis, Department of Electrical and Computer Engineering, University of Toronto

Yao Y-P (2012) Sufficient conditions for output regulation on metric and banach spaces - a set-theoretic approach. MASc thesis, Department of Electrical and Computer Engineering, University of Toronto

Yi S (2004) Hierarchical supervision with nonblocking. MASc thesis, Department of Electrical and Computer Engineering, University of Toronto

Yuen C-Y (1992) Control synthesis for timed discrete event systems. MASc thesis, Department of Electrical Engineering, University of Toronto

Zhang X-Q (1997) Control of Boolean discrete-event systems. MASc thesis, Department of Electrical and Computer Engineering, University of Toronto

Zhang Y-Q (1996) Software for state-event observation theory and its application to supervisory control. MASc thesis, Department of Electrical and Computer Engineering, University of Toronto

Zhu L (1993) Control theory of stochastic discrete event systems. MASc thesis, Department of Electrical and Computer Engineering, University of Toronto

Index

Symbols
C-observable, 279
I-fold $L_m(\mathbf{G})$-observer, 188
I-fold nonconflicting, 188
L_i-observer, 266
$L_m(\mathbf{G})$-observer, 185
L_m-observer, 305
L_{voc}-observer, 230
ρ-supported, 83

A
Activity
  transition, 412
  transition graph, *see* transition, graph, activity
Admissible, 301
  agent, 246–249
AGV, 118, 163, 184, 202
Alphabet, 45, 55–57, 70, 71, 86, 87, 92, 93, 99, 106, 118, 120, 121, 158, 208, 210, 258, 303, 307, 308, 349, 365, 380, 412, 417, 419, 420, 423–425
  event label
    high-level, 210
  input, 214, 215
  output, 55, 56, 208, 214
    extended, 211
  sub-, 155, 158, 210
  transition label, 70
Ambient, 279
Arden's rule, 67–69
ATG, *see* transition, graph, activity
Automaton, 48, 49, 55, 57–60, 61, 84, 93, 109, 148, 193, 215, 358, 361, 402, 411, 429
  canonical, 49
    Moore, 245
  controllable, 111, 113, 136, 148–150, 152, 153, 158–160, 440
  coreachable, 61, 62, 88, 89, 111, 136
  Moore, 55, 56
  nonblocking, 61, 62, 89, 101–104, 107, 111, 121, 148, 149, 194, 280, 287, 303, 308
  non-coreachable, 89
  product, 58, 148
  reachable, 57, 58, 61, 62, 88, 89, 93, 94, 110, 111, 121, 136, 148
  timed, 443
  trim, *see* DES,trim
  uncontrollable, 432

B
Behavior
  admissible, 114
  closed, 61, 63, 64, 70, 88, 101, 103, 120, 121, 126, 159, 240, 243, 296, 307, 342, 349, 350, 373, 380, 393, 416–418, 421, 424, 427, 431, 436
  closed-loop, 110, 217
  controlled, 120, 127, 141, 195, 217, 287, 306, 347, 355, 404, 406, 407, 432
  marked, 61, 63, 64, 70, 78, 88, 101, 103, 114, 120, 121, 159, 228, 229, 240, 243, 280, 282, 307, 342, 416–418, 421, 424, 427, 436
  optimal controlled, 141, 403
  uncontrolled, 207, 421
Behaviorally equivalent, 150
Big Factory, 154

W. M. Wonham and K. Cai, *Supervisory Control of Discrete-Event Systems*, Communications and Control Engineering,
https://doi.org/10.1007/978-3-319-77452-7

Bijective, 12, 16, 52, 55
Bisimulation, 337
Blocking, 65, 91–93, 96, 97, 150, 152, 159, 194, 197, 198, 246, 305, 307, 308
sequence, 403, 408, 409
set, 246
state, 404

C
Canonical
automaton, *see* automaton, canonical
factorization, 13
projection, *see* projection, canonical
recognizer, 48–57, 58, 66, 210
Cat and Mouse, 137
Catenation, *see* string, catenation
Catenative, 90
Causal, 70
Causal output mapping, 69
Cell, 8, 10, 51, 52, 55, 58, 72, 73
Chains of regular languages, 78
Channel, 291
as encoder, 295
Closed
behavior, *see* behavior, closed
language, 48, 51, 52, 61, 101–107, 113, 208, 216–221, 235, 236, 247, 262–266, 271, 272, 275, 276, 280–282, 295–301, 313, 342, 345, 349–351, 424–429
sublanguage, 208, 228, 262, 264, 273, 275, 282, 297
Closure, 100, 159, 262, 264, 265, 296, 421
algebraic property, 261
boolean, 367
language, 48
prefix, 48, 101, 208, 231, 234
transitive, 11
Codomain, 12
Compatible, 349
Complexity reduction, 335
Composition, 22
Congruence, 15, 20
coarse, 20
Congruence of a dynamic system, 15
Conjunction, 148
Consistency, 223
property, 229
Control cover, 203
Control pattern, 101
Control VDES, 392
Convergence, 78
Coordinating modular supervisor, 167
Coordination
efficient, 331
Coordinator, 181, 331
coordinators, 196
Coreachable state subset, *see* state, coreachable, subset
Coset, 7

D
Deadly Embrace, 150
Decentralized modular synthesis, 147
DES, 86
complexity, 334
controllable, *see* automaton, controllable
isomorphic, 121, 152, 159, 160, 197
trim, 61, 62, 88, 89, 93, 100, 111, 113, 114, 117, 148–150, 153, 159, 262, 440, 441
vocalized, 213
Descending chain, 78
Determinism
structural, 321
Dining Philosophers, 194, 202
Disabled-event map
high-level, 216
low-level, 216
Distributed control, 291
Distributed supervision, 200
Domain, 12, 213
Downward continuous, 19
DSFBC, 350, 386
balanced, 350, 351, 382
optimal, 382, 393, 394
Dynamic state feedback control, *see* DSFBC
Dynamic system, 15

E
Elements
comparable, 1, 2
indistinguishable, 11
Empty sequence, *see* string, empty
Encoder, 293
Encoding, 294
Endangered Pedestrian, 430
ϵ, *see* string, empty
Event
blocking, *see* blocking
control-enabled, 121
controllable, 87, 88, 101, 112, 115–120, 122, 125–127, 137, 192, 194, 207, 210, 211, 213, 215, 217, 219–221, 226, 227, 240, 243, 257, 258, 276, 297, 299, 302,

303, 311, 346, 347, 358, 383, 386, 388, 391, 404, 406, 421, 423
high-level, 215
low-level, 216
delayable, 423
disable, 87, 112, 114–117, 121, 122, 125–127, 144, 152, 156, 158, 159, 193, 210, 215, 216, 219, 220, 222, 226, 246, 248, 258, 276, 286, 302, 303, 311, 313, 343, 353, 356–358, 371, 379, 391, 406, 407, 411, 414, 420, 421, 423
mechanism, 112
enable, 87, 94, 101, 122, 126, 148, 215, 313, 341, 343, 347, 355–357, 363, 367, 368, 371, 379, 386, 411, 414, 415, 417, 423
enablement, 380
enablement condition, 363, 380, 400
forced, 114, 125–129, 411, 425, 433, 438, 439, 443
forcible, 126
forcing, 126–128
hiding, 197
label, 65, 86, 88, 91, 207, 210, 257, 412, 421, 434
observable, 92, 258, 273, 299, 302, 303, 306, 308, 310, 311, 313
permanently enabled, 399
preempt, 126–128
preemptible, 126, 127
state-enabled, 121
uncontrollable, 87, 101, 102, 106, 119–121, 126, 127, 137, 192, 197, 207, 210, 213, 226–228, 297, 308, 343, 345, 369, 371, 379, 391, 393, 394, 404, 407, 423, 425
unobservable, 264, 265, 273, 276, 285, 286, 305–307, 311, 313, 406, 407
Event set, 311, 366, 382, 383
active, 271
controllable, 423
inactive, 271
uncontrollable, 423
Event, flag, 96
Exponentiation, 22

F

Factorization
canonical, *see* canonical, factorization
Feasible
supervisory control, *see* supervisory, control, feasible
Fiber, 12
Fixpoint, 18
Flags, 96
FTG, *see* transition, graph
Function
inverse image, 12, 90, 259
partial, 60

G

Gas Burner, 353
Generator, 60–66, 69, 70, 72, 73, 75, 84–86, 89, 94, 95, 100, 109, 111, 118, 148, 150, 197, 210, 225, 243, 258, 310, 358, 404, 414
controlled, 86
coreachable, 62
deterministic, 64, 65
Mealy, 73, 74
Moore, 208
Nerode, *see* Nerode, generator
nonblocking, 62, 113
nondeterministic, 64, 65
product, 112
reachable, 62, 63
Guideway, 310

H

Hasse diagram, 4, 11, 105
Hierarchical
aggregation, 77
consistency, 205, 207, 208, 220, 223, 224, 228, 235, 245
high-level, 249
low-level, 219
control, 77, 205, 210
structure, 205, 207
loop, 206, 215, 216
supervisory control, *see* supervisory, control, hierarchical
Hierarchical aggregation, 69

I

Iff, 2
Image, 12, 14, 210, 224, 230, 236
inverse, function, 12, 90, 259
Implementation, 109
Inadmissible
agent, 246–249
Independent functions, 22
Induced, 16
Infimal closed observable sublanguages, 295

Interrupts, 123

J
Join, 2–6, 8, 9, 17, 47, 105

K
Kanban, 96
Kernel
coarse, 13
equivalence, 12, 13, 73
fine, 13
Kleene's Theorem, 66

L
Language, 45–46
ambient, 279
closed-loop, 117, 121, 190, 215, 217
controllable, 101–104, 106, 107, 110, 111, 113, 121, 122, 148, 149, 208, 215–221, 223, 224, 234, 236, 247, 276, 281–283, 297–301, 303, 304, 311, 346, 350, 399, 425–429, 440
closed, 106
high-level, 219, 224, 228
locally, 234, 235
low-level, 217, 218, 224
supremal, 108, 116, 124, 155, 197, 208, 237, 432, 436, 438
empty, 46, 66, 261
legal, 118, 151
marked, 342
nonblocking, 121
nonconflicting, *see* nonconflict, language
normal, 261–265, 275, 276, 301, 303–305, 311, 313
optimal, 262
supremal, 262, 264
not normal, 313
observable, 257, 272, 273, 275, 276, 281–283, 295–299, 303, 311, 313
paranormal, 264, 265
prefix, 102
prefix-closed, 70, 71, 208, 375
uncontrollable, 102, 121, 122, 215, 222, 431
unobservable, 296
Lattice, 1, 4–6, 6, 10, 18, 47, 105, 106, 262, 340
complete, 5, 6, 8, 10, 19, 340
distributive, 5
modular, 5
product, 6
semilattice, 17
upper, 105
sublattice, 6, 16, 47, 105, 106
complete, 262
Lin-Brandt formula, 263
Livelock, 195
Localizable, 190
Local observability, 291
Loop-freeness, 371, 375

M
Machine, 60
Manufacturing Workcell, 402
Marked
activity, 412
L-, 106, 107, 113, 428, 429
L_m-, *see* marked, L-
string, *see* string, marked
Marker state, *see* state, marker
Marking, 260
Mealy output map, 73
Meet, 2–6, 8, 9, 47, 73
Memory, 349, 381
Message passing, 122
Mine Pump, supervisory control of, *see* Appendix, 361
MNFSC, 282, 283
MNSC, 103, 108–111, 113, 427, 429
Monoid, 363
morphism, 363
multiplicative, *see* multiplicative, monoid
Monolithic, 148
Monotone, 18
Multiplicative
monoid, 46
semigroup with identity, 46
Mutual exclusion, 136

N
Nerode
cell, 48, 56, 72
equivalence, *see* relation, Nerode equivalence
generator, 72–74
Neutralization System, 354
NFSC, 281, 282, 301
Nonanticipative, 70
Nonblocking
control, 152
DES, *see* automaton, nonblocking

feasible supervisory control, *see* NFSC
hierarchical supervision, 228–239
predicate, 351
property, 117, 148, 190, 195, 208, 296, 351
SFBC, 351
optimal, 351
supervisor, *see* nonblocking, supervisory control
supervisory control, 103, 107, 110, 120, 153, 191, 192, 302, 391, 424–429, 440
Nonconflict
language, 113, 148, 149, 190, 262, 303–305, 441
predicate, 351
property, 113
Nondeterminism, 315, 321
structural, 321
Nondeterministic dynamic system, 17
Nondeterministic generator, 64
Nondeterministic transition structure, 67
Normality, *see* property, normality
NSC, *see* nonblocking, supervisory control

O
Observability, 271
checking, 276
relative, 279
strong, 274
Observable, 20, 271, 272
relatively, 279
Observation, 20
partial, 257
Observational equivalence, 337
Observation-consistent (o.c.), 277
Observer, 20, 202, 230, 236, 266
natural
state-size reduction under, 309
natural, for **G**, 310
Observer property, 315
OCC, *see* output control consistent
Oscillator, 36
Output
controllable, 220, 246
globally, 233
map, 55, 73, 75, 211
Mealy, 73
uncontrollable, 246
Output control consistency, 213–215, 223, 228, 233, 240
strict, 220, 222, 228, 235, 243, 245
Output control consistent, 215, 216, 218, 220, 223, 226, 228, 245, 249
strict, 220, 221, 223, 227, 228, 236, 238, 239, 245–247
Output language, 71
Output map, 20

P
Parametrized internal model, 33
Paranormal, 264
Partition, 7–12, 15, 16, 20, 46, 47, 54, 55, 87, 96, 207, 210, 245, 261, 264, 412, 413, 428
binary, 55
coarse, 8, 10, 74
fine, 8, 10, 20, 46, 47, 55, 56, 78
partition, 10
refine, 8, 48, 58
Path, 15, 17
Permutable, 22
Petri net, 361, 401, 402, 443
Piston Rod Robotic Assembly Cell, 404
Poset, 1–4, 4, 6, 8, 73, 104
product, 2
Predicate, 340–343, 347, 349, 350, 352, 353, 367–369, 378, 381–383, 390, 392
controllable, 343–347, 349, 351, 369, 370
false, *see* predicate, strongest
family, 343, 351
linear, 367, 368, 375, 376
linear dynamic, 381
strongest, 340, 343
transformer, 341, 345
true, *see* predicate, weakest
uncontrollable, 344
weakest, 340, 345
Prefix, *see* string, prefix
closure, *see* closure,prefix
-preserving, 56, 208
Preimage, 224
Prioritized synchronous product, 95
Production Network, 386
Projection, 63, 148, 194, 197, 261, 267, 273, 276, 300, 305, 307, 311
canonical, 12–15, 49, 58
catenative, 90
natural, 75, 90–92, 95, 258, 265, 266, 417
to obtain TATG from TTG, 417
Property
admissible, 283, 301, 302
composition, 51
controllability, 101, 102, 111, 122, 218, 222, 224, 233, 237, 238, 264, 282, 298,

304, 342–345, 362, 368, 370, 394, 398, 411
local controllability, 230, 231, 233, 234, 237
normality, 257, 263, 276, 300–304, 311
observability, 257, 271, 273, 276, 282, 283, 298, 300, 302, 303, 311, 342, 352, 362
reachability, 346, 368
relative observability, 257
TDES controllability, 425, 426, 428, 440, 443
Pyramidal hierarchy, 207

Q
Quasi-congruence, 17, 315

R
Reachability
guaranteed, 129
predicate, 341, 343
set, 353
tree, 70, 71, 211–213, 220, 230, 231, 237, 240, 243, 247, 341, 350, 424
Reachable state subset, *see* state, reachable, subset
Recognizer, 48, 57, 59–61, 64, 68, 103, 113, 427
canonical, *see* canonical, recognizer
deterministic, 69
minimal-state, 64
nondeterministic, 68
Reconfiguration, 130
Reduction, 260
Regular expressions, 66
Relabel, 138
Relation
antisymmetric, 1, 8
binary, 1, 7, 11, 91, 271
equivalence, 1, 7–12 20, 47, 56, 57, 71, 73, 209, 210, 244, 352
refine, 74, 77, 273
Nerode equivalence, 46–48, 55, 57, 58, 63, 72, 76
coarse, 49
fine, 47
reduced, 260
reflexive, 1, 7, 8, 11, 271
right congruence, 47, 50, 55, 72–74, 76, 78, 209, 213, 258
coarse, 47, 55, 56
reduced, 260
refinement, 209, 213
symmetric, 7, 11, 271
tolerance, 11, 271
transitive, 1, 7, 8, 11
Relationship
absorption, 5
consistency, 5
Relative observability, 279
Reporter, 70
Represents, 89

S
SCOP supervisory control and observation problem, 301, 305–307
Seedlist, 330
Selfloop, 65, 69, 78, 96, 100, 119, 155, 158, 191, 219, 363, 379, 431, 437
Semantics, 61, 88, 414
Semilattice, *see* lattice, semilattice
SFBC, 343, 344, 346, 347, 350–352, 368, 370, 371, 375, 376, 386, 393
balanced, 347, 350
modular, 347, 351
nonblocking, *see* nonblocking, SFBC
optimal, 347, 376, 382
Shuffle product, 93
Σ, *see* alphabet
Silent
node, 211, 213, 220, 222, 240, 242–246
output symbol, 70, 209, 240
path, 211, 220, 240, 246, 247
silent, 237
silent path
uncontrollable, 244
symbol, 56, 209
transition, 98
Simultaneous events, 366
Small Factory, 99, 153, 197, 306
SOCC, *see* output control consistent, strict
Specification
language, 86, 104, 113, 114, 155, 228, 301, 303, 308, 310, 349, 350, 394
State, 50, 53, 60
coreachable, 61, 62, 89, 113
subset, 88
description, 86
dump, 48, 51–54, 60, 62, 99, 120
entrance, 77, 86, 415
exit, 64, 86
initial, 15, 17, 20, 21, 51, 52, 55, 57, 60, 61, 68, 70, 86, 87, 94, 98, 100, 137, 143, 144, 194, 207, 211, 235, 240, 241, 353, 358, 365, 368, 391, 413

marker, 51–53, 57, 60, 62, 86, 87, 94, 98–100, 113, 118, 119, 137, 143, 144, 152, 191–194, 207, 227, 241, 349, 351, 391, 413, 438
non-coreachable, 62, 65, 119
nonreachable, 62
nonvocalizable, 244, 245
output, 213
output map, 209
reachability, 342
reachable, 57, 58, 61, 64, 89, 113
subset, 88
root, 70, 211, 213, 220, 222, 230, 244, 245, 247
set, 86
transition graph, *see* transition, graph
vocal, 137
vocalization, 230
vocalize, 221, 244, 245
State feedback control, *see* SFBC
String, 45–48, 51, 54–57, 59, 61, 64, 68, 70, 71, 73, 77, 86, 87, 90, 93, 94, 101, 103, 106, 113, 119, 159, 191, 192, 208, 209, 211–214, 216–218, 220, 222, 230, 234, 236, 237, 247, 258, 267, 271–273, 276, 280–282, 298, 300, 305, 313, 342, 362, 368, 369, 383, 399, 414, 416, 417, 421, 424, 427, 429, 431, 432
catenation, 45–47, 66, 76, 264
empty, 45–48, 50–53, 55, 56, 60, 64, 66–68, 70–72, 74–76, 88, 89, 91, 93, 101, 102, 106, 107, 113, 114, 138, 192, 208, 209, 212, 218, 222, 230, 233, 234, 258, 264, 265, 272, 273, 296, 298–300, 302, 304, 345, 424, 426
infinite, 416
length, 46, 50
look-alike, 273
marked, 61, 94, 101, 191, 236, 300, 438
prefix, 48, 101, 113
uncontrollable, 216, 244
Strongly (**G**, P)-observable, 274
Strong observability, 275
Strong quasi-congruence, 329
Structural boundedness, 401
Structurally bounded, 364
Structurally deterministic, 321
Structurally nondeterministic, 321
Subalphabet, *see* alphabet, sub-
Sublattice, *see* lattice, sublattice
Subpredicate, 340
Subset construction, 64
Subsystems
uncontrollable, 123
Successive approximation, 78
Supervisor, 111, 114, 116, 117, 120, 122, 124, 142, 147, 152, 156, 191, 257, 258, 276, 282, 285, 308, 352, 406, 424, 439, 441
candidate, 117
centralized, 148, 155, 158, 160
component, 148, 158
conjunction, 148–150
decentralized, 147, 148, 159, 160
design, 114
feasible, 310, 407
global, 194
high-level, 224
language, 111
localization, 199, 203
modular, 147, 153, 192, 194
monolithic, *see* supervisor, centralized
optimal, 115, 116, 401
optimal global, 151
optimal low-level, 225
proper, 111, 113, 114, 116, 117, 148–150, 153, 155, 156, 160, 191, 436, 441
reduction, 141–142
state, 144
supremal, 405
synthesis, 143
Supervisory
action, 103, 114, 147, 148, 156, 190, 282, 311, 427
control, 101–104, 107, 109, 110, 112, 118, 120, 121, 127, 137, 142, 202, 215, 257, 262, 276, 280, 297, 300, 302, 313, 358, 402, 408, 411, 421, 423, 424, 426, 427, 429, 443
feasible, 280, 282, 297
hierarchical, 239
controller, 210
modules, 148
structure, 210
Supported, 82
Surjective, 12, 13, 49, 75
Synchronously nonconflicting, 266
Synchronous product, 89
System
uncontrollable, 375

T

Tail map, 70
TATG, *see* transition, graph, timed activity

TCT
- allevents, 121, 193
- complement, 99, 120
- condat, 111, 115–117, 119, 121, 122, 153, 158, 159, 286, 306, 311
- create, 213, 310
- hiconsis, 223, 227
- higen, 213, 214, 223, 226
- isomorph, 118, 154, 306
- meet, 93, 94, 96, 100, 113, 115–117, 121, 148, 149, 151, 152, 154, 155, 158, 160, 225, 306, 311, 402, 431, 436, 437, 440
- minstate, 59, 63, 98, 121, 306
- mutex, 136, 310
- natobs, 330
- nonconflict, 113, 116, 117, 119, 121, 149, 150, 152–311
- outconsis, 214
- project, 69, 95, 97, 98, 120, 127, 197, 198, 258, 259, 264, 305–307, 311, 418, 419
- relabel, 194
- selfloop, 94, 96, 117, 119, 126, 127, 158, 197, 198, 227, 242, 259, 305, 306, 311
- (s)observ, 277
- supcon, 111, 114, 115, 118–121, 123, 127, 136, 141, 150, 151, 155, 158, 194, 197, 198, 305, 306, 311, 403, 436, 438, 439
- supconrobs, 291
- supnorm, 263, 264, 305, 306, 310
- suprobs, 280
- supscop, 309
- sup(s)qc, 329
- sync, 90–93, 96, 99, 118, 119, 122, 127, 136, 151, 154, 157, 158, 194, 197, 198, 225, 306, 407, 421
- trim, 88, 100, 116–118, 153, 159
- vocalize, 226, 227, 244–246, 248, 249

TDES, 411, 414–418, 421, 424, 428, 430, 431, 436, 438, 439, 442
- composition, 420
- controllability, 421–428
- generalized, 436
- modular supervision, 439
- transition, 421, 422
  - structure, 416

Three Cooks, 120

Tick, 412–421, 423–427, 430–432, 437–441
- preemptive, 425

Tickcount, 414

Timed
- activity
  - transition graph, *see* transition, graph, timed activity
- transition graph, *see* transition, graph, timed

Timed automata, 443

Timed manufacturing cell, 433

Toolmaker, Parable of the, 15

Toy, 98

Trajectory, 20

Transfer Line, 157, 198, 224, 235, 239

Transition, 52, 59–61, 64, 67, 71, 72, 74, 77, 86, 87, 96, 98, 99, 111, 118–120, 122, 126, 137, 143, 144, 155, 158, 159, 192–194, 221, 226, 242, 243, 246, 247, 249, 285, 310, 311, 341, 353, 356, 365, 369, 381, 387, 393, 405, 414, 416–421, 425, 436–439
- action, 71, 77, 384
- activity, 418, 430
- arrow, 87
- constraint, 141
- diagram, 225
- function, 50, 55, 57, 60, 61, 64, 70, 71, 76, 78, 86, 94, 99, 109, 111, 113, 148, 207, 343, 362, 387, 413
- graph, 52–55, 60, 64, 66, 68, 86, 87, 115, 124, 219, 240, 245–248, 273, 311, 414
  - activity, 414, 416, 418, 421, 422, 430–434, 437
  - timed, 414, 417, 421, 430–435
  - timed activity, 414, 417, 418
- invariant, 367
- label, 87, 98, 211, 240, 412
- map, 208
- matrix, 353
- model, 353
- rule, 356
  - timed, 420
- structure, 50, 60, 99, 114–116, 125, 155, 210, 213, 214, 220, 240, 242, 249, 311, 339, 353, 354, 414, 416, 419, 420
  - activity, 416, 420, 425
  - refinement, 220
  - timed, 416, 420
- timed, 430
- tracking, 75

Transition function, 20

Tree, 70, 71, 211, 212, 243–245, 247, 424
- extended, 212
- new-, 244, 245
- sub-, 244, 245

TTG, *see* transition, graph, timed

U
Uncertainty sets, 277
Uncontrollable subsystems, 123
Unreliable Machine (deadlock avoidance), 407
Upward continuous, 19

V
VDES I mplementation (VDESI), 393
Vector Discrete-Event System (VDES), 362
Vocal
  node, 211–213, 220, 222, 226, 230, 231, 235, 237, 238, 240–248

W
Weakly ρ-supported, 82
Weakly supported, 79
Witness, 292
Word, 48
Workcell, 119

MIX
Papier aus verantwortungsvollen Quellen
Paper from responsible sources
FSC® C105338

If you have any concerns about our products,
you can contact us on
ProductSafety@springernature.com

In case Publisher is established outside the EU,
the EU authorized representative is:
Springer Nature Customer Service Center GmbH
Europaplatz 3, 69115 Heidelberg, Germany

Printed by Libri Plureos GmbH
in Hamburg, Germany